Environmental Stress and Amelioration in Livestock Production

Veerasamy Sejian · S. M. K. Naqvi
Thaddeus Ezeji · Jeffrey Lakritz
Rattan Lal
Editors

Environmental Stress and Amelioration in Livestock Production

Springer

Editors
Veerasamy Sejian
Division of Physiology and Biochemistry
Central Sheep and Wool Research Institute
Avikanagar, Via-Jaipur
Rajasthan 304501
India

S. M. K. Naqvi
Division of Physiology and Biochemistry
Central Sheep and Wool Research Institute
Avikanagar, Via-Jaipur
Rajasthan 304501
India

Thaddeus Ezeji
Department of Animal Sciences
The Ohio State University and OARDC
Wooster, OH 44691
USA

Jeffrey Lakritz
Department of Veterinary Clinical Science
The Ohio State University
Columbus, OH 43210
USA

Rattan Lal
Carbon Management and Sequestration
Centre, School of Environment and Natural
Resources
The Ohio State University
Columbus, OH 43210
USA

ISBN 978-3-642-29204-0 ISBN 978-3-642-29205-7 (eBook)
DOI 10.1007/978-3-642-29205-7
Springer Heidelberg New York Dordrecht London

Library of Congress Control Number: 2012937461

Printed on acid-free paper

Springer is part of Springer Science+Business Media (www.springer.com)

Preface

Homeostasis by definition is the tendency of a system, especially the physiological system of higher animals, to maintain internal stability, through coordinated response of its components to any perturbation or stimulus that would tend to disturb its normal condition or function. Animals interact with the environment in a dynamic manner. Changes in the environment, in other words stress, trigger an equal response from the animals which is known as adaptation. Thus, adaptation is an ongoing process that helps an animal to adjust to the environment. Stress, to an animal, can be physical including climatic parameters such as heat, cold, relative humidity, and changes in oxygen tension with altitude. Alternately, stress can also be physiological due to an infection, nutritional deficiency, and metabolic diseases. Whatever the etiology of stress, an animal always strives to establish a balance that would enhance its survival. However, just survival does not suffice. To be successfully adapted, an animal must survive, produce, and reproduce. Therefore, depending on its physiological status, the effect of stress on an animal varies. Although there are innate adaptive mechanisms that help the animal to overcome stress and establish homeostasis, these adaptive mechanisms can be limiting especially when the animal is being raised for production. Stress can have a serious impact on the health of the animal which in turn can reduce feed intake, feed efficiency, milk or meat production, and fertility. Therefore, as a manager, the farmer has an essential role to play in ameliorating stress in animals. It is precisely in this context that this volume is specifically devoted to several adaptive strategies which elaborate how judicious management practices can help an animal adjust and adapt to any changes in its environment. This in turn helps maintain animal health and productivity.

This volume is specifically prepared by a team of multidisciplinary scientists to be a valuable reference material for researchers as the primary target group for this compendium. In addition, the material contained in this volume is also relevant to teaching undergraduates, graduates, and other professionals involved in livestock production. Given the importance of livestock to the global economy, there is a strong need for a world class reference material on sustainable management of livestock in diverse ecoregions. With uncertain climate involving unpredictable extreme events (e.g., heat, drought, infectious disease), environmental stresses are becoming the most crucial factors affecting livestock productivity. Reference materials pertaining to stress physiology of livestock are scanty and obsolete. By addressing systematically and comprehensively all aspects of environmental stresses and livestock productivity, this volume is a useful tool for graduates and undergraduates in understanding the various intricacies of stress physiology. Further, scholars involved in research concerning livestock and its welfare can make use of this book for conducting high quality research in the field of adaptation physiology of livestock. In addition, this volume can also guide livestock researchers in identifying researchable priorities in adaptation physiology. With information and case studies collated and synthesized by professionals working in diversified ecological zones, the volume attempts to study the influence of the environment on livestock production across global biomes.

This 17-chapter compendium provides readers with an insight into the major stress factors that livestock are exposed to and their influence on livestock physiology and production. An attempt is also made to discuss the innate adaptive mechanisms that animals exhibit to counteract the adverse effects of stress. In addition to the adaptive mechanisms, several management and feeding practices have also been established as tested methods for reduction of stress effects. This book also highlights the challenges the livestock industry faces in maintaining the delicate balance between animal welfare and production. Therefore, this book is a comprehensive resource for researchers to understand stress, stress management, and livestock productivity.

The contributors of the various chapters are world class professionals with vast experience in the chosen field supported by several peer reviewed publications. The Editorial Committee takes this opportunity to thank all the contributors from different parts of the world for their dedication in preparing these chapters, for their prompt and timely response, and for sharing their knowledge and experience with others. The efforts of many others, all of those that cannot be individually listed, were also very pertinent in completing this relevant and important volume.

15 February 2012

V. Sejian
S. M. K. Naqvi
T. Ezeji
J. Lakritz
Rattan Lal

Contents

Contributors

Saumya Bahadur Division of Physiology and Biochemistry, Central Sheep and Wool Research Institute, Avikanagar, Via-Jaipur, Rajasthan 304501, India

Priya Balasubramanian Department of Pharmacology and Toxicology, College of Veterinary Medicine, Michigan State University, East Lansing, MI 48824, USA

Lance H. Baumgard Department of Animal Science, Iowa State University, Ames, IA 50011, USA, e-mail: baumgard@iastate.edu

Vijay Kumar Bharti Environmental Physiology and Toxicology Laboratory, DIHAR, Defense Research and Development Organization, Leh Ladakh 901205, India

Rebbeca L. Boddicker Department of Animal Science, Iowa State University, Ames, IA 50011, USA

Ashish Chopra Division of Animal Genetics and Breeding, Central Sheep and Wool Research Institute, Avikanagar, Via-Jaipur, Rajasthan 304501, India

Robert J. Collier Department of Physiological Sciences, University of Arizona, Tucson, AZ 85721, USA, e-mail: rcollier@ag.arizona.edu

André Martinho de Almeida Instituto de Investigação Científica Tropical, Lisbon, Portugal, e-mail: aalmeida@fmv.utl.pt

Meenambigai Dharmaraj Department of Animal Biotechnology, Tamil Nadu Veterinary and Animal Sciences University, Chennai 600007, India

Thaddeus Ezeji Department of Animal Sciences, The Ohio State University and OARDC, Wooster, OH 44691, USA

Nicholas K. Gabler Department of Animal Science, Iowa State University, Ames, IA 50011, USA

Miriam Gallardo Instituto de Patobiología INTA Castelar, 1712 Hurlingham, Buenos Aires, Argentina

John B. Gaughan School of Agriculture and Food Sciences, The University of Queensland, Gatton 4343, Australia, e-mail: j.gaughan@uq.edu.au

Kifle Gebremedhin Department of Biological and Environmental Engineering, Cornell University Ithaca, Ithaca, NY 48824, USA

Gopal R. Gowane Division of Animal Genetics and Breeding, Central Sheep and Wool Research Institute, Avikanagar, Via-Jaipur, Rajasthan 304501, India

Manuela Mendes Guerra Escola Superior de Hotelaria e Turismo do Estoril, Estoril, Portugal

Aileen F. Keating Department of Animal Science, Iowa State University, Ames, IA 50011, USA

Narayanan Krishnaswamy Division of Veterinary Obstetrics and Gynaecology, Indian Veterinary Research Institute, Mukteswar, Nainital 263138, Uttarakhand, India

Davendra Kumar Division of Physiology and Biochemistry, Central Sheep and Wool Research Institute, Avikanagar, Via-Jaipur, Rajasthan 304501, India

Kamal Kumar Division of Physiology and Biochemistry, Central Sheep and Wool Research Institute, Avikanagar, Via-Jaipur, Rajasthan 304501, India

Jeffrey Lakritz Department of Veterinary Clinical Science, The Ohio State University, Columbus, OH 43210, USA

Rattan Lal Carbon Management and Sequestration Centre, School of Environment and Natural Resources, The Ohio State University, Columbus, OH 43210, USA

Elsa Lamy Instituto de Ciências Agrárias e Ambientais Mediterrânicas, Universidade de Évora, Évora, Portugal; Escola Superior de Hotelaria e Turismo do Estoril, Estoril, Portugal

Sangeeta Lenka Division of Soil Physics, Indian Institute of Soil Science, Nabibagh, Berasia Road, Bhopal, Madhya Pradesh 462038, India

Antoni R. Macko Department of Physiological Sciences, University of Arizona, Tucson, AZ 85721, USA

Vijai P. Maurya Division of Physiology and Climatology, Indian Veterinary Research Institute, Izatnagar, Bareilly, Uttar Pradesh 243122, India, e-mail: vijaipmaurya@gmail.com

Onagbesan Okanlawon Mohammed Department of Animal Physiology, University of Agriculture, Abeokuta, Nigeria

Puliyur S. MohanKumar Departments of Pathobiology and Diagnostic Investigation, College of Veterinary Medicine, Michigan State University, East Lansing, MI 48824, USA

Sheba M. J. MohanKumar Department of Pharmacology and Toxicology, College of Veterinary Medicine, Michigan State University, East Lansing, MI 48824, USA, e-mail: mohankumrs@cvm.msu.edu

S. M. K. Naqvi Division of Physiology and Biochemistry, Central Sheep and Wool Research Institute, Avikanagar, Via-Jaipur, Rajasthan 304501, India, e-mail: naqvismk@yahoo.co.in

Soumen Naskar National Research Centre on Pig, Rani, Guwahati, Assam 781131, India, e-mail: snrana@gmail.com

Abioja Monsuru Oladimeji Department of Animal Physiology, University of Agriculture, Abeokuta, Nigeria

Daramola James Olamitibo Department of Animal Physiology, University of Agriculture, Abeokuta, Nigeria, e-mail: daramolajames2003@yahoo.com

Chandan Paswan Division of Animal Genetics and Breeding, Central Sheep and Wool Research Institute, Avikanagar, Via-Jaipur, Rajasthan 304501, India

Rajani Kr. Paul Division of Physiology and Biochemistry, Central Sheep and Wool Research Institute, Avikanagar, Via-Jaipur, Rajasthan 304501, India

Leslie Leo L. Prince Division of Animal Genetics and Breeding, Central Sheep and Wool Research Institute, Avikanagar, Via-Jaipur, Rajasthan 304501, India

Michelle L. Rhoads Department of Animal Science, Virginia Tech, Blacksburg, VA 24061, USA

Robert P. Rhoads Department of Animal Science, Virginia Tech, Blacksburg, VA 24061, USA

Jason W. Ross Department of Animal Science, Iowa State University, Ames, IA 50011, USA

Kajal Sankar Roy National Institute of Animal Nutrition and Physiology, Adugodi, Bangalore, Karnataka 560030, India

Elvira Sales-Baptista Instituto de Ciências Agrárias e Ambientais Mediterrânicas, Universidade de Évora, Évora, Portugal; Departamento de Zootecnia, Universidade de Évora, Évora, Portugal

Vijay Kumar Saxena Division of Physiology and Biochemistry, Central Sheep and Wool Research Institute, Avikanagar, Via-Jaipur, Rajasthan 304501, India, e-mail: drvijaysaxena@gmail.com

Veerasamy Sejian Adaptation Physiology Laboratory, Division of Physiology and Biochemistry, Central Sheep and Wool Research Institute, Avikanagar, Via-Jaipur, Rajasthan 304501, India, e-mail: drsejian@gmail.com

Kailash C. Sharma Division of Physiology and Biochemistry, Central Sheep and Wool Research Institute, Avikanagar, Via-Jaipur, Rajasthan 304501, India

Indu Shekhawat Division of Physiology and Biochemistry, Central Sheep and Wool Research Institute, Avikanagar, Via-Jaipur, Rajasthan 304501, India

Anoop Kumar Singh Division of Physiology and Biochemistry, Central Sheep and Wool Research Institute, Avikanagar, Via-Jaipur, Rajasthan 304501, India

Gyanendra Singh Division of Physiology and Climatology, Indian Veterinary Research Institute, Izatnagar, Bareilly, Uttar Pradesh 243122, India

Nira Manik Soren Division of Animal Nutrition, Central Sheep and Wool Research Institute, Avikanagar, Via-Jaipur, Rajasthan 304501, India, e-mail: drmanik75@gmail.com

Rajendra Swaroop Srivastava Division of Physiology and Climatology, Indian Veterinary Research Institute, Izatnagar, Bareilly, Uttar Pradesh 243122, India

Victor Ujor Department of Animal Sciences, The Ohio State University and OARDC, Wooster, OH 44691, USA

Silvia Valtorta Instituto de Patobiología INTA Castelar, 1712 Hurlingham, Buenos Aires, Argentina

Sofia van Harten Instituto de Investigação Científica Tropical, Lisbon, Portugal

Abbreviations

°C	Degree Centigrade
°F	Degree Fahrenheit
11-HSD	11-hydoxysteroid dehydrogenase
3β HSD	3-beta-hydroxysteroid dehydrogenase
5-HT	5-hydroxy tryptamine
AAMD	Aromatic-amino-acid-decarboxylase
Ach	Acetylcholine
ACTH	Adrenocorticotrophic hormone
ADG	Average daily gain
ADP	Adenosine di phosphate
AFS	Australian Friesian Sahiwal
AI	Artificial insemination
ALT	Alanine amino transferase
AMZ	Australian Milking Zebu
AnGR	Animal genetic resources
ANPP	Aboveground net primary productivity
ARD	Average relative deviations
AST	Aspartate amino transferase
ATP	Adenosine tri-phosphate
ATPase	Adenosine tri-phosphatase
AVP	Arginine vasopressin
BAT	Brown adipose tissue
BCS	Body condition scoring
BDNF	Brain-derived neurotrophic factor
BMPs	Best management practices
BNST	Bed nucleus of the stria terminalis
bST	Bovine somatotrophin
BW	Body weight
CA	Carbonic anhydrase
Caco-2	Human epithelial colorectal adenocarcinoma cells
CBD	Convention on biological diversity
CC	Climate change

CCK	Cholecystokinin
CD14	Cluster of differentiation14
CDC	Center for disease control
CER	Common environment specific response genes
CFCs	Chlorofluorocarbons
CGRFA	Commission on genetic resources for food and agriculture
CH_4	Methane
CK	Creatine phospho kinase
CL	Corpus luteum
cm	Centimetres
CNS	Central nervous system
CO_2	Carbon dioxide
cox2	Cyclo-oxygenase-2
CPRs	Common property resources
CR	Conception rate
CRF	Corticotropin releasing factor
CRH	Corticotrophin-releasing hormone
d	Day
DA	Dopamine
DBB	Diagonal band of Broca
DBRP	Destruction box recognizing protein
Dd	Distance downward
DE	Digestible energy
Dh	Distance travelled horizontal
dl	Decilitre
DM	Dry matter
DMI	Dry matter intake
DNA	Deoxyribonucleic acid
DNMTs	DNA MethylTransferases
Dw	Dwarfism
EAT	Effective ambient temperature
EB	Energy balance
ECF	East coast fever
ECl	Energy cost for moving on level
ECsl	Energy cost for moving on slope
ECT	Evaporative critical temperature
ECw	Energy cost of walking
EE	Energy expenditure
EI	External insulation
EOPs	Endogenous opioid peptides
EPI	Epinephrine
ER	Estrogen receptor
ERK	Extracellular-signal-regulated kinases
FAnGR	Farm animal genetic resources
FAO	Food and Agriculture Organization

FCE	Feed conversion efficiency
FCR	Feed conversion ratio
FEC	Faecal egg count
FFA	Free fatty acids
F gene	Frizzle gene
FMD	Foot and mouth disease
FR	Free radicals
FSH	Follicle stimulating hormone
FSSF	Farm system simulation framework
g	Gram
G6P	glucose -6-phosphatase
GABA	Gamma amino butyric acid
GDP	Gross domestic product
GE	Gross energy
GGA1	Golgi-localized, gamma adaptin protein
GH	Growth hormone
GHG	Green house gas
GHRH	Growth hormone releasing hormone
GIP	Gastric inhibitory polypeptide
GIS	Geographic information system
GIT	GastroIntestinal tract
GLP-1	Glucagon-like peptide-1
GLUT-1	Glucose transporter 1
GLUT2	Glucose transporter 2
GnRH	Gonadotropin releasing hormone
GPS	Global positioning system
GPT	Glutamic pyruvic transaminase
GPX	Glutathione peroxidase
GR	Glutathione reductase
GSH	Growth stimulating hormone
GVBD	Germinal vesicle breakdown
GVP	Genetic variation pool
GWP	Global warming potential
GxE	Genotype by Environment interaction
h	Hours
H_2	Hydrogen ions
HAL	Halothane (*Hal*) *Genotypes*
HAS2	Hyaluronan synthase 2
HATs	Histone Acyl Transferases
Hb	Hemoglobin
hCG	Human chorionic gonadotrophin
HDACs	Histone deacetylases
HHAA	Hypothalamo-hypophyseal-adrenal axis

HIOMT	Hydroxyindole-O-methyltransferase
HLI	Heat load index
HO-1	Heme oxygenase-1
HP	Heat production
HPA	Hypothalamo-pituitary-adrenal
HPG	Hypothalamo-pituitary-gonadal
HPst	Heat production in standing
HPw	Heat production in walking
HR	Heart rate
hrs	Hours
HS	Heat stress
HSE	Heat shock element
HSE-BF	Heat shock element binding factor
HSF	Heat shock factors
HSP	Heat shock proteins
HSP 70	Heat shock protein 70
ICTP	International Centre for Theoretical Physics
IFSM	Integrated farm system model
IGF-1	Insulin like growth factor-1
IL-1	Interleukin-1
IL-2	Interleukin-2
IPCC	Intergovernmental panel on climate change
IUCN	International Union for the Conservation of Nature and Natural Resource's
IUGR	Intra uterine growth restriction
IVDMD	In vitro dry matter disappearance
IVF	In vitro fertilization
IVM	In vitro maturation
JNK	C-Jun NH_2-terminal kinase
Kcal	Kilo calories
kDa	Kilo daltons
kg	Kilogram
KISS	Keep it simple to be sustainable
L	Litre
LBP	Lipopolysaccharide Binding Protein
LCMS	*Leymus Chinensis* Meadow Steppe
LCT	Lower critical temperature
LD	Lethal dose
LDH	Lactic dehydrogenase
LH	Luteinizing hormone
LPS	Lipopolysaccharides
MA-BE	Marker assisted breeding value estimation
MAPK	Mitogen activated protein kinase
MC	Melanocortin
MCH	Melanin concentrating hormone

MD	Marek's disease
ME	Metabolizable energy
Me3His	3-methyl histidyne
MEm	Metabolizable energy for maintenance
MHC	Major histocompatibility complex
MIG	Management-intensive grazing
min	Minutes
miRNA	Mitochondrial RNA
MLC	Myosin light chain
MLCK	Myosin light chain kinase
MNBs	Molasses nutrient blocks
MPA	Medial preoptic area
mRNA	Messenger RNA
mtDNA	Mitochondrial DNA
MY	Milk yield
N_2O	Nitrous oxide
NAT	N-acetyl-transferase
NDF	Neutral detergent fiber
NE	Norepinephrine
NEB	Negative energy balance
NEFA	Non-esterified fatty acids
NEm	Maintenance energy
NEp	Productive energy
NH_3	Ammonia
NK	Natural killer
NMDA	*N*-methyl D-aspartate
NOS	Nitric oxide synthase
NPP	Net primary production
NPY	Neuropeptide Y
OIE	Office International Epizootics
OVLT	Organum vasculosum lamina terminalis
p85	phosphoinositide-3 kinase
PBMC	Peripheral blood mononuclear cell
PCR	Polymerase chain reaction
PCV	Packed cell volume
PEPCK	Phosphoenolpyruvate carboxy kinase
PGE2	Prostaglandin E2
PI	Phagocytosis index
PI-IUGR	Placental insufficiency and intrauterine growth restriction
PlGF	Placental growth factor
PNMT	Phenylethanolamine *N*-methyltransferase
POF	Premature ovarian failure
POMC	Pro-opiomelanocortin
PP	Pineal proteins
PR	Pulse rate

PRPs	Proline-rich proteins
PSE	Pale, soft and exudative
PSMs	Plant secondary metabolites
PSS	Porcine stress syndrome
PTGR	Post transcriptional gene regulation
PTGS	Prostaglandin-endoperoxide synthase
PUN	Plasma urea nitrogen
PVN	Paraventricular nucleus
PYY	Peptide YY
QTL	Quantitative trait loci
RBC	Red blood corpuscles
REV	Relative economic value
RFI	Residual feed intake
RH	Relative humidity
RNA	Ribonucleic acid
ROS	Reactive oxygen species
rpm	Revolution per minute
RR	Respiration rate
rRNA	Ribosomal ribonucleic acid
RT	Rectal temperature
RT-PCR	Real-time polymerase chain reaction
SA	Surface area
SARA	Subacute rumen acidosis
SCC	Somatic cell count
SF6	Sulfur hexafluoride
SGLT1	Sodium-dependent glucose co-transporter
sHSPs	Small heat shock proteins
SIT	Sterile insect technique
SLC2A3	Reduced expression of glucose transporter 3
SLC5A1	Sodium/glucose co-transporter 1
SNPs	Single-nucleotide polymorphisms
SNS	Sympathetic nervous system
SOC	Soil organic carbon
SOD	Superoxide dismutase
SR	Stocking rate
SRBC	Sheep red blood cells
STRE	Stress response elements
SWL	Seasonal weight loss
T_3	Tri-iodo-thyronine
T_4	Thyroxine
TBE	Tick born encephalitis
TBSPs	Tannin-binding salivary proteins
TDI	Tunica dartos indices
TDS	Total dissolved solids
TER	Trans-epithelial electrical resistance

TH	Tryptophan hydroxylase
THI	Temperature humidity index
TI	Tissue insulation
TLR4	Toll like receptor 4
TMR	Total maintenance ration
TNF-α	Tumor necrosis factor alpha
TNZ	Thermoneutral zone
TRH	Thyrotropin releasing hormone
TSH	Thyroid stimulating hormone
TUNEL	Terminal deoxynucleotidyl transferase dUTP nick end labeling
UCP-1	Uncoupling protein-1
UCT	Upper critical temperature
UNFCCC	United Nations Framework Convention on Climate Change
USEPA	United States Environmental Protection Agency
VEGF	Vascular endothelial growth factor
VFA	Volatile fatty acids
VFI	Voluntary feed intake
Vit	Vitamins
VMH	Ventromedial hypothalamus
WBC	White blood corpuscles
WFM	Whole farm model
WHO	World Health Organization
WUE	Water use efficiency
ZO-1, ZO-2	Zonula occudens-1 and 2
α-MSH	α-melanocyte stimulating hormone
μg	Micro gram
μm	Micro meter

Chapter 1
Introduction

Veerasamy Sejian

Abstract Animals live in complex environments in which they are constantly confronted with short- and long-term changes due to a wide range of factors, such as environmental temperature, photoperiod, geographical location, nutrition, and socio-sexual signals. Homeostasis, the state of relative physiological stability in an organism, is a prerequisite to survive. Despite changes in environmental conditions, many living species have the ability to maintain their homeostasis within fixed limits by means of a set of specific innate repertoire of counter regulatory behavioral and physiological mechanisms. When the individual innate and acquired repertoire of counter regulatory mechanisms are overridden by environmental or internal perturbations a state of stress is reached and the 'stress responsive systems' are activated. The 'stress system' consists of neuroanatomical and functional structures that produce the behavioral, physiological, and biochemical changes directed toward maintaining homeostasis, when threatened. The environment surrounding livestock plays a significant role in influencing their productivity. Among the environmental variables affecting livestock, heat stress seems to be one of the most intriguing factors making difficult animal production in many of the world areas. Though the animals live in a complex world but researchers most often study the influence of only one stress at a time since comprehensive, balanced multifactorial experiments are technically difficult to manage, analyze, and interpret. There is, in general, a strong relationship between agro-climatic conditions, population density, cropping systems, and livestock production. Rangelands are the largest land use systems on the Earth. They predominate in semi-arid tropical areas of the world. These pastoral systems are those in which people depend entirely on livestock for their livelihoods. The key constraints of arid and semi-arid tropical environment are their low biomass

V. Sejian (✉)
Division of Physiology and Biochemistry, Central Sheep and Wool Research Institute, Avikanagar, Jaipur, Rajasthan 304501, India
e-mail: drsejian@gmail.com

V. Sejian et al. (eds.), *Environmental Stress and Amelioration in Livestock Production*, DOI: 10.1007/978-3-642-29205-7_1,

productivity, high climatic variability, and limited availability of water. All these constraints make these regions difficult for sustainable livestock production. Research agendas need to take into account the trade-offs and synergies arising from these livestock population in tropical environments so that the poor are able to reap the multiple benefits provided by these ecosystems.

Keywords Adaptation • Climate change • Environmental stress • GHGs • Grazing Ruminants • Livestock Economy • Mitigation • Multiple stress • Production • Reproduction • Stress mitigation

Contents

Livestock have been an integral part of the human civilization from times immemorial (FAO 2006). Humans have lived and learned to control domesticated animals for more than 10,000 years (Dib 2010). Over these years, human behavior changed and wild animals with potential to be explored became tamed. Livestock provide a diverse number of products that promote quality of human life, such as wool, skin, meat, milk, eggs, among others. They have also been used for transportation, labor and traction, companionship, hunting along with other activities for necessity or recreation of human.

As time progressed, mankind invented and discovered new technologies and ways to live life comfortably and leisurely. Increase in population created several challenges, such as food, shelter, clothing, etc. which in turn presented the need of synthetic food production, increased demand for land area, and excessive textile manufacturing and hence the need for industrialization. In the course of fulfilling his ever-increasing demands, mankind caused drastic changes in land use through deforestation and cultivation. Along with combustion of fossil fuels, emissions of greenhouse gases (GHGs) have and will alter the climatic patterns. The industrialization, technological advent, and population explosion together brought about drastic changes in the environment such that it has reached alarming levels. The adverse consequences of industrialization, e.g., pollution, air; water; soil and noise, deforestation, etc. are leading to a gradual increase in atmospheric temperature, increase in extreme events, such as droughts, increased incidences of natural calamities, melting of glaciers, rise in sea-level, ozone layer depletion, among others. Ecosystems including plants, animals, and birds cannot adjust and adapt to rapid changes in temperature, precipitation, and other extreme climatic events.

Animals live in complex environment in which they are constantly confronted with short- and long-term changes due to a wide range of factors, such as environmental temperature, photoperiod, geographical location, nutrition, and socio-sexual signals (Kleemann and Walker 2005; Ali and Hayder 2008). Homeostasis, the state of relative physiological stability in an organism, is a prerequisite to survive. Despite changes in environmental conditions, many living species have the ability to maintain their homeostasis within fixed limits by means of a set of specific innate repertoire of counter regulatory behavioral and physiological mechanisms. When the individual innate and acquired repertoire of counter regulatory mechanisms are overridden by environmental or internal perturbations a state of stress is reached and the 'stress responsive systems' are activated (Johnson et al. 1992; Karman 2003). The 'stress system' consists of neuroanatomical and functional structures that produce the behavioral, physiological, and biochemical changes directed toward maintaining homeostasis when it is threatened (Karman 2003).

There is, in general, a strong relationship between agro-climatic conditions, population density, cropping systems, and livestock production (Marai et al. 2007; Ali and Hayder 2008). Rangelands are the largest land use system on the Earth. They predominate in semi-arid tropical areas of the world. In some of these pastoral systems, people depend entirely on livestock for their livelihoods. The key constraints of arid and semi-arid tropical environment are their low biomass productivity, high climatic variability, and limited availability of water (Sejian et al. 2010a; Maurya et al. 2010). All these constraints exacerbate the challenges of sustainable livestock production in the rangeland ecosystems. Thus, research agendas must take into account the trade-offs and synergies arising from these tropical environments so that the resource poor land managers and farmers can harvest the multiple benefits provided by these ecosystems.

Seasonal variations in climatic conditions impact the availability of feed in the livestock. Over and above the effect of seasonal variations that cause considerable economic hardship, it is the unforeseen and unexpected periods of inclement and severe weather conditions which exacerbate the gross economic losses. In addition to mortalities of livestock associated with severe climatic conditions, reductions in reproductive and productive performances generate sizeable economic setback. Impressive advances in research have been made to assess the impact of climatic stressors on the physiological and dynamic responses of livestock. At the same time livestock managers and farmers continue to search for management options which can alleviate and reduce the effects of severe weather on livestock performance and productivity.

1.1 Economic Importance of Livestock

Livestock sector includes animal husbandry, dairy, and fishery. It plays a significant role in national and international economy, and in socio-economic development. In developing countries and emerging economies alike, it has a critical role in contributing to the rural economy by supplementing family incomes and generating gainful employment, particularly among the landless laborers, small and marginal farmers, and women (FAO 2009). Environmental factors affecting livestock productivity are discussed below.

1.1.1 Heat Stress in Livestock

Environmental stresses reduce the productivity and health of livestock resulting in significant economic losses. Livestock productivity is thought to be affected by many factors including, biomass productivity, photoperiod, geographical location, age, breed, nutrient availability, water availability, management practices, environmental conditions so on, and so forth (Khalifa 2003). Of all these factors, environmental condition influencing livestock productivity is of major concern (Shelton 2000; Koubková et al. 2002). Among the environmental variables, heat stress seems to be the most detrimental factor affecting livestock production (Rivington et al. 2009). Heat stress can cause a significant financial burden to livestock producers by decreasing milk and meat production, decreasing reproductive efficiency, and adversely affecting livestock health. Heat stress is a significant issue for livestock grazing in the tropics and subtropics and the effects of heat stress may aggravate with prospects of global warming caused by the accelerating emission of GHGs. Heat stress affects animal performance and productivity at all stages of the life cycle. The adverse effect of heat stress in livestock may be attributed to repartitioning of energy necessary for maintenance of homeothermy. The impact encompasses decreased growth, reduced milk yield, decreased reproduction, increased susceptibility to diseases, and delayed initiation of lactation. Heat stress in cattle reduces feed intake and growth, and in extreme cases can cause death, resulting in substantial revenue loss to producers (Brown-brandl et al. 2005). Heat stress also negatively affects reproductive function (Maurya et al. 2004; Sejian et al. 2011b). Infact reproductive inefficiency is one of the most costly production-limiting problems facing the livestock industry. Heat stress causes infertility in farm animals and this represents a major source of economic loss to the farmers. Reproductive processes in both the male and female are sensitive to environment. As a general rule, increased temperature decreases ovulation rates, shortens duration of estrus, decreases fertility of males, and increases rate of embryonic mortality. Given the associated economic losses through heat stress in several livestock species, additional research is needed on the interactions among heat stress, nutritional requirements, immunological status, and the overall livestock performance.

1.1.2 Economic Losses by Heat Stress

The livestock sector is socially, culturally, and politically significant both nationally and globally. It accounts for 40% of the world's agricultural gross domestic product (GDP). It employs 1.3 billion people, and creates livelihoods for one billion of the world's population living in poverty (Gaughan et al. 2010). Global demand for livestock products is expected to double during the first half of this century (FAO 2009), as a result of the growing human population, and its growing affluence. Hence it is of utmost importance to concentrate on improving the productivity of livestock to meet the growing needs of human population.

Although most livestock possess well-developed mechanisms of thermoregulation, they may not be able to adequately maintain homeothermy under heat stress, especially the high producing animals (Gaughan et al. 2010). Indeed, the hyperthermia negatively affects any form of productivity, regardless of breed, and stage of adaptation (Marai et al. 1999). An understanding of the control of body temperature in livestock under heat loads, and the relationship of this to productivity must come from an approach in which the animal is viewed in relation to both its thermal and nutritive environments (Sejian et al. 2010a). Exposure to elevated ambient temperature evokes a series of drastic changes in the livestock biological functions that include depression in feed intake efficiency and utilization; disturbances in metabolism of water and protein; and alteration in energy, and mineral balances, enzymatic reactions, hormonal secretions, and blood metabolites. Such physiological changes lead to a low live body weight and impaired reproduction, i.e., depression in age at puberty, reproductive activity, and fertility (Marai et al. 2009). As a result the production potential of the livestock species are directly under threat leading to severe economic losses.

1.2 Stress and Reproduction

Reproductive axis is one plane where stress effects are most pronounced and have gross economic impact. Stress activates systems which influence reproduction at hypothalamus, pituitary, or gonads levels. The reproductive axis is inhibited at all levels; steroidogenesis is directly inhibited at both ovaries and testes. The principle target is the GnRH neuron activity thus affecting the GnRH secretion into the hypophyseal portal blood. Stress can also affect the gonadotrophic cell responsiveness to GnRH. Glucocorticoids are critical to mediate inhibitory effect on reproduction. Environmental stresses affect the estrous behavior, embryo production, birth weights of lambs, placental size, and function and foetal growth rate.

Several factors affect the reproductive performance of farm animals, among which the physical environment and nutrition play a significant role (Gaughan et al. 2009). Proper nutrition supports mediocre biological types to reach their genetic potential, and may even alleviate the negative effects of a harsh physical environment. Poor nutrition on the other hand, may not only exacerbate detrimental environmental effects, but also reduce performance to below the genetic potential. In other words, nutritional factors appear quite important in terms of their direct effects on reproduction, and the potential to moderate the effects of other factors. Most reproductive responses to environmental factors are coordinated at the brain level, where all external and internal inputs ultimately converge into a final common pathway that controls the secretion of gonadotrophin-releasing hormone (GnRH). In turn, this neurohormone controls the secretion of gonadotrophins, the pituitary hormones that determine the activity of the reproductive axis (Martin et al. 2004). Reproductive fitness may be regarded as the most important criteria for studying or evaluating of animal adaptation. Body systems activated by stress influence reproduction by altering the activities of the hypothalamus, pituitary gland, or gonads. Reproduction processes in animals may be impacted during heat exposure, and glucocorticoids are paramount in mediating the inhibitory effects of stress on reproduction (Kornmatitsuk et al. 2008).

1.3 Significance of Optimum Nutrition to Livestock Production

Livestock grazing in hot semi-arid environment is prone to extreme fluctuations in the quantity and quality of feed throughout the year (Martin et al. 2004). Quality of feed, and thus, nutrition is one of the main factors affecting ovulation rate and sexual activity in livestock (Vinoles et al. 2005; Forcada and Albecia 2006). Nutrition modulates reproductive endocrine functions in many species including livestock (Martin et al. 2004; Sejian et al. 2011a). Further, undernutrition affects reproductive functions in ruminants at different levels of the hypothalamus-pituitary–gonadal axis (Robinson 1996; Boland et al. 2001; Chadio et al. 2007). Thermal stress and feed scarcity are the major predisposing factors for the low productivity of ewes under hot semi-arid environment (Martin et al. 2004). High ambient temperature augments the effort by livestock to dissipate body heat, resulting in increased rate of respiration, body temperature, heart beat, and water consumption (Marai et al. 2000). Increased body temperature and respiration rate are the most important signs of heat stress in livestock (Al-Haidary 2004). Further, increase in body temperature is associated with marked reduction in feed intake, redistribution in blood flow, and changes in endocrine functions that negatively affect the productive and reproductive performance of livestock (Averos et al. 2008). In addition, exposure of livestock to elevated temperature decreases body weight, growth rate, and body total solids (Marai et al. 2007). The depleted body condition during periods of energy deficiency also reduces heat tolerance (Minka and Ayo 2009).

1.4 Climate Change and Multiple Stresses Concept

In the present changing climate scenario, there are numerous stresses other than the heat stress which constraint the livestock and have severe consequences on their production (Sejian et al. 2010b). The projected climate change (CC) seriously hampers the pasture availability especially during the period of frequent drought in summer. Thus, livestock suffer from drastic nutrition deficiency. Both the quantity and the quality of the available pastures are affected during extreme environmental conditions (Gaughan et al. 2009). Further, with the changing climate, animals have to walk long distances in search of pastures. This locomotory activity also puts the livestock species under enormous stress (Gustafson et al. 1993; Sejian et al. 2011a). The majority of domesticated ruminants are raised solely or partially in semi-extensive or extensive production systems in which most nutrients are derived from grazed forage. Grazing is associated with daily activities considerably different than for confined animals such as time spent for eating and distances traveled. These activities result in greater energy expenditure (EE) than in confinement, which can limit energy available for maintenance and production. The grazing animals in the tropical areas usually have access to poor quality food available at lower densities per unit area, and to counter such hardship, animals increase their grazing time and disperse widely. Hence its not only the heat stress that needs to be counteracted but the nutrition and walking stress are also of great concern. Though the animals live in a complex world, researchers most often study the influence of only one stress factor at a time. Comprehensive, balanced, and multifactorial experiments are technically difficult to manage, analyze, and interpret (Blanc et al. 2001). When exposed to one stress at a time, animals can effectively counter it based on their stored body reserves and without altering the productive functions (Sejian et al. 2011b). However, if they are exposed to more than one stress at a time, the summated effects of the different stressors might prove detrimental to these animals. Such a response is attributed to animal's inability to cope with the combined effects of different stressors simultaneously (Sejian et al. 2010b; Sejian et al. 2011b). In such a case, the animal's body reserves are not sufficient to effectively counter multiple environmental stressors. As a result their adaptive capabilities are hampered and the animals struggle to maintain normal homeothermy (Sejian et al. 2010b). Moberg (2000) hypothesized that when animals are exposed to only one stress, they may not require the diversion of biological resources needed for other functions. If, however, two of these stressors occur simultaneously, the total cost may have a severe impact on other biological functions. Thus, normal basal functions are drastically affected which jeopardizes production.

1.5 Ameliorative Measures to Counter Environmental Stresses

This volume focuses on developing suitable ameliorative strategies which must be given due consideration to minimize economic losses incurred through impact of environmental stresses on livestock production. Several chapters are specifically devoted to the ameliorative strategies. Specific focus is on the managemental and nutritional strategies which must be adopted to prevent environmental stresses affecting livestock production. Further, an emerging theme of the probable role of pineal gland in relieving heat stress by its endocrine secretion is also addressed. Besides being a neuroendocrine transducer of cyclic photic input, pineal gland also impacts seasonal changes in reproductive capability of many animal species. Apart from this, it has influence on extra reproductive processes, such as pineal-adrenal; pineal-thytorid, and pineal-immune planes. It has antistress and tranquilizing effect, via melatonin. Melatonin has marked effect on several adrenal-cortex secretions and functions during the stress. Other pineal peptides have also been identified having antithermal stress property. Pineal plays an important role in heat stress amelioration in livestock (Sejian et al. 2008; Sejian and Srivastava 2009; 2010a). Pineal gland through its secretions, melatonin and other pineal proteins, was able to reduce heat stress in ruminant livestock (Darul and Kruczynska 2004; Sejian and Srivastava 2010b; Sejian et al. 2011c). Thus interrelated, both adrenal and pineal glands help animal to cope with stressful environment (Sejian and Srivastava 2009, 2010a, b).

Multidisciplinary approaches are required to counter environmental stresses influence on livestock production. There are varieties of options pertaining to animal nutrition, housing, and animal health (Collier et al. 2003). Some of the biotechnological options may also be used to reduce environmental stresses. However, it is important to understand the livestock responses to environment, analyze them, in order to design modifications of nutritional and environmental management thereby improving animal comfort and performance (Sejian and Naqvi 2011). An amelioration strategy must be cost-effective, suitable to the target agro-ecological zone, and provide high returns to farmers implementing it. Nutritional manipulation is one of the principal ways to optimize production during extreme environmental conditions (Martin et al. 2004; Scaramuzzi et al. 2006). Improved nutrition is an important tool to enhance ovulation rate and accentuate the overall reproductive efficiency especially in low input systems prevalent in arid, and semi-arid tropical environment (Archer et al. 2002; Scaramuzzi et al. 2006).

1.6 Adaptive Mechanisms of Livestock

The third section of the volume describes several mechanisms by which the livestock tries to adapt to an adverse environment. It addresses the basic principles that are involved in livestock adaptation to the environmental stresses. It focuses on the neuroendocrine mechanisms that control the process of livestock adaptation, and also deliberates the molecular mechanisms involved in making an animal adaptable to a specific environment. The genetic basis and its significance for livestock adaptation are also addressed in this section. Finally, efforts are made to incorporate information pertaining to identifying genes involved in thermal tolerance of livestock. This section provides an insight into how adaptations are controlled involving various systems of the body including nervous, endocrine, genetic, and molecular level control.

Animals in different parts of the world face different types of environmental factors in the form of temperature, solar radiation, photoperiod, humidity, geographical location, nutrition, and socio-economic signals. Homeostasis is referred to as the relative physiological activity in an organism critical to survival. Regardless of the changes in the environmental conditions, living species attempt to maintain constant core body temperature within a range through a definite set of regulatory behavioral and physiological mechanisms. Each species, breed, or animal category, correlated with its physiological state, has a comfort zone, in which the energy expenditure of the animal is minimal, constant, and independent of environmental temperature. Outside of this zone, the animal experiences stress to maintain homeothermy. Because maintaining homeothermy requires extra energy to thermoregulate, less energy is available for production processes (Nardone et al. 2006). Thus, animals modifies its behavior, especially feeding, physiological, and metabolic functions and the quantity and quality of its production. The extents to which they are able to adapt are limited by physiological (genetic) constraints (Alhidary et al. 2012). Responses of animals vary according to the type of thermal challenge: short-term adaptive changes in behavioral, physiological, and immunological functions (survival-oriented) are the initial responses to acute events, while longer term challenges impact performance-oriented responses (Gaughan et al. 2010). When environmental conditions change, an animal's ability to cope (or adapt) to the new conditions is determined by its ability to maintain performance and oxidative metabolism (Gaughan et al. 2010). The stress response is influenced by a number of factors including: species, breed, previous exposure, health status, level of performance, body condition, mental state, and age. Neuroendocrine responses to stress play an integral role in the maintenance of homeostasis in livestock. There are substantial evidences which suggests that neuroendocrine responses varies with the type of stressor and are specific and graded, rather than 'all or none'. While acute responses bring about survival; chronic responses may result in morbidity and mortality. Both of these responses are integrated via a network of mutual interplay between immune system, central nervous system, and the endocrine system. Infact, it is a network that exists between nervous and endocrine system which coordinate this stress response. An important component of this network is the hypothalamo-pituitary-adrenal (HPA) axis; it consists of 3 components: corticotrophin releasing hormone (CRH) neurons in the hypothalamus, corticotrophs in the anterior pituitary, and the adrenal cortex (Bernabucci et al. 2010). The HPA axis is a critical part of this mesh which is activated by the release of several neurotransmitters and hormones. Further, understanding the cellular dynamics behind the short- and long-term adaptation in the tropical animals is useful in developing mitigatory measures for improving the productivity (Collier et al. 2006). Genetic selection has been a traditional method to reduce effects of environment on livestock by development of animals that are genetically adapted to hot climates. Despite the strong knowledge base about the physiological aspects, the effects of heat stress at the cellular and genetic level are not clearly understood (Basirico et al. 2011). It is the cellular/molecular level at which stress also has its deleterious effects. Thus, the adaptive response is observed at cellular level as well and an insight into the molecular/cellular mechanism of stress relieve is important. As a result of stress, there is an increased number of nonnative conformational proteins with anomalous folding. Heat shock proteins, as we know, are evolutionarily conserved and many of them act as regulator of protein folding and structural functions of proteins. There is presence of common environment-specific response genes, making 18–38% of the genome. These genes induce expression of classical heat shock proteins, osmotic stress protectants, protein degradation enzyme, etc.

Genetic adaptability/improvement is an evolutionary process. Evolution is defined as a ongoing process of adaptation of population of organisms to the changing geological, biological, and climatic environment. Due to innumerable combinations of environmental dynamics, animals have a range of genetic types that can counter a variety of climatic, locational, biological, or other conditions. Any population must therefore be genetically heterogenous to be able to withstand the challenges of the environmental stressors. It is this concept that forms the basis of genetic improvement and which is critical to the livestock farming. Indeed, the livelihood of billions of families around the world depends on the economic returns from the livestock sector.

Functional genomics research is providing new knowledge about the impact of heat stress on livestock production and reproduction. Using functional genomics to identify genes that are regulated up or down during a stressful event can lead to the identification of animals that are genetically superior for coping with stress and toward the creation of therapeutic drugs and treatments that target affected genes (Collier et al. 2006). Given the complexity of the traits related to adaptation to tropical environments, the discovery of genes controlling these traits is a very difficult task. One obvious approach of identifying genes associated with acclimation to thermal stress is to utilize gene expression microarrays in models of thermal acclimation to identify changes in gene expression during acute and chronic thermal stress (Collier et al. 2008). Further, gene knockout models in single cells also allow for better delineation of the cellular metabolic machinery required to acclimate to thermal stress. With the development of molecular biotechnologies, new opportunities are available to characterize gene expression and identify key cellular responses to heat stress. These new tools enable to improve the accuracy and the efficiency of selection for heat tolerance. Epigenetic regulation of gene expression and thermal imprinting of the genome could also be an efficient method to improve thermal tolerance.

1.7 Livestock and Climate Change

The final section of the volume addresses the impact of climate change on livestock production, the latter is adversely affected by detrimental effects of extreme climatic conditions (Gaughan et al. 2009). Climate change is projected to be a major threat to the survival of many species, ecosystems and viability and sustainability of livestock production systems in many parts of the world (King 2004; Frankham 2005; Hulme 2005). Even if to varying extents, all productive traits of livestock are affected by climate. Heat associated with high humidity or drought represents the most stressful constraint for animal (Nardone et al. 2006).

The animals respond to these events by decreasing the feed intake, physiological responses, and production responses like reduction in reproductive efficiency and milk production etc. Consequently, adaptation to and mitigation of detrimental effects of extreme climates have played a major role in combating the adverse impact in livestock production. Whereas the animals can adapt to the hot climate, the response mechanisms are helpful to survival but detrimental to performance. Hence formulating mitigation strategies incorporating all requirements of livestock is the essential need of the hour to optimize productivity in livestock farms.

One of biggest challenges facing animal science is to increase the production in the context of climate change. Animal performance may be limited by adverse climatic focus. The economic impact of climate changes in relation to livestock production are widely reported (St-Pierre et al. 2003; Rosenweig et al. 2007), and several losses are predicted if current management systems are not modified to reflect the shift in climate. Animals that have evolved to survive in adverse conditions are endowed with the following characteristics: high resistance to stress, low metabolic rate, low fecundity, long lives, behavioral differences, late maturing, smaller mature size, and slow rate of development (Hansen 2004; Gaughan et al. 2010). Therefore, selection or use of animals (often indigenous breeds) that are adapted to adverse climates may have lower productivity than those selected for less stressful climates. Hence, it is necessary to select livestock, and use livestock systems (e.g., pasture management) on the basis of projected climatic conditions (Gaughan et al. 2010).

Like people, livestock are both the cause and the victim of the CC. The last section of the book describes the various impacts of CC on livestock production and identifies several mitigation strategies available to reduce the enteric methane (CH_4) emission from livestock. There is a growing interest in decreasing the potential threat of CC by reducing emissions of GHGs into the atmosphere (Moss et al. 2000; Sejian et al. 2011d). The scientific evidence of anthropogenic interference with the climate system through GHG emissions has led to worldwide research on assessing impacts that could result from potential CC associated with GHG accumulation (Sejian et al. 2011d). As ecosystems are sensitive to CC, it is necessary to examine the likely impacts of climate change on various components within ecosystems to provide a comprehensive understanding of the long-term effects. While carbon dioxide (CO_2) receives the most attention as a factor relative to CC, there are other gases to consider, including CH_4, nitrous oxide (N_2O) and chlorofluorocarbons (CFCs). Agricultural activities contribute significantly to global GHG emissions, namely CO_2, CH_4, N_2O, and ammonia (NH_3), which are major GHGs contributing to the radiative forcing (IPCC 2001). The global release of CH_4 from agricultural sources accounts for two-thirds of the anthropogenic CH_4 sources (Moss et al. 2000). These sources include rice paddies, enteric fermentation by ruminants (enteric CH_4), biomass burning, and animal wastes (Sejian et al. 2011d). CH_4 is a potent GHG and its release into the atmosphere is directly linked with animal agriculture, particularly ruminant production (Beauchemin and McGinn 2005). It has a global warming potential (GWP) 25 times more potent that of CO_2, making CH_4 one of the most important GHG because of its stronger molar absorption coefficient for infrared radiation and its longer residence time in the atmosphere (Wuebbles and Hayhoe 2002; Forster et al. 2007). In fact, livestock are produced throughout the world, and are an important agricultural product in virtually every country. Globally, ruminant livestock are responsible for about 85 Tg of the 550 Tg CH_4 released annually (Sejian et al. 2011d). Ruminant animals, particularly cattle, buffalo, sheep, goat, and camels produce significant amounts of CH_4 under the anaerobic conditions present as part of their normal digestive processes. This microbial fermentation process, referred to as 'enteric fermentation', produces CH_4 as a by-product which is released mainly through eructation and normal respiration, and small quantities as flatus (Lassey 2007; Chhabra et al. 2009; Sejian and Saumya 2011).

There is mounting awareness worldwide of the necessity to protect the environment by reducing the emission of GHGs. Offering relatively fewer cost-effective options than other sectors, such as energy, transport and buildings, agriculture has not yet been a major player in the reduction of GHG emissions (UNFCCC 2008). Agriculture and livestock are nevertheless poised to play a greater role in post-2012 climate agreements (UNFCCC 2008), and indeed wide ranging policy actions are inevitable (McAlpine et al. 2009). Adapting to CC and reducing GHG emission may require significant changes in production technology and farming systems that could affect productivity. Several opportunities exist for reducing CH_4 emissions from enteric fermentation in ruminant animals (Sejian and Indu 2011). To be considered viable, these emissions reduction strategies must be consistent with the continued economic viability of the producer, and must accommodate cultural factors that affect livestock ownership and management (Sejian et al. 2011d). With high GWP of CH_4, livestock producers can be ideal environmental custodians and adopt environment-friendly practices encumbering the enteric emissions. While meeting consumer demands for livestock production, animal welfare, and social responsibilities encounter issues that challenge the environmental domains. These issues must be recognized in order to be dealt with and resolved. Therefore, in considering ethical animal production practices, special consideration must be given to the impacts of the system on the environment.

1.8 Concluding Remarks

This volume entitled "Environmental Stress and Amelioration in Livestock Production" is targeting primarily the researchers involved in improving livestock production under the changing climate scenario. This state-of-the-knowledge compendium pertains to environmental influence on livestock production. This volume is unique because it addresses numerous issues facing researchers around the world. The information presented is specifically relevant to the impact of environmental stresses to livestock production, and adaptation and mitigation strategies to counter such environmental extremes. World class researchers around the world representing different agro-climatic zones have contributed their knowledge and experiences in this volume. The information presented will enhance understanding of improving livestock production under different agro-climatic zones, especially under the changing climate scenarios.

References

Al-Haidary AA (2004) Physiological responses of Naimey sheep to heat stress challenge under semi-arid environments. Int J Agr Biol 2:307–309

Alhidary IA, Shini S, Al.Jassim RAM, Gaughan JB (2012) Physiological responses of Australian Merino wethers exposed to high heat load. J Anim Sci 90:212–220

Ali A, Hayder M (2008) Seasonal variation of reproductive performance, foetal development and progesterone concentration of sheep in the subtropics. Rep Dom Anim 43:730–734

Archer ZA, Rhind SM, Findlay PA, Kyle CE, Thomas L, Marie M, Adam CL (2002) Contrasting effects of different levels of food intake and adiposity on LH secretion and hypothalamic gene expression in sheep. J Endocrinol 175:383–393

Averos A, Martin S, Riu M, Serratosa J, Gosalvez LF (2008) Stress response of extensively reared young bulls being transported to growing-finishing farm under Spanish summer commercial conditions. Life Sci 119:174–182

Basiricò L, Morera P, Primi V, Lacetera N, Nardone A, Bernabucci U (2011) Cellular thermotolerance is associated with heat shock protein 70.1 genetic polymorphisms in Holstein lactating cows. Cell Stress Chaperones 16(4):441–448

Beauchemin KA, McGinn SM (2005) Methane emissions from feedlot cattle fed barley or corn diets. J Anim Sci 83:653–661

Bernabucci U, Lacetera N, Baumgard LH, Rhoads RP, Ronchi B, Nardone A (2010) Metabolic and hormonal adaptations to heat stress in domesticated ruminants. Animal 4:1167–1183

Blanc F, Martin GB, Bocquier F (2001) Modelling reproduction in farm animals: a review. Reprod Nutr Dev 13:337–353

Boland MP, Lonergan P, O'Callaghan D (2001) Effect of nutrition on endocrine parameters, ovarian physiology, oocyte and embryo development. Theriogenology 55:1323–1340

Brown-Brandl TM, Eigenberg RA, Hahn GL, Nienaber JA, Mader TL, Spiers DE, Parkhurst AM (2005) Analyses of thermoregulatory responses of feeder cattle exposed to simulated heat waves. Int J Biometeorol 49:285–296

Chadio SE, Kotsampasi B, Papadomichelakis G, Deligeorgis S, Kalogiannis D, Menegatos D, Zerwas G (2007) Impact of maternal undernutrition on the hypothalamic –pitutary-adrenal axis responsiveness in the sheep at different ages postnatal. J Endocrinol 192:495–503

Chhabra A, Manjunath KR, Panigrahy S, Parihar JS (2009) Spatial pattern of methane emissions from Indian livestock. Current Sci 96(5):683–689

Collier RJ, Coppola C, Wolfgram A (2003) Novel approaches for the alleviation of climatic stress in farm animals. In: Interactions between Climate and Animal Production. EAAP Technical Series No. 7. Wageningen Academic Publishers, Wageningen

Collier RJ, Stiening CM, Pollard BC, Van Baale MJ, Baumgard LH, Gentry PC, Coussens PM (2006) Use of gene expression microarrays for evaluating environmental stress tolerance at the cellular level in cattle. J Anim Sci 84((E Suppl)):E1–E13

Collier RJ, Collier JL, Rhoads RP, Baumgard LH (2008) Genes involved in the bovine heat stress response. J Dairy Sci 91:445–454

Darul K, Kruczynska H (2004) Changes in selected blood metabolites associated with Melatonin administration in Dairy goats. Folia Biologica (Krakow) 52(3–4):239–241

Dib MG (2010) Strategies for beef cattle adaptation to finishing diets, ractopamine hydrochloride utilization and mature size genetic selection. Master's Thesis in Animal Science, University of Nebraska at Lincoln, USA, pp 1–105

FAO (2006) World agriculture: towards 2030/2050, Interim Report. Rome, Italy
FAO (2009) State of food and agriculture—Livestock in the balance. Rome, FAO
Forcada F, Albecia AJ (2006) The effect of nutrition on the seasonality of reproduction in Ewes. Reprod Nutr Dev 46:355–365
Forster P, Ramaswamy V, Artaxo P, Berntsen T, Betts R, Fahey DW, Haywood J, Lean J, Lowe DC, Myhre G, Nganga J, Prinn R, Raga GMS, Van Dorland R (2007) Changes in atmospheric constituents and in radiative forcing. In: Solomon S et al (eds) Climate change 2007: the physical science basis. Contribution of working group I to the fourth assessment report of the intergovernmental panel on climate change. Cambridge University Press, Cambridge
Frankham R (2005) Stress and adaptation in conservation genetics. J Evol Biol 18:750–755
Gaughan J, Lacetera N, Valtorta SE, Khalifa HH, Hahn L, Mader T (2009) Response of domestic animals to climate challenges. In: Ebi KL, Burton I, McGregr GR (eds) Biometeorology for adaptation to climate variability and change. Springer, Dordrecht, pp 131–170
Gaughan JB, Mader TL, Holt SM, Sullivan ML, Hahn GL (2010) Assessing the heat tolerance of 17 beef cattle genotypes. Int J Biometeorol 54(6):617–627
Gustafson GM, Luthman J, Burstedt E (1993) Effect of daily exercise on performance, feed efficiency and energy balance of tied dairy cows. Acta Agric Scand 43:219–227
Hansen PJ (2004) Physiological and cellular adaptations of zebu cattle to thermal stress. Anim Reprod Sci 82–83:349–360
Hulme PH (2005) Adapting to climate change: is there scope for ecological management in the face of a global threat. J Appl Ecol 42:784–794
IPCC (Intergovernmental Panel on Climate Change) (2001) Climate change 2001: the scientific basis. Intergovernment panel on climate change. Cambridge, UK, Cambridge University Press
Johnson EO, Kamilaris TC, Chrousos GP, Gold PW (1992) Mechanisms of stress: a dynamic overview of hormonal and behavioral homeostasis. Neurosci Biobehav Rev 16(2):115–130
Karman AG (2003) Neroendocrine adaptation to stress in pigs.CRH and vasopressin in the paraventricular nucleaus. PhD Thesis, Department of animal Science, Wageningen University, The Netherlands
Khalifa HH (2003) Bioclimatology and adaptation of farm animals in a changing climate. In: Interactions between climate and animal production. EAAP Tech Ser 7:15–29
King DA (2004) Climate change science: adapt, mitigate, or ignore? Science 302:176–177
Kleemann DO, Walker SK (2005) Fertility in south Australian commercial Merino flocks: relationships between reproductive traits and environmental cues. Theriogenology 63:2416–2433
Kornmatitsuk B, Chantaraprateep P, Kornmatitsuk S, Kindahl H (2008) Different types of postpartum luteal activity affected by the exposure of heat stress and subsequent reproductive performance in Holstein lactating cows. Reprod Dom Anim 43:515–519
Koubková M, Knížková I, Kunc P, Härtlová H, Flusser J, Doležal O (2002) Influence of high environmental temperatures and evaporative cooling on some physiological, hematological and biochemical parameters in high-yielding dairy cows. Czech J Anim Sci 47(8):309–318
Lassey KR (2007) Livestock methane emission: from the individual grazing animal through national inventories to the global methane cycle. Agric For Meteorol 142:120–132
Marai IFM, Habeeb AAM, Farghaly HM (1999) Productive, physiological and biochemical changes in imported and locally born Friesian and Holstein lactating cows under hot summer conditions of Egypt. Trop Anim Health Prod 31:233–243
Marai IFM, Bhagat LB, Shalaby TH, Abdel-Hafez MA (2000) Fattening performances, some behavioral traits and physiological reactions of male lambs fed concentrates mixture alone with or without natural clay, under hot summer of Egypt. Ann Arid Zone 39(4):449–460
Marai IFM, El-Darawany AA, Fadiel A, Abdel-Hafez MAM (2007) Physiological traits as affected by heat stress in sheep—A review. Small Rumin Res 71:1–12
Marai IFM, Daader AH, Soliman AM, El-Menshawy SMS (2009) Nongenetic factors affecting growth and reproduction traits of buffaloes under dry management housing (in sub-tropical environment) in Egypt. Livest Res Rural Dev 21 (3):1–13
Martin GB, Rogar J, Blache D (2004) Nutritional and environmental effects on reproduction in small ruminants. Reprod Fertil Dev 16:491–501

Maurya VP, Naqvi SMK, Mittal JP (2004) Effect of dietary energy level on physiological responses and reproductive performance in Malpura sheep in hot semi-arid region of India. Small Rumin Res 55:117–122

Maurya VP, Sejian V, Kumar D, Naqvi SMK (2010) Effect of induced body condition score differences on sexual behavior, scrotal measurements, semen attributes, and endocrine responses in Malpura rams under hot semi-arid environment. J Anim Physiol Anim Nutri 94:e308–e317. doi:10.1111/j.1439-0396.2010.01012.x

McAlpine CA, Etter A, Fearnside PM, Seabrook L, Laurance WF (2009) Increasing world consumption of beef as a driver of regional and global change: A call for policy action based on evidence from Queensland (Australia), Colombia and Brazil. Global Environ Change Human Policy Dimensions 19(1):21–23

Minka NS, Ayo JO (2009) Physiological responses of food animals to road transportation tress. Afr J Biotechnol 8(25):7415–7427

Moberg GP (2000) Biological responses to stress. Implications for animal welfare. In: Moberg GP, Mench JP (eds) Biology of animal stress. CAB International, Wallingford

Moss AR, Jounany JP, Neevbold J (2000) Methane production by ruminants: its contribution to global warming. Ann Zootech 49:231–253

Nardone A, Ronchi B, Lacetera N, Bernabucci U (2006) Climatic effects on productive traits in livestock. Vet Res Commun 30(Suppl 1):75–81

Rivington M, Matthews KB, Buchan K, Miller D, Russell G (2009) Investigating climate change impacts and adaptation options using integrated assessment methods. Aspect Appl Biol 93:85–92

Robinson JJ (1996) Nutrition and reproduction. Anim Reprod Sci 42:25–34

Rosenzweig C, Casassa G, Karoly DJ, Imeson A, Liu C, Menzel A, Rawlins S, Root TL, Seguin B, Tryjanowski P (2007) Assessment of observed changes and responses in natural and managed systems. Climate change 2007: impacts, adaptation and vulnerability. Contribution of working group II to the fourth assessment report of the intergovernmental panel on climate change. Cambridge University Press, Cambridge, UK, pp 79–13

Scaramuzzi RJ, Cambel BK, Downin JA, Kendall NR, Khalid M, Gutierrez M, Somchit A (2006) A review of the effect of supplementary nutrition in the ewe on the concentration of reproductive and metabolic hormones and the mechanism that regulate folliculogenesis and ovulation rate. Reprod Nutr Dev 46:339–354

Sejian V, Indu S (2011) Salient mitigation strategies to reduce enteric methane emission from livestock. In: Sejian V, Naqvi SMK, Bhatt RS, Karim SA (eds) NAIP sponsored national training manual on *Carbon sequestration, carbon trading and climate change* Division of Physiology and Biochemistry. Central Sheep and Wool Research Institute Avikanagar, Rajasthan, India, pp 118–131

Sejian V, Naqvi SMK (2011) Climate change and sheep production: concept of multiple stresses, adaptation and mitigation strategies to sustain production. In: Sahoo A, Sankhyan SK, Swarnkar CP, Shinde AK, Karim SA (eds) Trends in small ruminant production: perspectives and prospects, Satish Serial Publishing House, Azadpur, Delhi, pp 137–167

Sejian V, Saumya B (2011) Enteric methane emissions in livestock: contributors, prediction, estimations and repercussion. In: Sejian V, Naqvi SMK, Bhatt RS, Karim SA (eds) NAIP Sponsored national training manual on *Carbon sequestration, carbon trading and climate change*. Division of Physiology and Biochemistry, Central Sheep and Wool Research Institute, Avikanagar, Rajasthan, India, pp 68–80

Sejian V, Srivastava RS (2009) Effects of melatonin on adrenal cortical functions of Indian goats under thermal stress. Veterinary Med Int. doi:10.4061/2010/348919

Sejian V, Srivastava RS (2010a) Pineal-adrenal-immune system relationship under thermal stress: effect on physiological, endocrine and non-specific immune response in goats. J Physiol Biochem 66(4):339–349

Sejian V, Srivastava RS (2010b) Effects of pineal proteins on biochemical profile, enzyme profile and non-specific immune response of Indian goats under thermal stress. Anim Prod Res Adv 6(1):1–6

Sejian V, Lal R, Lakritz J, Ezeji T (2011a) Measurement and prediction of enteric methane emission. Int J Biometeorol 55:1–16

Sejian V, Maurya VP, Naqvi SMK (2010a) Adaptive capability as indicated by endocrine and biochemical responses of Malpura ewes subjected to combined stresses (thermal and nutritional) under semi-arid tropical environment. Int J Biometeorol 54:653–661

Sejian V, Maurya VP, Naqvi SMK (2010b) Adaptability and growth of Malpura ewes subjected to thermal and nutritional stress. Trop Anim Health Prod 42:1763–1770

Sejian V, Maurya VP, Naqvi SMK (2011b) Effect of walking stress on growth, physiological adaptability and endocrine responses in Malpura ewes under semi-arid tropical environment. Int J Biometeorol. doi:10.1007/s00484-011-0420-y

Sejian V, Maurya VP, Naqvi SMK (2011c) Effect of thermal, nutritional and combined (thermal and nutritional) stresses on growth and reproductive performance of Malpura ewes under semi-arid tropical environment. J Anim Physiol Anim Nutr 95:252–258

Sejian V, Maurya VP, Naqvi SMK, Kumar D, Joshi A (2010c) Effect of induced body condition score differences on physiological response, productive and reproductive performance of Malpura ewes kept in a hot, semi-arid environment. J Anim Physiol Anim Nutr 94(2): 154–161

Sejian V, Srivastava RS, Varshney VP (2011d) Effect of melatonin on mineral and enzyme profile in chemically adrenalectomized goats under thermal stress. Indian Vet J 88(2):18–20

Sejian V, Srivastava RS, Varshney VP (2008) Pineal-adrenal relationship: modulating effects of glucocorticoids on pineal function to ameliorate thermal-stress in Goats. Asian Aust J Anim Sci 21:988–994

Shelton M (2000) Reproductive perfoprmance of sheep exposed to hot environments. In: Malik RC, Razaqque MA, al-Nasser AY (eds) Sheep production in hot and arid zones. Published by the Kuwait Institute for Scientific Research, pp 155–162

St-Pierre NR, Cobanov B, Schnitkey G (2003) Economic losses from heat stress by US livestock industries. J Dairy Sci 86(E Suppl):E52–E77

UNFCCC (2008) Challenges and opportunities for mitigation in the agricultural sector: technical paper. United Nation Framework Convention on Climate Change. Online at http://unfccc.int/resource/docs/2008/tp/08.pdf

Vinoles C, Forsberg M, Martin GB, Cajarville C, Repetto J, Meikle A (2005) Short term nutritional supplementation of ewes in low condition affects follicle development due to an increase in glucose and metabolic hormones. Reproduction 129:299–309

Wuebbles DJ, Hayhoe K (2002) Atmospheric methane and global change. Earth Sci Rev 57:117–210

Part I
Stress and Livestock Productivity

Chapter 2
Factors Influencing Livestock Productivity

Elsa Lamy, Sofia van Harten, Elvira Sales-Baptista, Maria Manuela Mendes Guerra and André Martinho de Almeida

Abstract Numerous factors affect livestock production and productivity. In this chapter we will address those that are of paramount importance: climate, nutrition, and health aspects. In initial section we will review the effects of climate on livestock productivity and provide examples of livestock adaptation to climate constraints such as the *Bos indicus* cattle breeds adapted to hot weather and the fat tailed sheep, particularly adapted to arid conditions. In subsequent sections we address the influence of diseases and parasitism on livestock production and provide specific case studies on how diseases and parasites conditions affect livestock productivity and how domestic animals have adapted in order to cope with them. Finally, we describe two major nutrition-related factors affecting livestock productivity: seasonal weight loss and the browsing vs. grazing abilities in ruminants at the level of the oral cavity. In all section, case studies are provided as examples of specific adaptations to these problems.

Keywords Livestock productivity · Climate · Health · Nutrition

---

E. Lamy · E. Sales-Baptista
Instituto de Ciências Agrárias e Ambientais Mediterrânicas,
Universidade de Évora, Évora, Portugal

S. van Harten · A. M. de Almeida (✉)
Centro de Veterinária e Zootecnia FMV, Instituto de Investigação Científica Tropical,
Av Universidade Técnica, 1300-477 Lisbon, Portugal
e-mail: aalmeida@fmv.utl.pt

E. Lamy · M. M. M. Guerra
Escola Superior de Hotelaria e Turismo do Estoril, Estoril, Portugal

E. Sales-Baptista
Departamento de Zootecnia, Universidade de Évora, Évora, Portugal

V. Sejian et al. (eds.), *Environmental Stress and Amelioration in Livestock Production*, DOI: 10.1007/978-3-642-29205-7_2,

Contents

2.1 Introduction

Livestock systems occupy about 30% of the planet's ice-free terrestrial surface area and this sector is increasingly organized in long market chains employing approximately 1.3 billion people globally and directly supporting the livelihoods of 600 million smallholding farmers in the developing countries (Thornton 2010). Livestock production is therefore a key component of world agriculture. In fact, throughout the world, human populations largely depend on domestic animals for a multitude of purposes, essentially the production of meat, fat, milk, and other dairy products, eggs and fibers like wool or cashmere as well as other purposes such as transport, draft, and provision of fertilizers, especially in developing countries. Additionally, in some parts of the world West Africa, for example, farm animals particularly cattle are symbolic of an individual's wealth and status. Herders in Scandinavian Lapland largely depend on reindeer (*Rangifer tarandus*), (Ulvevadet and Hausner 2011) as the major source of protein, the Pacific Islanders largely depend on pig as the only source of meat while Andean cultures of South America depends on the Llama (*Lama glama*) for transport, milk, meat, and hides production (Markemann and Valle Zárate 2010). Although less conspicuous, the situation is analogous to the developed countries like Australia where merino sheep wool production is very advanced or United States and Canada with sophisticated cattle ranching, meat, and dairy production. Brazil and Argentina are examples of emerging countries with advanced meat and dairy production industry.

In such a context, livestock productivity is therefore of utmost importance as breeder's income, livelihoods, and ultimately the survival of entire populations and cultures relying on animal production. Numerous factors affect livestock productivity which is the main focus area of this book and also of the present chapter. Most of them will be dealt separately throughout this book. In the present chapter we will give a brief introduction to those factors that have the most significant and frequently limiting effects: Climate, Diseases and Parasites, and Nutrition.

The latter will be divided into two sections, the first deals with the effect of seasonal weight loss (SWL), the major drawbacks to animal production in tropical and Mediterranean climates, and the second details on adaptation of ruminants to different diets and particularly the difference between browsers and grazers in the adaptations to poor quality diets. In all sections examples are provided on how such factors impact livestock productivity. Furthermore, case studies depicting specific adaptation levels of farm animals to stress and other environmental factors are also provided.

2.2 Climate Influence on Livestock Productivity

Of all the factors influencing livestock production, climate, and location are undoubtedly the most significant. In fact, climatology characteristics such as ambient temperature and rainfall patterns have great influence on pasture and food resources availability cycle throughout the year, and types of disease and parasite outbreaks among animal populations.

In the literature, there are numerous climate classifications types. The most widely used is the Köppen classification that roughly divides world climates under the following categories: Tropical (Group A), Dry (Group B), Temperate (Group C), Continental (Group D), Polar (Group E), and Alpine (Group H). Generally speaking, it is assumed that animal production is a vital economic activity in all categories with the exception of polar climates. These climate types pose important constraints to animal production, for example, uneven distribution of rainfall during rainy and dry (tropical), prolonged dry periods of several years (Dry), long and extremely cold winters (Temperate), long hot summers (Continental), or winters with a significant amount of snowfall (Alpine). Domestic animals that have been selected in such climate conditions have developed strategies in order to cope with factors associated with climate variability either by maintaining body temperature (homeostasis) under high or low environmental temperatures through a broad range of physiological responses; or by adapting to seasonal nutrition scarcity, in particular SWL through physiological and behavioral adaptations or by possessing the ability to tolerate endemic diseases and parasites that severely limit animal performance.

Tropical climate (Group A) is usually divided into 3 subgroups (Tropical rainforest, Monsoon and Tropical wet and dry, or Savannah). All three are characterized by dry and rainy seasons with different durations according to the geographical location, among other factors. In general, the rainy season is characterized by high ambient temperatures and humidity. To cope with such conditions, farm animals have developed several physiological strategies such as higher surface of skin area, localized fat depots, or behavioral strategies like late afternoon grazing and the search for shade. In cattle, such combination of strategies is extraordinarily evident and is therefore an interesting case study.

Fig. 2.1 A *B. indicus* (zebu) bull indigenous to India. Contrary to European cattle (*B. taurus*), zebu cattle are particularly adapted to hot environments. Such trait is mainly achieved through a number of physiological adaptations such as the characteristic hump and presence of appendages, localized fat deposition, and a light coat color

Domestic cattle comprise two species, the European cattle (*Bos taurus*) and the Zebu or Indian cattle (*Bos indicus*), presented in Fig. 2.1. *B. taurus* was bred from the auroch, a wild bovid species existing in Eurasia until the nineteenth century and presently spread globally due to human settlements. On the contrary, *B. indicus* evolved in the Indian subcontinent but later spread to Africa, and to the Americas, the Pacific Islands and Australia around 300 years ago. It is generally accepted that both species share a common ancestor that at some point around 100,000 years ago diverged (Hansen 2004). Both species are however very similar, produce fertile hybrids and in fact numerous composite breeds of the two species exist. *B. indicus* cattle are particularly adapted to tropical environments, particularly to hot environments. Several studies on the comparison of zebu breeds and *B. taurus* breeds when subjected to high ambient temperatures are available in the literature (Barros et al. 2006; Collier et al. 2006, 2008), particularly originating from the United States, Brazil, and Australia, either under field conditions or more frequently in controlled environmental chambers (Barros et al. 2006) or gene expression studies (Collier et al. 2006, 2008). Additionally, zebu adaptation to heat stress has been thoroughly reviewed by Hansen (2004). To avoid increase in body temperature and maximize heat loss when animals are subjected to high ambient temperatures, the amount of heat produced by the body must equal to the amount dissipated to the surrounding environment. *B. indicus* rely on a combination of six strategies in order to accomplish this objective: (1) increasing surface area per unit of body weight; (2) increasing temperature gradient between animal and air; (3) increasing conduction of heat from the body core to the skin; (4) decreasing solar radiation reflection; (5) increasing metabolic rate and feed intake, and (6) adjusting cellular mechanisms.

Such a combination of strategies is achieved by a number of physiological abilities, as reviewed by Hansen (2004) and Berman (2011). Metabolic rate may be defined as the amount of energy produced by unit of surface area. *B. taurus* tend to have higher metabolic rates than *B. indicus*, as the latter are lighter, have proportionally smaller internal organs and tend to have larger body surface area. Zebu cattle also have a greater ability to produce sweat. In particular, in comparison to

B. taurus breeds, Zebu cattle have a higher density of larger and more superficially located sweat glands. Increases in sweat production allows zebu to dissipate more heat through sweating than *B. taurus*. A second adaptation lays in a higher resistance to outer influx of heat flow from the environment. This is achieved chiefly through the coat cover of *B. indicus* tending to be lighter in color, sleeker, and shinier in contrast to the darker, denser, and typically wooly coating of European cattle. Skin appendages are also an important and recognizable feature of *B. indicus* cattle. The main function of appendages is to increase the surface area of the skin therefore specifically contributing to the already mentioned lowering of the metabolic rate and the increase in the presence of sweat glands, hence reducing the amount of heat produced by the body and increasing heat dissipation. A very visible characteristic of all Zebu breeds is the cervo-thoracic hump. In *B. indicus*, hump fat allows decreased body fat deposition throughout the body, resulting in increased heat dissipation. Additionally, the hump is an additional appendage that contributes to the increase of body surface area. Zebu cattle reputedly utilize a series of mechanisms at the cellular and molecular level resulting in a greater adaptation to high environmental temperatures. Such mechanisms are poorly defined at present, however phenomena such as higher lymphocyte in in vitro of Zebu cattle when exposed to higher ambient temperatures in comparison to European cattle. *B. indicus* are undoubtedly an interesting case study regarding adaptation to high ambient temperatures that make them the cattle type of choice for most of the extensive production systems in the tropics and subtropics. They are known for having higher tolerance to tick infestations and tick-borne diseases and also have greater capacity to digest fodders with high dietary fiber content when compared to *B. taurus*. Nevertheless, zebus are also regarded by breeders as animals with poorer productive performance with lower meat tenderness that significantly affects access to several export markets. Milk production is also considered to be poor with a persistent lactation. Finally, short estrus and long pre-pubertal periods and a characteristic difficult temperament rendering difficult herd management. Nevertheless, it is generally accepted that the ability to withstand and produce under hot temperatures clearly surpasses the above-mentioned inconveniences.

Group B (Dry climates) include deserts and steppe-like climates, encompassing large areas of all continents. Large areas in Africa, the Middle East and Central Asia have this type of climate which is characterized by rain seasonality and the existence of periods of prolonged droughts. In these vast areas, sheep production relies heavily on indigenous breeds with distinct anatomy namely fat tail and fat rump (see Fig. 2.2). These breeds are the main ovine genetic resource in the Middle East, North Africa, Iran, Pakistan, Central Asian Republics, China, Mongolia, East and Southern Africa and have important populations in agricultural export countries like Brazil and Australia (Almeida 2011a). With regards to international diffusion, only two fat tailed breeds may be classified as international: the Awassi breed originating from several Middle Eastern countries that was improved in Israel and dissipated to the rest of the world as a selected dairy breed (Gootwine 2011). The Karakul (*Astrakhan*) breed from the former Soviet

warming and climate change on livestock productivity is the likely increase in incidence of tropical diseases and parasitism in temperate and Mediterranean climates. The occurrence of bluetongue in Southern Europe (Barros et al. 2007) and sporadic cases in the UK (Szmaragd et al. 2010) that requires specific control measures is frequently presented as an example of disease outbreaks related to climate change. It may therefore be inferred that cattle production in temperate climates will also shift toward genotypes with higher tolerance to diseases and parasites, with particular relevance to tick-borne diseases.

2.3 Diseases and Parasites: Influence on Livestock Productivity

Diseases and parasites are among the most severe factors that impact livestock production and productivity. Animal diseases have great impact on food supplies, trade and commerce, and human health globally. The last few decades have seen a general reduction in the burden of livestock diseases. Such reduction is the direct result of the availability and effectiveness of drugs and vaccines, as well as improvements in diagnostic technologies (Pearson 2006; Thornton 2010). Future disease trends are likely to be effectively managed by disease surveillance and control technologies. At the same time, new diseases have emerged and will continue to spread by the international movement of animals and animal products, such as avian influenza H5N1. This disease has caused considerable global concern about the potential for a change in host species from poultry to man and an emerging global pandemic of human influenza (Murray 2006; Thornton 2010).

From the point of view of producers, livestock diseases are essentially an economic problem. Diseases that reduce production, productivity, and profitability are associated with the cost of their treatment, disruption of local markets, international trade, and exacerbate poverty on rural, local, and regional communities. At the biological level, pathogens compete for the productive potential of animals and reduce the share that can be captured for human purposes (FAO 2009; Rushton 2009).

Livestock diseases can cause direct losses (deaths, stunting, reduced fertility, and changes in herd structure) and indirect losses (additional costs for drugs and vaccines, added labor costs and profit losses due to denied access to better markets and use of suboptimal production technology) in revenue (Rushton 2009).

Large ruminants are generally regarded as the most important domestic livestock species in the world. The importance is demonstrated by the list of products they provide. In developed countries, their contributions are mainly restricted to commercial products such as meat and milk. In developing countries they are a source of food, particularly protein for human diets, and they provide income, employment, transport, can serve as a store of wealth, and provide draft power and organic fertilizer for crop production (Perry et al. 2005; Rushton 2009). The optimization of animal production in these regions is therefore of paramount importance.

Many diseases affect livestock productivity. The added complication is that they can also cause disease in humans, such as *Trypanosomiasis*, *Salmonellosis*, and *Brucellosis*. Many animal diseases are not zoonotic, however they result in severe economic hardship (Rushton 2009). In Africa there are several tropical parasitic livestock diseases such as the tsetse-transmitted *Trypanosomiasis*, with a sub continental scale of distribution that poses a major burden on cattle farming and rural livelihoods. Increased production costs associated to *Trypanosomiasis* is expected to be compensated by reduction in productivity losses, but this may not be the case if animal healthcare services are of poor quality and the treatment is not applied correctly. This is a serious problem in many developing countries, where veterinary services are scarce. Livestock in developing countries is exposed to a range of diseases that affect productivity; it is a serious hurdle when it infects draft animals during the plowing season, limiting their ability to work. This reduces farmers' incomes from renting out draft animals and causes a reduction in the area of land that can be planted with staple food crops. Similarly, *Salmonellosis* and *Brucellosis* have also a detrimental effect of animal production.

The selected diseases subsequently described are those that in our opinion present significant economic impacts on animal production in developed and developing countries with recurring consequences. These consequences will be briefly described in the following paragraphs. In later section, we will address specific case studies on how some diseases and parasites impact livestock productivity. We will also refer to how specific animal breeds have developed mechanisms of adaptations to those diseases and parasites.

Foot-and-mouth disease (FMD) is a highly contagious, clinically acute, vesiculating viral disease of cloven-hoofed animals like domesticated ruminants, pigs, and more than 70 wildlife species. FMD virus is a non-enveloped icosahedral virus, member of the Picornaviridae family that contains RNA of around 8.4 kb (Alexandersen et al. 2003; NRC 2005). The most common mechanism of spread is by direct contact, ingestion, or inhalation of the virus from contagious animals or innate objects, such as contaminated vehicles, clothing, feed, or water (Alexandersen et al. 2003; NRC 2005). The disease is characterized by fever and blister-like lesions followed by erosions on the gastrointestinal tracts, the teats, and between the skin areas of the hooves (NRC 2005). FMD often leaves affected animals debilitated and livestock herds experience severe losses in production of meat and milk (Kitching et al. 2007; NRC 2005). The main effects of FMD on animal production are abortions, reduction in milk yields (increases the probability of mastitis due to damage to the teats), lameness in animals, weight loss (animals are unable to eat), and increases mortality in young animals (NRC 2005). These losses are most pronounced in intensive cattle and pig production systems. This disease is also a major constraint to international trade in live animals and their products because it cannot take place between a FMD-infected country and FMD-free country. The presence of FMD can also affect the export of fresh fruit and vegetables to FMD-free countries (Rushton 2009). In much of Africa and Asia, FMD is endemic and remains a perpetual obstacle to the export of meat and other livestock products.

Salmonellosis is one of the most commonly reported zoonotic diseases in humans in Europe (Kangas et al. 2007). *Salmonella* is not one disease or one causal agent. More than 2,300 serovars of *Salmonella enterica* have been identified and classified (Heithoff et al. 2008). While chickens can acquire *Salmonella* from the poultry house environment, feed, rodents, or insects or through direct contact between infected and uninfected birds (horizontal transmission), many *Salmonella* serotypes are egg transmitted (Dorea et al. 2010). Clinical manifestations of human and animal *Salmonellosis* range from self-limiting gastroenteritis to severe bacteremia and typhoid fever (Heithoff et al. 2008). The production losses or problems due to *Salmonella* go from abortions in large and small ruminants to diarrhea-associated morbidity and mortality of young animals of large and small ruminants and pigs, loss of young chicks, lower egg production by hens and turkeys, depending upon the causative agent (*Salmonella pullorum* and *Salmonella gallinarum* in hens and *Salmonella arizona* in turkeys) and adult mortality. These disease problems have many associated costs such as the treatment of sick animals, destocking of farms and higher feed costs due to poor feed conversion (Rushton 2009) that are all well described. To prevent foodborne *Salmonellosis*, various control strategies have been designed in different countries, namely eradication campaigns at farm level, improved food processing and education campaigns for households (Rushton 2009).

Brucellosis is one of the most important bacterial zoonoses worldwide and in particular in developing countries where this disease may have important economic, veterinary, and public health consequences. It is caused by *Brucellae* (Gram-negative, facultative intracellular bacteria) and it affects humans, large and small ruminants, and pigs. The main pathogenic species worldwide are *Brucellae abortus*, responsible for bovine brucellosis; *Brucellae melitensis*, the main etiologic agent of ovine and caprine brucellosis; and *Brucellae suis*, responsible for swine brucellosis. The disease can be spread from livestock to humans (that show similar symptoms such as abortions) via the consumption of untreated milk or milk products or by direct contact with diseased animals (Abdoel et al. 2008; Godfroid et al. 2010; NRC 2005). It is an important reproductive disease causing abortions in the later stages of pregnancy. The main losses are abortions, loss of milk production, and weak calves. This disease has high costs related with its control, such as costs with diagnostic tests, vaccination, the slaughter of positive animals, and incentives to encourage farmers to eradicate brucellosis at herd level (Rushton 2009). Brucellosis in cattle is effectively controlled and in many industrialized or developed countries eradicated. The disease in small ruminants has a considerable public health risk due to the production of soft cheeses from unpasteurized milk. Brucellosis in pigs also carries a public health risk and causes losses in production (Rushton 2009).

Trypanosomiasis is a disease that affects a range of species and its impact is different in different regions of the world. This disease is caused by parasitic protozoa of the genus *Trypanosoma*. In Africa, trypanosomes are mainly transmitted by the tsetse fly of the *Glossina* species. The disease is caused by *Trypanosoma congolense*, *Trypanosoma brucei brucei*, and *Trypanosoma vivax*

(Courtin et al. 2008; Naessens 2006; NRC 2005). The main interest in this disease has been Africa, which has all the trypanosome protozoas, including the trypanosomes that cause sleeping sickness. The more common symptoms for animal *Trypanosomiasis* are hyperthermia, anemia, rapid weight loss, mucous pallor, miscarriage, 'petering out', pica, splenomegaly, cachexia, and death (Courtin et al. 2008). Trypanosomiasis in Africa accounts for heavy losses due to mortality and morbidity in cattle and other species. Public health issues as cattle can harbor the trypanosome that causes sleeping sickness in humans and can also move it from one place to another is also a preeminent issue (Rushton 2009).

East Coast fever (ECF) is a disease caused by a tick-borne intracellular protozoan parasite (*Theileria parva*) transmitted by *Rhipicephalus appendiculatus* ticks. It infects T and B lymphocytes of cattle. The disease is prevalent in eastern, central, and southern Africa and has a high mortality and morbidity rate in cattle (Cicek et al. 2009; Geysen et al. 1999; Katzer et al. 2010; Rushton 2009). Efforts to control ECF are largely based on the use of acaricides to control tick population, but this approach is increasingly being compromised by the emergence of acaricide resistance in the vector tick populations. Although drugs are available to treat the disease, they are expensive and require an early diagnosis to be effective (Katzer et al. 2010).

Bluetongue is a non-contagious hemorrhagic disease of small ruminants and pigs caused by an arbovirus transmitted via Culicoides species of biting midges (Noad and Roy 2009; NRC 2005; Willgert et al. 2011). Clinical signs and lesions include cell injury and necrosis, consumptive coagulopathy (white-tailed deer), pulmonary edema, and vascular thrombosis (MacLachlan 2004). In many countries, it is present without clinical outbreaks but these may occur when climate changes encourage insect vectors of the disease to move to regions outside their original distribution. It is endemic in many tropical regions of the world and clinical symptoms are rare in indigenous animals, but are problems for the introduction of exotic animals. Therefore, this disease has a considerable economic impact in terms of restricting trade of live animals and adoption of exotic animals in countries which have endemic bluetongue. It presents minor losses in endemic areas due to reproductive problems and mortality in situations where clinical symptoms occur (Rushton 2009).

Many different agents and factors can cause diseases of livestock. Animal disease agents can have a major impact on livestock health and production and the zoonotic agents can also cause human as well as animal diseases. New, more cost-effective approaches for delivering animal health services are critical to poverty reduction processes, such as vaccines, which are critical technologies for the prevention of infectious diseases. Many animal diseases prevalent in the developing world do not occur in the developed countries. Of particular importance are the tsetse-transmitted trypanosomiasis and the tick-borne East Coast Fever in Africa, for which safe and effective vaccines do not exist. These are complicated infections, but because their distributions are restricted to developing countries and the risk of their spreading beyond Africa is minimal, the research investment on these diseases has been very limited. However, for tropical diseases there has been some development in the study of adapted breeds to some of these diseases, so as to diminish the costs related to prevention and treatment and also to search alternatives to existing drugs.

2.3.1 Diseases and Parasites: Influence on Livestock Productivity: Case Studies of Adaptation

Adaptability of an animal can be defined as the ability to survive and reproduce within a defined environment or the degree to which an organism, population or species can remain/become adapted to a wide range of environments by physiological or genetic means (Mirkena et al. 2010).

In livestock, genetic diversity with respect to disease resistance is important given that disease-causing organisms evolve continuously and develop resistance to drugs. Today, modern livestock production is highly dependent upon the use of antimicrobial compounds and antihelmintics to control diseases and parasites. Natural resistance to diseases and parasites is not usually selected while breeding animals (Gillespie and Flanders 2010). If a new disease occurs in a country, animals with a narrow genetic base may all be affected whereas in genetically diverse livestock, the chances that some animals survive, when others die, increase (Mirkena et al. 2010). Approximately one billion cattle, most of which are in the tropics, are at risk from various tick species, tick-borne diseases and worms, all of which can cause significant production losses (Frisch 1999). Some native livestock are less affected by these challenges than imported ones. In tsetse fly infested areas of Africa, indigenous cattle have developed tolerance to Trypanosomiasis disorders, whereas those imported from non-endemic areas die if not managed appropriately and treated with chemicals. Similarly, local cattle, sheep and goats in West Africa are resistant to heartwater, a deadly disease for imported animals or crossbreds (Mirkena et al. 2010). Cattle are dipped in dip tanks or directly sprayed with acaricides to prevent deaths due to anemia, to minimize losses in bodyweight due to infestation and to protect against tick-borne diseases (Jonsson 2006).

To control ticks, strategic programs have been recommended based on applications of acaricides in spring when the number of ticks on cattle is low and the proportion of ticks in the parasitic phase is high. This results in a strong effect on the size of subsequent generations (Jonsson 2006). However, acaricide use has produced acaricide resistant ticks, leading to a continuing need for the development of new drugs. All of these strategies are not cost effective. So, a search for alternative methods of control was needed, taking into account public concerns over the safety of chemical residues in livestock products and the undesirable effects of these chemicals in the general environment (Frisch 1999).

For ticks, an existent efficacious antigen, Bm86, is the basis of two commercial vaccines, directed against the cattle tick *Rhipicephalus* (*Boophilus*) *microplus* (Hope et al. 2010). These vaccines have so far proven to be the most effective alternative control method for spread of ticks in cattle. However, an integrated approach utilizing vaccination and high host resistance will be more cost effective in controlling ticks than vaccination alone, particularly in extensive pastoral systems (Frisch 1999).

The main method of *Boophilus microplus* control is the use of tick-resistant *B. indicus* breeds (Frisch 1999). *B. indicus* breeds have evolved a relatively stable relationship with ticks and, except in unusual circumstances, are not greatly affected by continuous exposure to ticks. However, almost all of the resistance found in different breeds is the result of natural, not deliberate selection. Nevertheless, due to low productivity of these breeds, their population size is slowly decreasing mostly due to crossbreeding with exotic breeds for increased production (Wambura et al. 1998). The resistance of the crossbreed is, however, directly related to the proportion of resistant breed in the cross. Thus, in tick-infested regions, crossing between "resistant" tropical breeds and "susceptible" temperate breeds will culminate in the need to use other control methods if production losses are to be minimized (Frisch 1999). Disease resistance is arguably one of the important traits possessed by these indigenous breeds and is an important attribute of livestock in low-input livestock production systems. Such traits, if identified, are useful in breed improvement programs involving crossbreeding of productive exotic breeds with indigenous breeds (Jonsson 2006).

Trypanotolerance is the capacity of certain West-African, *B. taurus* breeds of cattle to remain productive and gain weight after trypanosome infection (Naessens 2006). This infection is caused by parasitic protozoa of the genus *Trypanosoma*. In Africa, trypanosomes are mainly transmitted by the tsetse flies of the *Glossina* species. The disease is caused by *T. congolense*, *T. b. brucei*, and *T. vivax* and occurs in 37 countries resulting in a risk to approximately 60 million cattle living over 7–9 million km^2. It is estimated livestock producers and consumers lose $1.34 billion annually to trypanosomiasis in Africa. The main economic losses attributed to animal trypanosomes are related to cattle mortality and morbidity, diagnosis and treatment costs, the reduction in meat and milk production by the infected animals, and the reduction of livestock production areas. Human African trypanosomiasis, or sleeping sickness, a disease caused by *Trypanosoma brucei gambiense* and *Trypanosoma brucei rhodesiense* parasites, remains a major public health problem and occurs in 36 countries of sub-Saharan Africa. Cattle are an epidemiologically important reservoir for the human-infective parasite *T. b. rhodesiense* (Courtin et al. 2008; Delespaux et al. 2008; Naessens 2006).

African trypanosomes are unique for being able to multiply and survive in the blood of their mammalian hosts. Trypanosomes elude antibody attack by sporadically varying their surface glycoprotein, forcing the host to mount a new cycle of antibody production each time a new variant appears. In this way, the parasite manages to survive and increase its chances of transmission by tsetse or biting flies. Unfortunately for the host, the disease often leads to fatal outcomes (Naessens 2006).

Trypanosomiasis is controlled either by controlling the vector (tsetse fly) or by controlling the parasite, or a combination of both. Over the years, a large arsenal of vector-control tools has been developed with eventual impacts ranging from reduction of tsetse fly populations to total eradication. Targets and traps, and use of insecticides on animals have been effective in controlling tsetse fly populations locally and have been used extensively in agricultural settings (Aksoy 2003; Delespaux et al. 2008).

Nevertheless, the control of animal trypanosomiasis in poor rural communities will continue to rely on the use of trypanocidal drugs. This is not surprising considering the private nature of such treatments and the difficulties in controlling tsetse flies (Delespaux et al. 2008).

An alternative approach with extensive coverage is the sterile insect technique (SIT) which has been successfully applied in the control of several important pests such as the screwworm fly and the Mediterranean fruit fly. SIT is a genetic population suppression approach and involves sustained, systematic releases of irradiated sterile male insects among the wild population. Males are sterilized by irradiation and then taken to the selected area and released. As females only mate once, by continually releasing sterile males in large numbers, the reproductive capacity of the target population is progressively reduced until the population is eliminated. However, this technique involves high costs associated with its implementation. Despite this counterpart, a recent successful eradication of *Glossina austeni* from the island of Zanzibar by SIT has demonstrated the feasibility and applicability of this technology in integrated tsetse control programs (Aksoy 2003; Vreysen et al. 2000).

The decreasing efficacy of available trypanocidal drugs and the difficulties of sustaining tsetse control increase the imperative need to enhance trypanotolerance through selective breeding, either within breeds or through cross-breeding (d'Ieteren, 1998). Trypanotolerance occurs in some African bovine breeds (*B. taurus*) such as longhorn (N'Dama) and shorthorn (Baoule or West-African Shorthorn) cattle. They have the ability to control the development of the disease, unlike zebu and exotic taurine breeds. *B. taurus* breeds are tolerant to both *T. vivax* and *T. congolense*, with a higher degree of resistance to *T. vivax*. These breeds have a relatively high capability to reduce parasitaemia waves and related anemia which are the main pathogenic effects. They also have lower mortality, superior weight gain and better reproductive performance than more susceptible *B. indicus* breeds (Courtin et al. 2008; Naessens 2006).

Resistance to infections with endoparasites is defined as the initiation and maintenance of responses provoked in the host to suppress the establishment of parasites and/or eliminate parasite burdens (Baker and Gray 2004). Ruminant diseases caused by gastrointestinal nematode parasite infections are the diseases with the greatest impact on animal health and productivity. Ovine hemonchosis is an endemic helminth disease of considerable economic importance in tropical and sub-tropical regions of the world. Production losses result from depression in food intake, increase in the loss of endogenous proteins, reduced efficiency of use of food energy for tissue deposition, and impairment of bone growth in sheep (Bishop and Morris 2007; Mugambi et al. 1997; Van Houtert and Sykes 1996). Due to a growing antihelmintic resistance, there is a need to find alternative and sustainable control strategies (Magona et al. 2011). Most of the breeds identified as being relatively resistant are indigenous. This might reflect the fact that these breeds have been under natural selection for resistance for many centuries with no anti-helmintic treatment (Baker and Gray 2004). Sheep breeds resistant and/or resilient to endoparasites (predominantly *Haemonchus contortus*) include the East African Red Maasai, the Florida Native and Gulf Native in the USA, Barbados blackbelly and the St. Croix from the Caribbean. The Indonesian Thin Tail sheep have been shown to be resistant to the liver fluke *Fasciola gigantica*. The Small East African goat in Kenya and the Alpine goat in France can be included in the endoparasites-resistant goat breeds (Baker and Gray 2004).

It is generally hypothesized that differences in host resistance relates to selection for a better immune response against gastrointestinal nematodes, which affect different stages of the parasite's life cycle (Hoste and Torres-Acosta 2011). The major histocompatibility complex has been implicated as a determinant of host resistance and/or sensitivity to gastrointestinal parasitism in several species. Mucosal humoral responses to parasites have been involved in mechanisms that restrict parasite growth and mediate the expulsion of worms (Lee et al. 2011).

Mastitis resistance Mastitis is a complex disease which can be defined as an inflammation of the mammary gland resulting from the introduction and multiplication of pathogenic microorganisms in the mammary gland (Heringstad et al. 2000).

Mastitis causes major economic losses through reduction in milk yield and waste because milk from infected cows is unfit for consumption. Mastitis is a major cause of premature culling and it is the most common reason for antibiotic use in lactating dairy cattle. The prevention and treatment of mastitis present a serious hurdle to producers and it is always the primary concern of the dairy industry (Heringstad et al. 2000; Rainard and Riollet 2006). Vaccination against mastitis has long been an active field of research, but for the time being, the panoply of mastitis vaccines is neither well stocked nor very efficient (Rainard and Riollet 2006). Another approach to the control of mastitis is the selection of more resistant animals. Genetic improvement of mastitis resistance may reduce the need for treatment and, consequently, reduce the use of antibiotics with concomitant reduction in chemical residues in dairy products (Heringstad et al. 2000; Rainard and Riollet 2006).

Breeding for increased resistance to mastitis can be performed by direct selection corresponding to the diagnosis of infection (bacteriology, observation of clinical cases), by indirect selection using traits genetically correlated to mastitis or by a combination of both. The most commonly used indirect measures have so far been focused on milk somatic cell counts (SCC) to predict the bacterial status of udders (Heringstad et al. 2000; Rupp and Boichard 2003). Increase in milk SCC mainly corresponds to an afflux of white blood cells that come from the bloodstream into the milk to fight infection in the udder. Therefore, SCC is closely related to the magnitude of the inflammatory process (Rupp and Boichard 2003).

Scandinavian countries were the first to consider udder health in their breeding objectives for dairy cattle. In the last decade, many other countries similarly modified their breeding objectives for dairy cattle and sheep in response to the increasing consumer's concern for better animal health and food quality, and also to maximize profitability by reducing production costs (Rupp and Boichard 2007).

Marked differences between breeds may also be found regarding tolerance to mastitits. Dairy breeds originating from eastern France (*Montbéliarde*, *Abondance*) or central Europe (*Simmentaler*, Brown Swiss) have lower SCC and clinical mastitis frequency than Holstein. Within breed, genetic variability is quite large (Rupp and Boichard 2003). Based on progress in understanding genetic basis of host's defense mechanisms, new phenotypes and genes may emerge to target key components of resistance of the udder gland, and potentially control resistance to various pathogens and environment-pathogen interactions (Rupp and Boichard 2007).

Considerable work has been made in the last decade in understanding immune mechanisms and identifying genes that play key role in the mammary gland defenses, but the function is highly complex and is still a large field for investigation. Studies combining different field approaches (genetics, QTL characterizations, immunology), including technology such as transcriptomics and proteomics, may be promising to better understand genetic basis of udder health, to predict long-term responses to selection, and to develop new tools and strategies for genetic improvement of udder health (Rupp and Boichard 2007).

Innate immunity is a target of choice for selection against infectious diseases. Natural resistance to a number of vectors and vector-borne microorganisms exists in ruminants and varies with breed. Thus, breeding schemes based on these characteristics are encouraged since they can contribute to a reduction in the use of chemicals, to an increase in the effectiveness of drugs (as naturally resistant animals respond better to treatment), to a reduction in the risk and incidence of drug-resistant strains of pathogens and vectors and lower the cost of animal production through a diminished pathological impact, and, consequently a reduced use of drugs.

2.4 Nutritional Influences on Livestock Productivity: The Problem of SWL

As discussed earlier, the rain pattern during the year strongly conditions livestock production systems through pasture development and disease and parasites outbreaks, therefore influencing animal production systems, productivity. Tropical and Mediterranean climates are characterized by the existence of a season of varied duration, when rainfall is scanty or non-prevalent. Such season is termed dry season in the tropics and summer in Mediterranean climates. During rainy season pastures are available in higher quantities and show good nutritional quality whereas dry season's pastures have poor nutritional quality with high fiber and low protein contents (Butterworth 1984), which often results in SWL. A schematic representation of the effects of SWL on animal productivity is presented in Fig. 2.4.

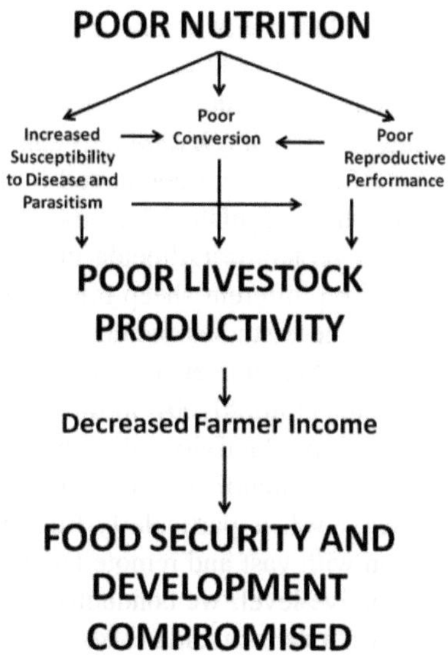

Fig. 2.4 Schematic representation of the effects of poor nutrition caused by SWL on farm animal productivity and food security

To mitigate this problem, animal production systems in these regions resort to strategies such as transhumance in pastoral societies or to supplementation in developed countries. Animals that evolved in such production systems tend to have physiological adaptations that enable them to cope with SWL. In this section we discuss the deleterious effect of SWL and strategies for mitigating it to enhance livestock productivity. Such strategies rely essentially on transhumance or cattle and human population migrations, supplementation with concentrates and minerals and determining the physiological adaptations to nutritional stress displayed by farm animal breeds that were selected in regions where SWL is a severe problem.

In tropical countries, lack of feed supplementation during the dry season is frequent in extensive or traditional management systems. This situation leads to a problem of SWL of approximately 20–40% of the body weight at the onset of the dry season. This fact has been reported by several authors (Preston and Leng 1987; Clariget et al. 1998). Hence, pasture shortage significantly affects animal production in tropical nations such as Mali (Wilson 1987), Brazil (Abdalla et al. 1999), the Philippines (Alejandrino et al. 1999), South Africa (Lusweti 2000; Almeida et al. 2006a, 2007), or Guinea-Bissau (de Almeida and Cardoso 2008a, b). SWL has therefore great impact on all aspects of animal production. The productive and physiological impact of SWL has been a major area of research activities for our research group over the last 12 years in laboratory and farm animals.

Our research group has studied production standards and productivity on one of the most important goat breeds of the world, the Boer goat. The results obtained from these studies clearly illustrate the deleterious effect of SWL at the level of growth, carcass characteristics (Almeida et al. 2006a), carcass minerals (Almeida et al. 2006b) several reproduction indicators in the male goat (Almeida

of these techniques as tools for animal selection is envisaged to be of paramount interest in the twenty-first century (Fadiel et al. 2005), mitigating the adverse effects of SWL on livestock productivity.

To study such breed differences, we started with two rabbit breeds subjected to control and food restriction conditions: wild rabbits (not selected to increased meat productivity) and the well-known meat producer New Zealand white (the most common domesticated breed). Studies focused on male reproductive performance (Carvalho et al. 2009), liver regulatory enzymes (van Harten and Cardoso 2010), and the skeletal muscle proteome (Almeida et al. 2010a). The objective of using the rabbit as a mammalian model would allow the extrapolation of the results to other conventional farm animal species such as swine, cattle, or sheep.

In these studies, a 20% weight reduction was induced in the two above-mentioned rabbit breeds: New Zealand white, a selected meat producer (*Oryctolagus cuniculus cuniculus*) and the Iberian wild rabbits (*Oryctolagus cuniculus algirus*). Regarding the proteomics experiment (Almeida et al. 2010a), the determination of the differential protein expression in the *gastrocnemius* muscle within control (ad libitum) and restricted diet experimental animal groups, using techniques of two-dimensional gel electrophoresis and mass spectrometry-based identification was carried out. Results show that spots identified as energy metabolism L-lactate dehydrogenase, adenylate kinase, β enolase and α enolase, fructose bisphosphate aldolase A, and glyceraldehyde 3-phosphate dehydrogenase enzymes are differentially expressed, in restricted diet experimental animal groups. These enzymes are available to be further tested as relevant biomarkers of weight loss and putative objects of manipulation as a selection tool toward increasing tolerance to weight loss. Similar reasoning could be applied to spots corresponding to important structural proteins tropomyosin β chain and troponin I. Additionally, a spot identified as mitochondrial import stimulation factor seems of capital interest as a marker of undernutrition and a possible object of further studies aiming to better understand its physiological role. In parallel, we conducted a study on the Hepatic glycolytic, lipidic, and protein regulatory enzyme activity at the transcriptional and metabolite levels (van Harten and Cardoso 2010). Insulin-like growth factor (IGF-1), triiodothyronine, and cortisol were also evaluated. In the glycolytic pathways, the New Zealand control rabbits showed a higher phosphofructokinase and pyruvate kinase activity level when compared to the wild rabbit, while the latter group showed a higher expression of glycogen synthase, although with less glycogen content. In the nitrogen metabolism, our results showed a lower activity level of glutamate dehydrogenase in Wild Rabbits when subjected to food restriction. The lipid metabolism results showed that although Wild Rabbits had a significantly higher mRNA hepatic lipase, non-esterified fatty acid levels were similar between the experimental groups. New Zealand rabbits exhibited a better glycemia control and greater energy substrate availability leading to enhanced productivities in which triiodothyronine and IGF-1 played a relevant role.

Results obtained in both studies in conjunction with the productive data led us to conduct further studies aimed at understanding breed differences in sheep as a consequence of the effect of SWL. Further trials were therefore conducted in sheep with very diverse genetic background and varying levels of adaptation to SWL: the Australian Merino (Animals of European origin and less tolerant to SWL), the Damara, a fat tailed sheep from Southern Africa with high levels of adaptation to nutritional stress and the Dorper, a composite breed of mixed European and African origin with an intermediate level of adaptation to nutritional stress.

These studies focused essentially on both productive trials, as well as on proteomics-based comparison of the hepatic tissue exploring breed differences. Scanlon et al. (2008) and Kilminster et al. (2008) studied the effects of food restriction (15% live weight decrease) on growth and carcass characteristics of pure Damara, Dorper, and Australian Merino ram lambs. Contrary to our expectation, results indicate that food restriction affected all breeds similarly, as weight lost proportional to initial body weight (Scanlon et al. 2008). As expected, Dorper rams had higher proportions of weight gains than Damara and Merinos. In addition, it was found that differences in carcass and loin meat characteristics were seen between breeds and between feeding levels. Dorper and Damara lambs had a higher dressing percentage than Merino lambs and this might partially be explained by differences in fleece weight. Dorper lambs had heavier carcasses generally when fed to gain weight than the Damara and Merino lambs. Meat from Damara lambs was darker in color than that from other breeds, an indication of differences in metabolic rate. Feeding to lose weight reduced carcass weight and fatness but not dressing percentage. Meat from lambs fed to lose weight tended to be lighter and less red in color than meat from lambs fed to gain weight.

Additionally, we conducted experiments on the effect of SWL on liver proteome profiles in the same sheep breeds (Almeida et al. 2010b). At the end of the afore-said trial, animals were euthanized and liver sampled. Total liver protein was extracted; quantified and used in two-dimensional gel electrophoresis analysis. After gel analysis, a total of 67 spots were selected for identification using MALDI-TOF/TOF. Results obtained reflect the extreme complexity of the liver when compared to tissues previously studied on the SWL issue, particularly the skeletal muscle in two different rabbit genotypes (Almeida et al. 2010a), and also a high conservation of protein expression levels in this tissue as a consequence of weight loss. The study suggests that proteins such as glutathione S-transferase, triose phosphate isomerase, phosphoglycerate mutase, carbonic anhydrase, carbonyl reductase, and a stress protein similar to heat shock protein could be considered as biomarker for the tolerance to weight loss in sheep. To the best of our knowledge, this work was the first proteomics-based approach to the study of protein expression profiles at the liver level in underfed sheep. The study was also enriched by the possibility of comparing three different breeds of sheep that show apparent different tolerance levels to weight loss, distinct metabolism type, and breed origins.

SWL is the main constraint to animal productivity in tropical and Mediterranean regions, the results obtained in the aforementioned studies, allowed us to quantify the effects of SWL on ruminant production and productivity, primarily at an experimental level. Further studies were conducted on numerous physiological and biochemical aspects of effects of SWL and different levels of adaptation on animals of commercial interest such as the rabbit and the sheep. These strategies will enhance animal breeding and development of effective selection tools for obtaining genotypes with higher tolerance to SWL and improved productivity.

2.4.1 Nutritional Influences on Livestock Productivity: Adaptation to Poor Quality Diets in Ruminants at the Oral Cavity Level: Browsers vs. Grazers

In this section, we will explore the differences between ruminants consuming varied types of foliage (grass vs. shrubs; i.e., browser vs. grazers) primarily at the level of the adaptations that are found at the oral cavity. The coexistence of species in the same areas, without competing is possible due to the differentiation of ecological niches. Support for this statement concerns the ruminant feeding types proposed by Hofmann (1989): browsers or concentrate selectors, grazers, and intermediate feeders. This classification was based on the relation between the functional–anatomical and histological characteristics and the different chemical and physical properties of the respective food sources. In general, grazers have a highly developed fermentation system enabling them to digest fibrous fractions of plants with high amounts of cellulose and lignin, such as monocotyledons. On the other hand, browsers are animals that select for fresh, juicy foliage, forbs and other dicotyledonous matter, highly digestible, relatively rich in energy and protein, and low in fiber. In the between are the intermediate feeder type that behaves as browsers or grazers seasonally, according to the available vegetation species. One of the main sources of diet variation among these Hofman's ecophysiological feeding types is the level of plant secondary metabolites (PSMs) (e.g. tannins) present in diet, which require detoxification mechanisms within consumers. Browsers are faced with diets with high levels of these compounds, followed by intermediate feeders and ultimately by grazers. Although some authors criticized (reviewed in Pérez-Barberia et al. 2005) some of Hofman's assumptions, this continues to be a widely accepted classification scheme.

Since ruminants are faced with variable levels of nutrient and PSMs in plants, they must be able to differentiate foods with different compositions. They also need to associate the sensory properties of foods, namely palatability, with the metabolic consequences of eating them, in order to increase diet quality and maintain productivity. The role of oral cavity in food perception is extremely important in the exploratory behavior. Learned associations formed between the sensory properties of the food perceived in the oral cavity and its post-ingestion effects allow individuals to modify feeding behaviors (Green et al. 1984; Provenza 1995; Ralphs 1997). In fact, preferences are likely indicative of underlying physiological adaptations promoting further behavioral, physiological, and ultimately genetic differences between the species. For example, different ruminant species with different tolerance levels for PSMs react differently to diet sensorial cues.

With regards to ruminant diets, the main sensorial characteristics perceived at the mouth level and responsible for the acceptance or rejection of food items are bitter taste and astringency. Glendinning (1994) hypothesized that mammals in different trophic groups have evolved different strategies for coping with the unpredictable bitterness/toxicity relationship of food items. Accordingly, bitter taste thresholds and bitter rejection response is dependent on the relative occurrence of bitter and potentially toxic compounds in a species regular diet. Lower thresholds of bitter detection would reduce drastically the chances of toxic food ingestion. However, for species which have to deal with higher levels of PSMs, which are regularly associated to these sensory characteristic, the presence of much higher bitter taste sensitivity would not be advantageous at the expense of limiting drastically the range of potential foods.

In all the physiological oral processes, saliva is central. Whole saliva constitutes the fluid that bathes the oral cavity. It originates from major and minor salivary glands and from gingival crevicular sulcus, the latter of which is an area located between teeth and marginal free gingival. This fluid has the main role of protection and maintenance of the upper part of the digestive tract, acting through lubrication, buffering action, maintenance of tooth and mucosal integrity, antibacterial and antiviral activity. Moreover, it can play several roles in feeding behavior, through taste perception and assistance in the processes of food ingestion and digestion (Mese and Matsuo 2007).

The salivary glands of ruminants, which rely on foregut fermentation for digestion, differ from other mammalian salivary glands (Steward et al. 1996). The amounts of saliva produced are extremely high as it is an important source of fluid to rumen [e.g. one sheep (*O. aries*) producing at least 15 L/day (Kay 1960)]. The ruminant parotid saliva, with a pH of 8.2, is unusually rich in mineral ions, particularly sodium, phosphate, and bicarbonate (McDougall 1948), which are involved in providing a buffered medium for ruminal fermentation. Ruminant parotid serous glands are the main facilitator in the digestive processes during feeding and rumination, whereas submandibular, sublingual, and other minor mucous salivary glands are confined to lubricate the mouth and esophagus (Carr 1984).

Mouth being the first entry level for food, saliva plays an important physiological role in protecting ruminants against external and internal milieus. Secretion and composition of saliva is mainly under autonomic nervous system control, allowing a rapid adjustment to changes in dietary conditions. The role of saliva in ruminant dietary adaptation can be viewed as a consequence of two main inter-related functions: importance of saliva in food perception and palatability, and the direct interaction between salivary proteins and PSMs, acting as a defense mechanism against negative effects of their ingestion.

Animals possessing different dietary habits concomitantly secrete different saliva volumes, buffer capacity, and protein composition (Sales Baptista et al. 2009). Browsers have been referred as having a greater secretion of thin serous saliva, followed by intermediate feeders, being grazers the ones with the lowest volume of secretions (Hofmann 1989). Saliva protein composition varies considerably among species, and also reflects their diverse diets. Animals using identical feeding niches may exibit similarities in their saliva protein composition, whereas the presence of particular proteins appears to be specific for particular feeding niches. In sheep and goats we observed that salivary proteins are expressed at different levels, between the species, what appears to be related to food consumption (Lamy et al. 2009). This issue will be further developed.

Salivary proteins can interact with taste substances, modulating taste perception and, indirectly, dietary habits. Although not present in saliva of ruminants, alpha amylase is a salivary protein suggested to be important in food perception. This protein initiates the digestion of starch in the mouth and this may have some influence on taste of carbohydrates (Becerra et al. 2003). Other salivary enzyme suggested to interact with food components, changing their original taste, is lingual lipase, which is secreted by the von Ebner's minor salivary glands. This protein can break down dietary triglycerides to fatty acids and other small molecules, which, in turn, can stimulate taste receptors and result in fat perception (Kawai and Fushiki 2003). Moreover, its expression levels change according to the levels of dietary fat (Armand et al. 1990). Von Ebner's gland protein, abundantly expressed in the small von Ebner's salivary glands of the tongue, was known to bind to lipophilic molecules some of which have bitter taste, influencing their perception (Gurkan and Bradley 1988). The salivary carbonic anhydrase (CA) VI has also been associated with taste sensitivity. One of the ways this protein contributes to taste function seems to be by protecting taste receptor cells (Leinonen et al. 2001). CA VI expression was recently demonstrated in both sheep and goat saliva, although our results suggest differences between the two species in the expression of some isoforms (Lamy et al. 2009). However, it is difficult to make conclusions on how this relates to species different ingestive behavior without further studies. Although not studied in ruminants, salivary cystatins are also proteins differently expressed based on diet composition, with some researchers associating them to an aversive oral stimuli such as pungency and astringency (Katsukawa et al. 2002; Dinnella et al. 2010). The involvement of salivary proteins in ruminant taste perception, however, is a very interesting subject for research that still needs to be further investigated.

Animal species may have to deal with distinct levels of PSMs present in their diet differently during the year, as these compounds vary among plants and are influenced by climatic conditions. Among the PSMs, tannins have a major importance in ruminant food choice and nutrition. Tannins are a class of PSMs with a high capacity to bind proteins, polysaccharides, carbohydrates, and other macromolecules. Tannins may form stable complexes that tend to precipitate (Lu and Bennick 1998), which can result in the anti-nutritional property usually attributed to these compounds. The levels of dietary tannins influence food selection, and may even result in food rejection attributable to either their astringent properties or detrimental post-ingestion effects (Iason 2005). High tannin levels reduce acceptability of plants by cattle, sheep, and goats

(Perevolotsky et al. 1993). There are several studies reporting salivary proteins as a countermeasure against the potential negative effects of tannins. Such countermeasure role is essentially due to the high affinity that some classes present for these PSMs, resulting in the formation of complexes, which change the way these phytochemicals are perceived in the mouth and the way they act through the digestive tract. A large number of researchers (reviewed in Shimada 2006) link the occurrence of such tannin-binding salivary proteins (TBSPs) to the levels of tannins present in the individual's regular diet: species with low tannin content in their natural forage, such as grazers, have little or none of such salivary proteins, whereas browsers having a diet rich in tannins throughout the year present TBSPs constitutively. Species in which the levels of tannins in the diet change seasonally can adapt by producing higher amounts of TBSPs only when consuming tannin-rich diets (Austin et al. 1989; Robbins et al. 1987; Fickel et al. 1998; Lamy et al. 2011).

The most studied TBSPs are the proline-rich proteins (PRPs). PRPs have a high capacity to bind tannins, and the complexes formed appear to be stable across pH ranges of the digestive tract, allowing tannins to pass intact through it and to be excreted (Bennick 2002). It was suggested that the presence of these proteins might override the negative effects of tannins on palatability, and consequently on feed intake, improving the utilization of plants containing such compounds (Glendinning 1992). Besides PRPs, other salivary proteins show affinity for tannins, are salivary histatins (Wroblewski et al. 2001), and salivary amylase (Zajácz et al. 2007; da Costa et al. 2008; Lamy et al. 2010). However, these proteins seem to be absent from ruminant saliva, and as such they do not contribute as a defense medium for these animals.

Until now, there has been some controversy about the presence of TBSPs in sheep and goat saliva. Whereas some researchers suggest that sheep and goat saliva does not have a great ability to bind tannins (Pérez-Maldonado et al. 1995), others point to the possibility of their presence in both species (Vaithiyanathan et al. 2001). There is more speculation about the occurrence of TBSPs in goat than in sheep. According to the "niche theory" (i.e., the relative position of a species in its ecosystem), goats produce TBSPs since their diet is based in tannin-rich browse (Kababya et al. 1998; Landau et al. 2002). Gilboa (cited by Silanikove et al. 1996) found that the parotid saliva of goats was relatively rich in proline (6.5%), glutamine (16.5%), and glycine (6.1%). This author also observed that the concentration of parotid saliva of goat fed tannin-rich diet was higher as compared to goats maintained on diets low in these PSMs. On the other hand, other researchers (e.g. Distel and Provenza 1991) did not detect TBSPs in the saliva of goats even when fed a tannin-rich diet.

In the last few years, our research has focused on studies concerning the influence of saliva in tannin-rich diet consumption by ruminants. We observed differences between the salivary proteome of sheep and goats (Lamy et al. 2008; Lamy et al. 2009). These two ruminant species, in Mediterranean areas, are frequently found together in pastures and although sharing the same areas, they do not compete, since they select plants and/or plant parts with different characteristics, namely the levels of tannins (Gong et al. 1996). Accordingly, we suggest that the different salivary protein profile may be related to this difference in ingestive behavior. In fact, when subjected to condensed tannin-enriched diets changes in the expression profiles of some proteins were observed (Lamy et al. 2011). By using proteomic approaches, it was observed that the consumption of quebracho tannins resulted in the increase in the expression of the proteins cytoplasmic actin 1 and annexin A1, in both species. We do not know whether these proteins may act as TBSPs, or whether they are only the consequence of an increased salivary gland function induced by tannins. For example the increase of actin 1, which is a protein from the cytoskeleton, may be related to the "apocrine-like" mode of salivary secretion present in ruminants (Stolte and Ito 1996) and to a potential increased salivation rate resulting from tannin consumption. Additionally, annexin A1 is a protein already shown to be increased in humans after tasting bitter/sour substances (Neyraud et al. 2006) and a relation of this protein to sensory properties of tanniniferous foods is not to be discarded. However, to our knowledge, an association of this protein to tannin consumption has not been reported in the literature.

A recent study (Hanovice-Ziony et al. 2010) also failed to identify potential proteins that bind tannic acid or quebracho tannins in goat saliva. However, these researchers observed that goat mixed saliva possesses a considerable affinity to tannins. A possible (although less explored) explanation offered for this affinity of goats mixed saliva to tannins is the potential role of salivary mucins. Van Soest (1994) had reported that goats may secrete higher levels of salivary mucins than sheep or cattle, and for that reason they could be more tolerant to high levels of dietary tannins. However, these hypothesis need to be further explored, since the role of salivary mucins in astringency perception is not completely clear.

In conclusion, several reasons seem to indicate that the oral cavity characteristics, namely of saliva are important in ruminant adaptation to diet. Although in other species saliva has drawn attention in the last years, in ruminants it has been less studied. However, there are evidences pointing to the importance of thoroughly investigating ruminant saliva interaction with diet in order to better understand ruminant food choices, which may allow improvements in nutritive management and therefore on ruminant productivity.

2.5 Final Considerations

This chapter was an attempt to address the most significant factors affecting livestock productivity: Climate, Health, and Nutrition. We have made an effort to analyze the factors individually. In our opinion these are extremely complex factors that are difficult to quantify, particularly in extensive production systems. Nevertheless, other important factors also exist. Such factors are most of the time human influenced and include, among other aspects, livestock selection, a tool that has been in practise since the Neolithic period and that has achieved a high level of

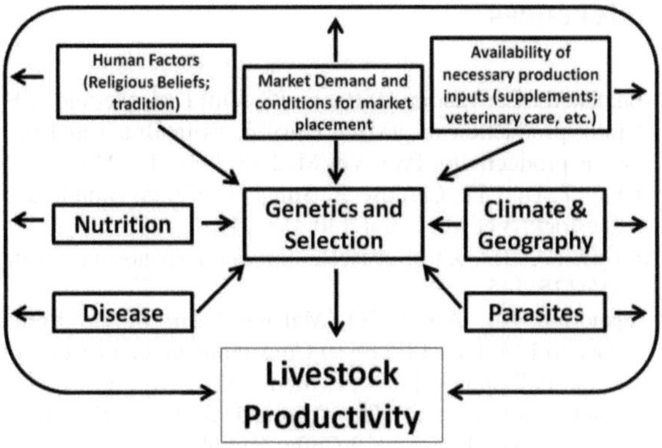

Fig. 2.5 Major factors influencing livestock productivity and their interconnection. Preponderant factors (nutrition, climate, diseases, and parasites are shown with a larger font size

sophistication over the last 200 years. Other man-made factors include religion and cultural beliefs that strongly influence choices in animal production species or breeds. Finally, economical constraints, marketing choices, and access to infrastructures such as roads, harbors, and railways are also strong factors that have great impacts on livestock welfare and productivity. In Fig. 2.5, we present a schematic representation of the major conditionings of farm animal productivity.

To minimize the effect of all the above-mentioned adverse factors in farm animal productivity, it is essential to design mitigation strategies at the local, regional, national, and transnational level. In our opinion, it is vital that such strategies would focus on the study and use of local genetic resources showing a high level of adaptation to the most significant issue for that specific region, either climate, disease, or nutrition-induced. Such studies have necessarily to be directed towards the full comprehension of the local breed's genetic background and production ability, as well as the search for markers of tolerance to those limiting factors. Such markers would be of great importance and use in the definition of selection strategies and objectives to increase livestock productivity, with special reference to developing countries. Such a strong research input would allow decreasing the need for costly production factors like disease or environmental control costs, as well as feed supplementation. A similar strategy would need to be thought at the local, regional, national, and transnational level and would require significant cooperation between North and Southern countries, state and private organizations, aiming to improve the productivity of the farming systems, therefore improving farmers' incomes, standard of living and ultimately, the sustainable development of whole communities. This would be an outstanding contribution to the fulfillment of the Millennium Development Goals established by the United Nations.

Acknowledgments Authors acknowledge funding from the *Fundação para a Ciência e a Tecnologia* of the Ministry of Science and Technology and Higher Education (Lisbon, Portugal) in the form of Post-Doctoral grants (SFRH/BPD/34751/2007—S. van Harten and SFRH/BPD/63240/2009—E. Lamy) and Research contract by the Ciência 2007 program (André M. Almeida).

References

Abdalla AL, Louvandini H, Bueno IC, Vitti DM, Meireles CF, Gennari SM (1999) Constraints to milk production in grazing dairy cows in Brazil and management strategies for improving their productivity. Prev Vet Med 38(2–3):217–230

Abdoel T, Dias IT, Cardoso R, Smits HL (2008) Simple and rapid field tests for brucellosis in livestock. Vet Microbiol 130:312–319

Aksoy S (2003) Control of tsetse flies and trypanosomes using molecular genetics. Vet Parasitol 115:125–145

Alejandrino AL, Assad CO, Malabayabasm B, De Vera AC, Herrera MS, Deocaris CC, Ignacio LM, Palo LP (1999) Constraints on dairy cattle productivity at the smallholder level in the Philippines. Prev Vet Med 38(2–3):167–178

Alexandersen S, Zhang Z, Donaldson AI, Garland AJM (2003) The pathogenesis and diagnosis of foot-and-mouth disease. J Comp Pathol 129:1–36

Almeida AM (2011a) *By endurance we conquer* fat tailed sheep in the 21st century. Trop Anim Health Prod 43:1233–1235

Almeida AM, Carvalho MR, Franco CF, Francisco R, Kilminster T, Scanlon T, Greef J, Milton J, Oldham C, Cardoso LA, Coelho AV (2010b) Weight loss and the hepatic proteome in african fat tailed, european and intermediate domestic sheep breeds. In: Marko-Varga G, Simões T (eds) 4th EuPA scientific meeting—A proteomics odyssey towards next decades. EUPA—European Proteomics Association, Estoril (Portugal)

Almeida AM (2011b) The Damara in the context of Southern Africa fat tailed sheep breeds. Trop Anim Health Prod 43:1427–1441

Almeida AM, Campos A, Francisco R, van Harten S, Cardoso LA, Coelho AV (2010a) Proteomic investigation of the effects of weight loss in the gastrocnemius muscle of wild and New Zealand white rabbits via 2D-electrophoresis and MALDI-TOF MS. Anim Genetics 41: 260–272

Almeida AM, Schwalbach LM, de Waal HO, Greyling JPC, Cardoso LA (2006a) A study of mineral status in the carcass of Boer Goat bucks: influence of supplementation. Arch Zootecnia (Córdoba, Spain) 55:313–316

Almeida AM, Schwalbach LM, de Waal HO, Greyling JP, Cardoso LA (2006b) The effect of supplementation on productive performance of Boer goat bucks fed winter veld hay. Trop Anim Health Prod 38:443–449

Almeida AM, Schwalbach LM, deWaal HO, Greyling JPC, Cardoso LA (2004) Serum amino acid and myofibrillar protein profiles in Boer Goat bucks following undernutrition. Small Rum Res 55:141–147

Almeida AM, Schwalbach LMJ, Cardoso LA, Greyling JPC (2007) Scrotal, testicular and semen characteristics of young Boer bucks fed winter veld hay: the effect of nutritional supplementation. Small Rum Res 73:216–220

Armand M, Borel P, Cara L, Senft M, Chautan M, Lafont H, Lairon D (1990) Adaptation of lingual lipase to dietary fat in rats. J Nutr 120:1148–1156

Austin PJ, Suchar LA, Robbins CT, Hagerman AE (1989) Tannins-binding proteins in saliva of deer and their absence in saliva of sheep and cattle. J Chem Ecol 15:1335–1347

Baker RL, Gray GD (2004) Appropriate breeds and breeding schemes for sheep and goats in the tropics. In: Sani RA, Gray GD, Baker RL (eds) Worm control for small ruminants in tropical Asia. ACIAR, Canberra

Barros CM, Pegorer MF, Vasconcelos JL, Eberhardt BG, Monteiro FM (2006) Importance of sperm genotype (*indicus* versus *taurus*) for fertility and embryonic development at elevated temperatures. Theriogenol 65:210–218

Barros SC, Ramos F, Luís TM, Vaz A, Duarte M, Henriques M, Cruz B, Fevereiro M (2007) Molecular epidemiology of bluetongue virus in Portugal during 2004–2006 outbreak. Vet Microbiol 124(1–2):25–34

Becerra L, Soares RV, Bruno LS, Siqueira CC, Oppenheim FG, Offner GD, Troxler RF (2003) Patterns of secretion of mucins and non-mucin glycoproteins in human submandibular/sublingual secretion. Arch Oral Biol 48:147–154

Bennick A (2002) Interaction of plant polyphenols with salivary proteins. Crit Rev Oral Biol Med 13:184–196

Berman A (2011) Invited review: are adaptations present to support dairy cattle productivity in warm climates? J Dairy Sci 94(5):2147–2158

Bishop SC, Morris CA (2007) Genetics of disease resistance in sheep and goats. Small Rum Res 70:48–59

Black PF, Murray JG, Nunn MJ (2008) Managing animal disease risk in Australia: the impact of climate change. Rev Sci Tech 27(2):563–580

Butterworth MH (1984) Beef cattle nutrition and tropical pastures. Longman, London (England)

Carr DH (1984) The regulation of parotid and submandibular salivary secretion in sheep. Q J Exp Physiol 69:589–597

Carvalho M, Mateus L, Afonso F, Van Harten S, Cardoso LA, Redmer DA, Ferreira-Dias G (2009) Testicular angiogenic activity in response to food restriction in rabbits. Reproduction 137(3):509–515

Cicek H, Cicec H, Eser M, Tandogan M (2009) Current status of ruminant theileriosis and its economical impact in Turkey. Turkiye Parazitol Derg 33:273–279

Clariget RP, Forsberg M, Rodriguez-Martinez H (1998) Seasonal variation in live weight, testes size, Testosterone, LH secretion, Melatonin and Thyroxine in Merino and Corriedale rams in SubTropical climate. Acta Vet Scand 39:35–47

Collier RJ, Collier JL, Rhoads RP, Baumgard LH (2008) Invited review: genes involved in the bovine heat stress response. J Dairy Sci 91:445–454

Collier RJ, Stiening CM, Pollard BC, Van Baale MJ, Baumgard LH, Gentry PC, Coussens PM (2006) Use of gene expression microarrays for evaluating environmental stress tolerance at the cellular level in cattle. J Anim Sci 84((E. Suppl.)):E1–E13

Courtin D, Berthier D, Thevenon S, Dayo G-K, Garcia A, Bucheton B (2008) Host genetics in African trypanosomiasis. Infection Genet Evol 8:229–238

da Costa G, Lamy E, Capela e Silva F, Andersen J, Sales-Baptista E, Coelho AV (2008) Salivary amylase induction by tannin-enriched diets as a possible countermeasure against tannins. J Chem Ecol 34:376–387

de Almeida AM, Cardoso LA (2008a) Animal production and genetic resources in Guinea Bissau: I—Northern Cacheu province. Trop Anim Health Prod 40:529–536

de Almeida AM, Cardoso LA (2008b) Animal production and genetic resources in Guinea Bissau: II—Tombali province. Trop Anim Health Prod 40:537–543

Delespaux V, Geysen D, Van den Bossche P, Geerts S (2008) Molecular tools for the rapid detection of drug resistance in animal trypanosomes. Trends Parasitol 24:236–242

d'Ieteren G, Authié E, Wissocq N, Murray M (1998) Trypanotolerance, an option for sustainable livestock production in areas at risk from Trypanosomosis. Rev Sci Tech 17:154–175

Dinnella C, Recchia A, Vincenzi S, Tuorila H, Monteleone E (2010) Temporary modification of salivary protein profile and individual responses to repeated phenolic astringent stimuli. Chem Senses 35:75–85

Distel RA, Provenza FD (1991) Experience early in life affects voluntary intake of blackbrush by goats. J Chem Ecol 17:431–450

Dorea FC, Cole DJ, Hofacre C, Zamperini K, Mathis D, Doyle MP, Lee MD, Maurer JJ (2010) Effect of *Salmonella* vaccination of breeder chickens on contamination of broiler chicken carcasses in integrated poultry operations. Appl Environ Microbiol 76:7820–7825

Fadiel A, Anidi I, Eichenbaum D (2005) Farm animal genetics and informatics: an update. Nucl Acids Res 33:6308–6318

FAO (2009) The state of food and agriculture. Livestock in the balance. FAO, Rome

Fickel J, Göritz F, Joest BA, Hildebrandt T, Hofmann RR, Breves G (1998) Analysis of parotid and mixed saliva in roe deer. J Comp Physiol B 168:257–264

Forman S, Hungerford N, Yamakawa M, Yanase T, Tsai HJ, Joo YS, Yang DK, Nha JJ (2008) Climate change impacts and risks for animal health in Asia. Rev Sci Tech 27(2):581–597
Frisch JE (1999) Towards a permanent solution for controlling cattle ticks. Int J Parasitol 29: 57–71
Geysen D, Bishop R, Skilton R, Dolan TT, Morzaria S (1999) Molecular epidemiology of *Theileria parva* in the field. Trop Med Int Health 4:A21–A27
Gillespie JR, Flanders FB (2010) Modern livestock and poultry production, 8th edn. Cengage Learning, New York
Glendinning JI (1994) Is the bitter rejection response always adaptative? Physiol Behav 56:1217–1227
Glendinning JI (1992) Effect of salivary proline-rich proteins on ingestive reponses to tannic acid in mice. Chem Senses 17:85–116
Godfroid J, Nielsen K, Saegerman C (2010) Diagnosis of *Brucellosis* in livestock and wildlife. Croat Med J 51:296–305
Gong Y, Hodgson J, Lambert MG, Gordon IL (1996) Short-term ingestive behaviour of sheep and goats grazing grasses and legumes. 1. Comparison of bite weight, bite rate, and bite dimensions for forages at two stages of maturity. New Zealand J Agri Res 39:63–73
Gootwine E (2011) Mini review: breeding Awassi and Assaf sheep for diverse management conditions. Trop Anim Health Prod 43:1289–1296
Green GC, Elwin RL, Mottershead BE, Keogh RG, Lynch JJ (1984) Long term effects of early experience to supplementary feeding in sheep. Proc Aust Soc Anim Prod 15:373–375
Gurkan S, Bradley RM (1988) Secretions of von Ebner's glands influence responses from taste buds in rat circunvallate papilla. Chem Senses 13:655–661
Hanovice-Ziony M, Gollop N, Landau SY, Ungar ED, Muklada H, Glasser TA, Perevolotsky A, Walker JW (2010) No major role for binding by salivary proteins as a defence against dietary tannins in Mediterranean goats. J Chem Ecol 36:736–743
Hansen PJ (2004) Physiological and cellular adaptations of zebu cattle to thermal stress. Anim Reproduc Sci 82–83:349–360
Heithoff DM, Shimp WR, Lau PW, Badie G, Enioutina EY, Daynes RA, Byrne BA, House JK, Mahan MJ (2008) Human *Salmonella* clinical isolates distinct from those of animal origin. Appl Environ Microbiol 74:1757–1766
Heringstad B, Klemetsdal G, Ruane J (2000) Selection for mastitis resistance in dairy cattle: a review with focus on the situation in the Nordic countries. Livestock Prod Sci 64:95–106
Hofmann RR (1989) Evolutionary steps of ecophysiological adaptation and diversification of ruminants. A comparative view of their digestive systems. Oecologia 78:443–457
Hope M, Jiang X, Gough J, Cadogan L, Josh P, Jonsson N, Willadsen P (2010) Experimental vaccination of sheep and cattle against tick infestation using recombinant 5′-nucleotidase. Parasitol Immunol 32:135–142
Hoste H, Torres-Acosta JFJ (2011) Non chemical control of helminths in ruminants: adapting solutions for changing worms in a changing world. Vet Parasitol 180:144–154
Iason G (2005) The role of plant secondary metabolites in mammalian herbivory: ecological perspectives. Proc Nutr Soc 64:123–131
Jonsson NN (2006) The productivity effects of cattle tick (*Boophilus microplus*) infestation on cattle, with particular reference to *Bos indicus* cattle and their crosses. Vet Parasitol 137:1–10
Kababya D, Perevolotsky A, Bruckental I, Landau S (1998) Selection of diets by dual-purpose Mamber goats in Mediterranean woodland. J Agric Sci 131:221–228
Kangas S, Lyytikainen T, Peltola J, Ranta J, Maijala R (2007) Costs of two alternative *Salmonella* control policies in Finnish broiler production. Acta Vet Scandinavica 49:35
Katsukawa H, Shang Y, Nakashima K, Yang KH, Sugita D, Mishima K, Nakata M, Ninomiya Y, Sugimura T (2002) Salivary cystatins influence ingestion of capsaicin-containing diets in the rat. Life Sci 71:457–467
Katzer F, Ngugi D, Walker AR, McKeever DJ (2010) Genotypic diversity, a survival strategy for the apicomplexan parasite *Theileria parva*. Vet Parasitol 167:236–243

Kawai T, Fushiki T (2003) Importance of lipolysis in oral cavity for orosensory detection of fat. Am J Physiol 285:R447–R454

Kay RNB (1960) The rate of flow and composition of various salivary secretions in sheep and calves. J Physiol 150:515–537

Kilminster T, Almeida AM, Greef J, Milton J, Oldham C, Scanlon T (2008) Are Damara and Dorper sheep better adapted than merinos to nutritional stress?—Carcass attributes. Agribusiness Livestock Updates: 42–43

Kitching P, Hammond J, Jeggo M, Charleston B, Paton D, Rodriguez L, Heckert R (2007) Global FMD control–Is it an option? Vaccine 25:5660–5664

Lamy E, da Costa G, Capela e Silva F, Potes J, Coelho AV, Sales Baptista E (2008) Comparison of electrophoretic protein profiles from sheep and goat parotid saliva. J Ecol 34:388–397

Lamy E, da Costa G, Santos R, Capela e Silva F, Potes J, Pereira A, Coelho AV, Sales Baptista E (2009) Sheep and goat saliva proteome analysis: a useful tool for ingestive behavior research? Physiol Behav 98:393–401

Lamy E, da Costa G, Santos R, Capela e Silva F, Potes J, Pereira A, Coelho AV, Sales Baptista E (2011) Effect of condensed tannin ingestion in sheep and goat parotid saliva proteome. J Anim Physiol Anim Nutr (Berl) 95:304–312

Lamy E, Graça G, da Costa G, Franco C, Capela e Silva F, Sales Baptista E, Coelho AV (2010) Changes in mouse whole saliva soluble proteome induced by tannin-enriched diet. Proteome Sci 8:65

Landau SY, Perevolotsky A, Kababya D, Silanikove N, Nitsan R, Baram H, Provenza FD (2002) Polyethylene glycol affects goats' feeding behaviour in a tannin-rich environment. J Range Manag 55:598–603

Lee CY, Munyard KA, Gregg K, Wetherall JD, Stear MJ, Groth DM (2011) The Influence of MHC and Immunoglobulins A and E on Host Resistance to Gastrointestinal Nematodes in Sheep. J Parasitol Res 2011:101848

Leinonen J, Parkkila S, Kaunisto K, Koivunen P, Rajaniemi H (2001) Secretion of carbonic anhydrase isoenzyme VI (CA VI) from human and rat lingual serous von Ebner's glands. J Histochem Cytochem 49:657–662

Lu Y, Bennick A (1998) Interaction of tannin with human salivary proline-rich proteins. Arch Oral Biol 43:717–728

Lundie RS (2011) The genetics of colour in fat-tailed sheep: a review. Trop Anim Health Prod 43:1245–1265

Lusweti EC (2000) The performance of the Nguni Afrikander and Bonsmara cattle breeds in developing areas of Southern Africa. South Afr J Anim Sci 30(S1):28–29

MacLachlan NJ (2004) Bluetongue: pathogenesis and duration of viraemia. Vet Ital 40:462–467

Magona J, Walubengo J, Kabi F (2011) Response of Nkedi Zebu and Ankole cattle to tick infestation and natural tick-borne, helminth and trypanosome infections in Uganda. Trop Anim Health Prod 43:1019–1033

Markemann A, Valle Zárate A (2010) Traditional llama husbandry and breeding management in the Ayopaya region. Bolivia Trop Anim Health Prod 42(1):79–87

McDougall EI (1948) Studies on ruminant saliva. 1. The composition and output of sheep's saliva. Bichem J 43:99–109

Mese H, Matsuo R (2007) Salivary secretion, taste and hyposalivation. J Oral Rehabil 34: 711–723

Miraglia M, Marvin HJ, Kleter GA, Battilani P, Brera C, Coni E, Cubadda F, Croci L, De Santis B, Dekkers S, Filippi L, Hutjes RW, Noordam MY, Pisante M, Piva G, Prandini A, Toti L, van den Born GJ, Vespermann A (2009) Climate change and food safety: an emerging issue with special focus on Europe. Food Chem Toxicol 47:1009–1021

Mirkena T, Duguma G, Haile A, Tibbo M, Okeyo AM, Wurzinger M, Sölkner J (2010) Genetics of adaptation in domestic farm animals: a review. Livestock Sci 132:1–12

Mugambi JM, Bain RK, Wanyangu SW, Ihiga MA, Duncan JL, Murray M, Stear MJ (1997) Resistance of four sheep breeds to natural and subsequent artificial *Haemonchus contortus* infection. Vet Parasitol 69:265–273

Murray N (2006) International trade and the spread of animal diseases: assessing the risks. Vet Italiana 42:319–336

Naessens J (2006) Bovine trypanotolerance: a natural ability to prevent severe anaemia and haemophagocytic syndrome? Int J Parasitol 36:521–528

Neyraud E, Sayd T, Morzel M, Dransfield E (2006) Proteomic analysis of human whole and parotid salivas following stimulation by different tastes. J Proteome Res 5:2474–2480

Noad R, Roy P (2009) Bluetongue vaccines. Vaccine 27:D86–D89

NRC (2005) Animal health at the crossroads. National Research Council/National Academy of Sciences, Washington, D.C

Pearson JE (2006) Worldwide risks of animal diseases: introduction. Vet Italiana 42:293

Perevolotsky A, Brosh A, Ehrlich O, Gutman M, Henkin Z, Holzer Z (1993) Nutritional value of common oak (*Quercus calliprinos*) browse as fodder for goats: experimental results in ecological perspective. Small Rumin Res 11:95–106

Pérez-Barberia FJ, Elston DA, Gordon IJ, Illius AW (2005) The evolution of phylogenetic differences in the efficiency of digestion in ruminants. Proc R Soc Lond B 271:1085–1090

Pérez-Maldonado RA, Norton BW, Kerven GL (1995) Factors affecting in vitro formation of tannin protein complexes. J Sci Food Agric 69:291–298

Perry BN, Sones AP, Stevens KC (2005) An appropriate level of risk: balancing the need for safe livestock products with fair market access for the poor (Pro-Poor Livestock Policy Initiative (PPLPI)). http://www.fao.org/ag/againfo/programmes/en/pplpi/docarc/execsumm_wp23.pdf. Accessed 21 Aug 2011

Pinto J, Bonacic C, Hamilton-West C, Romero J, Lubroth J (2008) Climate change and animal diseases in South America. Rev Sci Tech 27(2):599–613

Pourlis AF (2011) A review of morphological characteristics relating to the production and reproduction of fat-tailed sheep breeds. Trop Anim Health Prod 43:1267–1287

Preston TR, Leng RA (1987) Matching ruminant production systems with available resources in the tropics and sub-tropics, 1st edn. Penambul Books, Armindale (New South Wales, Australia)

Provenza FD (1995) Postingestive feedback as an elementary determinant of food preference and intake in ruminants. J Rang Manag 48:2–17

Rainard P, Riollet C (2006) Innate immunity of the bovine mammary gland. Vet Res 37:369–400

Ralphs MH (1997) Persistence of aversions to larkspur in naïve and native cattle. J Rang Manag 50:367–370

Robbins CT, Hanley TA, Hagerman AE, Hjeljord O, Baker DL, Schwartz CC, Mautz WW (1987) Role of tannins in defending plants against ruminants: reduction in protein availability. Ecology 68:98–107

Rupp R, Boichard D (2003) Genetics of resistance to mastitis in dairy cattle. Vet Res 34:671–688

Rupp R, Boichard D (2007) Genetics of udder health in dairy ruminants. 2007 EAAP meeting, Dublin. http://www.eaap.org/Previous_Annual_Meetings/2007Dublin/Papers/S34_02_Rupp.pdf. Accessed 22 Aug 2011

Rushton J (2009) The economics of animal health and production. CAB International, Oxford

Sales-Baptista E, Lamy E, Mau M, Capela e Silva F, Coelho AV (2009) Variation in salivary protein composition related to feeding behaviour and its ecological implications. In: Zhang W, Liu H (eds) Behavioural and chemical ecology. Nova Science Publishers, Haupagge

Scanlon T, Almeida AM, Greef J, Kilminster T, Milton J Oldham C (2008) Are Damara and Dorper sheep better adapted than merinos to nutritional stress?—Growth rates. Agribusiness Livestock Updates: 40–41

Shimada T (2006) Salivary proteins as a defence against dietary tannins. J Chem Ecol 32: 1149–1163

Silanikove N, Gilboa N, Perevolotsky A, Nitsan Z (1996) Goats fed tannin-containing leaves do not exhibit toxic syndromes. Small Rumin Res 21:195–201

Steward MC, Poronnik P, Cook DI (1996) Bicarbonate transport in sheep parotid secretory cells. J Physiol 494:819–830

Stolte M, Ito S (1996) A comparative ultrastructural study of the parotid gland acinar cells of nine wild ruminant species (mammalian, artiodactyla). Eur J Morphol 34:79–85

Szmaragd C, Wilson AJ, Carpenter S, Wood JL, Mellor PS, Gubbins S (2010) The spread of bluetongue virus serotype 8 in Great Britain and its control by vaccination. PLoS One 22 5(2):e9353

Thornton PK (2010) Livestock production: recent trends, future prospects. Philos Trans R Soc B Biol Sci 365:2853–2867

Udo HM, Budisatria IG (2011) Fat-tailed sheep in Indonesia; an essential resource for smallholders. Trop Anim Health Prod 43:1411–1418

Ulvevadet B, Hausner VH (2011) Incentives and regulations to reconcile conservation and development: thirty years of governance of the Sami pastoral ecosystem in Finnmark Norway. J Environ Manage 92(10):2794–2802

Vaithiyanathan S, Mishra JP, Sheikh Q, Kumar R (2001) Salivary glands tannins binding proteins of sheep and goat. Indian J Anim Sci 71:1131–1134

Van den Bossche P, Coetzer JA (2008) Climate change and animal health in Africa. Rev Sci Tech 27(2):551–562

van Harten S, Almeida AM, Morais Z, Schwalbach LM, Greyling JP, de Waal HO, Cardoso LA (2003) Free fatty acids and fatty acids of triacylglycerols profiles in muscle and plasma of fed and underfed Boer goats. Nutr Res 23:1447–1452

van Harten S, Cardoso LA (2010) Feed restriction and genetic selection on the expression and activity of metabolism regulatory enzymes in rabbits. Animal 4:1873–1883

Van Houtert MFJ, Sykes AR (1996) Implications of nutrition for the ability of ruminants to withstand gastrointestinal nematode infections. Int J Parasitol 26:1151–1167

Van Soest PJ (1994) Nutritional ecology of the ruminant, 2nd edn. Cornell University Press, New York

Vreysen MJBS, Khalfan M, Ali MY, Abdulla AM, Zhu ZR, Juma KG, Dyck VA, Msangi AR, Mkonyi PA, Feldmann HU (2000) Glossina austeni (Diptera: Glossinidae) eradicated on the Island of Unguja, Zanzibar, using the sterile insect technique. J Econ Entomol 93:123–135

Wambura PN, Gwakisa PS, Silayo RS, Rugaimukamu EA (1998) Breed-associated resistance to tick infestation in *Bos indicus* and their crosses with *Bos taurus*. Vet Parasitol 77:63–70

Willgert KJE, Schroedle B, Schwermer H (2011) Spatial analysis of bluetongue cases and vaccination of Swiss cattle in 2008 and 2009. Geosp Health 5:227–237

Wilson RT (1987) Livestock production in Central Mali: factors influencing growth and live weight in agro-pastoral cattle. Trop Anim Health Prod 19(2):103–114

Wroblewski K, Muhandiram R, Chakrabartty A, Bennick A (2001) The molecular interaction of human salivary histatins with polyphenolic compounds. Eur J Biochem 268:4384–4397

Zajácz A, Gyémánt G, Vittori N, Kandra L (2007) Aleppo tannin: structural analysis and salivary amylase inhibition. Carbohydr Res 342:717–723

Chapter 3
Heat Stress Impact on Livestock Production

James Olamitibo Daramola, Monsuru Oladimeji Abioja and Okanlawon Mohammed Onagbesan

Abstract The effects of heat stress on several aspects of animal production are well documented. Heat stress results from the animal's inability to dissipate sufficient heat to maintain homeothermy. High ambient temperature, relative humidity, and radiant energy compromise the ability of animals to dissipate heat. As a result, there is an increase in body temperature, which in turn initiates compensatory and adaptive mechanisms to re-establish homeothermy and homeostasis. Heat stress could affect animal production and well-being, especially because of increase in air temperature. Heat stress is very common and on the increase particularly in the tropics. There is considerable research evidence that shows significant decline in animal performance when subjected to heat stress. Heat stress inflicts heavy economic losses on livestock production. The effects of heat stress is evident in feed consumption, production efficiency in terms of milk yield or weight gain per unit of feed energy, growth rate, egg production, and reproductive efficiency. The physiologic mechanisms underlying the action of heat stress on the decline of production performance of domestic animals have not been fully investigated. Heat stress requires further investigation, and the elucidation of the mechanisms may facilitate adoption of comprehensive preventive and control measures to combat heat stress in domestic animals. This chapter examines heat stress and its negative impacts on livestock production. It elucidates the general negative effects of heat stress on physiologic and production parameters of domestic livestock. The mechanisms involved when animals are subjected to heat stress and impacts of heat stress on domestic animals are emphasized. An understanding of these mechanisms may result in the development of improved techniques for enhancing livestock productivity in tropical environments.

Keywords Heat Stress · Livestock · Production

J. O. Daramola (✉) · M. O. Abioja · O. M. Onagbesan
Department of Animal Physiology, University of Agriculture, Abeokuta, Nigeria
e-mail: daramolajames2003@yahoo.com

V. Sejian et al. (eds.), *Environmental Stress and Amelioration in Livestock Production*, DOI: 10.1007/978-3-642-29205-7_3,

Contents

3.1 Introduction

Domesticated animals, contribute to agricultural and economic development through provision of much needed protein of a nation; in particular, those countries in which the populations are growing faster than the economy. Animal protein requirements are met through domestic animal sources. The present supply of protein from domestic animals is currently inadequate to meet the protein requirement of the world's growing population because of many factors. One paramount and currently overriding factor is heat stress affecting the performance of domestic animals. It has been reported that domestic animals kept and managed in harsh environments are exposed to stress (Adeloye and Daramola 2004). Under extreme climate conditions, reduced ability of the animal to dissipate environmental heat results in significant heat stress during at least part of the year. Animals that are managed in these states loose condition, and often fail to perform when compared to animals not undergoing heat stress. Heat stress is one pressure placed upon an animal resulting in an increased need for the animal to dissipate excess body heat, requiring the animal to modify it's behavior to reduce this added stress. Figure 3.1 describes the details of the stress response in animals.

Physiologic changes do occur when animals are stressed but paramount and fundamental are loss of weight, reduced production, and death (Adeloye and Daramola 2004). Changes occurring in the animal as a result of heat stress include elevated body temperature, respiration rates, increased maintenance energy requirement, decreased efficiency of nutrient utilization, decreased dry matter (DM) intake, decreased milk production, and reproductive performance. The severity of heat stress experienced by an animal depends on actual temperature and humidity, length of the heat stress period, degree of night cooling that occurs, ventilation and air flow and level of production. The numerous physiologic mechanisms for coping with heat stress have been reported (Blackshaw and Blackshaw 1994). Sweating, elevated respiratory rates, vasodilation with increased

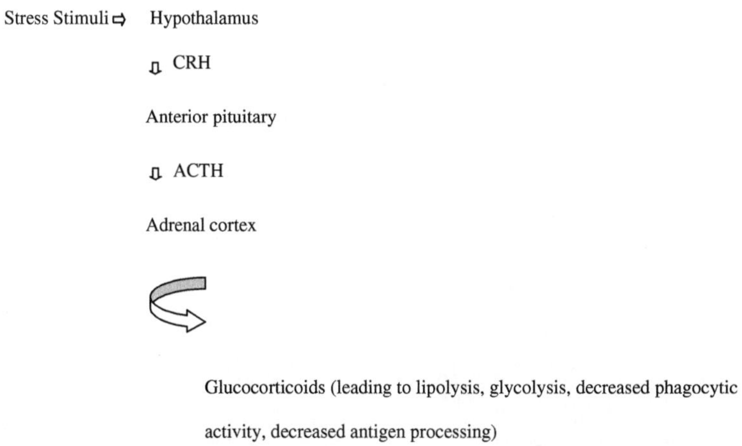

Fig. 3.1 Description of the stress cycle. The figure describes the stress cycle starting from the stress stimulus on hypothalamus to release CRH. CRH acts on the anterior pituitary and release ACTH. ACTH acts on adrenal cortex to release glucocorticoids to elicit the biologic actions to relieve stress (*Source* Adeloye and Daramola 2004)

blood flow to skin surface, reduced metabolic rate, decreased DM intake, and altered water metabolism are the physiologic responses that have negative impact on the production performance of livestock (West 1999). This review intends to integrate information documented on impacts of heat stress on the performance of domestic animals. The review identifies the mechanisms involved when animals are subjected to heat stress and knowledge of animal responses to heat stress may be used to evaluate the impacts of heat stress in animal production. Figure 3.2 describes the impact of heat stress on the productive parameters in livestock.

3.2 Feed Intake and Nutrient Utilization

Temperature levels outside the thermal comfort zone will increase body maintenance requirements for animals. Thermal comfort zone is a range in ambient temperature in which body temperature is possible and an animal need not change the metabolic rate. Consequently, efficiency of utilization of energy for production is decreased. Lowered efficiency occurs because animals eat less, thereby requiring more time to achieve a given level of weight and an appreciable amount of the energy consumed is siphoned off for the labor exerted to mitigate heat stress. The effect of heat stress on nutrient intake and utilization are well-documented in literature (Shanklin 1963; Attebery and Johnson 1969; Warren et al. 1974; Igono et al. 1985; Mallonee et al. 1985; Holter et al. 1996;). Studies have found that there is a significant negative correlation between Temperature-Humidity Index and DM intake (Shanklin 1963). However, differences exist for heat tolerance of different

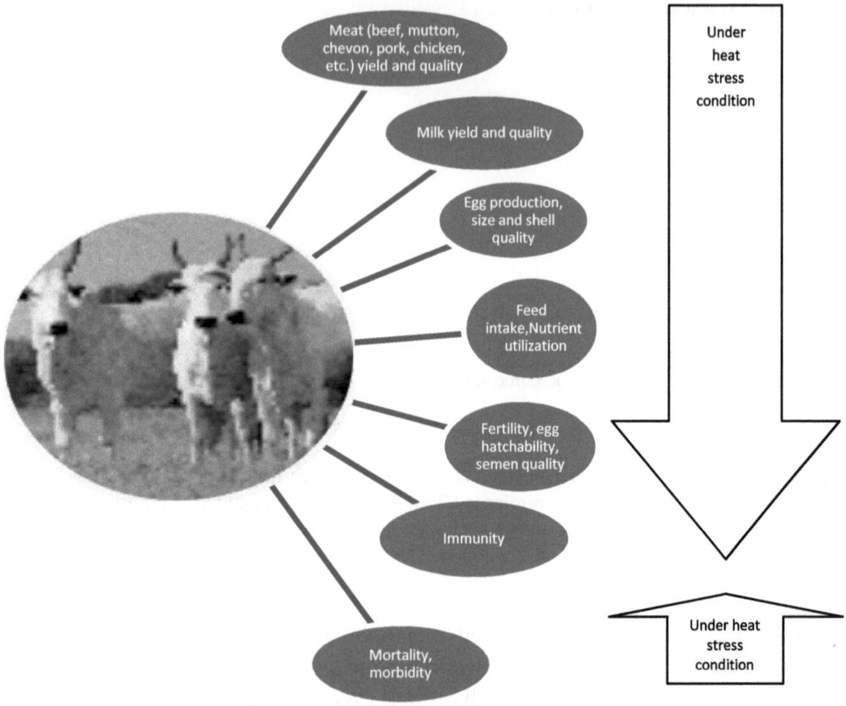

Fig. 3.2 Impact of heat stress on livestock production parameters. The figure describes the effect of heat stress on meat production, milk yield and quality, egg production, feed intake, nutrition utilization, fertility, semen quality, egg hatchability, immunity, mortality, and morbidity

breeds; and sweating capacity and metabolic rates are implicated in these variations (Blackshaw and Blackshaw 1994). Holter et al. (1997) reported a reduction in DM intake in Jersey cows when minimum Temperature-Humidity Index exceeded 56 and further reduced until Temperature-Humidity Index reached 72. During heat stress, DM intake is reduced to 22% for multiparous and 6% for primiparous dairy cows because of smaller body size and lower metabolic rate in primiparous cows (Igono et al. 1985). Although reduced DM intake and heat generated during ruminal fermentation and body metabolism aid in maintaining heat balance, heat stress caused by increased environmental temperature elevates the respiratory rate and water intake (Mallonee et al. 1985), reduces the gut motility, rumination, ruminal contractions (Attebery and Johnson 1969) and depresses appetite (Warren et al. 1974) by directly affecting the appetite center of the hypothalamus (Baile and Forbes 1974).

Feed consumption, feed quality, nutrient composition, rates of passage of digesta and volumes of ruminal and postruminal digestive organs are various factors that affect digestibility (Ellis et al. 1984). At high temperature, decreased feed intake evokes increased digestion by decreasing the passage of digesta and increasing the ruminal volume (Lippke 1975). These physiologic alterations are more helpful for animals consuming higher forage diets. Peripheral vasodilation and central vasoconstriction cause reduced blood flow to the ruminant forestomach (Engelhardt et al. 1977). This in turn decreases the portal vein blood flow hence, reducing nutrient absorption (McGuire et al. 1989).

In periods of heat stress, the risks of acidosis are increased. Factors that can contribute to rumen acidosis problems are: decreased DM intake with lower proportion of forage and higher levels of fermentable carbohydrates, decreased rumination, and decreased saliva to the gut—a source of bicarbonate—with a reduction of its buffering power due to increased carbon dioxide (CO_2) expelled (Bach et al. 2007). Rumen pH also decreases which impairs fiber digestion efficiency: rumen fibrolytic bacteria are the most affected when rumen pH drops below 6.0 (Bach et al. 2007). All of these factors contribute to decreasing feed utilization during heat stress.

In poultry, reduction in feed intake consequentially results in deficiency of essential nutrients. Growth rate is reduced in broilers when environmental temperature rises because energy obtained from the small amount of feed consumed is expended in panting. The result is that birds have lower final body weight after normal feeding period with higher environmental temperatures. Often, the broiler chickens reach market weight of less than 4 kg at day 84 instead of day 42. Ain Baziz et al. (1990) observed that in heat-stressed birds, body weight gain reduced more than feed intake, because part of the metabolizable energy intake is used for heat dissipation, resulting in impaired feed conversion ratio. Heat stress, also leads to lower feed efficiency. More feed is required to produce a unit weight of chicken. Consequently, longer times are required to reach market weight in broiler chickens reared under heat stress conditions. These negative effects are found to be greater in poultry with a high potential for growth rate (Cahaner et al. 1998; Settar et al. 1999). High thermal loads may depress enterocyte proliferation, reduction in villus mass and dry weight per unit length of jejunum (Sahin et al. 2001). A number of morphologic and physiologic changes occur in the gastrointestinal tracts of chickens exposed to heat stress. There is a reduction in gut mobility and depression of gastrointestinal blood flow (Wolfenson et al. 2000). Combinations of all these effects, lead to reduction in digestibility of feed and nutrient absorption from the intestines. It is not known whether high ambient temperature has impacts on the secretion and efficiency of digestive enzymes. However, Sahin et al. (2001) showed that plasma triiodothyronine (T_3) and thyroxine (T_4), which are important growth promoters in animals, are adversely affected in heat-stressed broiler chickens. Heat stress in domestic birds elicits secretion and increase in plasma levels of corticosteroids, while that of plasma proteins decreases with marked increase in blood glucose concentrations. Therefore, the inhibition of growth and production in heat-stressed broiler birds may be engineered by stress hormones, especially corticosteroids. As suggested by these observations, it is likely that combined effects of reduced feed consumption, increased energy cost required for heat dissipation, altered metabolic and gut physiologic processes cause the reduction in weight gains.

The consequences of heat exposure during rearing and during transportation to slaughterhouses generally involve an increase in mortality, reduced meat yield, and quality. Seasonal heat stress has been reported to accelerate postmortem glycolytic metabolism leading to biochemical changes in muscle and the production of pale, soft exudative meat characteristics in chickens. These detrimental effects are exacerbated in older birds (Sandercock et al. 2001). Ante and postmortem muscle metabolism is affected by stress reactions prior to slaughter. The rate and extent of glycogen breakdown, pH decline, and drip loss are also influenced. Terlouw (2004) stated that the effect is principally due to variations in adenosine-triphosphatase (ATPase) activity and muscle glycogen reserve. But acute heat stress appeared to have no effect upon breast meat color in broilers (Sandercock et al. 2001). No change in pH of broiler meat of acutely heat-stressed chickens was also reported. Terlouw (2004) indicated that production of meat with normal pH does not necessarily mean that animals have not been stressed. Lu et al. (2007) attributed high mortality and decreased growth, carcass, and breast muscle yield during heat exposure and indicated that heat stress did indeed cause physiologic stress in broilers, even though the resulting meat had normal pH.

3.3 Body Growth

Growth, the increase in live body mass, is under genetic and environmental regulation. The environmental factors that influence body growth include available nutrients, hormones, and enzymes as well as ambient temperatures (Bar and Radde 2009). Adequate and organized interaction of vascular growth factors and their receptors are required for placental development. When subjected to environmental heat stress at early stages of placental development, impaired placental vascular development occurs in part, due to lower levels of vascular growth factors in the tissues. Exposure to chronic heat stress has been observed to lower the circulating placental hormone concentrations due to impaired trophoblast cell development (Regnault et al. 2000).

The noticeable effects of heat stress on growth performance are the results of a decrease in anabolic activity caused by reduced voluntary feed intake, and increase in tissue catabolism. It is well-established that heat stress negatively impacts growth rate in swine. Although reduced feed intake undoubtedly plays a significant role in this reduction, studies in laboratory animals and other non-swine species indicate that muscle growth also is affected by heat stress—related alterations in muscle physiology (Kamanga-Sollo et al. 2011). Cattle, goats, and sheep are less sensitive to the effects of temperature than swine and poultry. Provided feeding is adequate, growth rate is not appreciably affected until the ambient temperature increases above the thermal comfort zone. When animal is subjected to temperatures outside of the comfort zone for a given animal size, the amount of feed required per unit of gain rises markedly. From the time of hatching until about 4 weeks of age, chicks have a narrow comfort range (32–34°C), efficiency of gain is reduced if the chicks are kept at a high temperature, growth rate will be almost ceased by the 7 or 8th week. While cold is a limiting factor in early stages of animal development, heat is most limiting in latter stages.

It is well-established that heat stress negatively impacts growth rate in animals. Although reduced feed intake undoubtedly plays a significant role in this reduction, studies in laboratory animals indicate muscle growth also is affected by heat stress-related alterations in muscle physiology. There is now emerging evidence that heat shock proteins (hsp), produced in response to heat stress and other types of cellular stress, may play important roles in regulating rate and efficiency of muscle growth. Stress, whether thermal or otherwise causes denaturing of proteins in the cells. During heat stress, the synthesis of hsp is increased while synthesis of other proteins is downregulated. So, the proteins that should have helped as building blocks are denatured or not synthesized at all. Because muscle satellite cells play crucial roles in postnatal muscle growth, the effects of heat stress on rates of satellite cell proliferation, protein synthesis, and protein degradation play an important role in determining the rate and extent of muscle growth.

Chronic exposure of growing pigs to a high ambient temperature is associated with enhanced lipid metabolism in the liver and the adipose tissue (Kouba et al. 2001). As a consequence, plasma triglyceride uptake and storage is facilitated in the adipose tissues, which results in greater fatness (Kouba et al. 2001). Increased fatness in long-term heat-exposed pigs was accompanied by the changes in the distribution of adipose tissues: a shift of body fat toward internal sites (Le Dividich et al. 1998), an increased weight of flare fat, and increased ratio of flare fat:back fat + flare fat (Kouba et al. 2001). The change in fat distribution in these heat-exposed pigs would appear to increase heat loss and represents an adaptation to high ambient temperature (Le Dividich et al. 1998; Kouba et al. 2001). Heat-exposed chickens also exhibit enhanced fat deposition (Ain Baziz et al. 1990).

Feed efficiency under hot conditions differs somewhat between mammals and birds. Feed to weight gain ratio is enhanced in hot conditions in chickens (Ain Baziz et al. 1990). On the other hand, an improvement in feed efficiency is often observed in rats and pigs under heat exposure (Rinaldo and Le dividich 1991). Stress reactions prior to slaughter may influence ante and postmortem muscle metabolism, and consequently, the rate and extent of glycogen breakdown, pH decline, and drip loss. Seasonal heat stress accelerates postmortem metabolism and biochemical changes in the muscle, which produces a faster pH decline, lower ultimate pH, and higher lightness values in turkey meat (McKee and Sams 1997). McKee and Sams (1997) and Lu et al. (2007) showed that chronic heat stress increased the lightness in muscle. The impact of stress response on meat quality is not inevitable. Terlouw (2004) indicated that production of meat with normal ultimate pH does not necessarily mean that animals have not been stressed.

3.4 Milk Production

During heat stress cows exhibit reduced feed intake, decreased activity, increased respiratory rate, and increased peripheral blood flow in sweating. When heat stress is experienced close to calving, an additional negative side effect is reduced cow's ability to produce high quality colostrum and impaired transfer of maternal IgG's to colostrum (West 2003). Levels of milk yield are sensitive to temperature conditions because temperature affects the feed intake of lactating cow. Temperature level at which significant depressions in milk yield occur depends on the humidity conditions, level of production, size of animal, and the breeds involved. At higher temperature above the comfort range, milk yield is depressed.

Heat stress is most detrimental to dairy cattle. Consequently reductions in feed consumption, milk production, and reproductive performance have been reported (Cavestany et al. 1985; Sharma et al. 1988; Bernabucci et al. 1999). Heat stress is a major factor contributing to low milk production and low fertility in lactating dairy cows. Various tissues are affected and their functions disrupted under heat stress (Wolfenson et al. 2000). High milk production is associated with high metabolic heat production. As a result, cows have to dissipate larger amounts of heat in order to maintain normothermia. At air temperatures of 27°C, under humid climates, the body temperature of lactating cows rises above normothermic values, and severe hyperthermia develops as air temperature rises (Wolfenson et al. 2000). The ambient temperatures rise to levels that induce hyperthermia in lactating cows under heat stress. The impact of heat stress is compounded by relatively low sweating rate in cattle (Berman and Wolfenson 1992).

The severity of heat stress is correlated to both ambient temperature and humidity level (Bach et al. 2007). The animal comfort is optimal, with a body temperature between 38.4 and 39.1°C. Above 25°C, and even 20°C for some authors, the cow suffers from heat stress: its health status and production performance are affected. Cows have ways to maintain thermal balance and regulate body temperature under high heat conditions. This involves favoring heat dispersion, in particular through evaporation, by increasing subcutaneous blood flow, panting, drooling, etc. These activities increase the maintenance energy needs of the animal by an estimated 20% at 35°C (Bach et al. 2007). In the case of the dairy cow, this means that part of its production energy will be redirected to thermal regulation. Also, rumination, which produces heat, decreases dramatically. Cows will tend to eat less during the day, but more often and in small quantities. They will tend to consume more feed at night when it is cooler, slug feed, sort feed and tend to choose feeds that produce less heat during digestion, choosing grains and proteins over forages.

3.5 Semen Production and Sperm Characteristics

Some physiologic traits that have direct bearing on quality of ejaculate are known to be affected when male animals are stressed due to handling, methods of ejaculation and elevated temperature during the time of semen collection (Marai et al. 1997). Among all climatic elements, temperature is the most important parameter affecting spermatogenesis. Skinner and Louw (1966) reported that high ambient temperature causes a sharp reduction in semen quality with many abnormal sperm cells. Exposure to heat stress is registered by the temperature-humidity index that includes both ambient temperature and relative humidity (LPHSI 1990; Marai et al. 2000). Heat stress is known to cause temporary interruption of sperm production, sperm motility, and secondary defects (Moreira et al. 2001).

Heat stress affects all phases of semen production in breeder cocks as reported in other species (Banks et al. 2005). Although limited high temperature stimulates testicular growth in the early phase and promotes increased semen volume and concentration, a subsequent rise lead to decreased semen quality and quantity with time (Obidi et al. 2008; McDaniel et al. 1996; Edens 1983). Serum calcium and phosphorus levels were observed to be significantly lowered in heat-stressed birds (McDaniel et al. 1995, 1996). Transient inward calcium ion currents whose density increased during spermatogenesis, from spermatogonia to early spermatids, have been observed (Hagiwara and Kawa 1984). The decrease in spermatogenesis due to inhibition of calcium and potassium ion exchange (Hagiwara and Kawa 1984; Schreiber et al. 1998), implies that distinct expression and non-inhibition of ion channels during spermatogenesis may enhance the excitation and differentiation of seminiferous epithelium (Hagiwara and Kawa 1984; Schreiber et al. 1998), as is characteristic of excitable tissues. Some of the ion channels regulating ion exchange during the preliminary stages of germinal cell differentiation end up in mature spermatozoa, determining their physiologic properties (Schreiber et al. 1998. The decrease in calcium ion level due to heat stress causes deleterious multiple effects on testicular function through inhibition of intracellular ion exchange (McDaniel et al. 1996).

In a study conducted to investigate the changes that might occur in spermiograms, blood and physiologic indices following successive electroejaculation (EE) during cold and hot periods of the day, progressive sperm motility, sperm concentration, and mass activity followed similar trend and the values deteriorated with respect to elevated temperatures during semen collection periods (Daramola and Adeloye 2010). Primary abnormalities increased with respect to elevated temperatures during semen collection periods (Daramola and Adeloye 2010). The authors observed that reduced ejaculate quality reflects stress stimuli arising from increase ambient temperature and physiologic traits in West African Dwarf (WAD) goat and reported it as the adaptive mechanism evolved to cope with stress arising from elevated temperatures. The proportion of the sperm abnormalities increased concurrently in response to increased ambient temperature intensity (Daramola and Adeloye 2010). The increase in morphologic abnormalities at high ambient temperature indicates that periods of collection have deleterious effect on the testes or epididymis, such as testicular degeneration (Daramola and Adeloye 2010). Skinner and Louw (1966) reported that high ambient temperature causes a sharp reduction in semen quality with many abnormal sperm cells. The rise in primary abnormalities therefore indicates that elevated ambient temperature results in the rapid release of immature spermatozoa (Skinner and Louw 1966; Daramola and Adeloye 2010).

The reproductive performance of the rooster is greatly depressed during environmental stress. In a simulated study on the effects of heat stress on fertility in broiler breeder roosters, McDaniel et al. (1996) showed that the broiler breeder contributed more to heat-induced infertility than the female. When the male broiler breeder was exposed to a temperature of 32°C, male fertility declined to 42% and in vivo sperm-egg penetration declined to 52%, compared to values obtained from males that were maintained at 21°C. This observation demonstrated a significant inhibition of the rooster's spermatozoa viability through qualitative and quantitative depression in semen characteristics, such as spermatozoa motility.

Heat stress may be responsible for the inhibition of osmotic equilibrium and ionic channels that are key elements in the interplay between spermatozoa, its environment, and the egg, thus disrupting spermatozoa cellular homeostasis, distorting spermatozoa behavior, and metabolic machinery (Darszon et al. 1999). In mammals, excessive levels of reactive oxygen species have been correlated with decreased sperm motility (Agarwal et al. 1994; Armstrong et al. 1999). The report of McDaniel et al. (1996) showed that semen characteristics, such as consistency, spermatozoa concentration, and seminal volume were depressed as a result of a decrease in seminiferous epithelial cell differentiation, which is manifested by environmental temperatures outside the zone of thermal comfort. Heat-induced infertility is mediated through any compromise in the fluidity and integrity of spermatozoa cell membranes as well as acrosomal and deoxyribonucleic acid damage (Surai 2000, 2002, 2010), and the inhibition of expression of hyaluronic acid binding sites as well as acrosomal integrity (Shamsuddin and Rodriguez-Martinez 1994; Morrell and Rodriguez-Martinez 2011). The differences found when breeder cocks were exposed to elevated ambient temperatures were not evident when the female birds alone were exposed to the same high ambient temperatures (McDaniel et al. 1996; Abd-Ellah 1995). Edens (1983) reported significant effects of ambient temperature on male fertility, which were evident within 12 h of challenge at a typical summer temperature of 29°C, although semen characteristics, such as semen volume, spermatozoa concentration, and percentage dead spermatozoa were unaffected by the heat treatment. This apparent lack of observable depreciation in semen characteristics obtained in the study of Edens (1983) suggests that roosters can adapt to short-term exposure to thermal stress. Thus, physiologic changes inimical (injurious/harmful) to testicular functions may not occur in short-term exposure to heat stress. The finding of Edens (1983) disagreed with those of McDaniel et al. (1995, 1996), who subjected roosters to a long-term heat exposure. The depression in in vivo sperm-egg penetration and fertility in heat-stressed roosters reported by McDaniel et al. (1995, 1996) may be due to a decrease in number of spermatozoa stored in the sperm host glands in the hen's reproductive tract (Bakst et al. 1994; Brillard 2003). In other words, a decrease in oviductal spermatozoa storage results in fewer spermatozoa cells available to bind, penetrate, and fertilize the egg in the infundibulum of the hen as documented by King et al. (2002).

In mammals, spermatozoa's binding with uterine epithelial cells is a strong index of spermatozoa viability and fertilizing capacity, implying that spermatozoa attachment to uterine epithelial cells is indicative of normal ultrastructure and mitochondrial membrane potential (Mburu et al. 1996; Taylor et al. 2008; Taylor et al. 2009). It is known that spermatozoa that have been bound temporarily to uterine epithelial cells can pass along the oviduct for fertilization (Taylor et al. 2008). In this context, it is reasonable to conclude that heat stress in the rooster retards or even prevents important physiologic mechanisms, such as sperm-uterine epithelial cells interaction, capacitation, acrosome reaction, and zonal vesicle binding, resulting in depressed fertility. This is, apparently, due to a depletion of endogenous antioxidant milieu in semen, leading to speedy exhaustion of spermatozoa energy reserves. On the other hand, it is likely that exposed spermatozoa are properly stored in the hen's oviduct, but their release was inhibited; thus, the spermatozoa were unable to bind and penetrate the ovum (Abd-Ellah 1995; King et al. 2002; Brillard 2003). It is worthy to note that roosters in pen-mated (natural mating) breeding system are known to reduce mating activity, and sexual arousal behavior (libido) is strongly impaired during heat stress, presumably through dehydration and alteration in secretion of sex hormones.

3.6 Immune Response and Endocrine System

Several studies have been conducted on the effects of heat stress due to high temperature on the immune responses of chickens, with variable results. Thaxton et al. (1968) demonstrated that high environmental temperatures (44.4–47.8°C) affect the development of specific immune responses in young chickens. These effects include the suppression of circulating white blood cells (Nathan et al. 1976; Heller et al. 1979) and an increase in the heterophil/lymphocyte ratio (H/L ratio) (Mogenet and Youbicier-Simo 1998), which are indicators of stress (Gross and Siegel 1983). Heat stress also reportedly causes a reduction in antibody production in young chickens (Zulkifi et al. 2000). On the other hand, Donker et al. (1990) found that heat exposure did not reduce antibody production, rather, significantly increased antibody titers were observed following heat exposure (Heller et al. 1979). The difference in these findings could be associated with age and breed. Regnier et al. (1980) suggested that heat-induced immunosuppression may depend on breed of bird and Kelley (1983) reported that effects on immune responses may depend on the length and intensity of the heat exposure.

Heat stress can negatively affect an animal's growth performance and the immune competence to some bacterial or viral infections (Goligorsky 2001). It has been reported that heat stress results in decrease of both primary and secondary lymphoid organs, profiles of circulating leukocytes, T cell in the blood, and antibody response to sheep red blood cells or against Newcastle disease (Davison et al. 1988; Liew et al. 2003).

The endocrine system involved in coordination of metabolism is substantially altered because of thermal stress (Beede and Collier 1986). The hormones associated with adaptation to heat stress are prolactin, growth hormone, thyroid hormones, glucocorticoids, mineralocorticoids, catecholamines, and antidiuretic hormone. Prolactin is vital for mammogenesis (Buttle et al. 1979), lactogenesis (Akers et al. 1981). Growth hormone is a calorigenic hormone produced from the anterior pituitary gland and does not function through a target gland but exerts its effects on almost all tissues of the body. Igono et al. (1988) reported that Growth hormone content in milk of low, medium, and high production groups declined when Temperature-Humidity Index exceeded 70. Plasma Growth hormone reductions that occurred with heat-stressed cows did not occur in thermoneutral conditions for cows fed restricted intakes that were similar to those consumed during heat stress (McGuire et al. 1989).

Although stress isn't the only reason that cortisol is secreted into the bloodstream, it has been termed "the stress hormone" because it's also secreted in higher levels during the body's response to stress, and is responsible for several stress-related changes in the body. Daramola and Adeloye (2010) and Daramola et al. (2011) reported increased cortisol level with respect to the semen collection periods and indicated that the increase observed in cortisol concentrations in the hot period of the day reflects stress stimuli due to elevated ambient temperature (Daramola and Adeloye 2010; Daramola et al. 2011). Apparently, the higher cortisol levels observed was attributed to stress caused by increase in ambient temperature (Daramola and Adeloye 2010), similar to the reports of Ortiz-de-Montellano et al. (2007) and therefore reflects stress stimuli due to elevated ambient temperature.

3.7 Blood Parameters

There is a great variation in the hematologic and biochemical parameters as observed between breeds of goats (Azab and Abdel-Maksoud 1999; Tambuwal et al. 2002). Seasonal variations in hematologic parameters of domestic chickens have also been reported (Oladele et al. 2003). Lowered packed cell volumes in domestic chickens have been observed during the hot-dry season, a period associated elevated ambient temperature as compared to other season (Oladele et al. 2003). Oladele et al. (2001) attributed the low values of hemoglobin and packed cell volume during the hot-dry season to heat and nutritional stress, which impairs the production of blood cells in birds and further observed significant correlation between the hot-dry season meteorological elements and packed cell volume, hemoglobin, and total protein in the chicken. Hemoglobin therefore seems to be highly responsive to fluctuations in ambient temperature, with a significant negative response to the deleterious effect of heat stress. Total protein values also demonstrate a significant and negative relationship with elevated ambient temperature (Oladele et al. 2003). These observations support the findings of Sahin et al. (2001), who demonstrated significant negative effects of heat stress on total serum proteins in broiler chickens. Sahin et al. (2001) also reported significant negative effects of heat stress (32°C) on serum concentration of some metabolites and minerals in broilers. Serum levels of thyroxine (T_4) and triiodothyronine (T_3) were significantly reduced due to high levels of adrenocorticotropic hormone in unsupplemented (vitamins E and A) broiler chickens, when compared to values obtained in birds that received antioxidant feed supplements (vitamins E and A). Serum calcium and phosphorus levels were observed to be significantly lowered in heat-stressed birds (McDaniel et al. 1995, 1996).

3.8 Egg Production

The effect of heat stress in laying birds is chronologic: laying flocks typically have a reduction in egg size, followed by lowered egg production, and reduced egg shell quality (Grieve 2003). Ambient temperatures influence reproductive function through alteration of feed intake. Reduction in feed intake resulting from heat stress is observed in layer chickens during hot-dry seasons (Simon 2003; Ayo et al. 1999) which ultimately leads to reduction in hen-day production. The observed decreases in voluntary feed intake by birds is attributed to physiologic responses to heat stress, aimed at reducing the excessive endogenous heat generated in the body due to feed metabolism (Simon 2003). The depression is attributed to an imbalance in calcium-estrogen relationship and lowered Haugh unit of the ovalbumin (Mahmoud et al. 1996). This implies that high environmental temperature depresses yolk size, ovalbumin consistency, and optimum calcium deposition within the egg shell.

Various authors reported that increased environmental temperatures affect egg production. Smith (2000) stated that the effect of ambient temperature on average egg weight appears to be cumulative. Thus, when birds are kept at 26°C, the mean egg weight increases by 1 g per week whereas when kept at 35°C, the average egg weight remains constant for a period of six months. North (1984) illustrated the effect of poultry house temperature on egg production, egg weight and feed consumed per egg as percentage of the optimum of 16°C. The author did not observe any change in egg production until a temperature of 24°C was reached. However a gradual decline in egg production was recorded as the temperature increased to 32°C.

Egg shell weight, shell thickness and specific gravity significantly declined in laying hens heat-stressed for 5 weeks (Mashaly et al. 2004). The heat-stressed laying flock often lays eggs with thinner shells because of acid–base disturbances in the blood (respiratory alkalosis). The higher blood pH caused by decreases in blood CO_2 concentrations reduce the amount of ionized calcium and bicarbonate in the blood. Ionized calcium is the form of calcium utilized by the shell gland in producing the egg shell (Grieve 2003). Increasing the amount of calcium in the diet does not correct this problem.

3.9 Incubation, Embryonic Development and Hatchability

The modern incubator is a simulated artificial design that mimics the mother-hen's role of providing fertile eggs with optimum environmental conditions (temperature and humidity) to stimulate embryonic development until hatching (French 1997). Deleterious effects of heat stress on the incubation of the avian embryo, hatchability, post-hatch development of chicks are well-documented (Romanoff 1972; Deeming and Ferguson 1991; French 1997; Hill 2001; Moraes et al. 2003, 2004; Lourens et al. 2007). Optimum environmental conditions are synonymous with incubation temperatures, which determine the efficiency of embryonic and post-hatch development of chicks (Lin et al. 2006; Romanoff 1972). In line with observation of Romanoff (1972), French (1997), Hill (2001) and Lourens et al. (2007) reported deleterious effects of heat stress on the incubation of the avian embryo, and showed extensive influences of temperature on chicks' embryo development, and that environmental temperature is the most critical factor in incubation efficiency. Wilson (1991), Lourens et al. (2005, 2007) and Moraes et al. (2003) confirmed the adverse effects of temperature on post-hatch development of chicks. It has been shown that a constant incubation temperature of 37.8°C, established as thermal homeostasis in the chick embryo (Lourens 2001), gave the best embryo development and hatchability (Lourens et al. 2007; Wilson 1991). Any marginal deviation from this fragile balance is detrimental to the developing embryo (Lourens et al. 2007). Thus, a constant high temperature of 38.9°C during incubation initially accelerates embryonic growth, utilization of nutrients and energy from the yolk and albumen reserves, but later decreases embryonic development as a result of limited metabolic process by insufficient exchange of oxygen (Lourens et al. 2005; Rahn et al. 1974).

Heat stress in the incubation process has been shown to have diverse detrimental influences on embryos. French (1994) observed deleterious effect of heat stress on embryo survival and showed that dead embryos occurred soon after subjecting them to heat stress, especially on days 7 and 19. This implies that embryos at these stages of development may be very sensitive to all types of stress, including heat stress, which could be related to the chorioallantoic membrane susceptibility to environmental stress. Increased embryonic death is, apparently, due to increased endogenous (metabolic) heat production (French 1994). The observation is in line with the findings of Lourens et al. (2005), who reported significant embryo mortality and, hence, lower hatchability in chicken eggs, when they were subjected to a high incubation temperature of 38.9°C. Apart from embryonic mortality, the quality of chicks from heat-stressed embryos has been reported to be adversely affected. Lourens et al. (2005) established depressed chick quality, lower percentage of first-grade chicks due to adverse effects of heat stress on chick quality and production.

Temperature is the most important factor in the incubation of any avian egg. An egg can, to a certain degree, compensate for various insults to its well-being except for temperature extremes. A 0.5°F change in incubator temperature can have a profound effect in overall performance of a group of eggs that incubate (Jeffrey et al. 2007). Extremes in temperature (high or low) can cause problems with embryo growth, or in many cases death. Eggs that are hatching early or very small chicks at hatch are an indication of incubator temperatures being too high (Jeffrey et al. 2007).

As water evaporates during incubation, it is replaced by gas molecules that form the air cell, which should occupy about 15% of the egg volume prior to hatching (Rahn et al. 1977) Increasing shell porosity and permeability will increase the oxygen uptake of the embryo until a maximum rate is reached, after which further increase in shell porosity and permeability have little effect (Burton and Tullett 1982). Therefore, oxygen availability is not a limiting factor for embryos within highly permeable, porous shells, but rather there is the danger of dehydration caused by elevated temperature (Burton and Tullett 1984). During incubation, eggs should ideally lose a quantity of water equal to 12% of their initial mass (Davis and Ackerman 1987). Water loss exceeding 20% of initial egg mass causes increased mortality and subsequent dehydration of the embryo, thus decreasing hatching success (Davis and Ackerman 1987).

Heat-stressed embryos have been observed to exhibit shorter face length and low lung weight, resulting in weaker chicks with high incidence of culled-out birds due to unsteady gait (Yalcin and Siegel 2003). An increase in environmental temperature may cause metabolizable energy to be diverted from growth and development to functions involved in homeothermy. High environmental temperatures reduce thyroid function and, consequently, metabolic rate, oxygen consumption, and growth rate (Romanoff et al. 1938; Moraes et al. 2003). Christensen et al. (2002) showed that the chick embryonic thyroid plays a major role in maturation of vital tissues during the final stages of in ovo life; the authors reiterated that the embryonic thyroid had a significant control of hatching times and survival rates of neonates. These reports support the fact that lower egg fertility and hatchability, retarded embryonic and post-hatch chick developments are due to consistent heat stress (Abioja 2010).

3.10 Conclusion

Heat stress occurs at the point where the animals cannot dissipate an adequate quantity of heat to maintain body thermal balance. High temperature, high humidity, and radiant energy (sunlight) are the major environmental factors that contribute to heat stress. Heat stress is very common and on the increase particularly in the tropics. There is considerable research evidence that shows significant decline in animal performance when subjected to heat stress. It inflicts heavy economic losses on livestock production. The effects of heat stress is evidenced in feed consumption, gross efficiency in terms of milk yield or weight gain per unit of feed energy, growth rate, milk production, egg production, and reproductive efficiency. The effects should be considered in future experiments, designed to elucidate the mechanism of heat stress on production efficiency in the domestic animals reared in the tropical and subtropical regions of the world. The physiologic mechanisms underlying the action of heat stress on the decline of production performance of domestic animals have not been fully investigated. This requires further investigation, and the elucidation of the mechanisms may facilitate the adoption of comprehensive preventive and control measures of combating heat stress in domestic animals.

References

Abd-Ellah AM (1995) Effect of ascorbic acid supplementation on performance of laying hens during hot summer months. Assiut Vet Med J 34:83–95
Abioja MO (2010) Temperature-humidity effects on egg fertility and evaluation of vitamin C and cold water on broiler growth in hot season. Ph.D Thesis University of Agriculture, Abeokuta, Nigeria, p 165
Adeloye AA, Daramola JO (2004). Improving living conditions of domestic animals in Nigeria: a review. In: Amalu UC, Gottwald FT (eds) Studies of sustainable agriculture and animal science in sub-Saharan Africa, vol 1. Support Africa International, pp 99–115
Agarwal A, Ikemoto I, Loughlin KR (1994) Relationship of sperm parameters with levels of reactive oxygen species in semen specimens. J Urol 152(1):107–110
Ain Baziz H, Geraert PA, Guillaumin S (1990) Effects of high temperature and dietary composition on growth, body composition and retention in broilers. In: Proceedings of 8th European poultry conference, vol 1. World's Poultry Science Association, Barcelona, Spain, pp 626–629
Akers RM, Bauman DE, Capuco AV, Goodman GT, Tucker HA (1981) Prolactin regulation of milk excretions and biochemical differentiation of mammary epithelial cells in periparturient cows. Endocrinology 109:23–30
Armstrong JS, Rajasekaran M, Chamulitrat W, Gatti P, Hellstrom WJ, Sikka SC (1999) Characterization of reactive oxygen species induced effects on human spermatozoa movement and energy metabolism. Free Radical Biol Med 26(7–8):869–880
Attebery JT, Johnson HD (1969) Effect of environmental temperature, controlled feeding and fasting on rumen motility. J Anim Sci 29:727
Ayo JO, Oladele SB, Ngam S, Fayomi A, Afolayan SB (1999) Diurnal fluctuations in rectal temperature of the Red Sokoto goat during the harmattan season. Res Vet Sci 66(1):7–9
Azab ME, Abdel-Maksoud HA (1999) Changes in some hematologic and biochemical parameters during pre-partum and post-partum periods in female Baladi goats. Small Ruminant Res 34:77–85
Bach A, Iglesias C, Devant M (2007) Daily rumen pH of loose-housed dairy cattle as affected by feeding pattern and live yeast supplementation. Anim Feed Sci Technol 136(1–2): 146–153
Baile CA, Forbes (1974) Control of feed intake and regulation of energy balance in ruminants. Physiol Rev 54:160
Bakst MR, Wishart G, Brillard JP (1994) Oviductal sperm selection, transport and storage in poultry. Poult Sci Rev 5:117–143
Banks S, King SA, Irvine DS, Saunders PTK (2005) Impact of a mild scrotal heat stress on DNA integrity in murine spermatozoa. Reproduction 129(4):505–514
Bar NS, Radde N (2009) Long-term prediction of fish growth under varying ambient temperature using a multiscale dynamic model. BMC Syst Biol 3: 107 http://www.biomedcentral.com/1752-0509/3/107
Beede DK, Collier RJ (1986) Potential nutritional strategies for intensively managed cattle during thermal stress. J Anim Sci 62:543–554

Berman A, Wolfenson D (1992) Environmental modifications to improve production and fertility. In: Van Horn HH, Wilcox CJ (eds) Large dairy herd management. American Dairy Science Association, Champaign, pp 126–134

Bernabucci U, Bani P, Ronchi B, Lacetera N, Nardone A (1999) Influence of short and long term exposure to a hot environment on rumen passage rate and diet digestibility by Friesian heifers. J Dairy Sci 82:967–973

Blackshaw JK, Blackshaw AW (1994) Heat stress in cattle and the effect of shade on production and behavior. Aust J Exp Agric 34:285–295

Brillard JP (2003) Practical aspects of fertility in poultry. World's Poult Sci J 59(4):441–446

Burton FG, Tullett SG (1982) A comparison of the effects of eggshell porosity on the respiration and growth of domestic fowl, duck and turkey embryos. Comp Biochem Physiol 75A: 167–174

Burton FG, Tullett SG (1984) The effects of egg weight and shell porosity on the growth and water balance of the chicken embryo. Comp Biochem Physiol 81A:377–385

Buttle HL, Cowie AT, Jones EA, Turvey A (1979) Mammary growth during pregnancy in hypophysectomized or bromocriptine-treated goats. J Endocr 80:343–351

Cahaner A, Deeb N, Yunis R, Lavi Y (1998) Reduced stress tolerance in fast growing broilers. In: Proceedings of 10th European poultry conference, vol 1. Jerusalem, Israel, pp 113–117

Cavestany D, El-Whisky AB, Foot RH (1985) Effect of season and high environmental temperature on fertility of Holstein cattle. J Dairy Sci 68:1471–1478

Christensen VL, Davis GS, Nestor KE (2002) Environmental incubation factors influence embryonic thyroid hormones. Poult Sci 81(4):442–450

Daramola JO, Adeloye AA (2010) Changes in spermiograms, biochemical and physiologic indices following successive electroejaculation during different periods of the day in west African Dwarf bucks. ASSET. Series A, 10 (1) (in press)

Daramola JO, Adeloye AA, Ife A Balogun, Yousuf MB, Olatunde AO, Abiona JA (2011) Changes in serum cortisol concentrations in west African Dwarf (WAD) bucks during electroejaculation. Niger J Anim Prod (in press)

Darszon V, Labarca P, Nishigaki T, Espinosa F (1999) Ion channels in sperm physiology. Physiol Rev 79(2):481–510

Davis TA, Ackerman RA (1987) Effects of increased water loss on growth and water content of the chick embryo. J Exp Zool 76(Suppl 1):357–364

Davison TF, Misson BH, Williamson RA, Rea J (1988) Effect of increased circulating corticosterone in the immature fowl on the blastogenic responses of peripheral blood lymphocytes. Dev Comp Immunol 12(1):131–144

Deeming DC, Ferguson MWJ (1991) Physiologic effects of incubation temperature on embryonic development in reptiles and birds. In: Deeming DC, Ferguson MJW (eds) Egg incubation. Cambridge University Press, Cambridge, pp 147–172

Donker RA, Nieuwland MG, Van der Zijpp AJ (1990) Heat stress influences on antibody production in chicken lines selected for high and low immune responsiveness. Poult Sci 69:599–607

Edens FW (1983) Effect of environmental stressors on male reproduction. Poult Sci 62(8): 1676–1689

Ellis WC, Matis JH, Pond KR, Lascano CE, Telford JP (1984) Dietary influences on flow rate and digestive capacity. In: Gilchrist FMC, Mackie RI (ed) Herbivore nutrition in the subtropics and tropics. The Science Press (PTY) Ltd., Craighill, South Africa, pp 269–293

Engelhardt W, Von Hales JRS (1977) Partition of capillary blood flow in rumen, reticulum and omasum of sheep. Amer J Physiol 232:E53

French N (1994) Effect of incubation temperature on the gross pathology and development of turkey embryos. Br Poult Sci 3(5):363–371

French NA (1997) Modeling incubation temperature: the effects of incubator design, embryonic development and egg size. Poult Sci 76(1):124–133

Goligorsky MS (2001) The concept of cellular "fight-or-flight" reaction to stress. Am J Physiol 280(4):F551–F561

Grieve D (2003) Heat stress in commercial layers and breeders. Tech Bull Hy-Line Int, Iowa, HLST 19(1):1–3
Gross WB, Siegel HS (1983) Evaluation of the heterophil/lymphocyte ratio as a measure of stress in chickens. Avian Dis 27:972–978
Hagiwara S, Kawa K (1984) Calcium and potassium currents in spermatogenic cells dissociated from rat seminiferous tubules. J Physiol 356:135–149
Heller ED, Nathan DB, Perek M (1979) Short heat stress as an immunostimulant in chicks. Avian Pathol 8:195–203
Hill D (2001) Chick length uniformity profiles as a field measurement of chick quality. Avian Poult Biol Rev 12:188
Holter JB, West JW, McGilliard ML, Pell AN (1996) Predicting ad libitum dry matter intake and yields of Jersey cows. J Dairy Sci 79:912–921
Holter JB, West JW, McGilliard ML (1997) Predicting adlibitum dry matter intake and yield of Holstein cows. J Dairy Sci 80:2188–2199
Igono MO, Steevens BJ, Shanklin MD, Johnson HD (1985) Spray cooling effects on milk production, milk and rectal temperatures of cows during a moderate temperature summer season. J Dairy Sci 68:979–985
Igono MO, Johnson HD, Steevens BJ, Shanklin MD (1988) Effect of season on milk temperature, milk growth hormone, prolactin and somatic cell counts of lactating cattle. Int J Biometeorol 32:194–200
Jeffrey JS, Martin GP, Fanguy RC (2007) The incubation of ratite eggs. http://www.thepoultrysite.com/articles/812/the-incubation-of-ratite-eggs. Accessed 5 Jan 2012
Kamanga-Sollo E, Pampusch MS, White ME, Hathaway MR, Dayton WR (2011) Effects of heat stress on proliferation, protein turnover, and abundance of heat shock protein messenger ribonucleic acid in cultured porcine muscle satellite cells. J Anim Sci 89(11):3473–3480. doi:10.2527/jas.2011-4123
Kelley KW (1983) Immunobiology of domestic animal as affected by hot and cold weather. Trans Am Soc Agric Eng 26:834–840
King LM, Brillard JP, Garrett WM, Bakst MR, Donoghue AM (2002) Segregation of spermatozoa within sperm storage tubules of fowl and turkey hens. Reproduction 123(1): 79–86
Kouba M, Hermier D, Le Dividich J (2001) Influence of a high ambient temperature on lipid metabolism in the growing pig. J Anim Sci 79:81–87
Le Dividich J, Noblet J, Herpin P, van Milgen J, Quiniou N (1998) Thermoregulation. In: Wiseman J, Vailez MA, Chadwick JP (eds) Progress in Pig Science. Nottingham Univ. Press, UK, pp 229–263
Liew PK, Zulkifli I, r-Bejo M, Omar AR, Israf DA (2003) Effects of early age feed restriction and heat conditioning on heat shock protein 70 expression, resistance to infectious bursal disease, and growth in male broiler chickens subjected to heat stress. Poult Sci 82(12):1879–1885
Lin H, Decuypere E, Buyse J (2006) Acute heat stress induces oxidative stress in broiler chickens. Comp Biochem Physiol 144(1):11–17
Lippke H (1975) Digestibility and volatile fatty acids in steers and wethers at 21 and 32 C ambient temperatures. J Dairy Sci 58:1860
Lourens A (2001) The importance of air velocity in incubation. World Poult 17(3):29–30
Lourens A, van den Brand H, Meijerhof R, Kemp B (2005) Effect of eggshell temperature during incubation on embryo development, hatchability, and posthatch development. Poult Sci 84(6): 914–920
Lourens A, van den Brand H, Heetkamp MJW, Meijerhof R, Kemp B (2007) Effects of eggshell temperature and oxygen concentration on embryo growth and metabolism during incubation. Poult Sci 86(10):2194–2199
LPHSI (Livestock, poultry heat stress indices) (1990) Agriculture engineering technology guide. Clemson University, Clemson
Lu Q, Wen J, Zhang H (2007) Effect of chronic heat exposure on fat deposition and meat quality in two genetic types of chicken. Poult Sci 86:1059–1064

Mahmoud KZ, Beck MM, Scheideler SE, Forman MF, Anderson KP, Kachman SD (1996) Acute high environmental temperature and calcium-estrogen relationships in the hen. Poult Sci 75(12):1555–1562

Mallonee PG, Beede DK, Collier RJ, Wilcox CJ (1985) Production and physiologic responses of dairy cows to varying dietary potassium during heat stress. J Dairy Sci 68:1479

Marai IFM, Shalaby TH, Bahgat LB, Abdel-Hafez MA (1997) Fattening of lambs on concentrates mixture diet alone without roughages or with addition of natural clay under subtropical conditions of Egypt. 2. Physiologic reactions. In: Proceeding of international conference on animal production and health, Cairo, Egypt, 2–4 Sept 1997

Marai IFM, Bahgat LB, Shalaby TH, Abdel-Hafez MA (2000) Fattering performance, some behavioral traits and physiologic reactions of male lambs fed concentrates mixture alone with or without nature clay under hot summer of Egypt. Ann Arid Zone 39(4):449–560

Mashaly MM, Hendricks GL, Kalama MA, Gehad AE, Abbas AO, Patterson PH (2004) Effect of heat stress on production parameters and immune responses of commercial laying hens. Poult Sci 83(6):889–894

Mburu JN, Einarsson S, Lundeheim N, Rodriguez-Martinez H (1996) Distribution, number and membrane integrity of spermatozoa in the pig oviduct in relation to spontaneous ovulation. Anim Reprod Sci 45(1–2):109–121

McDaniel CD, Bramwell RK, Wilson JL, Howarth B (1995) Fertility of male and female broiler breeders following exposure to elevated ambient temperatures. Poult Sci 74(6):1029–1038

McDaniel CD, Bramwell RK, Howarth B Jr (1996) The male's contribution to broiler breeder heat-induced infertility as determined by sperm-egg penetration and sperm storage within the hen's oviduct. Poult Sci 75(12):1546–1554

McGuire MA, Beede DK, DeLorenzo MA, Collier RJ, Thatcher LA, Israel CJ, Wilcox GB, Huntington CK, Reynolds RJ (1989) Effects of thermal stress and level of feed intake on portal plasma flow and net fluxes of metabolites in lactating Holstein cows. J Anim Sci 67: 1050–1060

McKee SR, Sams AR (1997) The effect of seasonal heat stress on rigor development and the incidence of pale, exudative turkey meat. Poult Sci 76:1616–1620

Mogenet LY, Youbicier-Simo BJ (19980 Determination of reliable biochemical parameters of heat stress, and application to the evaluation of medications: example of erythromycin E. In: Proceedings of 10th European poultry conference, Jerusalem, Israel, pp 538–541

Moraes VMB, Malheiros RD, Bruggeman V (2003) Effect of thermal conditioning during embryonic development on aspects of physiologic responses of broilers to heat stress. J Therm Biol 28(2):133–140

Moraes VMB, Malheiros RD, Bruggeman V, Collin A, Tona K, Van As P, Onagbesan OM, Buyse J, Decuypere E, Macari M (2004) The effect of timing of thermal conditioning during incubation on embryo physiologic parameters and its relatioship to thermotolerance in adult broiler chickens. J Therm Biol 29:55–61

Moreira PE, Moura AAA, Araujo AA (2001) Effects of scrotal insulation on testis size and semen criteria in santa Ines hariy sheep raised in the state of Ceara, north–east of Brazil. Rev Bras Zootec 30(6):1704–1711

Morrell JM, Rodriguez-Martinez H (2011) Practical applications of sperm selection techniques as a tool for improving reproductive efficiency. Vet Med Int 894767:9

Nathan DB, Heller ED, Perek M (1976) The effect of short heat stress upon leucocyte count, plasma corticosterone level, plasma and leukocyte ascorbic acid content. Br Poult Sci 17:481–485

North MO (1984) Commercial chicken production manual. Layer management on floors, 3rd edn. Avi publishing Company Inc. West Port, CT

Obidi JA, Onyeanusi BI, Rekwot PI, Ayo JO, Dzenda T (2008) Seasonal variations in seminal characteristics of shikabrown breeder cocks. Int J Poult Sci 7(12):1219–1223

Oladele SB, Ogundipc S, Ayo JO, Esievo KAN (2001) Effects of season and sex on packed cell volume, hemoglobin and total proteins of indigenous pigeons in Zaria, northern Nigeria. Veterinarski Arh 71(5):277–286

Oladele SB, Ogundipe S, Ayo JO, Esievo KAN (2003) Seasonal and species variations in erythrocyte osmotic fragility of indigeneous poultry species in Zaria, northern Guinea Savannah zone of Nigeria. Bull Anim Health Prod Afr 51:204–214

Ortiz-de-Montellano M, Galindo-Maldonadob F, Cavazos-Arizpec EO, Aguayo-Arceoc AM, Torres-Acostac JFJ, Orihuela A (2007) Effect of electro-ejaculation on the serum cortisol response of criollo goats (*Capra hircus*). Small Rumin Res 69:228–231

Rahn H, Paganelli CV, Ar A (1974) The avian egg: air cell gas tension, metabolism and incubation time. Respir Physiol 22(3):297–309

Rahn H, Ackerman RA, Paganelli CV (1977) Humidity in the avian nest and egg water loss during incubation. Physiol. Zool. 50:269–283. Altel concentrations of placental hormones during maximal placental growth in a model placental insufficiency. J. Endocrinol. 162 (3): 433-442

Regnault TRH, Orbus RJ, Battaglia FC, Wilkening RB, Anthony RV (2000) Altel concentrations of placental hormones during maximal placental growth in a model placental insufficiency. J Endocrinol 162(3):433–442

Regnier JA, Kelley KW, Gaskins CT (1980) Acute thermal stressors and synthesis of antibodies in chickens. Poult Sci 59:985–990

Rinaldo D, Le Dividich J (1991) Influence de température ambiante sur les performance de croissant du porc. Prod Anim 4:57–65

Romanoff AI (1972) Assimilation of avian yolk and albumen under normal and extreme incubating temperatures. In: Pathogenesis of the avian embryo. Wiley-Interscience, New York

Romanoff AI, Smith LL, Sullivan RA (1938) Biochemistry and biophysics of the developing hen's egg. 3. Influence of temperature. Memorandum Cornell Univ Agric Exp Stat 216:1–42

Sahin N, Sahin K, Küçük O (2001) Effects of vitamin E and vitamin A supplementation on performance, thyroid status and serum concentrations of some metabolites and minerals in broilers reared under heat stress (32°C). Vet Med 46(11–12):286–292

Sandercock DA, Hunter RR, Nute GR, Mitchel MA, Hocking PM (2001) Acute heat stress-induced alterations in blood acid-based status and skeletal muscle membrane in broiler chickens at two ages: implications for meat quality. Poult Sci 80:418–425

Schreiber M, Wei A, Yuan A, Gaut J, Saito M, Salkoff L (1998) Slo3, a novel pH-sensitive K channel from mammalian spermatocytes. J Biol Chem 273(6):3509–3516

Settar P, Yalcin S, Turkmut L, Ozkan S, Cahaner A (1999) Season by genotype interaction related to broiler growth rate and heat tolerance. Poult Sci 78:1353–1358

Shamsuddin M, Rodriguez-Martinez H (1994) A simple, non-traumatic swim-up method for the selection of spermatozoa for in vitro fertilization in the bovine. Anim Reprod Sci 36(1–2): 61–75

Shanklin MD (1963) Temperature-humidity effects including influence of acclimation in feed and water consumption of Holstein cattle. Univ Mo Res Bull, No 846

Sharma AK, Rodriguez LA, Wilcox CJ, Collier RJ, Bachman KC, Martin FG (1988) Interactions of climatic factors affecting milk yield and composition. J Dairy Sci 71:819–825

Simon MS (2003) Reducing heat stress problem. World Poult 19(3):16–17

Skinner JD, Louw GN (1966) Heat stress and spermatogenesis in *Bos indicus* and *Bos taurus* cattle. J Appl Physiol 21:1784–1790

Smith TW (2000) 4-H poultry manual. Mississippi State University Extension Service in Cooperation with USDA. Unit I. Publ. No. 255

Surai PF (2000) Effect of selenium and vitamin E content of the maternal diet on the antioxidant system of the yolk and the developing chick. Br Poult Sci 41(2):235–243

Surai PF (2002) Selenium in poultry nutrition 1. Antioxidant properties, deficiency and toxicity. World's Poult Sci J 58(3):333–347

Surai PF (2010) Natural antioxidants in poultry nutrition: new developments. In: Proceedings of the 16th European symposium on poultry nutrition, pp 669–675

Tambuwal FM, Agale BM, Bangana A (2002) Hematologic and biochemical values of apparently healthy red sokoto goats. In: Proceeding of 27th annual conference of Nigerian society for animal production (NSAP) FUTA, Akure, Nigeria, 17–21 Mar 2002, pp 50–53

Taylor U, Rath D, Zerbe H, Schuberth HJ (2008) Interaction of intact porcine spermatozoa with epithelial cells and neutrophilic granulocytes during uterine passage. Reprod Domest Anim 43(2):166–175

Taylor U, Schuberth HJ, Rath V, Michelmann HW, Sauter-Louis C, Zerbe H (2009) Influence of inseminate components on porcine leucocyte migration in vitro and in vivo after pre- and post-ovulatory insemination. Reprod Domest Anim 44(2):180–188

Terlouw C (2004) Stress reactions at slaughter and meat quality in pigs: genetic background and prior experience, a brief review of recent findings. Livest Prod Sci 37:91–98

Thaxton P, Sadler CR, Glick B (1968) Immune response of chickens following heat exposure or injections with ACTH. Poult Sci 47:264–266

Warren WP, Martz FA, Asay KH, Hilderbrand ES, Payne CG, Vogt JR (1974) Digestibility and rate of passage by steers fed tall fescue, alfalfa and orchardgrass hay in 18 and 32 C ambient temperatures. J Anim Sci 39:93

West JW (1999) Nutritional strategies for managing the heat-stressed dairy cow. J Anim Sci 77:21–35

West JW (2003) Effects of heat stress on production in dairy cattle. J Dairy Sci 86:2131–2144

Wilson HR (1991) Physiologic requirements of the developing embryo: temperature and turning. In: Tullet SG (ed) Avian Incubation. Poultry science symposium, vol 22. Butterworths-Heinnemann, London, UK, pp 145–156

Wolfenson D, Roth Z, Meidan R (2000) Impaired reproduction in heat-stressed cattle: basic and applied aspects. Anim Reprod Sci 60–61:535–547

Yalcin S, Siegel PB (2003) Exposure to cold or heat during incubation on developmental stability of broiler embryos. Poult Sci 82(9):1388–1392

Zulkifi I, Norma MT, Israf DA, Omar AR (2000) The effect of early age feed restriction on subsequent response to high environmental temperatures in female broiler chickens. Poult Sci 79:1401–1407

Taylor U, Rath D, Zerbe H, Schuberth HJ (2008) Interaction of intact porcine spermatozoa with epithelial cells and neutrophilic granulocytes during uterine passage. Reprod Domest Anim 43:166–175
Taylor [illegible]

Chapter 4
Walking Stress Influence on Livestock Production

Vijai P. Maurya, Veerasamy Sejian, Kamal Kumar, Gyanendra Singh and S. M. K. Naqvi

Abstract The food available to grazing animals, particularly during the dry season, in the tropics, is often of very low quality and, in addition, is frequently available at low densities per unit area. Grazing ruminants attempt to adapt to these adverse conditions by increasing the time for which they graze each day and also by dispersing more widely. However, the time for which animals can graze may be limited by solar radiation and fly irritation during the day. Depending on management conditions, livestock are required to walk long distances. When walking is restricted (1–3 km/day), animal performance is generally not reduced. However, under certain conditions (scarce or hilly pasture) the distances walked by livestock can be substantially greater. Unusual physical activity is considered a stress factor in all species since it induces neuroendocrine and metabolic changes which in turn alter the physiological responses, endocrine and enzymes' release status and productivity in animals. In addition, there are both breed and inter-species differences in locomotory efficiency as a result of morphological, physiological and behavioural adaptations in livestock. The significant changes in physiological responses, adrenal and thyroid hormone concentrations after subjecting the livestock species to walking stress shows that they are capable to adapt to long-distance walking and adrenal and thyroid gland hormones play a significant role in such adaptation. Though, while trying to adapt to long-distance walking scenario in search of food, animals compromise their productive performance. This is reflected as significant reduction in growth, milk and reproductive performance of different livestock species. The reason for this low production

V. P. Maurya (✉) · G. Singh
Division of Physiology and Climatology, Indian Veterinary Research Institute,
Izatnagar, Bareilly 243122, India
e-mail: vijaipmaurya@gmail.com

V. Sejian · K. Kumar · S. M. K. Naqvi
Division of Physiology and Biochemistry, Central Sheep and Wool Research Institute,
Avikanagar, Via-Jaipur, Rajasthan 304501, India

V. Sejian et al. (eds.), *Environmental Stress and Amelioration in Livestock Production*, DOI: 10.1007/978-3-642-29205-7_4,

could be a major proportion of the energy is shifted to combating the walking stress as any counter measures need energy source for its implementation. The repercussions of walking or work on production performances have been studied by various researchers and reported that livestock energy requirement increases significantly during walking exercise. Further, most of the increase in energy expenditure of physical activity results from grazing and locomotion costs. The energy expenditure of locomotion during grazing contributes significantly to the energy requirement of animals in free-living conditions and must be included for accurate evaluation of the energy needs of the grazing animal.

Keywords Adaptation · Environmental stress · Energy · Livestock · Production · Walking

Contents

4.1 Introduction

Walking stress of animals wholly depends upon the management system used to maintain them. When walking is restricted, the performance of animals is generally not affected (Lamb et al. 1979; Gustafson et al. 1993). However, under certain conditions (scarce or hilly pasture) the distances walked by dairy animals can be substantially greater (Berhan et al. 2006). Walking or exercise has adverse consequences on animal productivity and tends to increase energy requirements of animals (Henning 1987; Lawrence and Stibbards 1990). Semi-arid environments are one of the major agro-ecological zones of the tropics. There is, in general, a strong relationship between agro-climatic conditions, population density, cropping systems and livestock production. Rangelands are the largest land use system on earth (FAO 2001). They predominate in semi-arid tropical areas of the world. These pastoral systems are those in which people depend entirely on livestock for their livelihoods. The key constraints of arid and semi-arid tropical environment are their low biomass productivity, high climatic variability and scarcity of water (Sejian et al. 2010a; Maurya et al. 2010a). All these constraints make these regions difficult for sustainable livestock production. Research agenda need to take into account the trade-offs and synergies arising from these tropical environments so that the poor are able to reap the multiple benefits provided by these ecosystems. The food available to grazing animals in developing countries, and particularly in the dry season in the tropics, is often very low in quality and, in addition, is available at low densities per unit area. The long-distance walking in search of water and feed creates negative energy balance in the animals which also affects body condition score of animals (Maurya et al. 2010b). In general, walking stress is more pronounced in the animals living in regions of low rain fall resulting in scarce availability of vegetation in the pasture as compared to the pasture where the rain fall is sufficient.

Walking results in a greater demand for nutrients, to fuel the increased metabolic rate of contracting muscle. The balance and metabolic destination of these nutrients are determined principally by the work rate, but level of fitness, diet and genetic potential also play a role. The major tissues which adapt to maintain nutrient balance are the gastrointestinal tract, liver, adipose tissue and skeletal muscle. Most part of the year under tropical environment, animals have to walk long distances in search of water, and are usually replenished once in two to three days. Nutrients consumed by ruminants are mostly utilised for maintenance and production. Maintenance energy is used to maintain body temperature, is lost as heat of fermentation, or is used for work (e.g. walking). The distance cattle walk daily varies both within days and between days on individual farms and is generally related to pasture availability and/or accessibility. Conversely, the duration for which animals can graze may be restricted by solar radiation and fly irritation (Manteca and Smith 1994). In arid and semi-arid tropical climate, depending on management conditions, livestock are bound to walk long distances for fodder and to replenish water.

Walking stress induces neuroendocrine and metabolic changes (Bruckmaier and Blum 1992). The neuroendocrine stress response is not only the result of metabolic demands but is also associated to the psychological aspects of the situation (Voigt et al. 1990). The neuroendocrine changes are characterised by the stimulation of the hypothalamic-pituitary-adrenal (HPA) axis with an increase in the circulating levels of adrenocorticotropic hormone (ACTH) and adrenal corticosteroids (Mormbde 1988). This hyperactivity can be explained by metabolic demands induced by the physical exercise and the essential role of glucocorticoids in energy metabolism (Richard and Rivest 1988). This metabolic action is favoured by sympathetic nervous system (SNS) activity which facilitates the necessary metabolic and cardiovascular adjustments (Verde and Gascon 1987).

Restricted walking (1–3 km/day), usually does not affect animal performance (Anderson et al. 1979; Lamb et al. 1979; Gustafson et al. 1993). However, certain geographical areas (scarce or hilly pasture) require animals to cover significantly longer distances. Studies have been conducted to explore the outcome of walking or work, in terms of energy requirements and production performances, in sheep

(Henning 1987; Lachica and Aguilera 2003). In scarcity of forage or when grazing on a hilly terrain, nutritional requirements of livestock increase as walking is a major activity (Mendez et al. 1996). During dry seasons, animals lose weight (Pagot 1993) and reduction in milk yield is been observed (Matthewman et al. 1993) in animals walking long distances in search of food and water. Henning (1987) reported increased energy demand due to walking exercise in sheep. Most of the increase in energy expenditure (EE) of physical activity results due to grazing and locomotion (Lachica et al. 1997). The EE of locomotion contributes significantly to the energy requirement of animals in free-living conditions and must be included for accurate evaluation of the energy needs of the grazing animal (Berhan et al. 2006). Further, breed differences have been defined in livestock in response to walking stress (D'Hour et al. 1994; Coulon and Garcl 1996). In addition, published data suggest that there are interspecies differences in locomotory efficiency as a result of morphological, physiological and behavioral adaptations (Farley et al. 1993; Lachica and Aguilera 2005a, b). Sheep under hot semi-arid environment is mostly reared in extensive system. The productive potential of sheep in these areas is influenced by the exposure to harsh climatic factors. In addition to thermal stress and feed scarcity, the animals need to walk longer distances for grazing (Maurya et al. 2004; Naqvi et al. 1991). These stresses lead to alteration in the process of homeostasis and metabolism. This chapter aims to build a clear understanding about the effect of walking stress on livestock productivity.

4.2 Significance of Livestock Walking

As is known, the food quality and densities per unit area of the food obtainable to grazing animals, is low in developing countries, particularly in the dry season in the tropics. To adapt to such adversities, grazing ruminants increase grazing times each day and walk longer distances. Solar radiation and fly irritation may, limit the duration of grazing (Manteca and Smith 1994). Depending on management conditions, livestock are required to walk long distances. It has been established that restricted walking (1–3 km/day), does not affect animal performance (Anderson et al. 1979; Lamb et al. 1979; Gustafson et al. 1993). We have already discussed that the EE due to locomotion activities is a significant contributor to the energy demand of animals in free-living conditions and must be included for accurate evaluation of the energy needs of the grazing animal. This is supported by the finding of Henning (1987) reporting increased energy demands through walking exercise in sheep. Further it was suggested that the spring like tendons of large mammals can potentially store more elastic energy than those of smaller mammals because their disproportionately stronger muscles can impose higher tendon stresses. Farley et al. (1993) using dogs, goats, horses and red kangaroos, found that the stiffness of the leg spring is nearly independent of speed $\times$ body mass. In goats and horses, the stiffness is increased 25 and 29%, respectively.

4.3 Effect of Walking on Energy Balance in Dairy Cow

The majority of domesticated ruminants are raised solely or partially in semi-extensive or extensive production systems in which most nutrients are derived from grazed forage. Grazing is associated with daily activities considerably different than for confined animals, such as time spent eating and distances travelled (Lachica and Aguilera 2003). These activities result in greater EE than in confinement, which can limit energy available for maintenance and production. Perhaps because of the difficulty of these studies, there is relatively little known regarding factors influencing the grazing activity energy expense by ruminants. (D'Hour et al. 1994). As such, there are, at present, no universally accepted methods of prediction of EE while grazing. For available methods pertaining to goats, NRC (1981) recommended the addition of 25% of the suggested metabolizable energy (ME) requirement for maintenance (MEm) with light activity, 50% with semi-arid rangeland and slightly hilly conditions and 75% with sparsely vegetated rangeland or mountainous transhumance pasture. Although Coop and Hill (1962) reported a higher grazing energy activity cost of 92% of MEm for sheep grazing perennial ryegrass-white clover pastures. AFRC (1998) recommended estimating the activity energy expense of grazing for goats from additional costs above confinement attributable to horizontal and vertical distances travelled and number of changes in position, based on specific activity costs of sheep and goats on a treadmill placed at different slopes (ARC 1980; Lachica et al. 1997). CSIRO (1990) presented a prediction equation for cattle, sheep and goats with independent variables of DM digestibility, terrain score, availability of green or total forage and BW. Because of the close relationship between grazing time and EE, Sahlu et al. (2004) proposed prediction based primarily on time spent grazing and walking, in addition to consider the influence of herbage digestibility, distance travelled and terrain ruggedness or topography. One of the conditions impacting aforementioned factors used to predict the activity energy cost of grazing is nutrient demand of the animal, and thus, forage intake (Fierro and Bryant 1990). Forage availability can influence both grazing time and the nutritive value of ingested forage (Seman et al. 1991; Krysl and Hess 1993; Herselman et al. 1999). As forage availability decreases, bite size declines, which results in at least partially compensatory changes in grazing time and rate of biting (Davies and Southey 2001). Decreased forage quality also increases time spent in ingestive mastication (Sahlu et al. 1989; Lachica and Aguilera 2003). Heart rate (HR) is known to be related to oxygen consumption and heat production, and thus, has been used as an indirect measure of EE for cattle and sheep (Brosh et al. 2002; Barkai et al. 2002). To do so, it is necessary to determine the quantity of heat produced or EE per heart beat, which can vary among individual animals (Brosh et al. 1998). However, Osuji (1974) reviewed literature indicating a close relationship between times spent grazing and the activity energy cost of grazing, and noted physiological processes contributing to this relationship such as skeletal muscle work for locomotion and energy use by the gastrointestinal tract and liver.

The increasing number of steps and presumably distance travelled with increasing Stocking Rate (SR) probably contributed to increasing EE as SR rose. In fact, distance travelled is a primary input of some systems used to predict the grazing activity energy cost (CSIRO 1990; AFRC 1998; NRC 2001; Animut et al. 2005).

Energy is the main limiting factor in animal production and its availability affects the animal's adaptation to environment, behaviour and feeding strategy. Small ruminants adapt well to environmental conditions prevailing in arid lands, being able to obtain an adequate diet even when forage is scarce and to feed over rugged and otherwise inaccessible terrain. There is little information in the literature concerning their energy requirements compared with bovine species, although small ruminants provide an important source of income, particularly in countries around the Mediterranean basin, and in other areas with dry climatic conditions. It is well-known that grazing animals expend more energy than ones in confinement, basically for two reasons. In the stall, the food is offered in a partially processed form, which makes it easier to be ingested and, second, the animal does not have to move to search for food. High maintenance requirements of sheep on pastures might be due to increased cost of body movement at pasture, especially the cost of walking and harvesting the herbage.

The usual procedure for estimating the total EE of grazing animals is the factorial method. It is based on the Hess's Law: "The heat produced in a chemical reaction is always the same, regardless of whether the process went directly or proceeded through a number of intermediate steps". That is, the total energy expended by an animal is equal to the partial EE of each one of the component processes or activities. The energy cost of each activity can be quantified by calorimetry and, thereafter, the total extra energy expended in the free-ranging animal is calculated by summation. Most of the energy required by the grazing animal is due to increased muscular efforts, mainly walking and eating, while the contributions of other activities are usually considered negligible (Garcia-Belenguer et al. 1996). An accurate assessment of the energy costs corresponding to such activities has a vital importance, as well as the methods to be applied in the field to identify and record those activities and their corresponding energy expenditure. Energy expenditure associated with basal metabolism plus heat increment of feeding usually constitutes the greatest portion of the daily budget of domestic animals (Gustafson et al. 1993). Lack of a generalisation as to what component dominates is to be expected because different environmental factors regulate the heat increment of feeding and the activity/thermoregulatory costs. In the field these factors are poorly studied and models draw heavily on studies with penned animals.

4.3.1 Calorimetric Methods to Estimate Energy Costs

The usual method to estimate the energy cost of locomotion has been by the combined use of calorimetry and treadmills. The common calorimeters used have been indirect, that is, with measurement of the gas exchange that is associated with the oxidation of energy substrates and calculation of heat production (HP) from the stoichometry of substrates, oxidation. Most of them operate following the open-circuit principle. Some of the studies have used masks or hoods made of transparent material (normally acrylic plates) enclosing only the head of the animal, but preferably the animal is confined in a respirometry chamber. Masks and head hoods permit rapid response times on relatively short periods but keep the animal under a comparatively less comfortable, "not natural" condition. Respirometry chambers allow the scientist to install the treadmill inside so the animal feels "more free". Nevertheless, this methodology presents some disadvantages, as the personnel have no close control over the animal and the time required for the system to reach a "steady state" is usually much longer due to the volume of air of the chamber that has to pass through the recording devices before they allow an accurate measurement of the gas exchange due to the physical activity of the confined animal. The immediate consequence is that the length of time to reach steady state depends on the characteristics of the chamber and/or type of experiment (flow rate, volume of the chamber versus animal size, level and period of physical exercise, etc. Nienaber et al. (1985) reported on difficulties encountered for accurate partitioning of EE into fasting heat production and heat production (HP) related to physical activity in experiments on fasted calves, sheep and pigs designed to measure the increase in HP due to physical activity by means of open-circuit respiration chambers. Undoubtedly, the so-called "confinement system" has definitive advantages, as it relies only on accurate analysis of gas composition. It is best suited and sufficiently accurate to fast response applications for measurement of gaseous exchange of animals confined for a short period of time in a respiration chamber. Comparisons of day-to-day variations in HP determined in goats showed no statistical difference between the confinement and the open-circuit system, even though data obtained by the confinement approach corresponded to short periods of gaseous measurement and required extrapolation to 1 h (Lachica and Aguilera 2005b).

Once the physical exercise starts, O_2 consumption (or HP) approaches a constant value after a certain period; therefore, it is necessary for an extended period of time to achieve a steady state. In the same way, when the work that is performed is severe there is an oxygen debt, i.e., the animal's O_2 consumption remains elevated for a while after the cessation of the work because of the recovery of the muscle to the prework level of lactic acid. In other words, lactic acid is then oxidised and creatine is phosphorylated. Consequently, the total HP attributable to the work done is the EE during the exercise period plus the energy expended during the recovery until the prework level is achieved. For practical purposes, it is important to have in mind that when the exercise is severe, in the calculation of the energy cost of the physical activity the extra HP over the preexercise period continues after work ceases. Full adaptations of the animals to experimental conditions are vital for the success of experiments (Green et al. 2009). For some weeks before starting the experimental protocol, the person in charge of work at the chambers lab should also feed the animals and spend time with them ("for the animal to become accustomed to them"). After this stage, the training process with the treadmill in the chamber can be initiated, first with the animal tethered to the front area of the treadmill and, after some days, when it walks steadily without dragging against its tether, and then with no tether at all. Only an obstacle (a chain, plate, etc.) at the rear side of the treadmill will be enough to keep the animal on it. Boyne et al. (1981) in experiments with sheep and cattle observed huge variations in EE caused by the tension voluntarily exerted in the tether between the animal and the treadmill, which varied with speed and gradient up to a maximum value of approximately 0.04 J/m walked per gram of tension or negative load. The overall effect is an overestimation of the energetic efficiency of the muscular activity. The presence of another animal seen by the one on the treadmill greatly lessens chance of bias.

4.3.2 Mathematical Method to Estimate Energy Cost

Two different procedures are used to estimate the energy cost of locomotion:

(a) By subtracting the value obtained for HP (J) or O_2 consumption (mL) of the animal while standing at rest from that measured during walking (subtraction approach):

$$\mathrm{ECw} = \mathrm{HPw} - \mathrm{HPst\ BW} \times \mathrm{Dt} \tag{4.1}$$

where ECw is the energy cost of walking (J kg − 1BWm − 1), and HPw and HPst are the HP (J) while walking and standing, respectively, BW the animal body weight (kg) and Dt is distance travelled (m).

The net energy cost of vertical movement on ascent, i.e., the energy expended in raising 1 kg BW one vertical metre (ECup, J kg − 1 BWm − 1), is calculated as:

$$\mathrm{ECup} = \mathrm{ECsl} - \mathrm{Ecl}\ \sin\alpha \tag{4.2}$$

where ECsl and ECl are the energy cost of walking (J kg − 1 BWm − 1) on slope and on the level, respectively, and sin α is the fraction of a metre ascended per metre travelled.

The efficiency with which the animal performs the work of walking on positive slopes is calculated as the ratio of 9.81 (joules potential energy per kg BW raised 1 m) to ECup and expressed as a percentage (Lachica and Aguilera 2005a, b). The net EE of vertical movement on descent, i.e., the energy recovered upon lowering 1 kg BW1 m (ECdown, J kg − 1 BWm − 1), is estimated as:

$$\mathrm{ECdown} = \mathrm{EC1} - \mathrm{Ecsl}\ \sin\alpha \tag{4.3}$$

where ECl and ECsl have the same meaning as in (2) and sin α is the fraction of a metre descended per metre travelled. The efficiency of the recovery of potential energy while walking on negative slopes is calculated by dividing ECdown by 9.81, and is expressed as a percentage.

(b) The energy cost of locomotion within slopes is estimated from the coefficient of the linear regression of HP (J kg − 1 BWh − 1) or O_2 consumption (ml kg − 1 BWh − 1) on distance travelled (Dt, m) (regression approach). The energy cost of vertical ascent and the energy recovered on vertical descent is calculated by multiple regression equations of HP (J kg − 1 BWh − 1) or O_2 consumption (ml kg − 1 BWh − 1) on distance travelled horizontal (Dh, m) and vertically in ascent (Du, m) or descent (Dd, m), respectively, using the following approach:

Du = distance upward, equals 0 otherwise;

Dd = distance downward, equals 0 otherwise (Lachica and Aguilera 2005a, b)

Estimates of the grazing activity energy cost (MEa) for goats vary tremendously, ranging from 0 to 1 × the ME requirement for maintenance (MEm) of confined goats (Lachica and Aguilera 2003). AFRC (1998) suggested that MEa for stall fed goats is 10% of fasting heat production [10% of 315 kJ/kg body weight (BW)0.75], with additional costs for horizontal movement [3.5 J/(kg BW × m)], vertical movement [28 J/(kg BW × m)], standing [0.417 kJ/(kg BW × h)] and position change [0.26 kJ/(kg BW × number of changes)], and then application of the efficiency of ME used for maintenance. These values were based primarily on reports of ARC (1980) with sheep and of Lachica et al. (1997) with goats on a treadmill placed at different slopes. Although some empirical methods to predict MEa rely only on energy for movement, time spent grazing is quite closely related to total heat production or EE regardless of distance travelled (Osuji 1974). How magnitudes of effects of locomotion and physiological processes associated with feed ingestion are compared is unknown. Such information could be of value in developing simple means of predicting MEa based on measures such as distance travelled, terrain, grazing and walking times, diet quality and (or) forage mass. Heart rate holds promise as a simple and inexpensive indirect means of evaluating EE or heat production while walking and (or) grazing. However, in order to use HR in this manner, it is necessary to know whether the ratio of EE to HR varies among HR in the range observed while grazing (Brosh et al. 2002) For example, the ratio has been shown relatively constant in cattle on pasture without extreme exercise or heat load (Aharoni et al. 2003). Use of HR to estimate EE of goats has not been extensively studied (Berhan et al. 2006). The EE of locomotion contributes significantly to the energy requirement of animals in free-living conditions and must be included for accurate evaluation of the energy needs of the grazing animal (Lachica et al. 1997). The usual and more reliable procedure for estimating free-living EE is the factorial method, whereby EE is calculated from calorimetric determination of the energy cost of various activities. Most of the increase in EE of physical activity results from grazing and locomotion costs, whereas the contribution of other activities is usually considered to be negligible. Published data suggest that there are interspecies differences in locomotory efficiency as a result of morphological, physiological and behavioural adaptations (Lachica et al. 1997). The net energy cost of locomotion, on the level, of a Granadina goat with an average body weight of 34–95 kg was 3–35 J/kg BW per m (Lachica et al. 1997). In sheep, Clapperton (1964) and Brockway and Boyne

(1980) reported values of 2–47, 2–83 and 2–30 J/kg BW per m, respectively; Lawrence and Stibbards (1990) obtained, in cattle, net energy expenditures of 209, 1–54 and 1–91 J/kg BW per m, respectively.

It is well documented that the energy cost of horizontal locomotion decreases with increasing body weight. It is generally recognised that the net energy cost of up slope locomotion is higher than that for moving on the level due to the energy expended to work against gravity, whereas during down slope travel potential energy is recovered as kinetic energy, leading to a decrease in EE relative to horizontal costs.

Energy cost of locomotion (ECW, J/kg BW per m) as a function of slope (SI, %):

$$\begin{aligned} \mathrm{ECW} = 3{-}39\ (\mathrm{SE\,i.022})\, \mathrm{e}0063\ (\mathrm{SE}\,00031)\, \mathrm{sl},\ \mathrm{r}\ 0 \\ -\ 996;\mathrm{residual\ SD}\,0{-}049;\ \mathrm{n}\ 5. \end{aligned} \tag{4.4}$$

The mean values (n 6) of the energy cost of walking, calculated for each goat by separation of the horizontal (Dh) and vertical (Du and Dd) components by multiple regression of HP (J/kg BW per h) v. the horizontal and vertical distances travelled (m) in ascent or descent, were those given by the average equation:

$$\begin{aligned} \mathrm{HP} = 6724\,(\mathrm{SE}\,301) + 3 - 31\,(\mathrm{SE}\,0 - 148)\,\mathrm{Dh} + 31 - 7\,(\mathrm{SE}\ 1 - 59)\,\mathrm{Du} \\ -\ 13 - 2\,(\mathrm{SE}\ 1 - 33)\,\mathrm{Dd}, \end{aligned} \tag{4.5}$$

where the regression coefficents of Dh, Du and Dd indicate values for the net energy cost (J/kg BW per m) for horizontal (ECh) and vertical locomotion on ascent (ECU) and on descent (ECd) respectively.

4.4 Effect of Walking Stress on Growth and Feeding Behavior

Body condition scoring (BCS) is a simple, non-invasive, time-saving and beneficial technique to rank animals according to their body reserve by sight and touch. This technique is integrated in contemporary reproductive programmes to accelerate productive performance in the animals. The scope of BCS continues to expand as the management strategies are constantly modifying to affect productivity and profitability. The size and frame of animals also play an important role in working ability of animals (Kartiarso Martin and Teleni 1989). When quality of forage is low, the animal's body condition score decreases as the energy is diverted towards energy required to carry out muscular exercise in the process of walking. Due to climatic and management condition of the farmers, sometimes dairy animals are required to walk long distances (Lachica and Aguilera 2005a, b). While walking, grazing or between grazing paddocks and drinking water, animals expend energy (Bailey et al. 1996). The distance cattle walk daily varies both within days and between days on individual farms and is generally related to pasture availability and/or accessibility

(Garcia-Belenguer et al. 1996). Distance traveled by grazing cattle is determined by a combination of several factors and management decisions. Grazing behaviour under these conditions will be determined, among other factors, by pasture quality and its availability. Sometimes there is reduction in vegetation and a cow is being supplemented to achieve energy equilibrium and increase production and profit (Garcia-Belenguer et al. 1996). In general, economics plays a major role in this decision. During supplementation, the cost effectiveness of supplementation should be considered.

Increases in energy requirements are linked with the distance of walking and the increases in energy requirements were not compensated by increase in forage intake. On the contrary, cows having walked, ingested less forage than cows at rest due to the reduced access time, fatigue engendered by the walk may be responsible for these differences Metz (1984). During the process of long-distance walking the animal generally mobilises body energy reserves, as indicated by the increases in plasma non esterified fatty acids (NEFA) content just after walking (Pearson and Archibald 1989; D'Hour et al. 1994; Animut and Chandler 1996). Winugroho (1993) indicated that during walking stress or work, the animals' feed intake was inconsistent depending on the supply of nutrients (feed, body reserves) and on the intensity of work. The work or walking stress increases heat load and decreases gut motility and rate of passage that could reduce appetite and intake. The energy cost for a 450 kg cow fed poor quality diet is about 1.12 J/kg live weight (Matthewman and Dijikman 1993). The energy cost of walking in cattle is widely reported as approximately 2 J/m//kg live weight (Lawrence and Stibbards 1990). The difference in energy cost may be due to differences in muscle and skeletal arrangements within the different animal breeds. Osuji (1974) indicated a close relationship between time spent grazing and the energy cost of grazing, and noted physiological processes contributing to this relationship such as skeletal muscle work for locomotion and energy use by the gastrointestinal tract and liver. The increasing number of steps and presumably distance travelled with increasing stocking rate, probably contributes to increasing energy expenditures. Zerbini and Asamenew (1991) in Ethopia, found that during the early lactation, work did not affect the milk production but it has a profound effect on the body weight of animals. They established that after three months of postpartum, a working cow lost 26 kg of body weight as compared to nonworking cows.

In a recent study on the influence of walking stress on feed intake and body weight in sheep, Sejian et al. (2011) reported that walking stress significantly decreases feed intake which is reflected in lower body weight. They further reported that average daily gain was also reduced significantly as the exposure to walking stress is prolonged Sejian et al. (2011). In order to maintain the uniformity of feed and water availability, Sejian et al. (2011) have used face masks for the walking stress sheep to prevent them from grazing. Figure 4.1 describes the walking stress sheep with their face masks. Coulon et al. (1998) also reported significant decrease in both body weight and feed intake in cows subjected to walking stress. In addition, Sejian et al. (2011) also reported significant increase in water intake. The increase of water intake might have occurred to compensate for

Fig. 4.1 Pictorial representation of walking stress sheep. Picture representing the walking stress Malpura ewes in a controlled study. The animals are with indigenously developed face mask to prevent grazing

total body water deficits, caused by the increase of evaporation through the respiratory tract and skin surface when these ewes were subjected to walking stress. Further, animals under walking stress consume more water to counteract the exercise stress. The high water intake in walking stress animals could be the adaptive mechanism exhibited by these animals to combat walking stress. The study establishes the importance of providing optimum nutrition to counteract walking stress; one of the major predisposing factors for the low productivity of animals particularly small ruminants in tropical environment. The nutritional deficiency affects growth of sheep but when nutrition is not a limitation sheep can counter environmental extremes effectively under semi-arid tropical environments (Sejian et al. 2010b).

4.5 Physiological Adaptability of Livestock to Walking Stress

Generally livestock species use their physiological responses as their primary means to adapt to a particular environment (Sejian et al. 2010d). On subjecting livestock species to environmental stresses, their physiological responses show significant variations as compared to that exhibited when they were at thermo-neutral zone (Sejian et al. 2010d). During periods of increased locomotory activity, it is essential that an adequate supply of oxygen be delivered to the working muscles if aerobic metabolism is to be maintained. Further, respiration rates and rectal temperatures have been shown to be good indicators of the stress condition in farm animals (Daramola and Adeloye 2009; Sejian et al. 2010c). Sejian et al. (2011) reported significant increases in respiration rate and rectal temperature in ewes subjected to walking stress by allowing them to walk for 14 KM a day. Similar exercise-induced increase in RR and RT were reported in goats and cows (Kasa et al. 1995; Animut and Chandler 1996; Coulon and Pradel 1997). Anderson et al. (1979) reported that if cows were allowed to walk more than 5 km, it imposed severe stress to the animals and they established this based on the significant changes in the physiological responses of these cows. Coulon et al. (1998) reported significant increase in rectal temperature in cow after 5 km walking daily for grazing. Berhan et al. (2006) demonstrated that pulse rate is an effective means to evaluate heat production during walking in cows. However, Sejian et al. (2011) did not find any significant influence of walking stress on pulse rate in sheep indicating the species difference.

4.6 Blood Biochemical Responses to Walking Stress

Generally Hb and PCV are considered to be good indicators of stress in farm animals (McManus et al. 2009). Sejian et al. (2011) reported a significant increase in both haemoglobin (Hb) and packed cell volume (PCV) in Malpura ewes after subjecting them to walking stress. The increased Hb and PCV in walking stress ewes could be to increase oxygen carrying capacity of blood to support rigorous muscular activity in these ewes. Garcia-Belenguer et al. (1996) also reported similar results of walking stress-induced increase in Hb and PCV in cows. Further, reduced availability of water during exercise stress may cause haemo-concentration leading to increase in both Hb and PCV. It seems, however, that the oxygen-carrying capacity of blood and aerobic work capacity, as reflected by the level of hemoglobin and PCV, increase with conditioning and improved fitness levels (Jones 1989). According to this theory, the higher levels of hematological parameters (PCV, RBC) found in the calves as compared to the cows might be related with a better performance of these young animals to muscular effort (Jones 1989). Garcia-Belenguer and Mormbde (1993) reported an increase in the WBC and the cortisol level in the blood of cattle, occurring just after initiation of walking. These changes were far larger in calves than in cows. Snow et al. (1983) observed an increase in lymphocyte numbers associated with splenic contraction as a normal response to the stress of exercise in horse. Three to four hours after exercise, an increase in neutrophils occurred which was associated with an increase in plasma cortisol levels at the end of exercise.

There are also reports indicating low blood glucose after subjecting the animals to walking stress (Veissier et al. 2008; Sejian et al. 2011). The decrease in plasma glucose in walking stress animals could be due to the mobilization of energy and the exhaustion of glycogen reserves (Payne and Payne 1987; Wasserman et al. 1989). In general, at rest, skeletal muscle of sheep uses considerable amounts of blood glucose, sufficient to account for half of the energy metabolism. As the level of exercise increases, the liver releases additional glucose which in turn is utilized by the exercising muscle. Importantly, the rate of glucose release by liver and the uptake by muscle remained sustained and matched throughout the mild walking period, resulting in no tendency for hypoglycemia (Pethick 1993). During the walking stress, glucocorticoids are being released, which induce increases in blood glucose levels via gluconeogenesis and glycogenolysis (Kaneko 1989), a decrease was observed in the animals immediately after exercise, which could be due to the mobilization of energy and the exhaustion of the glycogen reserves (Payne and Payne 1987; Wasserman et al. 1989). However, Gustafson et al. (1993) reported that walking exercise did not influence plasma glucose in cow.

The plasma protein, cholesterol and urea were also not altered significantly after walking stress. One of the principal functions of cortisol, the principal acclimation (stress relieving) hormone in ruminants, is to stimulate hepatic gluconeogenesis leading to conversion of noncarbohydrate substances into glucose to meet the energy demand during stressful conditions (Sejian et al. 2010d; Sejian and Srivastava 2010). Although the cortisol levels rose significantly, suggesting walking group ewes are under stress, it did not cause significant alterations in serum protein, cholesterol and urea concentrations. The higher plasma AST and ALT could serve as indicators of muscle injury (Payne and Payne 1987; Garcia-Belenguer et al. 1996) due to walking. However, AST and ALT are not specific for muscle in large animals.

These observations suggest that the animals were able to meet their energy requirement during walking stress from their reserve pools. Thus walking stress initiates a sequence of metabolic responses. First, the animal increases the level of oxidative metabolism in the muscle and spleenic contraction increases the number of circulating erythrocytes to provide more oxygen. Second, as exercise proceeds, muscle metabolism changes from an aerobic to anaerobic type resulting in lactic acid production. This alters the local environment of muscle cells, affecting cell membrane permeability and allowing the escape of cell enzymes such as creatine phospho kinase (CK), lactic dehydrogenase (LDH), aspartate amino transferase (AST) and alanine amino transferase (ALT) into blood (Garcia-Belenguer et al. 1996). All these enzymes are intracellular and increase in the circulation when muscle cell damage occurs. If the exercise continues excessively, the muscle changes become progressive requiring longer recovery times.

4.7 Endocrine Responses of Livestock to Exercise Stress

In a study conducted to establish the effect of walking stress on Malpura ewes, Sejian et al. (2011) reported reductions in thyroid hormones. They attributed this reduced thyroid hormones concentrations in walking ewes to the reduced metabolic activity of these ewes to suppress heat production. This is a common phenomenon exhibited by livestock species to prevent increase in heat load during stress (Sejian et al. 2010d). The significant reduction of T_3 and T_4 as compared to normal range as described by Kramer (2000) in sheep shows that this could be the adaptive mechanism of these ewes to walking stress. Depression of thyroid function during stress is a part of the process of metabolic adaptation by which the heat production may consequently be maintained at low level (Sano et al. 1979; Sejian et al. 2010d). It is an established fact that energy needs increase tremendously during exercise in livestock (Veissier et al. 2008). Exercise stress during walking increases the EE thereby altering the metabolic balance. In ruminants, release of cortisol in blood is linked to metabolic balance, with increased cortisol concentrations when needs exceed energy intake (Ward et al. 1992; Samuelsson et al. 1996). The increased metabolic need during walking stimulates the release of ACTH from anterior pituitary which inturn stimulates adrenal cortex to release cortisol. Cortisol stimulates hepatic gluconeogenesis to meet the energy demand (Lachica and Aguilera 2005a, b). Cortisol is the principle stress-relieving hormones in sheep and Sejian and Srivastava (2010) reported a significant increase in plasma cortisol level in walking stress ewes. Similar walking stress-induced increase in plasma cortisol concentration were reported in cows and calves by Garcia-Belenguer et al. (1996).

4.8 Effect of Walking Stress on Reproductive Efficiency of Animals

The distance covered by pastured animals during walk depends on many factors including size of the grazing area, the amount of grass available, the proximity of drinking water and management strategies (Anderson et al. 1977; Arnold and Dudzinski 1978; Anderson and Urghart 1986). When animals are required to forage over large areas, the breed's walking ability should be taken into account (Bibe and Vissac 1979). The long-distance walking in the search of food and water requires more energy which inturn affects the productive and reproductive efficiency of animals (Osuji 1974).

Animal traction is influenced by numerous mechanical, biological, management and socio-economic factors. Biological factors such as sex, animal condition, size, nutrition, weight, pregnancy, lactation status, temperature and general health of the animal are important in determining work performance (Matthewman and Dijikman 1993; Bartholomew et al. 1994). Body weight losses greater than 15% have been reported to impair ovarian activity in female buffaloes (Teleni et al. 1989). Work carried out in early lactation could delay return to oestrus. Zerbini et al. (1992) found out that in working cows even after supplementation, the conception rate delayed significantly. It is unlikely that the cessation of cyclic activity in working cows was a result of direct competition for nutrients between the ovary and other tissues. It is possible that the depletion of body reserve nutrients to certain critical levels may signal metabolic controls to switch off non-vital processes such as ovarian function. Nutritional constraints, combined with draught work activities, could cause a loss in body condition and lead to postpartum anestrus (Jabbar 1983). Zerbini and Asamenew (1991) reported that crossbred cow may also be used for work but they should be provided sufficient energy, as low energy input may lead to a low calf output and eventually in a complete stop of reproductive functions (Zerbini et al. 1992) with loss of production and cow value.

4.9 Effect of Walking Stress on Milk Yield on Animals

Generally animals walk long distance in search of feed and it is presumed that at least the animals walk twice a day to the pasture and back to the barn. This means that the animals at distant pastures have higher energy requirement for walking. In cows, the increased energy demand for walking 2 km/d is around 5% of maintenance corresponding to approximately 0.5 kg of milk according to NRC (2001). Sporndly and Wredle (2004) reported that with shorter walking distances cows had a higher milking frequency and milk yields than the cows in distant pasture because these cows got few hours to gaze per day which in turn decreased their feed intake and milk yields as well. The reports in the literature show a variable effect of work on milk production. Jabbar (1983) in Bangladesh suggested a fall in the milk yield when cows are used for drought. Goe (1983) reported that on work days cows can show a 10–20% decrease in milk yield. In such situations, lactating cows deal with a shortage of nutrients created through exercise by restricting secretion of protein and lactose whilst maintaining fat output. Coulon et al. (1986), Gordon (1984) further reported that the effect of walking on milk yield persists even during postexperiment periods in cow. Therefore, walking likely alters the synthesising capacity of the udder in an irreversible manner.

The walking stress also affects the somatic cell count (SCC) of the mammary gland. During the initial stage of walking the SCC increases drastically (Coulon and Pradel 1997). The increased SCC is associated with higher pH, lower milk lactose and elevated BSA and IgG1 concentrations. These effects reflect an increase in capillary permeability (Poutrel et al. 1983). In fact, it has been demonstrated that cows spend as much as 8 h per day walking, including time walking while grazing. In addition, walking time always exceeded the time of grazing by 10–15% (Ruckebusch and Bueno 1978). Coulon and Re'mond (1991), Coulon and Pradel (1997) have shown that the reduction in milk yield resulting from long walks (9–12 km) were significant, and varied according to animal type. The prolonged walking in the cows reduces the milk yield and changes composition of milk even in low producing animals (Coulon and Pradel 1997). The effect of walking is likely to be more marked for higher yielding cows. The reduction in the milk yield is thought to be due to the lack of additional forage intake by cows to compensate for the increased energy requirements induced by exercise. Stobbs and Brett (1974) have reported that milk and protein quantity decreased but the amount of fat produced did not vary probably because of a greater mobilization of lipid reserves. In fact, the walking cows consume less forage than did the sedentary cows because of restricted access time, which was not compensated by a higher intake rate or a sufficient increase in nocturnal intake (Coulon et al. 1998). Matthewman (1989) conducted experiments on 12 lactating and pregnant cows and allowed them to walk approximately 9 km/d for 3 weeks (maximum 15–17 h per week or 45–51 h over 3 weeks) at an average speed of 2.9 km/h. Milk yield was decreased by 7–14%. Milk fat yield (g/d) was not affected by exercise, but lactose and milk protein declined by the same proportion as milk yield.

4.10 Conclusion

It can be concluded from the findings of different workers that walking stress significantly alters the growth, physiological, haemato-biochemical responses as well as milk yield and its composition in animals undergoing these stresses. The significant changes in physiological responses, plasma cortisol, T_3 and T_4 show that the animals like sheep, goats and cattle are capable of adapting to long-distance walking and adrenal and thyroid hormones play a significant role in such adaptation. However, while trying to adapt to long- distance walking scenario in search of food, the productive performance of animals is greatly compromised which is observed in the significant reduction in growth, milk and reproductive performance of different livestock species. The low production could be attributed to the fact that a major proportion of the energy gets channelised in combating the walking stress as counter measures need energy source for its implementation. Agriculture is a major contributor to the economies of many countries; livestock are reared in tropical environments. In these hot environments during summer months, the animals need to walk long distances to graze. Hence, the findings from different studies discussed above have greater significance in terms of improving the performance of these animals under such environmental extremes.

Acknowledgment The authors are thankful to the senior research fellow, Ms.Saumya Bahadur, for her valuable help in assisting the preparation of this manuscript.

References

AFRC (1998) The nutrition of goats. CAB International, New York, pp 41–51

Aharoni Y, Brosh A, Kourilov P, Arieli A (2003) The variability of the ratio of oxygen consumption to heart rate in cattle and sheep at different hours of the day and under different heat load conditions. Livest Prod Sci 79:107–117

Anderson DM, Urghart NS (1986) Using digital pedometers to monitor travel of cows grazing arid rangeland. Appl Anim Behav Sci 16:11–23

Anderson MJ, Lamb RC, Mickelsen CH, Blake JT, Olsen JD, Arave CW (1977) Facility for exercising dairy cows. J Dairy Sci 60:1173

Anderson MJ, Lamb R, Walters JL (1979) Effect of prepartum exercise on feed intake and milk production of multiparous cows. J Dairy Sci 62:1420–1423

Animut G, Chandler KD (1996) Effects of exercise on mammary metabolism in the lactating ewe. Small Rumin Res 20:205–214

Animut G, Goetsch AL, Aiken GE, Puchala R, Detweiler G, Krehbiel CR, Merkel RC, Sahlu T, Dawson LJ, Johnson ZB, Gipson TA (2005) Grazing behavior and energy expenditure by sheep and goats co-grazing grass/forb pastures at three stocking rates. Small Rumin Res 59:191–201

ARC (1980) The nutrient requirements of ruminant livestock. Commonwealth Agricultural Bureaux, Farnham Royal 351p

Arnold GW, Dudzinski ML (1978) Ethology of free-ranging domestic animals. Elsevier, Amsterdam

Bailey DW, Gross JE, Laca EA, Rittenhouse LR, Coughenour MB, Swift DM, Sims PL (1996) Mechanisms that result in large herbivore grazing distribution patterns. J Range Mgmt 49:386–400
Barkai D, Landau S, Brosh A, Baram H, Molle G (2002) Estimation of energy intake from heart rate and energy expenditure in sheep under confinement or grazing conditions. Livest Prod Sci 73:237–246
Bartholomew PW, Khibe T, Little DA, Ba S (1994) Effect of change in body weight and condition during the dry season on capacity for work of draft oxen. Trop Anim Health Prod 25:50–58
Berhan T, Puchala R, Goetsch AL, Merkel RC (2006) Effects of walking speed and forage consumption on energy expenditure and heart rate by Alpine does. Small Rumin Res 63:119–124
Bibe B, Vissac B (1979) Amelioration gonotique et utilization du territoire. In: Utilisation par les ruminants des paturages d'altitude et parcours mdditerrandens, INRA Publication, Versailles, France
Boyne AW, Brockway JM, Ingram JF, Williams K (1981) Modification, by tractive loading, of the energy cost of working in sheep, cattle and man. J Physiol 315:303–316
Brockway JM, Boyne AW (1980) The energy cost for sheep of walking on gradients. In: Mount LE (ed) Energy metabolism of farm animals. European association for animal production publication no. 26, Butterworths, London, pp 449–453
Brosh A, Aharoni Y, Dejen AA, Wright D, Young B (1998) Estimation of energy expenditure from heart rate measurements in cattle maintained under different conditions. J Anim Sci 76:3054–3064
Brosh A, Aharoni Y, Holzer Z (2002) Energy expenditure estimation from heart rate: validation by long-term energy balance measurement in cows. Livest Prod Sci 77:287–299
Bruckmaier RM, Blum JW (1992) Responses of calves to treadmill exercise during P-adrenergic agonist administration. J Anim Sci 70:2809–28211
Clapperton JL (1964) The effect of walking upon the utilization of food by sheep. Br J Nut 18:39–46
Coop IE, Hill MK (1962) The energy requirements of sheep for maintenance and gain. II. Grazing sheep. J Agric Sci 58:187–199
Coulon JB, Garcl JP (1996) Aptitude a la marche de vachcs laitières de types g6n6tiques diff6rents :influence d'un exercice r6p6t6 sur la production laiti!re. Ann o Zhotec 45:349–355
Coulon JB, Pradel P (1997) Effect of walking on roughage intake and milk yield and composition of Montbe'liardes and Tarentaises dairy cows. Ann Zootech (Paris) 46:139–146
Coulon JB, Re′mond B (1991) Variations in milk output and milk protein content in response to the level of energy supply to the dairy cow: a review. Livest Prod Sci 29:31–47
Coulon JB, Garcl JP, Hoden A (1986) Evolution de la production et de la composition du lait a' la mise a' l'herbe. Bull. Tech. Ctr. Rech. Zootech. Vet. Theix Inst. Natl Rech Agron 66:23–29
Coulon JB, Pradel P, Cochard T, Poutrel B (1998) Effect of extreme walking conditions for dairy cows on milk yield, chemical composition, and somatic cell. Count J Dairy Sci 81:994–1003
CSIRO (1990) Feeding standards for Australian livestock: ruminants. Commonwealth Scientific and Industrial Research Organization, East Melbourne
D'Hour P, Hauwuy A, Coulon JB, Garel JP (1994) Walking and dairy cattle performance. Arur Zhoote 43:369–378
Daramola JO, Adeloye AA (2009) Physiological adaptation to humid tropics with special references to the West African Dwarf (WAD) goat. Trop Anim Health Prod 41:1005–1016
Davies HL, Southey IN (2001) Effects of grazing management and stocking rate on pasture production, ewe liveweight, ewe fertility and lamb growth on subterranean clover-based pasture in Western Australia. Aust J Exp Agric 41:161–168
FAO (2001) Pastoralism in the new millennium. FAO animal production and health paper 150, Rome, Italy
Farley CT, Glasheen J, McMahon TA (1993) Running springs speed and animal size. J Exp Biol 185:71–86

Fierro LC, Bryant FC (1990) Grazing activities and bioenergetics of sheep on native range in southern Peru. Small Rumin Res 3:135–146

Garcia-Belenguer S, Mormbde P (1993) Nuevo concepto de estres en ganaderia: psicobiologia y neuroendocrinolog'a de la adaptacin. Invest Agrar Prod Sanid Anim 8(87–1):100

Garcia-Belenguer S, Palacio J, Gascon M, Acena C, Revilla R, Mormede P (1996) Differences in the biological stress responses of two cattle breeds to walking up to mountain pastures in the pyrenees. Vet Res 27:515–526

Goe MR (1983) Current status of research on animal traction. World Anim Rev 45:2–17

Gordon FJ (1984) The effect of level of concentrate supplementation given with grass silage during the winter on the total performance of autumn-calving dairy cows. J Agric Sci (Camb) 102:163–179

Green JA, Halsey LG, Wilson RP, Frappell PB (2009) Estimating energy expenditure of animals using the accelerometry technique: activity, inactivity and comparison with the heart-rate technique. J Exp Biol 212:471–482

Gustafson GM, Luthman J, Burstedt E (1993) Effect of daily exercise on performance, feed efficiency and energy balance of tied dairy cows. Acta Agri Sci 43:219–227

Henning PH (1987) The effect of increased energy demand through walking exercise on intake and ruminal characteristics of sheep fed a roughage diet. J Agric Sci (Camb) 109:53–59

Herselman MJ, Hart SP, Sahlu T, Coleman SW, Goetsch AL (1999) Heat energy for growing goats and sheep grazing different pastures in the summer. J Anim Sci 77:1258–1265

Jabbar MA (1983) Effect of drought use of cows on fertility, milk production and consumption. In: Dvis CH, Treston TR, Jones WE (eds) Maximum live stock production from minimum land (1989) Equine sports medicine. Lea and Febiger, Philadelphia

Jones WE (1989) Equine sports medicine. Lea and Febiger, Philadelphia

Kaneko J (1989) Clinical biochemistry of domestic animals, 4th edn. Academic, San Diego

Kartiarso Martin D, Teleni E (1989) The pattern of utilization of body fat reserve by working cattle and buffalo. DAP Proj Bull 8:7–8

Kasa IW, Hill MK, Thwaites CJ, Baillie ND (1995) Physiological effects of exercise in male and female Saanen goats at the same body weight but different feed intake. Small Rumin Res 16:83–86

Kramer JW (2000) Normal hematology of cattle, sheep and goats. In: Feldman BF, Zinkl JG, Jain NC (eds) Schalm's veterinary henattology, 5th edn. Lippincott Williams & Wilkins, Philadelphia, pp 1075–1084

Krysl LJ, Hess BW (1993) Influence of supplementation on behavior of grazing cattle. J Anim Sci 71:2546–2555

Lachica M, Aguilera JF (2003) Estimation of energy needs in the free-ranging goat with particular reference to the assessment of its energy expenditure by the 13C-bicarbonate method. Small Rumin Res 49:303–318

Lachica M, Aguilera JF (2005a) Energy expenditure of walk in grassland for small ruminants. Small Rumin Res 59:105–121

Lachica M, Aguilera JF (2005b) Energy needs of the free-ranging goat. Small Rumin Res 2:111–125

Lachica M, Prieto C, Aguilera JF (1997) The energy costs of walking on the level and on negative and positive slopes in the granadina goat *(copra hircus)*. Br J Nut 77:73–81

Lamb RC, Barker BO, Anderson MJ, Walters JL (1979) Effects of forced exercise on two years old holstein heifers. J Dairy Sci 62:1791–1797

Lawrence PR, Stibbards RJ (1990) The energy cost of walking, carrying and pulling loads on flat surfaces by Brahman cattle and swamp buffalo. Anim Prod 50:29–39

Manteca X, Smith AJ (1994) Effects of poor forage conditions on the behaviour of grazing ruminants. Trop Anim Health Prod 26:129–138

Matthewman RW (1989) Effect of sustained exercised on milk yield, milk composition and blood metabolite concentrations in Hereford x Friesian cattle. PhD Dissertation, Centre for Tropical Veterinary Medicine, Roslin, Midlothian, EH25 9RG

Matthewman RW, Dijikman JT (1993) The nutrition of drought animals. J Agril Sci (Camb.) 121:297–306
Matthewman RW, Oldham JD, Horgan GW (1993) A note on the effect of sustained exercise on straw intake in lactating cattle. Anim Prod 57:491–494
Maurya VP, Naqvi SMK, Mittal JP (2004) Effect of dietary energy level on physiological responses and reproductive performance in Malpura sheep in hot semi-arid region of India. Small Rumin Res 55:117–122
Maurya VP, Prince LLL, Sejian V, Kumar D, Joshi A, Naqvi SMK (2010a) Relationship of body condition score with productive and reproductive efficiency of Avikalin ewes. Indian Vet J 87:189–190
Maurya VP, Sejian V, Kumar D, Naqvi SMK (2010b) Effect of induced body condition score differences on sexual behavior, scrotal measurements, semen attributes, and endocrine responses in Malpura rams under hot semi-arid environment. J Anim Phy Anim Nut 94: e308–e317
McManus C, Paluda GR, Louvandini H, Gugel R, Sasaki LCB, Paiva SR (2009) Heat tolerance in Brazilian sheep: physiological and blood parameters. Tropi Anim Health Prod 41:95–101
Mendez DG, Di Marco ON, Corva PM (1996) Energy expenditure of cattle walking on a flat terrain. Anim Sci 63:39–44
Metz JHM (1984) The reaction of cows to a short-term deprivation of lying. Appl Anim Behav Sci 13:301–307
Mormbde P (1988) Les réponses neuroendocriniennes de stress. Rec Med Vet 164:723–741
Naqvi SMK, Hooda OK, Saxena P (1991) Some plasma enzymes of sheep under thermal, nutritional and exercise stresses. Indian Vet J 68:1045–1047
Nienaber JA et al (1985) Performance of neonatal and newly weaned pigs as affected by temperature and diet. Trans ASAE 28(5):1626–1633
NRC (1981) Nutrient requirements of goats: angora, dairy and meat goats in temperate and tropical countries. National Academy Press, Washington
NRC (2001) Nutrient requirements of dairy cattle, 7th rev edn. National academic science, Washington
Osuji PO (1974) The physiology of eating and the energy expenditure of the ruminant at pasture. J Range Manage 27:437–443
Pagot J (1993) Animal production in the tropics and subtropics. McMillan Press Ltd, London
Payne JM, Payne S (1987) The metabolic profile test. Oxford University, Oxford
Pearson RA, Archibald RF (1989) Biochemical and haematological changes associated with short periods of work in draught oxen. Anim Prod 48:375–384
Pethick DW (1993) Carbohydrate and lipid oxidation during exercise. Aust J Agril Res 44:431–441
Poutrel B, Caffin JP, Rainard P (1983) Physiological and pathological factors influencing bovine serum albumin content of milk. J Dairy Sci 66:535–541
Richard D, Rivest S (1988) The role of exercise in thermogenesis and energy balance. Can J Physiol Pharmacol 67:402–409
Ruckebusch Y, Bueno L (1978) An analysis of ingestive behaviour and activity of cattle under field conditions. Applied Anim Etho 4:301–313
Sahlu T, Jung HG, Morris JG (1989) Influence of grazing pressure on energy cost of grazing by sheep on smooth Bromegrass. J Anim Sci 67:2098–2105
Sahlu T, Goetsch AL, Luo J, Nsahlai IV, Moore JE, Galyean ML, Owens FN, Ferrell CL, Johnson ZB (2004) Nutrient requirements of goats: developed equations, other considerations and future research to improve them. Small Rumin Res 53:191–219
Samuelsson B, Uvnas-Moberg K, Gorewit RC, Svennersten-Sjaunja K (1996) Profiles of the hormones somatostatin, gastrin, CCK, prolactin, growth hormone, oxytocin and cortisol. II. In dairy cows that are milked during food deprivation. Livest Prod Sci 46:57–64
Sano H, Takahashi K, Fujita M, Ambo K, Tsuda T (1979) Effect of environmental heat exposure on physiological responses, blood constituents and parameters of blood glucose metabolism in sheep. Tohoku J Agric Res 30(2):76–86

Sejian V, Srivastava RS (2010) Pineal-adrenal-immune system relationship under thermal stress: effect on physiological, endocrine and non-specific immune response in goats. J Physiol Biochem 66:339–349

Sejian V, Maurya VP, Naqvi SMK, Kumar D, Joshi A (2010a) Effect of induced body condition score differences on physiological response, productive and reproductive performances of Malpura ewes kept in a hot, semi-arid environment. J Anim Physiol Anim Nut 94:154–161

Sejian V, Maurya VP, Naqvi SMK (2010b) Effect of thermal, nutritional and combined (thermal and nutritional) stresses on growth and reproductive performance of Malpura ewes under semi-arid tropical environment. J Anim Phy Anim Nut. doi:10.1111/j.1439-0396.2010.01048.x

Sejian V, Maurya VP, Naqvi SMK (2010c) Adaptability and growth of Malpura ewes subjected to thermal and nutritional stress under semi-arid tropical environment. Tropi Anim Health Prod 42:1763–1770

Sejian V, Maurya VP, Naqvi SMK (2010d) Adaptive capability as indicated by endocrine and biochemical responses of Malpura ewes subjected to combined stresses (thermal and nutritional) under semi-arid tropical environment. Int J Biomet. doi:10.1007/s00484-010-0341-1

Sejian V, Maurya VP, Naqvi SMK (2011) Effect of walking stress on growth, physiological adaptability and endocrine responses in Malpura ewes under semi-arid tropical environment. Int J Biomet. doi:10.1007/s00484-011-0420-y

Seman DH, Frere MH, Stuedemann JA, Wilkinson SR (1991) Simulating the influence of stocking rate, sward height and density on steer productivity and grazing behavior. Agric Syst 37:165–181

Snow DH, Ricketts SW, Mason DK (1983) Haematological response to racing and training exercise in thoroughbred horses, with particular reference to leukocyte response. Equine Vet J 15:149–154

Sporndly E, Wredle E (2004) Automatic milking and grazing—effects of distance to pasture and level of supplements on milk yield and cow behavior. J Dairy Sci 87:1702–1712

Stobbs TH, Brett DJ (1974) Milk yield and composition of milk and blood as indicators of energy intake by Jersey cows. Aust J Agric Res 25:657–666

Teleni E, Boniface AN, Sutherland S, Entwistle KW (1989) The effect of depletion of body reserve nutrients on reproduction in Bos indicus cattle. Draught Anim Power Proj Bull 8:1–10. Aug 1989, James Cook University. Townsville, Australia

Veissier I, Andanson S, Dubroeucq H, Pomies D (2008) The motivation of cows to walk as thwarted by tethering. J Anim Sci 86:2723–2729

Verde MT, Gascon M (1987) Mecanismos de estrés en los animales domosticos. Med Vet 4:455–464

Voigt K, Ziegler M, Grunert-Fuchs M, Bickel U, Fehm-Wolfsdorf G (1990) Hormonal responses to exhausting physical exercise: the role of predictability and controllability of the situation. Psychoneuroendcrinol 15:173–184

Ward JR, Henricks DM, Jenkins TC, Bridges WC (1992) Serum hormone and metabolite concentrations in fasted young bulls and steers. Domest Anim Endocrinol 9:97–103

Wasserman DH, Williams PE, Lacy DB, Goldstein RE, Cherrington AD (1989) Exercise induced fall in insulin and hepatic carbohydrate metabolism during muscular work. Am J Physiol 256:500–509

Winugroho M (1993) Feeding draught animals in Indonesia. In: Pryor WJ (ed) Draught animal power in the Asian.Australasian region. A workshop in conjunction with 6th Asian. Australasian association of animal production societies, congress, 23–28 Nov, Bangkok, Thailand, ACIAR proceedings no. 46, p 93

Zerbini E, Asamenew G (1991) Potential of cow traction in the Ethiopian highlands farming system. Paper presented at the 4th national livestock improvement conference, 13–16 Nov 1991, held in Addis Ababa, Ethiopia

Zerbini E, Takele G, Alemu G, Abiye A (1992) Effect of draught work on lactation of F1 crossbred dairy cows, (preliminary results). In: Paper presented at the first workshop of the animal traction network for Eastern and Southern Africa (ATNESA) held 18–23 Jan 1992, Lusaka, Zambia

Chapter 5
Environmental Stresses and Livestock Reproduction

S. M. K. Naqvi, Davendra Kumar, Rajani. Kr. Paul and Veerasamy Sejian

Abstract Reproductive fitness may be regarded as the most important criteria for studying or evaluating animal adaptation. Body systems activated by stress are considered to influence reproduction by altering the activities of the hypothalamus, pituitary gland, or gonads. Activation of stress pathways may directly affect the activity of Gonadotropin-releasing hormone (GnRH) neurons within the hypothalamus or higher neural centers which in turn affects the synthesis or secretion of GnRH into the hypophysial portal blood. It is also possible that stress directly influences the responsiveness of gonadotrophin cells in the anterior pituitary gland via the action of GnRH. A further potential action of stress is to alter the feedback actions of sex steroids in the hypothalamus or pituitary and inhibin in the anterior pituitary gland. Reproduction processes in animals may be impacted during heat exposure and glucocorticoids are paramount in mediating the inhibitory effects of stress on reproduction. Glucocorticoids are capable of enhancing the negative feedback effects of estradiol and reducing the stimulation of GnRH receptor expression by estrogen. Glucocorticoids may also exert direct inhibitory effects on gonadal steroid secretion and sensitivity of target tissues to sex steroids. Heat stress (HS) influences estrous incidences and embryo production. The birth weights of lambs of heat stressed ewes are generally lower than the unstressed animals. This could be attributed to the fact that HS may cause a temporal impairment of placental size and function, resulting in a transient reduction in fetal growth rate. Secretion of the hormones regulating reproductive tract function may also be altered by HS. Further, HS can inhibit 3-beta-hydroxysteroid dehydrogenase (3β HSD) thereby minimizing progesterone secretion from luteal cells. Aromatase is an enzyme that converts androgens into estrogens and is present in

S. M. K. Naqvi (✉) · D. Kumar · R. Kr. Paul · V. Sejian
Division of Physiology and Biochemistry, Central Sheep and Wool Research Institute, Avikanagar, Jaipur, Rajasthan 304501, India
e-mail: naqvismk@yahoo.co.in

V. Sejian et al. (eds.), *Environmental Stress and Amelioration in Livestock Production*, DOI: 10.1007/978-3-642-29205-7_5,

the granulosa cells. By inhibiting the expression of this enzyme, HS may induce follicular atresia and consequently anestrus. Effects of steroid hormones on reproductive tract tissue could be reduced during exposure to HS due to increased synthesis of heat shock proteins (HSPs)—HSP 70 and HSP 90. Increased synthesis of HSP might alter assembly, transport, or binding activities of steroid receptors. Further, increased magnitude of these stresses will increase secretion of prostaglandin and reduce the secretion of interferon tau which affects the maternal recognition of pregnancy. In male, HS adversely affects spermatogenesis by inhibiting the proliferation of spermatocytes. This chapter will address the effect of environmental stresses on livestock reproduction.

Keywords Conception · Embryo · Fertilization · Heat stress · Lambing rate · Nutritional stress · Oocyte · Reproductive endocrinology

Contents

5.1 Introduction

Environmental stress is not limited to climatic factors but extends to nutrition, housing, and any stimuli that demand a response from the animal to adapt to new circumstances. Low energy and low or excessive protein levels in the diet are detrimental to reproduction. Likewise, high ambient temperatures, radiation, and humidity alter the intricate balance of endocrine profiles, leading to lower intensity of estrus behavior, anestrus, poor semen quality, embryonic death, and infertility. Various chemical substances present in the environment may lead to infertility in animals. Polluted soil, water, and air are major sources through which animals are exposed to xenobiotics. Substances such as pesticide, dioxins, and organic solvents which alter hormonal balance (endocrine disruptors) cause major damage to livestock reproduction. However, other inherent (genetic) and environmental (feed, climate, geographical location, diseases, etc.) factors also affect the reproductive efficiency of animals. In ruminants, for example sheep, the reproductive efficiency is determined by factors such as age of puberty, interlambing period, ovulation rate, estrus incidences, fertilization, embryo implantation, pregnancy, parturition, lactation, and mothering ability (Snowder 2008). A greater variability in various genotypes of animal has been reported for each factor (Safari et al. 2005).

Reproductive efficiency of animals generally improves the production and economic efficiency of any farm animal production system (Dickerson 1970). Animal farming has a great significance in human activities. Globally, livestock account for 40% of the world's agricultural gross domestic product (GDP) and generating employment for more than 1.3 billion people (FAO 2008, 2009). It also contributes significantly to the livelihood of people (about 1.0 billion) living below poverty level worldwide. Heat stress (HS) is known to have adverse impact on reproductive functions in animals. Disruption of follicular development, oocyte maturation, fertilization, embryogenesis and development, placental and fetus growth in female animals are major events caused by HS. These deleterious effects are manifested due to rise in the body temperature that could not be mitigated by inefficient thermolytic mechanism.

In order to maintain body function in steady state, homeotherms are required to maintain body temperature within a narrow range. Deviation from the set level of body temperature under stressful hot environment leads to interference with physiological events and consequently negative impacts on animal productivity. The maintenance of homeotherms is dependent upon the energy flow from animal to environment and vice versa. Effective ambient temperature (EAT) is a major environmental factor controlling the energy flow. Due to the fact that various factors influence the EAT viz. dry bulb temperature, wet bulb temperature, humidity, wind speed, heat radiation, contact surfaces, etc., no satisfactory measures have been developed so far to quantify EAT and hence ambient temperature is the most commonly used indicator. Sheep possess thick insulating boundary on the body and only about 10% of solar radiation received by fleeced sheep reaches to the skin. The exogenous (solar) heat load of the shorn sheep standing in sun may be about 5–6 times greater than its internal heat production (resting metabolic heat). Consequently, adaptation and mitigation of detrimental effects of extreme climates have played a major role in combating the climatic impact in livestock production (Khalifa 2003). Among the environmental variables affecting animals, HS and nutritional stress seem to be the more intriguing factors making animal production of many world areas difficult (Shelton 2000; Koubková et al. 2002).

Animals can adapt to the hot climate; nevertheless, the response mechanisms are helpful for survival but are detrimental to productive and reproductive performance. Hence, attempt will be made in this chapter to particularly address the effect of HS and nutrition stress on reproduction in livestock.

Animals live in complex environments in which they are constantly confronted with short- and long-term changes due to a wide range of factors such as environmental temperature, photoperiod, geographical location, nutrition, and socio-sexual signals. Environment plays an important role in influencing the reproductive performance of farm animals. Fertility of farm animals is affected by high ambient temperature, excess humidity, severe cold, and lesser access to drinking water. HS, due to high ambient temperature accompanied with excess humidity during summer months, causes infertility in most of the farm species. Dairy cattle are particularly susceptible to HS because of higher metabolic heat produced during milk production. Furthermore, high yielding cows are most sensitive to HS. Nutrition modulates reproductive functions in many species including sheep (Naqvi and Rai 1991). Failure to rebreed due to decline in average sheep flock fertility results in more serves per successful conception, extended lambing intervals, and increased culling (Naqvi et al. 2002). Genetics, management, and nutrition have all contributed to this decline in fertility. Nutritional deficiencies and imbalances are frequently implicated as an important cause of infertility in sheep (Naqvi et al. 2011).

5.2 Probable Mechanisms of Stress Affecting Reproduction

Figure 5.1 describes different possible mechanisms by which stress affects reproduction in livestock. These possible mechanisms are outlined below:

- The principal biological mechanism by which HS impacts on livestock reproduction is partly explained by reduced feed intake, and also includes altered endocrine status, reduction in rumination and nutrient absorption, and increased maintenance requirements resulting in a net decrease in nutrient/energy availability for reproduction
- Stress (heat, nutritional, pH, immunological, physiological, etc.) prevents the basic cell function by causing improper folding of proteins. To cope with this, expressions of HSPs like HSP70 are stimulated/upregulated. The basic function of HSPs is to help in 'proper folding' of proteins so that the three-dimensional structure of the protein is not compromised and thus maintaining its normal function. Further, the cell diverts its entire transcriptional and translational machinery to synthesize proteins that are required for maintaining 'house keeping functions' such as cellular respiration, ATP synthesis, excretion of protons (H^+ ions) at the expense of other functions of the cell, e.g., testosterone production. To combat stress, the Adrenocorticotropic hormone (ACTH)–Cortisol axis is activated and cortisol and other glucocorticoids are produced to enhance stress tolerance.

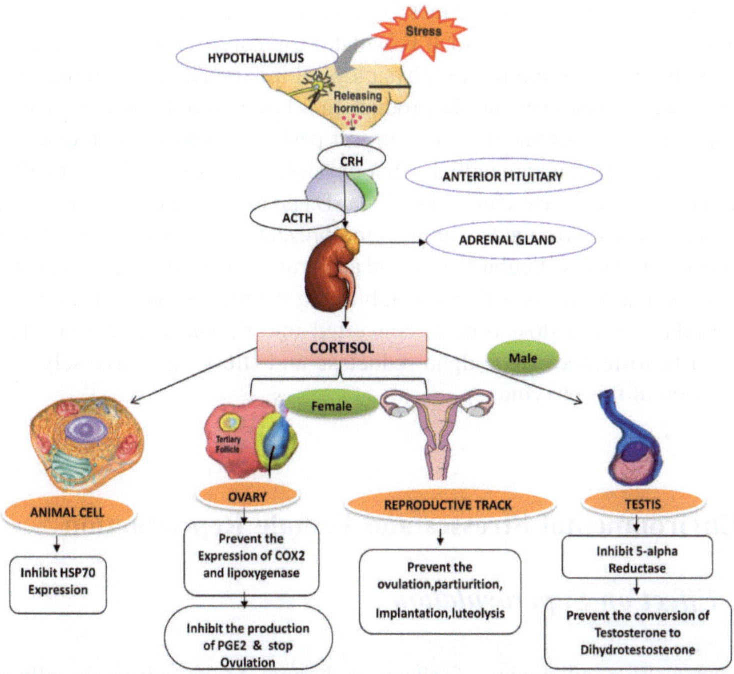

Fig. 5.1 Different mechanisms with which stress affects reproduction in livestock. Stress stimulus on hypothalamus produces CRH which in turn acts on anterior pituitary to release ACTH. ACTH atimulates cortisol release from adrenal cortex. Cortisol is the principal glucocorticoid which inhibits reproductive function by acting on animal cell, ovary, reproductive track in female and testis in male. Corticotrophin releasing hormone (*CRH*); Adrenocorticotropic hormone (*ACTH*), Cycloxigenase 2 (*COX2*); Prostaglandin E2 (*PGE2*)

- Glucocorticoids are anti-inflammatory. They prevent the expression of cyclo-oxygenase-2 (cox2) and lipoxygenase which gives rise to prostaglandins and leukotrienes, respectively. Luteinizing hormone (LH) surge induces cox2 expression and PGE2 production. PGE2 is essential for ovulation but may be inhibited by glucocorticoids.
- The growing fetus is under stress due to space and nutrient constraint. This fetal stress activates ACTH-Cortisol axis. The fetal cortisol enters maternal circulation and activates the expression of 17 alpha-hydroxylase enzyme in the placenta. This enzyme converts progesterone into estrogen. Declining progesterone levels with concomitant increase in estrogen levels induce parturition. Injection of glucocorticoids (dexa, beta, prednisone, triamcenalone, etc.) any time after first month of pregnancy induces abortion in sheep and goat. Glucocorticoids prevent the 'maturation of placentome'. Therefore, if parturition is induced with steroids, retained fetal membranes are common sequel.
- Cortisol 'stabilizes' the lysosomal membrane and thereby prevents the release of vasoactive substances (such as histamine, serotonin, heparin, substance P,

proteolytic enzymes) and subsequently prevents inflammatory response. Most of the common reproductive events such as ovulation, luteolysis, implantation, parturition, and involution of the uterus are 'physiological inflammation'. You may expect an adverse effect on these events. Reproduction is basically a 'luxurious phenomenon' and appropriate when the animal is in just perfect homeostasis. In case of severe stress, reproduction is typically the first physiological event to let go by the body.

- In female animals, stress can inhibit 3β HSD and minimize progesterone secretion from luteal cells. Aromatase, an enzyme present in the granulosa cells, may be inhibited to induce follicular atresia and anestrus. In male animals, stress adversely affects spermatogenesis perhaps by inhibiting the proliferation of spermatocytes. At cellular level, testosterone is converted into dihydrotestosterone (bioactive form of testosterone) by 5 alpha reductase and stress may adversely affect the expression of this enzyme.

5.3 Environmental Stresses and Female Reproduction

5.3.1 *Effect on Superovulation*

The reproductive efficiency of sheep is known to be adversely affected by hyperthermia (Thwaites 1971; Sawyer 1979). HS is also known to influence the superovulation response in sheep (Gordon 1997) and cattle (Hansen et al. 2001; Alfujairi et al. 1993; Gordon et al. 1987; Monty and Racowsky 1987) in a multiple ovulation and embryo transfer programme. Ewes exposed to HS produced relatively poor quality embryos when compared to ewes that were not exposed to HS. The results indicate that HS could adversely affect the quality of the embryos in Bharat Merino sheep reared in semi-arid tropics (Naqvi et al. 2004). The effect of HS on superovulation has been reported to vary in cattle. Monty and Racowsky (1987) reported no influence on superovulation response in Holstein cows, while adverse effects were observed by other workers (Gordon et al. 1987; Alfujairi et al. 1993). Alfujairi et al. (1993) reported a negative effect of hot summer on ovulation rate, total ova/embryos, and quality of embryos in cows. Similarly, Gordon et al. (1987) registered a highly significant difference in values recorded for Holstein cows treated for superovulation during midsummer and winter/spring.

5.3.2 *Estrus Intensity and Duration*

HS influence on estrus incidences is a well-established fact (Naqvi et al. 2004; Tabbaa et al. 2008). In general, duration and intensity of estrus in animals are reduced due to HS (Younas et al. 1993; Gwazdauskas et al. 1981). Exposure of Rambouillet cross ewes to severe HS from day 12 of the estrous cycle can extend the length of the cycle significantly (Dutt 1963). In addition, HS can influence the onset of estrus (Sejian et al. 2011). Exposure of ewes to high ambient temperatures 1.5–6 days prior to estrus has been reported to reduce estrus occurrence in ewes (Sawyer 1979). An alteration in the pulsatile release of LH and decrease in estrogen secretion is a potential reason for the delay of onset of estrus in ewes after exposure to HS. The normal GnRH release patterns (frequency and amplitude of LH pulses secreted from the pituitary) are reduced by exposure to HS (Dobson and Smith 2000). This results in abnormal ovarian functions and hence causes a delay in the LH surge. Furthermore, HS alters follicular development and dominance which leads to a decrease in estrogen secretion. Badinga et al. (1993) found follicular dominance to be altered in cows that were heat stressed during the first 8 days of the estrous cycle. However, Gangawar et al. (1965) also reported the duration of estrus to average 20 h in cows housed under cool conditions, compared to 11 and 14 h for cows reared in a hot psychrometric chamber and summer season, respectively. The intensity of estrus was also greater under cool than hot environmental conditions. The estrus response, fertilization rate, and neonatal survival may also decrease with HS (Mohamed 1974). The reduced estrus percentage and duration in livestock during summer months could be related to the high plasma progesterone concentration due to HS. Presumably, the longer estrus cycles were due to a slower rate of follicular maturation after corpus luteum (CL) regression. This statement agrees with Stewart and Oldham (1986), who reported that nutritional effect on ovulation rate seems to be more due to mechanisms that are confined to final stages of folliculogenesis rather than change in secretion of GnRH, LH, and follicle-stimulating hormone (FSH).

5.3.3 Sexual Behavior

HS reduces the normal manifestation of different sexual behaviors, which leads to decrease in productive potential of animals. The variation in sexual behavior pattern occurs in ewes during estrus, i.e., perceptivity (active search of ram) and receptivity (acceptance of mating attempts by ram) (Banks 1964). However, in Merino sheep sexual behavior occurs in the form of circling, tail fanning, head turning, standing, and approaching ram (Lynch et al. 1992).

5.3.4 Oocyte Maturation

Oocyte maturation process is a complex event and involves nuclear, cytoplasmic, and molecular maturation (Ferreira et al. 2009). Oocyte maturation, in vivo, begins with the resumption of meiotic process which is facilitated by cumulus cells. Calcium ions present in cumulus of oocyte brings about this nonhormone-mediated meiotic induction (Webb et al. 2002).

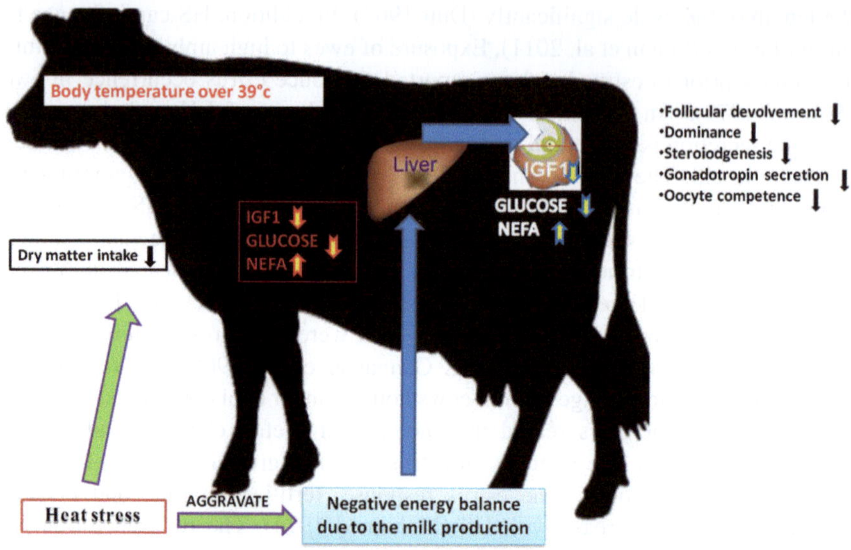

Fig. 5.2 Direct (hyperthermia) and indirect (negative energy balance associated with reduced dry matter intake) effects of HS (*green arrows*) on ovarian follicles, oocyte, and granulosa cells. Insulin-like growth factor-1 (*IGF1*), non-esterified fatty acids (*NEFA*). Metabolic parameters are altered in blood (*red text*) and have been reflected in the follicular fluid (*white text*) (*Source* Shehab-El-Deen 2011)

A sizeable population of livestock species in tropical and subtropical regions of the world are exposed to elevated ambient temperature and thus experience hyperthermia. HS, at the time of oocyte maturation (near estrus/breeding), is most susceptible to hyperthermic condition and lead to reduction in infertility (Cavestany et al. 1985; Putney et al. 1989). The magnitude of the effect of HS on fertility is dependent on the intensity and duration of HS on the animals (Cavestany et al. 1985; Barati et al. 2008). Ozawa et al. (2005) reported that HS in goats reduced circulating concentrations of estradiol and lowered follicular estradiol concentration, aromatase activity and LH receptor level, and delayed ovulation. Elevated temperature condition during in vitro culture of follicular cells of cattle reduced the steroid production (Bridges et al. 2005). Estradiol secretion in response to GnRH diminished in goats exposed to HS (Kanai et al. 1995). Figure 5.2 describes the direct and indirect effects of HS (green arrows) on ovarian follicles, oocyte, and granulosa cells.

5.3.5 *Nuclear Damage*

Many research findings suggested that the exposure of oocytes to elevated temperature induced DNA damage in oocytes prior to fertilization. Exposure of oocytes to

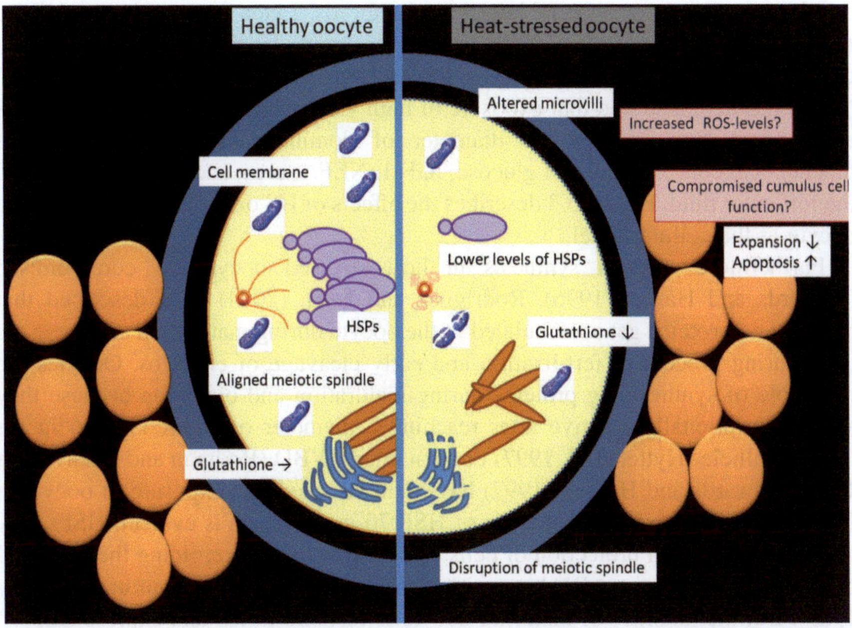

Fig. 5.3 Effects of HS on the organization of the oocyte at the cellular level (*Source*: Shehab-El-Deen 2011). *Left* an unstressed oocyte, *right* oocyte under heat stress. Heat stress leads to damage of the cytoskeleton and as such to poor alignment of the chromosomes (*brown* filaments). Organelles such as the Golgi and the endoplasmic reticulum (*blue*) become fragmented and disassemble. The number and integrity of mitochondria (*dark blue*) decreases. The synthesis of HSPs decreases (*purple*). Also, there are changes in the membrane morphology, aggregation of membrane proteins, and an increase in membrane fluidity. Together, all these effects stop cell growth and lead to cell-cycle arrest

elevated temperature (41.8°C) for 12 h reduces their ability to complete nuclear maturation and development after fertilization. Similarly, Roth and Hansen (2004) and Ju and Tseng (2004) reported that elevated temperature during maturation culture of oocytes induced DNA fragmentation and cytoskeleton disruption. When Porcine oocytes are exposed to temperatures 41.0 or 38.5°C (sham control) for 0, 0.5, 1.0, or 1.5 h, followed by culture for 44 h, meiotic competence was compromised. However, nuclear maturation can be improved with the use of antioxidants in heat stressed oocytes (Lawrence et al. 2004; Maya-Soriano et al. 2010).

5.3.6 Molecular Changes

Heat exposure of animals may alter the biochemical composition of follicles which indirectly affects the developmental competence of oocyte and quality of granulosa cells. Molecular changes occurring during oocyte meiotic maturation affect developmental competence of later stage embryos. Later stage embryos derived from the oocytes exposed to HS do not have optimum development competence as compared to that of non-stressed oocytes (Edwards et al. 2009). Shehab-el-Deen (2011) has demonstrated that exposure of high yielding dairy cows to summer HS early postpartum reduces the diameter of dominant follicle and alters its biochemical concentrations of glucose, IGF-1, NEFA, urea, and total cholesterol in the follicular fluid. Figure 5.3 describes the effects of HS on the organization of the oocyte at the cellular level.

Furthermore, protein synthesis in heat stressed oocytes is compromised (Edwards and Hansen 1996). Rodriguez and Farin (2004) have described that mRNA and proteins get accumulated in the oocyte during maturation and these are used during process of fertilization and early cleavage of embryos. Oocytes are incapable of synthesizing proteins during maturation and ovulation because they are transcriptionally inactive after reaching a diameter of about 110 μm in the tertiary follicle (Hyttel et al. 1997) or soon after GVBD (Rodman and Bachvarova 1976). Edwards and Hansen (1997) have demonstrated that exposure of oocyte to HS does not increase concentration of HSP 70 in cattle. As is evident, HSPs play an important role in protection of cells against HS through refolding the damaged proteins and stabilizing rRNA as a protection mechanism (Duncan and Hershey 1989). But the inability of maturing oocyte to respond to HS and express HSP 70 render them vulnerable except getting some protection by utilization of maternal RNA pools previously accumulated during oocyte growth for protein synthesis (Wassarman and Letourneau 1976). Further, HS conditions could have deleterious effects on oocyte growth, protein synthesis, and formation of transcripts required for subsequent embryonic development. The consequence of HS on the maturing oocyte can ultimately lead to a reduction in the capacity of an oocyte to be fertilized and to develop into a blastocyst.

5.3.7 Fertilization and Embryo Development

Lenz et al. (1983) and Edwards and Hansen (1996) performed in vitro maturation (IVM) and reported reduced maturation rate when bovine oocytes are exposed to 41°C. Subsequently, they also reported reduction in fertilization rate following in vitro fertilization (IVF) at 41°C. Rivera and Hansen (2001) showed that bovine oocytes fertilized at 41°C for 8 h had lower cleavage rates than did oocytes fertilized at 38.5°C in vitro. However, incubation at 40°C during IVF had no effect on cleavage rate and tended to increase the rate of oocytes forming blastocysts compared with oocytes fertilized at 38.5°C. This study mimicked rectal temperatures recorded during very hot mid-summer days in southeast Queensland, Australia. Sugiyama et al. (2007) recorded poor fertilization rates compared to the control group confirming that not only do high ambient temperature during IVF impairs fertilization, but also that on exposure to a maximum of 41°C for only 4 h is sufficient to adversely affect the outcome of IVF. Al-Katanani et al. (2002) collected oocytes from cows during the summer and reported that they have a reduced ability to develop to the blastocyst stage after IVF.

Most study suggests that early cleavages of embryos are sensitive to the exposure to elevated temperature (Putney et al. 1989). The adverse impact of HS declines as the embryo develops from 2-cell stage to morula and/or blastocyst (Ealy et al. 1993). The crossbred ewes (Bharat Merino) exposed to HS (40°C and 58.4% RH) of 6 h/d (10.00–16.00 h) for 4 weeks, yielded relatively poor quality embryos (Naqvi et al. 2004). In their study, the fertilization rate of the ewes was not affected by HS. Exposing Holstein cows to elevated ambient temperature (42°C for 10 h) during the periovulatory period increased the incidence of retarded and/or abnormal embryos (Putney et al. 1989). Similarly, Dutt (1963) reported that elevated ambient temperature 24 h prior to fertilization has no effect on the fertilization rate but increases the incidences of embryonic abnormalities. Similar results have been found in mice by Baumangartner and Chrisman (1987), who indicated that HS prior to ovulation did not interfere with oocyte fertilization but resulted in extensive developmental embryonic retardation.

The mechanism through which HS exerts influence on early embryos is not fully understood. Some evidence suggests increased production of reactive oxygen species (ROS) during preimplantation embryos following exposure to HS. Sakatani et al. (2004, 2008) reported increased production of ROS in cattle preimplantation embryos in response to elevated culture temperature. With the use of antioxidant, the negative effect of HS on the developing bovine embryos could be alleviated (Sakatani et al. 2008), but such effect was not reported by de Castroe Paula and Hansen (2008). Impaired function of oocytes and embryos were seen when they were exposed to heat during different stages of oocyte maturation and early embryo development in both in vivo and in vitro systems (Edwards et al. 2001; Naqvi et al. 2004). Naqvi et al. (2004) reported no affect on fertilization rate but found abnormal embryos from the Bharat Merino ewes exposed to HS during follicular phase. Further, Ealy et al. (1993) found that HS on day 1 after breeding decreased subsequent embryonic development. They further reported that HS on days 3, 5, or 7 after breeding, did not affect embryonic development. Therefore, the period of embryonic sensitivity to HS begins early during the development of the follicle and continues until about 1 day after breeding. Other studies based on in vitro culture system in cattle reported that the HS effect on embryo is stage dependent and zygote being most sensitive than the morula or blastocyst to high temperatures (Ealy et al. 1993; Edwards et al. 2001).

Exposure of bovine embryos to in vitro HS induces a range of effects including increased apoptosis (Paula-Lopes and Hansen 2002), increased expression of HSPs by porcine embryos (Bernardini et al. 2004), disruption of microtubule, and microfilaments (Rivera et al. 2004).

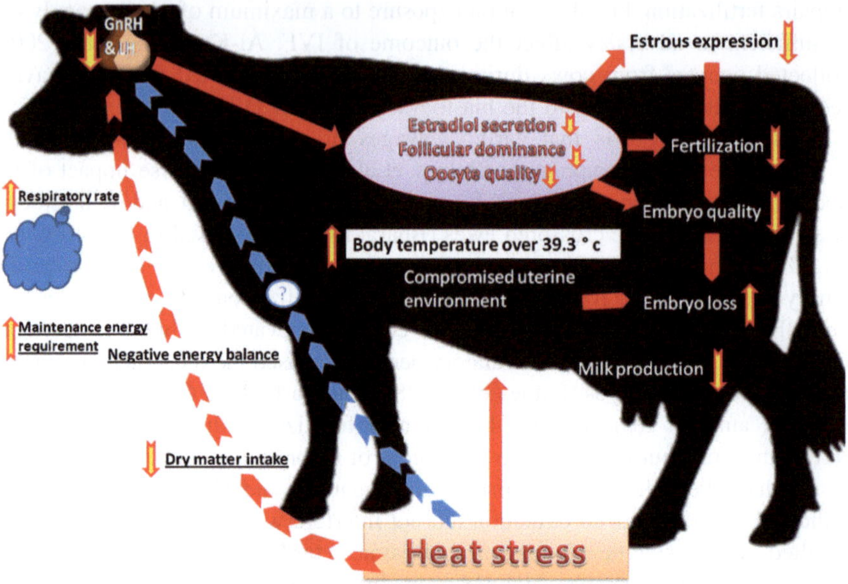

Fig. 5.4 A schematic description of the possible mechanisms with which HS may affect reproduction in the lactating dairy cow. HS can act in more than one way to reduce fertility in lactating dairy cows. HS can reduce dry matter intake and indirectly inhibit GnRH and LH secretion from the hypothalamo-pituitary system (*dashed red lines*). It is not clear if HS can directly affect the hypothalamo-pituitary system (*dashed blue line*) to reduce GnRH and LH secretion. HS can directly compromise the uterine environment (*solid lines*) to cause embryo loss and infertility (*Source* Shehab-El-Deen 2011)

HS at and immediately after the time of breeding results in lower conception rates. Heat stressed dairy cows tend to have a decrease in dry matter intake thus reducing the amount of energy in their diet. In order to maintain pregnancy, it is critical to sustain a healthy diet (Al-Katanani et al. 2002). Further, Du Preez et al. (1991) developed three regression models relating conception rate (CR) and mean monthly temperature-humidity index (THI) and reported that CR decreased with increasing THI during summer season. Chronic HS leads to delayed ovulation, low CR, as well as to a higher rate of abortions (Ben and Bouraoui 2009). Figure 5.4 represents the schematic description of the possible mechanisms generated by HS which may affect reproduction in the lactating dairy cow.

HS reduces the length and intensity of estrus behavior, modifies endocrine function, alters the oviductal and uterine environments, interrupts early embryonic development, and ultimately, lowers the conception rates of dairy cattle (Hansen 2009; Wolfenson 2009). Collier et al. (2006) reported a 10% reduction in CR when cows were mated during summer months. Similarly, HS can cause reduction in CR in sheep and goats, and sheep embryo is most susceptible (Thwaites 1971).

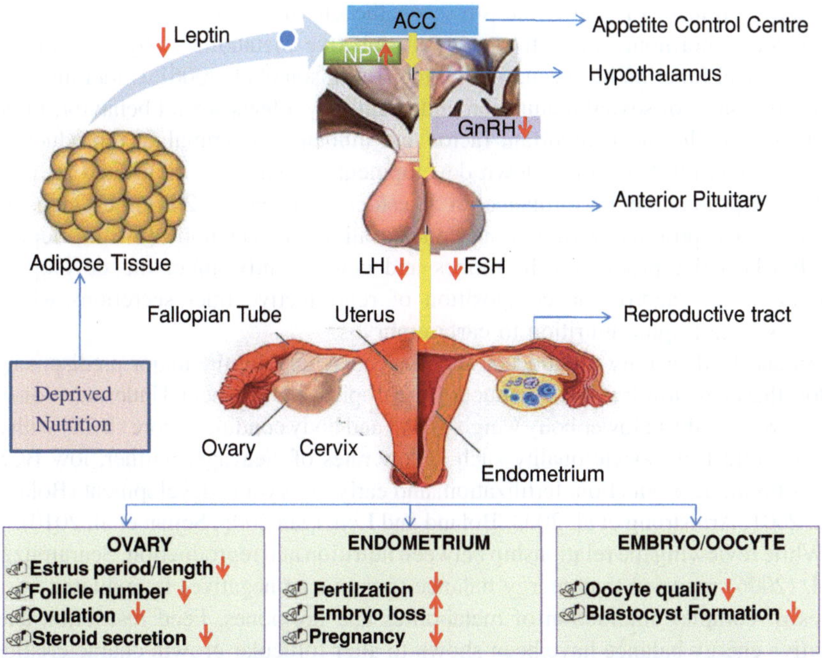

Fig. 5.5 Mechanism of nutritional stress affecting livestock reproduction. Deprived nutrition reduces blood leptin concentration. This stimulates the hypothalamus to reduce secretion of GnRH. This leads to reduced secretion of FSH and LH in the anterior pituitary. This leads to reduced sex steroids production in the reproductive tract and affects all the reproductive processes

5.4 Mechanisms of Nutritional Stress in Livestock Reproduction

Figure 5.5 describes the mechanism with which nutritional stress affects reproduction. Nutrient deficiency decreases the blood leptin concentration. This acts on the appetite control center in the brain and increases NPY which acts on the hypothalamus to decrease GnRH secretion. This decreased GnRH simultaneously leads to lower LH and FSH secretion from anterior pituitary. This in turn affects all the reproductive organs that control specific events of reproduction.

5.5 Effect of Nutritional Stress on Livestock Reproduction

Nutritional stress is an important environmental factor that influences ruminant fertility directly because it supplies specific nutrients required for oocyte development, ovulation, fertilization, embryo survival, and the establishment of pregnancy. Impact of nutritional stress on the circulating concentrations of the reproductive hormones and other nutrient-sensitive metabolites required for the physiological function has been highlighted (Robinson et al. 2006). Undernutrition delays the onset of sexual maturation and negatively affects sexual behavior. Food availability is the most important factor that influences mammalian reproduction because undernutrition slows down development of ovarian follicles and reduces lifetime reproductive performance in livestock (Rae et al. 2001). Changes in dietary intake promote variations in concentrations of metabolic [insulin, leptin, and IGF1] and reproductive hormones and consequently affect the developing ovarian follicle and/or the composition of reproductive tract secretions which provides histiotrophic nutrition to early embryos.

HS and feed scarcity often occur simultaneously and are the major predisposing factors that cause low livestock productivity in tropical environment. Undernutrition of donor ewes results in lower body weight (BW) and body condition score (BCS) with a negative effect on oocyte quality such as low rates of cleavage. Further, low BCS affects hormone production, fertilization, and early embryonic development (Boland et al. 2001; Armstrong et al. 2003; Boland and Lonergan 2005; Sejian et al. 2010).

While reviewing the relationship between nutrition and reproduction, Scaramuzzi et al. (2006) reported that energy balance (positive or negative) is regulated by a series of complex interaction of metabolites and hormones. Feed restriction and negative energy balance have been shown to alter follicular growth characteristics in cattle (Murphy et al. 1991) and sheep (Yaakub et al. 1997) and ultimately, estrus response. Restricted feeding (30% of ad lib intake) can significantly reduce the duration of estrus (15 vs 26 h) and increase the estrus interval (31.5 vs 18.6 days) in Malpura sheep (Maurya et al. 2004). The underlying mechanism for suppressed intensity of estrus and estrus duration has not been elucidated (Rhind 1992). However, it has been postulated that ovarian responses are influenced by the availability of nutrients, e.g., glucose and amino acids (Downing et al. 1995).

Optimal reproductive rates are essential for profitable sheep production (Vinoles et al. 2005). It has been postulated that nutrition is one of the main factors affecting ovulation rate and sexual activity in sheep (Forcada and Abecia 2006; Naqvi et al. 2011). Undernutrition during late pregnancy or in early postnatal life can irreversibly reduce the lambing rates of ewes (Gunn et al. 1995). Nutritional status has also been correlated with embryo survival in ewes and marked as key factor influencing efficiency in animal reproduction technology (Armstrong et al. 2003; Webb et al. 2004).

There are reports indicating the level of feeding affecting oocyte quality in ewes (Robinson et al. 2002). Ewes fed 50% of their maintenance requirements for two weeks had reduced expression of glucose transporter 3 (SLC2A3), sodium/glucose co-transporter 1 (SLC5A1), and Naþ/Kþ ATPase mRNA in oocytes, while expression of PTGS2, HAS2, and the leptin receptor in granulosa cells was increased (Pisani et al. 2008). Reduced expression of SLC2A3 is potentially relevant in the light of its significant role in post-implantation embryonic development (Schmidt et al. 2009).

Inadequate nutrition prior to calving results in emaciation of cows which delays the onset of estrual activity post calving. This delay influences the percent of cows available for breeding during the breeding season, thus reducing overall conception rates (Robinson et al. 2006; O'Callaghan et al. 2000). Lawson and Cahill (1983) postulated that variations in the physiological range of peripheral progesterone concentration due to management factors such as nutrition may induce asynchrony between the embryo and uterus resulting in failure of establishing pregnancy. Further, undernutrition in ewes before and after mating can increase embryonic mortality which consequently reduces the lambing rate (Rhind et al. 1989; Abecia et al. 2006).

5.6 Effect of Heat and Nutritional Stress on the Circulating Reproductive Hormones

Reproductive endocrinology plays a key role in livestock productivity during extreme climatic conditions. Endocrine responses to stress work toward suppressing productive functions such as growth and reproduction while favoring maintenance and survival (Rivest and Rivier 1995; Lindsay 1996). Heat and nutritional stress may cause infertility in farm animals. Notably, stress response involves the release of ACTH by the posterior pituitary gland and release of cortisol by the adrenal cortex (Minton 1994; Aoyama et al. 2003). High cortisol concentration in the system inhibits the pituitary response to GnRH (Dobson et al. 2001). However, cortisol decreases gonadal activity by reducing pulse frequency and amplitude of the LH released by the gonadotrophs (Dobson et al. 2001; Breen and Karsch 2004). The reduction of LH pulses in the luteal phase may induce follicular atresia. Suppression of LH release patterns during the follicular phase delays or inhibits the preovulatory LH surge and consequently disrupts the oocyte maturation and embryo quality (Mihm et al. 1994).

The association between HS and increased secretion of cortisol, the principal glucocorticoid hormone in small ruminants, is well documented (Ali and Hayder 2008; Sejian et al. 2008). Further, it is recognized that reproduction processes are influenced during heat exposure (Tilbrook et al. 2002; Naqvi et al. 2004; Kornmatitsuk et al. 2008). From these studies, it is pertinent to conclude that glucocorticoids are paramount in mediating the inhibitory effects of stress on reproduction. In support to this notion, various studies have shown that administration of natural or synthetic glucocorticoids can inhibit the secretion of the gonadotrophins in sheep (Juniewicz et al. 1987; Tilbrook et al. 2000). Further, glucocorticoids are capable of enhancing the negative feedback effects of estradiol and reducing the stimulation of GnRH receptor expression by estrogen (Adams et al. 1999; Daley et al. 1999). Glucocorticoids may also exert direct inhibitory effects on gonadal steroid secretion and sensitivity of target tissues to sex steroids (Magiakou et al. 1997). Inadequate nutrition delays or prevents the onset of puberty, interferes with normal cyclicity in the female, and results in hypogonadism (Sejian et al. 2010). This in turn affects all other reproductive processes.

Most reproductive responses to environmental factors are coordinated at brain level, where all external and internal inputs ultimately converge into a final common pathway that controls the secretion of GnRH. In turn, this hormone controls the secretion of gonadotrophins, the pituitary hormone, that determines the activity of reproductive axis. The effects of HS on LH concentration in peripheral blood plasma are inconsistent. Some studies report unchanged concentrations (Gwazdauskas et al. 1975; Howell et al. 1994). There are several researchers who reported increased plasma LH concentrations (Roman-Ponce et al. 1981) while others reported decreased concentrations (Gilad et al. 1993; Lee 1993) following HS. Plasma inhibin concentrations in summer are lower in heat stressed cows (Wolfenson et al. 1993) and in cyclic buffaloes, in India (Palta et al. 1997), perhaps reflecting reduced folliculogenesis since a significant proportion of plasma inhibin comes from small- and medium-sized follicles. Inhibin is an important factor in the regulation of FSH secretion (Findlay 1993; Kaneko et al. 1993). A clear inverse relationship between plasma FSH and immunoreactive inhibin concentrations was found throughout the oestrous cycle (Kaneko et al. 1997). The small amount of published information suggests that FSH is increased by HS and this may be due to decreased plasma inhibin production by compromised follicles.

Plasma estradiol concentrations are reduced by HS in dairy cows (Wilson et al. 1998), an effect that is consistent with decreased concentrations of LH. The mechanisms by which HS alters the concentrations of circulating reproductive hormones are not known. Some effects of HS may involve ACTH. HS can cause increased cortisol secretion (Elvinger et al. 1992; Sejian et al. 2011), and ACTH has been reported to block estradiol-induced sexual behavior (Hein and Allrich 1992). Decrease of estradiol concentration in the follicular fluid is more likely to occur after exposure to long-term, chronic (summer) HS than to acute HS. The effect of HS on plasma progesterone concentration is controversial. Wilson et al. (1998) found that HS had no effect on the plasma progesterone concentrations but that luteolysis was delayed. Several other studies have reported increased (Trout et al. 1998; Sejian et al. 2011), decreased (Younas et al. 1993; Ronchi et al. 2001) or unchanged (Guzeloglu et al. 2001) blood concentrations of this hormone during summer HS in dairy cows. These differences probably arise because of uncontrolled changes in other factors that affect blood progesterone concentrations.

It is generally accepted that nutrition modulates reproductive endocrine function in many species including sheep (Lindsay et al. 1993; Polkowska 1996). Kiyma et al. (2004) reported that serum concentrations of estradiol were lower in undernourished ewes. Similar results were reported in other species (Morin 1986; Otukongyong et al. 2000). Decreased concentration of estrogen may result from diminished ovarian follicular development caused by suppressed peripheral concentration of gonadotrophins (Gougeon 1996). Adams et al. (1994) contradicted this finding that undernourished ewes had lower plasma estradiol 17-β concentration. They established that food restriction is clearly associated with higher plasma concentration of estradiol 17-β concentration in ewes. Bell et al. (1989) reported that chronic HS exposure reduced plasma progesterone concentration in ewes. But Sejian et al. (2011) contradicted the above findings in which heat exposure significantly increased the plasma progesterone concentration. The level of nutrition and peripheral progesterone concentrations are inversely related (Parr 1992; Lozano et al. 1998) in ewes. This inverse relationship between the level of feed intake and plasma progesterone concentration was attributed to difference in metabolic clearance rate of progesterone (Parr et al. 1993). This view was further strengthened by the finding of Forcada and Albecia (2006) which states that difference in the rate of clearance rather than differences in secretion levels can explain the apparent inverse relationship between nutrition and peripheral progesterone concentrations in ewes. Plasma progesterone concentrations are determined by the differences between the rate of luteal production and the rate of hepatic metabolism. Both are affected by changes in dry matter intake. The increased plasma progesterone concentration in undernourished animals might be due to the limited extravascular pool in such animals with low body fat content (Lamond et al. 1972; Sejian et al. 2011). The reduced plasma progesterone concentration in the ad libitum fed ewes could be attributed to a consequence of higher metabolism of steroid by the liver (Parr et al. 1982).

5.7 Environmental Stresses and Male Reproduction

5.7.1 Scrotal and Testicular Morphology

Scrotal circumference and testicular consistency, tone, size, and weight are excellent indicators of sperm production capacity and spermatogenic functions. HS reduces these testicular measurements due to the degeneration of the germinal epithelium and partial atrophy in semniferous tubules. Mikelsen et al. (1981) recorded lowest scrotal circumference values during summer and the highest in autumn in rams. However, Hafez et al. (1955) reported that testes size of farm animals is not affected by seasonal changes. However, exposure to cold induces morphological, histopathological, and biochemical effects in rat testes (Blanco et al. 2007; El-Shahat et al. 2009).

5.7.2 Spermatogenesis

The testes of most mammalian species are located extra abdominally in the scrotum and function at a temperature that is a few degrees lower than normal body temperature. In addition, there is an intricate thermoregulatory system in the testis involving countercurrent exchange of heat from warm blood entering the testis and cool blood draining from the testis through an arterio-venous plexus called the pampiniform plexus. The degree of cooling is further controlled by two muscles: the tunica dartos in the scrotum that regulates scrotal surface area and the cremaster muscle that controls the position of the scrotum relative to the body. The lower intratesticular temperature is necessary for spermatogenesis and any disruption to thermoregulatory system of testis may cause problems in spermatogenesis. It can be observed when a local heat source is applied to the testis, the scrotum is insulated, the testes are internalized (i.e., cryptorchidism induced) or body temperature is raised because of fever or hot environment (Setchell 1998).

Exposure of testis to high temperature impairs spermatogenesis by elimination of spermatogonial germ cells in the seminiferous tubules and degeneration of sertoli and leydig cells. The heat damage in the testes is thought to be due to hypoxia causing oxidative stress and consequently germ cell apoptosis and DNA strand breaks (Perez-Crespo et al. 2008; Paul et al. 2008, 2009) mainly in pachytene spermatocytes and round spermatids (Lue et al. 2002). Spermatogonia are relatively resistant to heat compared to spermatocytes and spermatids, because of the fact that the number of spermatogonia remains unchanged and the morphological characteristics are less sensitive to heat exposure (Yin et al. 1997; Lue et al. 2000). Additionally, the testis is able to be repopulated with germ cells following a relative brief or mild temperature exposure (Lue et al. 1999; Yin et al. 1997).

Spermatogenic defects of HS are associated with decreased cytoplasmatic HSP60 immunoreactivity in spermatogonia (Werner et al. 1997). This decreased cytoplasmic HSP60 may negatively affect the mitotic proliferation of spermatogonia because of the fact that HSP60 is necessary for normal functioning of mitochondria. Normal spermatogonia proliferation continues to be drastically reduced for weeks even after the end of the heat treatment. The effects of heat on the spermatogonia seem to be dependent on the method, temperature, duration of heat application, and the livestock species. Exposure of rat testes to 43°C for 15 min causes only a slight increase of ''undetermined'' tubules 2, 8, and 26 days after heat exposure, whereas rats acclimatized to an environment of 35°C for 3 months showed 20% of severely affected tubular cross- sections. In llama, 38.2% of the tubular cross-sections showed ''no stage'' directly after the heat period, indicating that the llama is less able to stabilize an optimal temperature within the testes in elevated ambient heat conditions compared to the rat (Schwalm et al. 2007).

5.7.3 Seminal Attributes

Semen characteristics are not immediately affected by changes in testicular temperature because damaged spermatogenic cells do not enter ejaculates for some time after HS. In the bull, for example, where spermatogenesis takes about 61 days, alterations in semen occur about two weeks after HS and do not return to normal until up to eight weeks following the end of HS.

HS has a negative effect on semen attributes, such as sperm concentration, sperm motility, sperm viability, sperm morphology, and acrosome integrity. The seasonal infertility may be due, at least in part, to a combination of these parameters. Lower semen quality has been reported in bulls exposed to ambient temperatures over 27°C for as little as 6 h per day for several weeks. At 30°C semen quality was affected in 5 weeks, whereas at 38°C lower quality was observed in 2 weeks. In a study by Voglera et al. (1993) in bulls, the sperm motility starts declining on day 12 and reach lowest on day 15 after scrotal insulation for 48 h. Morphological changes first appear on day 12 and progress to peak on day 18 in a chronological sequence, i.e., tailless, (Days12–15); diadem, (Day 18); pyriform and nuclear vacuoles, (Day 21); knobbed acrosome, (Day 27); and Dag defect, (Day 30). Spermatozoa that appeared before 12 days of insulation were presumed to be in the epididymis or rete testes during scrotal insulation and the spermatozoa appeared after 12 days of insulation were presumed to be in spermatogenesis during scrotal insulation.

In rabbits, Marai et al. (2002) have reported that the HS did not have significant effect on reaction time, semen pH, sperm motility, percentages of dead sperm, sperm abnormalities, and acrosomal damage; however, ejaculate volume, sperm concentration, and total sperm output were significantly lower under HS. In seminal plasma, effects of HS were not significant on total protein, globulin, total lipid, cholesterol, creatinine, and alkaline phosphatase, while seminal plasma albumin, acid phosphatase were significantly lower, and GPT and GOT were significantly higher in HS conditions.

5.7.4 Sperm Capacitation and Fertilization

A satisfactory level of rams' fertility may be retained throughout the whole year, but in many instances, fertility is depressed when mating occurs during the hot months of the year (Hafez 1987). A high percentage of rams could be sterile during the summer time, especially under conditions of high humidity. Conception failure in ewes mated to heat stressed ram was related to a failure to fertilize than to embryonic mortality (Curtis 1983). Fertility of the rams is related to several phenomena: the ability to mate, sexual desire, sperm production and viability, and fertilizing capacity of ejaculated sperm, which are influenced by elevated ambient temperatures, as well as nutritional level.

The seasonal infertility may be due to early occurrence of the acrosome reaction in response to stimulus, possibly resulting from a decrease in acrosomal stabilizing proteins in the seminal plasma during summer (Murase et al. 2007). These changes may be modulated by heat/humidity stress and/or photoperiod-regulated testosterone. The decrease in seminal plasma and intracellular ion (Ca, Na^+ and Cl^-) concentrations due to high temperatures can also contribute to decreasing male fertility (Karaca et al. 2002).

There are a number of reports in the literature that suggest males exposed to HS can produce sperm which do not produce normal offspring in unexposed females.

Sperm produced by mice which had been exposed to a hot environment, bind to ova normally but are less able to fertilize in vivo and in vitro, even when motile sperms are selected by a swim-up procedure, and many of the resultant embryos do not develop normally.

In two different studies involving male mice exposed in a microclimate chamber to 36°C for 24 h Zhu and Setchell (2004) and Zhu et al. (2004) reported that there were no changes in the proportion of eggs showing two polar bodies if the mice were mated after subjecting them to HS on 7 or 35 days. However, the proportion of zygotes progressing to 2-cell stage by 34–39 h after mating was reduced when mating occurred 21 or 35 days after subjecting them to HS. Likewise, the proportion progressing to 4-cell or morulae by 61–65 h was reduced when mating occurred 7 or 21 days after HS. The proportion reaching blastocyst stage by 85–90 h after mating was reduced when mating occurred 7 or 21 days after HS, and the proportion of expanded blastocysts was reduced by mating on day 35. The proportion of abnormal embryos at 61–65 h and 85–90 h was increased by mating at 21 days post heat. When embryos were collected 25–28 h after human chorionic gonadotrophin (hCG) from mated superovulated females and cultured in vitro, development to 2-cell stage after 24 h of culture was normal with mating on day 3 or day 42, but reduced for days 7, 14, 21, 28, and 35; the 4-cell or morulae after 48 h culture, the 8-cell or blastocysts after 72 h, and blastocysts after 96 or 120 h culture were reduced with mating on the same days post heat. The percentage of abnormal embryos was increased after 48, 72, or 96 h of culture with mating on days 14–35, and the proportion of degenerating embryos after 72, 96, or 120 h culture was increased with mating on days 7–28, 7–35, and 3–35 post heat, respectively.

In another study, involving male mice exposed to 35°C for two periods of 12 h on consecutive days (Yaeram et al. 2006), the proportion of ova reaching 2-cell stage by 24 h after mating with superovulated females had fallen slightly when mating occurred 7 days post heat and appreciably with mating after 10 or 14 days. Using IVF, it was found that the same number of motile sperm from the epididymis of males heated 7, 10, or 14 days earlier were less effective in terms of the proportion of ova fertilized, with no effect 3 days after heating. Even when motile sperm were separated from immotile sperm in the sample by a swim-up procedure, a similar result was obtained. This was not due to a reduction in the number of sperm binding to the zona pellucida, but there were reductions in the proportion of ova with sperm in the perivitelline space and in the cytoplasm of the eggs.

In a different study, bulls were subjected to scrotal insulation for 48 h and semen samples were collected and cryopreserved 2 or 3 weeks later. Following IVF with swim-up sperm from these samples, there were decreased rates of sperm penetration, pronuclear formation (Walters et al. 2006), embryo cleavage, development, and blastocyst formation (Walters et al. 2005). The same bulls produced embryos with increased caspase activity after 8 days in culture, but there was no effect on apoptosis, as judged by percentage of cells positive in the TUNEL procedure. Paul et al. (2008) reported that IVF with sperm recovered from male mice in which the scrotum was heated to 42°C resulted in embryos with reduced ability to complete development. In addition, females mated to males exposed to scrotal heating had conceptuses with smaller fetal and placental weights compared with controls.

Data are equivocal as to whether ejaculated spermatozoa can be damaged by heat shock when deposited in the reproductive tract of a hyperthermic female. Culture of bull spermatozoa at 40°C did not alter fertilizing capability or the competence of the resultant embryos to develop to the blastocyst stage (Hendricks and Hansen 2009). In addition, they reported that ejaculated bull and stallion spermatozoa do not undergo apoptosis when cultured at temperatures characteristic of physiological hyperthermia. Nonetheless, there may be epigenetic changes in embryonic development associated with damage to the sperm in the reproductive tract. Insemination of rabbit done with sperm exposed to elevated temperature in vitro (Burfening and Ulberg 1968) or in the female reproductive tract (Howarth et al. 1965) resulted in reduced preimplantation as well as post-implantation survival. There is also evidence that X and Y spermatozoa are affected differentially by elevated temperature. The sex ratio of embryos was skewed toward female when female mice were bred to males experiencing scrotal heat treatment on the day of mating (Perez-Crespo et al. 2008). In contrast, incubation of sperm at 40°C for 4 h when compared with 38.5°C tended to reduce the proportion of embryos that were female following IVF (Hendricks and Hansen 2009).

5.7.5 Testosterone Concentration

Testes testosterone content fell from 1.1 to 0.4 μg/gm and spermatic vein plasma content from 8.2 to 1.9 μg/dl, when rams were exposed for 14 days to an average environmental temperature of 30°C (Curtis 1983). The lowest serum testosterone level was recorded during hot environmental conditions in Ossimi rams (El-Darawany 1999). Exposure of the intact rat to increased environmental temperatures is accompanied by decreased capacity of the testes to synthesize testosterone and consequently reduced serum testosterone concentration (Chap and Bedrack 1983). However, in developing ram lambs, direct exposure to high ambient temperature has no significant effect on testosterone production during non-breeding season (Rasooli et al. 2010). Short day-light stimulates the secretion of testosterone, FSH and LH in rams, while long daylight inhibits their secretion. Rams sexual activities peak occurs during the autumn breeding season and coincides with a sharp rise in plasma testosterone level. Then it declines in late winter, spring, and summer (Jainuden and Hafez 1987).

5.7.6 Expression Profile of Male Reproductive Genes During Heat Stress

It is widely accepted that changes in gene expression and in the activity of expressed proteins are an integral part of the cellular response to HS. Although the HSPs are perhaps the best-studied examples of genes whose expression is affected by heat shock, it has become apparent in recent years that HS also leads to induction of a substantial number of genes not traditionally considered to be HSPs.

Some of these genes are affected by a wide variety of stressors and probably represent a non-specific cellular response to stress, whereas others may eventually be found to be specific to certain types of stress.

The cellular response to HS characteristically includes an increase in thermotolerance (i.e., the ability to survive subsequent to more severe heat stresses) that is temporarily associated with increased expression of HSPs. Heat-induced changes in gene expression occur both during hyperthermia as well as after return to normothermia.

5.7.7 Heat Shock Protein Genes

HSPs were originally identified as proteins whose expression was markedly increased by heat shock (Lindquist 1986). Several HSPs are expressed even in unstressed cells and play important functions in normal cell physiology. The intensity and duration of the heat stimulus needed for HSP expression vary considerably from tissue to tissue. A typical in vitro exposure involves heating mammalian cells to 42–45°C for 20–60 min and then reverting to normothermic temperatures (37°C). Induction of HSP expression typically starts within minutes after the initiation of HS, with peak expression occurring up to several hours later. Importantly, several experiments have found that, during the period of hyperthermia and shortly thereafter, HSPs become the predominant proteins synthesized by cells (Lindquist 1986). Interestingly, most HSP genes lack introns (Lindquist 1986), which may facilitate their rapid expression and which may also help explain how they can be expressed in the presence of stressors (such as heat) that can interfere with RNA splicing.

Heat shock factors (HSFs) are transcription factors that regulate HSP expression through interaction with a specific DNA sequence in the promoter [the heat shock element (HSE)]. The HSE is a stretch of DNA, located in the promoter region of susceptible genes containing multiple sequential copies (adjacent and inverse) of the consensus pentanucleotide sequence 5′-nGAAn-3′ (Morimoto 1998) and has been found in both HSPs and in a number of other genes. Three HSFs have been identified in mammalian systems: HSF-1, HSF-2, and HSF-4 (Morimoto 1998). HSF-1 (Morimoto 1998) and HSF-2 (Mathew et al. 2001) are involved in the acute response to heat shock. HSF-2 also has a major function in controlling expression of genes important for embryonic development and maintenance of sperm production (Wang et al. 2003).

Before heat-induced activation, HSF-1 exists as a monomer localized to the cytoplasm. The initial stimulus for activation of HSF-1 appears to take place after exposure of hydrophobic domains of denatured proteins to HS. HSPs preferentially bind to denatured proteins, and hence activation of HSF-1 may occur as a result of competitive release of this transcription factor from HSPs when the concentration of denatured cytoplasmic proteins increases as a result of heat shock (Morimoto 1998). After activation by HS, HSF-1 is found primarily in the nucleus in trimeric form, concentrated (in human cell lines) in granules (Sarge et al. 1993). It is this activated, trimeric form of HSF-1 that binds to the HSE and is involved in increased HSP gene transcription during HS (Sarge et al. 1993).

Evidence has also indicated that HS induces tagging of HSF-1 with SUMO-1, a ubiquitin-like protein that is used by the cell to mark proteins for transport into different cellular compartments and to alter their activities (Hong et al. 2001). Importantly, in these experiments, HSF-1, in vitro, was incapable of binding DNA unless it had first acquired a SUMO-1 tag at lysine 298.

In addition to positive regulation of the HSE through HSF-1, evidence also exists for negative regulation of this promoter element in mice by means of a constitutively expressed protein known as the HSE binding factor (HSE-BF) (Liu et al. 1993; Kim et al. 1995). Thus, in addition to a phosphorylation state, the ability of HSF-1 to activate transcription may also be modulated by regulatory processes that affect the binding of HSE-BF to the HSE.

5.8 Conclusion

This chapter identifies the various reproductive responses that are sensitive to HS and nutrition stress. Evidently, HS can have profound effects on most of the aspects of reproductive function—male and female gamete formation and function, embryonic development, and fetal growth and development. HS challenges the reproductive performance of livestock through a variety of altered physiologic means including: altered follicular development, lowered estrus activity, and impaired embryonic development. The effect of heat environment on reproduction may take place through a direct action of hyperthermia upon the reproductive tissues or through an indirect manner due to lower nutrients intake, impairment of hypothalamic, pituitary, gonadal, and endometrial secretions. HS reduces reproductive efficiency of livestock through a variety of different mechanisms. The principal biological mechanism by which HS impacts on livestock reproduction is partly explained by reduced feed intake, and also includes altered endocrine status, reduction in rumination and nutrient absorption, and increased maintenance requirements resulting in a net decrease in nutrient/energy availability for reproduction. Food availability is the most important factor that influences mammalian reproduction. Adequate feeding has great influence on animal comfort and reproductive performance. Low energy and low or excessive protein levels in the diet are detrimental to reproduction. Undernutrition affects the development of reproductive axis, delaying the development of ovarian follicles and reducing lifetime reproductive performance in livestock. Undernutrition of livestock results in lower BW and BCS which has a negative effect on oocyte quality, which results in lower rates of cleavage, and numerous reproductive functions including hormone production, fertilization, and early embryonic development. Elevation of ambient temperature affects male reproductive functions deleteriously. Such phenomenon leads to testicular degeneration and reduces percentages of normal and fertile spermatozoa in the ejaculate of males. The ability of the male to mate and fertilize is also affected.

References

Abecia AJ, Sosa C, Forcada F, Meikle A (2006) The effect of undernutrition on the establishment of pregnancy in the ewe. Reprod Nut Dev 46:367–378

Adams NR, Abrondi JA, Briegel JR, Sanders MR (1994) Effect of diet on the clearance of estradiol-17β in the ewe. Biol Reprod 51:668–674

Adams TE, Sakurai H, Adams BM (1999) Effect of stress like concentrations of cortisol on estradiol dependent expression of gonadotrophin-releasing hormone receptor in orchidectomized sheep. Biol Reprod 60:164–168

Alfujairi MM, Albrahim RM, Elnouty ED (1993) Seasonal variations in superovulatory responses of Holstein cows treated with pregnant mare serum gonadotropin in Saudi Arabia. J Reprod Fertil 11:75

Ali A, Hayder M (2008) Seasonal variation of reproductive performance, foetal development and progesterone concentration of sheep in the subtropics. Reprod Domes Anim 43:730–734

Al-Katanani YM, Paula-Lopes FF, Hansen PJ (2002) Effect of season and exposure to heat stress on oocyte competence in Holstein cows. J Dairy Sci 85:390–396

Aoyama M, Negishi A, Abe A, Maejima Y, Sugita S (2003) Sex differences in stress responses to transportation in goats: effects of gonadal hormones. Anim Sci J 74:511–519

Armstrong DG, Gong JG, Webb R (2003) Interactions between nutrition and ovarian activity in cattle: physiological, cellular and molecular mechanisms. Reprod Suppl 61:403–414

Badinga L, Thatcher WW, Wilcox CJ (1993) Effect of environmental stress on follicular development and steroidogenesis in lactating Holstein cows. Theriogenology 39:797–810

Banks EM (1964) Some aspects of sexual behavior in domestic sheep. Ovis aries Behav 23:249

Barati F, Agung B, Wongsrikeao P, Taniguchi M, Nagai T, Otoi T (2008) Meiotic competence and DNA damage of porcine oocytes exposed to an elevated temperature. Theriogenology 69:767–772

Baumangartner AP, Chrisman CL (1987) Embryonic mortality caused by maternal heat stress during mouse oocyte maturation. Anim Reprod Sci 14:309–316

Bell AW, McBride BW, Slepetis R, Early RJ, Currie WB (1989) Chronic heat stress and prenatal development in sheep: I. Conceptus growth and maternal plasma hormones and metabolites. J Anim Sci 67:3289–3299

Ben Salem M, Bouraoui R (2009) Heat Stress in Tunisia: effects on dairy cows and potential means of alleviating it. South African J Anim Sci 39:256–259

Bernardini C, Fantinati P, Zannoni A, Forni M, Tamanini C, Bacci ML (2004) Expression of HSP70/HSC70 in swine blastocysts: effects of oxidative and heat stress. Mol Reprod Dev 69:303–307. doi:10.1002/mrd.20143

Blanco A, Moyano R, Vivo J, Flores-Acuña R, Molina A, Blanco C, Agüera E, Monterde JG (2007) Quantitative changes in the testicular structure in mice exposed to low doses of cadmium. Environ Toxicol Pharmacol 23:96–101

Boland MP, Lonergan P (2005) Effects of nutrition on fertility in dairy cows. Advances Dairy Tech 15:19–33

Boland MP, Lonergan P, O'Callaghan D (2001) Effect of nutrition on endocrine parameters, ovarian physiology, and oocyte and embryo development. Theriogenology 55:1323–1340

Breen KM, Karsch FJ (2004) Does cortisol inhibits pulsatile luteinizing hormone secretion at the hypothalamic or pituitary level? Endocrinology 145:692–698

Bridges PJ, Brusie MA, Fortune JE (2005) Elevated temperature (heat stress) in vitro reduces androstenedione and estradiol and increases progesterone secretion by follicular cells from bovine dominant follicles. Domest Anim Endocrinol 29:508–522

Burfening PJ, Ulberg LC (1968) Embryonic survival subsequent to culture of rabbit spermatozoa at 38 and 40 C. J Reprod Fertil 15:87–92

Cavestany D, El-Whishy AB, Foot RH (1985) Effect of season and high environmental temperature on fertility of Holstein cattle. J Dairy Sci 68:1471–1478

Chap Z, Bedrak E (1983) Interrelationship between pituitary-testicular axis activity and raised environmental temperature in the rat. J Endocrinol 97(2):193–200
Collier RJ, Dahl GE, Vanbaale MJ (2006) Major advances associated with environmental effects in dairy cattle. J Dairy Sci 89:1244–1253
Curtis SE (1983) Environmental management in animal agriculture. Iowa State University Press, Ames
Daley CD, Sakurai H, Adams BM, Adams TE (1999) Effect of stress like concentrations of cortisol on gonadotroph function in orchidectomized sheep. Biol Reprod 60:158–163
de Castroe Paula LA, Hansen PJ (2008) Modification of actions of heat shock on development and apoptosis of cultured preimplantation bovine embryos by oxygen concentration and dithiothreitol. Mol Reprod Dev 75:1338–1350
Dickerson GE (1970) Efficiency of animal production-molding the biological components. J Anim Sci 30:849–859
Dobson H, Smith RF (2000) What is stress and how does it affect fertility? Anim Reprod Sci 60:743–752
Dobson H, Tebble JE, Smith RF, Ward WR (2001) Is stress really all that important? Theriogenology 55:65–73
Downing JA, Joss J, Scarmuzzi RJ (1995) Ovulation rate and the concentration of gonadotrophins and metabolic hormones in ewes infused with glucose during the late luteal phase of the oestrous cycle. J Endocrinol 146:403–410
Du Preez JH, Terblanche SJ, Giesecke WH, Maree C, Welding MC (1991) Effect of heat stress on conception in a dairy herd model under South African conditions. Theriogenology 35:1039–1049
Duncan RF, Hershey JWB (1989) Protein-synthesis and protein-phosphorylation during heat-stress, recovery, and adaptation. J Cell Biol 109(4):1467–1481
Dutt RH (1963) Critical period for early embryo mortality in ewes exposed to high ambient temperature. J Anim Sci 22:713–719
Ealy AD, Drost M, Hansen PJ (1993) Developmental changes in embryonic resistance to adverse effects of maternal heat stress in cows. J Dairy Sci 76:2899–2905
Edwards J, Hansen P (1996) Elevated temperature increases heat shock protein 70 synthesis in bovine two-cell embryos and compromises function of maturing oocyte. Biol Reprod 55:340–346
Edwards JL, King WA, Kawarsky SJ, Ealy AD (2001) Response of early embryos to environmental insults: potential protective roles of HSP70 and glutathione. Theriogenology 55:209–223
Edwards JL, Hansen PJ (1997) Differential responses of bovine oocytes and preimplantation embryos to heat shock. Mol Reprod Dev 46(2):138–145
Edwards JL, Bogart AN, Rispoli LA, Saxton AM, Schrick FN (2009) Developmental competence of bovine embryos from heatstressed ova. J Dairy Sci 92:563–570
El-Darawany AA (1999) Improving semen quality of heat stressed rams in Egypt. Ind J Anim Sci 69:1020–1023
El-Shahat AE, Gabr A, Meki AR, Mehana ES (2009) Altered testicular morphology and oxidative stress induced by cadmium in experimental rats and protective effect of simultaneous green tea extract. Int J Morphol 27(3):757–764
Elvinger F, Natzke RP, Hansen PJ (1992) Interactions of heat stress and bovine somatotropin affecting physiology and immunology of lactating cows. J Dairy Sci 75:449–462
FAO (Food and Agricultural Organization) (2008). The State of Food and Agriculture2008, Rome
FAO (Food and Agricultural Organization) (2009). Rural Income Generating Activities database (available at www.fao.org/es/ESA/riga/english/index_en.htm)
Ferreira EM, Vireque AA, Adona PR, Meirelles FV, Ferriani RA, Navarro PAAS (2009) Cytoplasmic maturation of bovine oocytes: structural and biochemical modifications and acquisition of developmental competence. Theriogenology 71:836–848

Findlay JK (1993) An update on the roles of inhibin, activin, and follistatin as local regulators of folliculogenesis. Biol Reprod 48:15–23

Forcada F, Abecia AJ (2006) The effect of nutrition on the seasonality of reproduction in ewes. Reprod Nutr Dev 46:355–365

Gangawar PC, Branton C, Evans DL (1965) Reproductive and physiological responses Holstein heifers to control and natural climate. J Dairy Sci 48:222

Gilad E, Meidan R, Berman A, Graber Y, Wolfenson D (1993) Effect of heat stress on tonic and GnRH-induced gonadotrophin secretion in relation to concentration of oestradiol in plasma of cyclic cows. J Reprod Fertil 99:315–321

Gordon I (1997) Embryo transfer and associated techniques in sheep. In: Controlled Reproduction in Sheep and Goat. CAB International, Oxon

Gordon I, Boland MP, McGovern H, Lynn G (1987) Effect of season on superovulatory responses and embryo quality in Holstein cattle in Saudi Arabia. Theriogenology 27:231

Gougeon A (1996) Regulation of ovarian follicular development in primates: facts and hypotheses. Endocrine Rev 17:121–155

Gunn RG, Sim DA, Hunter EA (1995) Effects of nutrition in utero and in early-life on the subsequent lifetime reproductive-performance of Scottish Blackface ewes in two management-systems. J Anim Sci 60:223–230

Guzeloglu A, Ambrose JD, Kassa T, Diaz T, Thatcher MJ, Thatcher WW (2001) Long term follicular dynamicsand biochemical characteristics of dominant follicles in dairy cows subjected to acute heat stress. Anim Reprod Sci 66:15–34

Gwazdauskas FC, Wilcox CJ, Thatcher WW (1975) Environmental and managemental factors affecting conception rate in a subtropical climate. J Dairy Sci 58:88

Gwazdauskas FC, De Silva AWMV, Anderson GW, McGilliard ML, Lineweave JA (1981) Interrelationships with estrous behavior and conception in dairy cattle. J Dairy Sci 64: 2409–2418

Hafez ESE, Baderldin AL, Darwish YH (1955) Seasonal variations in semen characteristics of sheep in subtropics. J Agricul Sci 45:283–292

Hafez ESE (1987) Reproduction in farm animals, 5th edn. LEA and Febiger, Philadelphia

Hansen PJ, Drost M, Rivera RM, Paula Lopes FF, Al-Katanani YM, Krininger CE III, Chase CC Jr (2001) Adverse impact of heat stress on embryo production: causes and strategies for mitigation. Theriogenology 55:91–103

Hansen PJ (2009) Effects of heat stress on mammalian reproduction. Philos T R Soc B 364(1534):3341–3350

Hein KG, Allrich RD (1992) Influence of exogenous adrenocorticotropic hormone on estrous behavior in cattle. J Anim Sci 70:243–247

Hendricks KE, Hansen PJ (2009) Can programmed cell death be induced in post-ejaculatory bull and stallion spermatozoa? Theriogenology 71:1138–1146

Hong Y, Rogers R, Matunis MJ, Mayhew CN, Goodson ML, Park-Sarge OK, Sarge KD (2001) Regulation of heat shock transcription factor 1 by stress-induced SUMO-1 modification. J Biol Chem 276:40263–40267

Howarth B Jr, Alliston CW, Ulberg LC (1965) Importance of uterine environment on rabbit sperm prior to fertilization. J Anim Sci 24:1027–1032

Howell JL, Fuquay JW, Smith AE (1994) Corpus luteum growth and function in lactating Holstein cows during spring and summer. J Dairy Sci 77:735–739

Hyttel P, Fair T, Callesen H, Greve T (1997) Oocyte growth, capacitation and final maturation in cattle. Theriogenology 47(1):23–32

Jainuden MR, Hafez ESE (1987) Sheep and goats. In: Hafez ESE (ed) Reproduction in farm animals, 5th edn. LEA and Febiger, Philadelbhia

Ju JC, Tseng JK (2004) Nuclear and cytoskeletal alterations of in vitro matured porcine oocytes under hyperthermia. Mol Reprod Dev 68:125–133

Juniewicz PE, Johnson BH, Bolt DJ (1987) Effect of adrenal steroids on testosterone and luteinizing hormone secretion in the ram. J Androl 81:90–196

Kanai Y, Yagyu N, Shimizu T (1995) Hypogonadism in heat stressed goats: poor responsiveness of the ovary to the pulsatile LH stimulation induced by hourly injections of a small dose of GnRH. J Reprod Dev 41:133–139

Kaneko H, Nakanishi Y, Taya K, Kishi H, Watanabe G, Sasamoto S, Hasegawa Y (1993) Evidence that inhibin is an important factor in the regulation of FSH secretion during mid-luteal phase in cows. Endocrinology 136:35–42

Kaneko H, Taya K, Watanabe G, Noguchi J, Kikuchi K, Shimada A, Hasegawa Y (1997) Inhibin is involved in the suppression of FSH secretion in the growth phase of the dominant follicle during the early luteal phase in cows. Domes Anim Endocrinol 14:263–271

Karaca AG, Parker HM, Yeatman JB, McDaniel CD (2002) Role of seminal plasma in heat stress infertility of broiler breeder males. Poult Sci 81(12):1904–1909

Khalifa HH (2003) Bioclimatology and adaptation of farm animals in a changing climate. In: Interactions between climate and animal production, EAAP Tech Ser 7:15–29

Kim D, Ouyang H, Li GC (1995) Heat shock protein hsp70 accelerates the recovery of heat-shocked mammalian cells through its modulation of heat shock transcription factor HSF1. Proc Natl Acad Sci U S A 92:2126–2130

Kiyma Z, Alexander BM, Van Kirk EA, Murdoch WJ, Hallford DM, Moss GE (2004) Effect of feed restriction on reproductive and metabolic hormones in ewes. J Anim Sci 82:2548–2557

Kornmatitsuk B, Chantaraprateep P, Kornmatitsuk S, Kindahl H (2008) Different types of postpartum luteal activity affected by the exposure of heat stress and subsequent reproductive performance in Holstein lactating cows. Reprod Domes Anim 43:515–519

Koubkova M, Knizkova I, Kune P, Hartlova H, Flusser J, Dolezal O (2002) Influence of high environmental temperatures and evaporative cooling on some physiological, hematological and biochemical parameters in high-yielding dairy cows. Czech J Anim Sci 47:309–318

Lamond RD, Gaddy GR, Kennedy WS (1972) Influence of season and nutrition on luteal plasma progesterone in Rambuillet Ewes. J Anim Sci 34:626–629

Lawrence JL, Payton RR, Godkin JD, Saxton AM, Schrick FN, Edwards JL (2004) Retinol improves development of bovine oocytes compromised by heat stress during maturation. J Dairy Sci 87:2449–2454

Lawson RAS, Cahill LP (1983) Modification of the embryo-maternal relationship in ewes by progesterone treatment early in the oestrous cycle. J Reprod Fertil 67:473–475

Lee CN (1993) Environmental stress effect on bovine reproduction. Vet Clin North Am 9:263–273

Lenz RW, Ball GD, Leibfried ML, Ax RL, First NL (1983) In vitro maturation and fertilization of bovine oocytes are temperature dependent processes. Biol Reprod 29:173–179

Lindquist S (1986) The heat-shock response. Annu Rev Biochem 55:1151–1191

Lindsay DR (1996) Environment and reproductive behavior. Anim Reprod Sci 42:1–12

Lindsay DR, Martin GB, Williams IH (1993) Nutrition and reproduction. In: King GJ (ed) Reproduction in domesticated animals. World animal science series. Elsevier Science Publishers, Amsterdam, pp 459–491

Liu RY, Kim D, Yang SH, Li GC (1993) Dual control of heat shock response: involvement of a constitutive heat shock element-binding factor. Proc Natl Acad Sci U S A 90:3078–3082

Lozano JM, Albecia JA, Forcada F, Zarazaga L, Alfaro B (1998) Effect of under nutrition on the distribution of progesterone in the uterus of ewes during the luteal phase of the estrous cycle. Theriogenology 49(3):536–546

Lue YH, Lasley BL, Laughlin LS, Swerdloff RS, Sinha Hikim AP, Leung A et al (2002) Mild testicular hyperthermia induces profound transitorial spermatogenic suppression through increased germ cell apoptosis in adult Cynomolgus Monkeys (*Macaca fascicularis*). J Androl 23:799–805

Lue YH, Sinha Hikim AP, Swerdloff RS, Im P, Taing KS (1999) Single exposure to heat induces stage specific germ cell apoptosis in rats: role of intratesticular testosterone on stage specificity. Endocrinology 140:1709–1717

Lue YH, Sinha Hikim AP, Wang C, Im M, Leung A, Swerdloff RS (2000) Testicular heat exposure enhances the suppression of spermatogenesis by testosterone in rats: the ''two-hit'' approach to male contraceptive development. Endocrinology 141:1414–1424

Lynch JJ, Hinch GN, Adams DB (1992) The behaviour of sheep: biological principles and implications for production. C.A.B. International, Wallingford

Magiakou MA, Mastorakos G, Webster E, Chrousos GP (1997) The hypothalamic-pituitary-adrenal axis and the female reproductive system. Annals Newyork Acad Sci 816:42–56

Marai FM, Habeeb AAM, Gad AE (2002) Reproductive traits of male rabbits as affected by climatic conditions, in the subtropical environment of Egypt. Anim Sci 75:451–458

Mathew A, Mathur SK, Jolly C, Fox SG, Kim S, Morimoto RI (2001) Stress-specific activation and repression of heat shock factors 1 and 2. Mol Cell Biol 21:7163–7171

Maurya VP, Naqvi SMK, Mittal JP (2004) Effect of dietary energy level on physiological responses and reproductive performance of malpura sheep in the hot semi-arid regions of India. Small Rum Res 55:117–122

Maya-Soriano MJ, Andreu-Vazquez C, Lopez-Gatius F, Lopez-Bejar M (2010) Retinol improves in vitro Oocyte nuclear maturation under heat stress in dairy cows. Reprod Dom Anim 45:68

Mihm M, Baguisi A, Boland MP, Roche JF (1994) Association between the duration of dominance of the ovulatory follicle and pregnancy rate in beef heifers. J Reprod Fertil 102:123–130

Mikelsen WD, Paisley LG, Dahmen JJ (1981) The effect of semen on the scrotal circumference and sperm motility and morphology in rams. Theriogenology 16:45–51

Minton JE (1994) Function of the hypothalamic-pituitary-adrenal axis and the sympathetic nervous system in models of acute stress in domestic farm animals. J Anim Sci 72:1891–1898

Mohamed AA (1974) Physiological changes in the reproductive organ of buffalo from particular to conception. PhD Thesis, Faculty of Agriculture, Al Ajhar University, Egypt

Monty DE Jr, Racowsky C (1987) In vitro evaluation of early embryo viability and development in summer heat stressed, superovulated dairy cows. Theriogenology 28:451–465

Morimoto RI (1998) Regulation of the heat shock transcriptional response: cross talk between a family of heat shock factors, molecular chaperones, and negative regulators. Genes Dev 12:3788–3796

Morin LP (1986) Environment and hamster reproduction: responses to phase-specific starvation during estrous cycle. Am J Physiol 251:R663–R669

Murase T, Imaeda N, Yamada H, Miyazawa K (2007) Seasonal changes in semen characteristics, composition of seminal plasma and frequency of acrosome reaction induced by calcium and calcium ionophore A23187 in Large White boars. J Reprod Dev 53(4):853–865

Murphy MG, Enright WJ, Crowe MA, McConnell K, Spicer LJ, Boland MP, Roche JF (1991) Effect of dietary intake on pattern of growth of dominant follicles during the oestrus cycle in beef heifers. J Reprod Fertil 92:333–338

Naqvi SMK, Maurya VP, Gulyani R, Joshi A, Mittal JP (2004) The effect of heat stress on superovulatory response and embryo production in Bharat Merino ewes. Small Rum Res 55:57–63

Naqvi SMK, Rai AK (1991) Effect of feed restriction followed by real mentation on growth, physiological responses and blood metabolites of native and crossbred sheep. Ind J Anim Sci 61:629–631

Naqvi SMK, Soren NM, Karim SA (2011) Effect of concentrate supplementation on performance, ovarian response, and some biochemical profile of Malpura ewes. Trop Anim Health Prod 43:905–913

Naqvi SMK, Maurya VP, Joshi A, Sharma RC, Mittal JP (2002) Production of crossbred lambs through artificial insemination of non-prolific medium size Malpura and Avikalin ewes using fresh diluted semen of prolific micro size Garole rams. Asian-Aust J Anim Sci 15:633–636

O'Callaghan D, Yaakub H, Hyttel P, Spicer LJ, Boland MP (2000) Effect of nutrition and superovulation on oocyte morphology, follicular fluid composition and systemic hormone concentrations in ewes. J Reprod Fertil 18:303–313

Otukonyong EE, Okutani F, Takahashi S, Murata T, Morioka N, Kaba H, Higuchi T (2000) Effect of food deprivation and leptin repletion on the plasma levels of estrogen (E2) and NADPH-d reactivity in the ventromedial and arcuate nuclei of hypothalamus in the female rats. Brain Res 887:70–79

Ozawa M, Tabayashi D, Latief TA, Shimizu T, Oshima I, Kanai Y (2005) Alterations in follicular dynamics and steroidogenic abilities induced by heat stress during follicular recruitment in goats. Reproduction 129:621–630

Palta P, Mondal S, Prakash BS, Madan ML (1997) Peripheral inhibin levels in relation to climatic variations and stage of estrous cycle in buffalo (*Bubalus bubalis*). Theriogenology 47:989–995

Parr RA (1992) Nutrition-progesterone interactions during early pregnancy in sheep. Reprod Fertil Dev 4(3):297–300

Parr RA, Davis IF, Miles MA, Squires TJ (1993) Feed intake affects metabolic clearance rate of progesterone in sheep. Res Vet Sci 55:306–310

Parr RA, Cumming IA, Clark IJ (1982) Effect of maternal nutrition and plasma progesterone concentration on survival and growth of the sheep embryo in early gestation. J Agricul Sci (Cambridge) 98:39–46

Paul C, Murray AA, Spears N, Saunders PT (2008) A single, mild, transient scrotal heat stress causes DNA damage, subfertility and impairs formation of blastocysts in mice. Reproduction 136:73–84

Paul C, Teng S, Saunders PT (2009) A single, mild, transient scrotal heat stress causes hypoxia and oxidative stress in mouse testes, which induces germ cell death. Biol Reprod 80:913–919

Paula-Lopes FF, Hansen PJ (2002) Heat shock-induced apoptosis in preimplantation bovine embryos is a developmentally regulated phenomenon. Biol Reprod 66:1169–1177

Perez-Crespo M, Pintado B, Gutierrez-Adan A (2008) Scrotal heat stress effects on sperm viability, sperm DNA integrity, and the offspring sex ratio in mice. Mol Reprod Dev 75:40–47

Pisani LF, Antonini S, Pocar P, Ferrari S, Brevini TA, Rhind SM, Gandolfi F (2008) Effects of pre-mating nutrition on mRNA levels of developmentally relevant genes in sheep oocytes and granulosa cells. Reproduction 136:303–312

Polkowska J (1996) Stress and nutritional influences on GnRH and somatostatin neuronal systems in the ewe. Acta Neurobiol Exps 56:797–806

Putney DJ, Mullins S, Thatcher WW, Drost M, Gross TS (1989) Embryonic development in superovulated dairy cattle exposed to elevated ambient temperatures between the onset of estrus and insemination. Anim Reprod Sci 19:37–51

Rae MT, Palassio S, Kyle CE, Brooks AN, Lea RG, Miller DW, Rhind SM (2001) Effect of maternal undernutrition during pregnancy on early ovarian development and subsequent follicular development in sheep fetuses. Reproduction 122:915–922

Rasooli A, Jalali MT, Nouri M, Mohammadian B, Barati F (2010) Effects of chronic heat stress on testicular structures, serum testosterone and cortisol concentrations in developing lambs. Anim Reprod Sci 117(1–2):55–59

Rhind SM (1992) Nutrition: its effects on reproductive performance and its hormonal control in female sheep and goats. In: Speedy AW (ed) Progress in sheep and goat research. CAB International, UK, pp 25–51

Rhind SM, McKelvey WAC, McMillen SR, Gunn RG, Elston DA (1989) Effect of restricted food intake, before and/or after mating, on the reproductive performance of Greyface ewes. Anim Prod 48:149–155

Rivera RM, Hansen PJ (2001) Development of cultured bovine embryos after exposure to heat temperatures in the physiological range. Reproduction 121:107–115

Rivera RM, Kelley KL, Erdos GW, Hansen PJ (2004) Reorganization of microfilaments and microtubules by heat stress in two-cell bovine embryos. Biol Reprod 70:1852–1862

Rivest S, Rivier C (1995) The role of corticotrophin releasing factor and interleukin- 1 in the regulation of neurons controlling reproductive functions. Endocrine Rev 16:177–199

Robinson JJ, Ashworth CJ, Rooke JA, Mitchell LM, McEvoy TG (2006) Nutrition and fertility in ruminant livestock. Anim Feed Sci Technol 126:259–276

Robinson JJ, Rooke JA, McEvoy TG (2002) Nutrition for conception and pregnancy. In: Freer M, Dove H (eds) Sheep nutrition. CAB International, Wallingford, pp 189–211

Rodman TC, Bachvarova R (1976) RNA synthesis in preovulatory mouse oocytes. Cell Biol 70:251–257

Rodriguez KF, Farin CE (2004) Gene transcription and regulation of oocyte maturation. Reprod Fertil Dev 16(1–2):55–67

Roman-Ponce H, Thatcher WW, Wilcox CJ (1981) Hormonal interrelationships and physiological responses of lactating dairy cows to shade management system in a tropical environment. Theriogenology 16:139–154

Ronchi B, Stradaioli G, Verini Supplizi A, Bernabuci U, Lacetera N, Accorsi PA (2001) Influence of heat stress or feed restriction on plasma progesterone, oestradiol-17beta, LH, FSH, prolactin and cortisol in Holstein heifers. Livestock Prod Sci 68:231–241

Roth Z, Hansen PJ (2004) Involvement of apoptosis in disruption of developmental competence of bovine oocytes by heat shock during maturation. Biol Reprod 71:1898–1906

Safari E, Fogarty NM, Gilmour AR (2005) A review of genetic parameter estimates for wool, growth, meat and reproduction in sheep. Livest Prod Sci 92:271–289

Sakatani M, Kobayashi S, Takahashi M (2004) Effects of heat shock on in vitro development and intracellular oxidative state of bovine preimplantation embryos. Mol Reprod Dev 67:77–82

Sakatani M, Yamanaka K, Kobayashi S, Takahashi M (2008) Heat shock-derived reactive oxygen species induce embryonic mortality in vitro early stage bovine embryos. J Reprod Dev 54:496–501

Sarge KD, Murphy SP, Morimoto RI (1993) Activation of heat shock gene transcription by heat shock factor 1 involves oligomerization, acquisition of DNA-binding activity, and nuclear localization and can occur in the absence of stress. Mol Cell Biol 13:1392–1407

Sawyer GJ (1979) The influence to radiant heat load in Merino ewes. II. The relative effects of heating before and after insemination. Aust J Agric Res 30:1143–1149

Scaramuzzi RJ, Cambel BK, Downing JA, Kendall NR, Khalid M, Gutierrez M, Somchit A (2006) A review of the effect of supplementary nutrition in the ewe on the concentration of reproductive and metabolic hormones and the mechanism that regulate folliculogenesis and ovulation rate. Reprod Nut Dev 46:339–354

Schmidt S, Hommel A, Gawlik V, Augustin R, Junicke N, Florian S, Richter M, Walther DJ, Montag D, Joost HG, Schürmann A (2009) Essential role of glucose transporter GLUT3 for post-implantation embryonic development. J Endocrinol 200:23–33

Schwalm A, Gauly M, Erhardt G, Bergmann M (2007) Changes in testicular histology and sperm quality in llamas (Lama glama), following exposure to high ambient temperature. Theriogenology 67(8):1316–1323

Sejian V, Maurya VP, Naqvi SMK, Kumar D, Joshi A (2010) Effect of induced body condition score differences on physiological response, productive and reproductive performance of Malpura ewes kept in a hot, semi-arid environment. J Anim Physiol Anim Nut 94(2):154–161

Sejian V, Srivastava RS, Varshney VP (2008) Pineal-adrenal relationship: modulating effects of glucocorticoids on pineal function to ameliorate heat-stress in Goats. Asian-Austral J Anim Sci 21:988–994

Sejian V, Maurya VP, Naqvi SMK (2011) Effect of heat, nutritional and combined (heat and nutritional) stresses on growth and reproductive performance of Malpura ewes under semi-arid tropical environment. J Anim Physiol Anim Nut 95:252–258

Setchell BP (1998) The parkes lecture. Heat and the testis. J Reprod Fertil 114:179–194

Shehab-El-Deen MAMM (2011) Effects of metabolic stressors and high temperature on oocyte and embryo quality in high yielding dairy cows. PhD thesis submitted to Department of Reproduction Obstetrics and Herd Health, Salisburylaan 133, B-9820, Merelbeke

Shelton M (2000) Reproductive performance of sheep exposed to hot environments. In: Malik RC, Razzaque MA, Al-Nasser AY (eds) Sheep Production in Hot and Arid Zones. Published by the Kuwait Institute for Scientific Research, pp 155–162

Snowder GD (2008) Genetic improvement of overall reproductive success in sheep: a review. Arch Latinoam Prod Anim 16(1):32–40

Stewart R, Oldham CM (1986) Feeding lupins for 4 days during the luteal phase can increase the ovulation rate. In: Proceeding of the Australian society of animal production. Rev Reprod 16:367–370

Sugiyama S, Michael M, Nancy P, Majataba K, Mary Y (2007) Effects of increased ambient temperature during IVM and/or IVF on the in vitro development of Bovine Zygotes. Reprod Dom Anim 42(3):271–274

Tabbaa MJ, Alnimer MA, Shboul M, Titi HH (2008) Reproductive characterstics of Awassi rams. Anim Reprod 5(1):23–29

Thwaites CJ (1971) Short term heat stress and embryo mortality in sheep. Austral J Exp Agricul Anim Husbandry 11(50):265–267

Tilbrook AJ, Turne AI, Clark IJ (2000) Effect of stress on reproduction in non-rodent mammals: the role of glucocorticoids and sex differences. Rev Reprod 5:105–113

Tilbrook AJ, Turner AI, Clark IJ (2002) Stress and reproduction: central mechanisms and sex differences in non-rodent species. Stress 5(2):83–100

Trout JP, McDowell LR, Hansen PJ (1998) Characteristics of the oestrous cycle and antioxidant status of lactating Holstein cows exposed to stress. J Dairy Sci 81:1244–1250

Vinoles C, Forsberg M, Martin GB, Cajarville C, Repetto J, Meikle A (2005) Short term nutritional supplementation of ewes in low condition affects follicle development due to an increase in glucose and metabolic hormones. Reproduction 129:299–309

Voglera CJ, Bamea JH, DeJarnettea JM, McGilliarda ML, Saacke RG (1993) Effects of elevated testicular temperature on morphology characteristics of ejaculated spermatozoa in the bovine. Theriogenology 40(6):1207–1219

Walters AH, Saacke RG, Pearson RE, Gwazdauskas FC (2005) The incidence of apoptosis after IVF with morphologically abnormal bovine spermatozoa. Theriogeneology 64:1404–1421

Walters AH, Saacke RG, Pearson RE, Gwazdauskas FC (2006) Assessment of pronuclear formation following in vitro fertilization obtained after heat insulation of the testis. Theriogeneology 65:1016–1028

Wang G, Zhang J, Moskophidis D, Mivechi NF (2003) Targeted disruption of the heat shock transcription factor (*hsf*)-2 gene results in increased embryonic lethality, neuronal defects, and reduced spermatogenesis. Genesis 36:48–61

Wassarman PM, Letourneau GE (1976) RNA synthesis in fully grown mouse oocytes. Nature 361:73–74

Webb R, Garnsworthy PC, Gong JG, Armstrong DG (2004) Control of follicular growth: local interactions and nutritional influences. J Anim Sci 82:E63–E74

Webb RJ, Bains H, Cruttwell C, Carroll J (2002) Gap-junctional communication in mouse cumulus-oocyte complexes: implications for the mechanism of meiotic maturation. Reproduction 123:41–52

Werner A, Meinhardt A, Seitz J, Bergmann M (1997) Distribution of heat-shock protein immunoreactivity in testes of infertile men. Cell Tissue Res 288:539–544

Wilson SJ, Marion RS, Spain JN, Spiers DE, Keisler DH, Lucy MC (1998) Effect of controlled heat stress on ovarian function in dairy cattle: I. Lactating cows. J Dairy Sci 1:2124–2131

Wolfenson D, Bartol FF, Badinga L, Barros CM, Marple DN, Cummings K, Wolfe D, Lucy MC, Spencer TE, Thatcher WW (1993) Secretion of PGF2a and oxytocin during hyperthermia in cyclic and pregnant heifers. Theriogenology 39:1129–1141

Wolfenson D (2009) Impact of heat stress on production and fertility of dairy cattle. In: Proceedings 18th annual tri-state dairy nut conference, pp 55–59

Yaakub H, ÓCallaghan D, ÓDoherty JV, Hyttel P (1997) Effect of dietary intake on follicle numbers and oocyte morphology in unsuperovulated and superovulated ewes. Theriogenology 47:182

Yaeram J, Setchell BP, Maddocks S (2006) Effect of heat stress on the fertility of male mice in vivo and in vitro. Reprod Fertil Dev 18:647–653

Yin Y, Hawkins KL, Dewolf WC, Morgentaler A (1997) Heat stress causes testicular germ cell apoptosis in adult mice. J Androl 8:159–165

Younas M, Fuquay JW, Smith AE, Moore AB (1993) Estrus and endocrine responses of lactating Holsteins to forced ventilation during summer. J Dairy Sci 76:430–434

Zhu BK, Setchell BP (2004) Effects of paternal heat stress on the in vivo development of preimplantation embryos in the mouse. Reprod Nutr Dev 44:617–629

Zhu B, Walker SK, Oakey H, Setchell BP, Maddocks S (2004) Effect of paternal heat stress on the development in vitro of preimplantation embryos in the mouse. Andrologia 36:384–394

Chapter 6
Concept of Multiple Stresses and Its Significance on Livestock Productivity

Veerasamy Sejian, Vijai P. Maurya, Kailash C. Sharma and S. M. K. Naqvi

Abstract Animals reared in tropical environments are generally subjected to more than one stress at a time. This greatly influences the animal production and reproduction under such environmental conditions. Nearly all studies on the effect of environmental stress on farm animal productivity have generally implicated one stress at a time since comprehensive, balanced multifactorial experiments are technically difficult to manage, analyze, and interpret. Hence, few reports evaluating effects of multiple stresses on farm animals are available in the literature. However, researchers have described several hypothetical schemes highlighting how two stressors can synergistically influence normal physiologic functions in mammalian species. Thermal stress in livestock is aggravated when feed restriction is involved. These effects are often manifested in changes in the blood biochemical parameters, enzymes, and thyroid hormone levels in livestock. Generally, when nutrition is not compromised, livestock species cope with heat stress better. Further, several findings have shown that livestock species tolerate and adapt to nutritional stress more than thermal stress. However, when both stresses are present, severe impacts on all the biologic functions in the livestock have been observed. Hence, it may be pertinent to conclude that the combined effects of two stressors may have severe impact on biologic functions in livestock species. Further studies involving more than one stress at a time emphasized the importance of providing optimum nutrition to livestock for counteracting thermal

V. Sejian (✉) · K. C. Sharma · S. M. K. Naqvi
Adaptation Physiology Laboratory Division of Physiology and Biochemistry,
Central Sheep and Wool Research Institute, Avikanagar, Via-Jaipur,
Rajasthan 304501, India
e-mail: drsejian@gmail.com

V. P. Maurya
Division of Physiology and Climatology, Indian Veterinary Research Institute,
Izatnagar, Bareilly, Uttar Pradesh 243122, India

V. Sejian et al. (eds.), *Environmental Stress and Amelioration in Livestock Production*, DOI: 10.1007/978-3-642-29205-7_6,

stress during summer. These types of investigations will be instrumental in gaining a thorough understanding of the role of nutrition in mitigating adverse effects of environmental stress on livestock and how the knowledge garnered can be used to develop rational managerial strategies for optimizing productivity in livestock farms.

Keywords Combined stress · Heat stress · Multiple stress · Nutrition stress · Walking stress · Reproduction

Contents

6.1 Introduction

Animals live in complex environments where they are constantly confronted with short- and long-term changes due to a wide range of environmental factors such as temperature, photoperiod, geographic location, nutrition, and socio-sexual signals (Kleeman and Walker 2005; Ali and Hayder 2008). Although animals live in a complex world, yet researchers often study the influence of one stressor at a time since comprehensive, balanced multifactorial experiments are technically intricate to manage, analyze, and interpret (Blanc et al. 2001).

Semi-arid environments are one of the major agroecological zones of the tropics. There is a strong relationship between agro-climatic conditions, population density, cropping systems, and livestock production. Rangelands are the largest land use systems on earth. They predominate in semi-arid tropical areas of the world. In these pastoral systems people depend entirely on livestock for their livelihoods. The key constraints of arid and semi-arid tropical environments are their low biomass productivity, high climatic variability, and limited availability of water (Naqvi and Sejian 2010; Sejian et al. 2011). All these constraints make these regions difficult for sustainable livestock production. Research strategies and efforts need to take into account the tradeoffs and synergies arising from these tropical environments so that peasant farmers are able to obtain benefits provided by these ecosystems.

Livestock production is thought to be adversely affected by detrimental effects of extreme climatic conditions. Consequently, adaptation and mitigation of detrimental effects of extreme climates have played a major role in combating the climatic impact in livestock production (Khalifa 2003). Among the environmental variables affecting animals, heat stress seems to be one of the intriguing factors making animal production challenging in many geographic locations in the world (Shelton 2000; Koubková et al. 2002). Although animals can adapt to the hot climate, nevertheless the response mechanisms that ensures survival are also detrimental to performance (Rivington et al. 2009).

During stress in animals, various endocrine responses are involved in maintaining fitness. The front-line hormones to surmount stressful situations are the glucocorticoids and catecholamines. The secretion of glucocorticoids is the principal endocrine response to stress (Kannan et al. 2000). Currently, it appears that glucocorticosteroids provide an initial integrating signal, which in conjunction with other hormones and paracrine secretions, may determine specific behavioral, physiologic, and biochemical responses to different environmental conditions (Wingfield and Kitaysky 2002). The thyroid gland is one of the most sensitive organs to the ambient heat variation (Rasooli et al. 2004). Appropriate thyroid gland function and activity of thyroid hormone are considered crucial to sustain the productive performance in domestic animals (Todini 2007). When animals are exposed to heat stress, food ingestion is reduced and metabolism slows down causing a hypo-function of the thyroid gland (McManus et al. 2009). One of the principal functions of glucocorticoid under thermal stress is to control hepatic gluconeogenesis and insulin concentration. These endocrine changes directly control the energy status of animals to sustain production under extreme environmental conditions (Sejian et al. 2008a). Hence, measuring the metabolic hormones, such as thyroid hormone and insulin, gives clues about how an animal adapts to changing environments by altering its metabolic activity.

Sheep grazing in hot semi-arid environments face extreme fluctuations in the availability of quantity and quality of feed throughout the year (Martin et al. 2004). It has been postulated that nutrition is one of the main factors affecting ovulation rate and sexual activity in sheep (Vinoles et al. 2005; Forcada and Abecia 2006) and nutrition modulates reproductive endocrine functions in many species including sheep (Lindsay et al. 1993; Polkowska 1996; Martin et al. 2004). Furthermore, undernutrition affects reproductive function in ruminants at different levels of the hypothalamus-pituitary–gonadal axis (Robinson 1996; Boland et al. 2001; Chadio et al. 2007).

Thermal stress and feed scarcity are two major predisposing factors that cause low productivity of ewes in hot semi-arid environments (Naqvi and Hooda 1991; Martin et al. 2004). High ambient temperature augments the effort by sheep to dissipate body heat, resulting in increased rate of respiration, body temperature, heart beat, and water consumption (Marai et al. 2000). Increased body temperature and respiration rate (RR) are the most important signs of heat stress in sheep (Al-Haidary 2004). Further increase in body temperature is associated with marked reduction in feed intake, redistribution in blood flow, and changes in endocrine functions that negatively affect the productive and reproductive performance of sheep (Averos et al. 2008). In addition, exposure of sheep to elevated temperature results in a decrease in body weight, growth rate, and body total solids

(Marai et al. 2007). It has been reported that depleted body condition score during periods of energy deficiency leads to reduced energy reserves, which in turn reduces the heat tolerance capability of livestock (Minka and Ayo 2009).

Nurturing of animals in hot semi-arid environments exposes them to more than one stress at a time fettering animal production and reproduction (Naqvi and Sejian 2010). Studies investigating the effect of stress on farm animal productivity have generally focused on single stress at a time. Limited data are available in the literature on studies targeting the effects of multiple stresses on farm animals. Moberg (2000) put forth a supposition design signifying how two stressors, in conjunction, can manipulate normal body functions. He also insisted that when two stressors occur simultaneously, the total impact might be severe on biologic functions. However, the influence of thermal stress on biologic functions and productive responses, when coupled with long-term nutritional stress are not properly elucidated in livestock. In addition, despite the general awareness that energy demands vary in different seasons, only few quantitative data exist relating environment, nutrient need, and productive efficiency. This chapter will cover, in detail, information about more than one stress at a time in livestock including effects on various physiologic responses, growth parameters, blood metabolites, and endocrine responses. The chapter will also address the effect of multiple stresses on livestock production and the related research conducted in this area will be highlighted.

6.2 Environmental Stresses and Its Consequences on Livestock Economy

Global climate change is expected to alter temperature, precipitation, atmospheric carbon dioxide levels, and water availability in ways that will affect the productivity of crop and livestock systems (Hatfield et al. 2008). For livestock systems, climate change could affect the costs of production by altering the thermal environment of animals thereby affecting animal health, reproduction, and the efficiency by which livestock convert feed into retained products (especially meat and milk). Climatic changes could increase thermal stress for animals and thereby reduce animal production and profitability by lowering feed efficiency, milk production, and reproduction rates (Fuquay 1981; Morrison 1983; St-Pierre et al. 2003).

Climate changes, in addition, could impact the economic viability of livestock production systems worldwide. Environmental conditions directly affect mechanisms and rates of heat gain or loss in animals (NRC 1981). Lack of prior conditioning to weather events most often results in catastrophic losses in the domestic livestock industry (Mader 2003). The vulnerability of animals to weather is well known and their performance or even their survival is strongly influenced by weather conditions. Weather is a constraint on efficient livestock production systems (Hahn et al. 2001). Evaluation of the degree of constraint is a

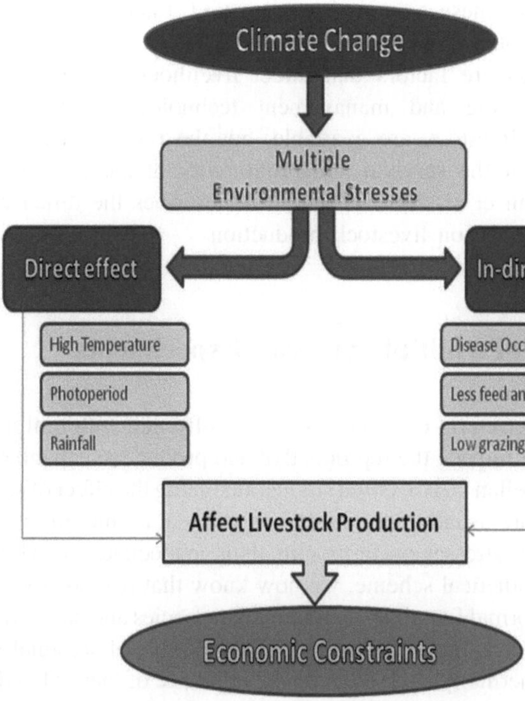

Fig. 6.1 Direct and indirect effects of climate change on livestock production

difficult but necessary task that must be conducted before selection of appropriate modifications in management or environment can be made. Environmental stress reduces the productivity and health of livestock resulting in significant economic losses. Heat stress affects animal performance and the productivity of dairy cows in all phases of production. Poor performance in calves and growing heifers stemmed from heat stress is due to repartitioning of energy necessary for maintenance of homeothermy. The outcomes include decreased growth, reduced reproduction, increased susceptibility to diseases, and ultimately delayed initiation of lactation. Dry matter intake and milk yield decrease as cows are exposed to ambient temperatures above the upper critical temperature of their comfort zone. Heat stress also negatively affects reproductive function (Amundson et al. 2006; Sprott et al. 2001). Normal estrus activity and fertility are disrupted in livestock during summer months. Economic losses are incurred by the livestock industries because farm animals are generally raised in locations and/or seasons where temperature conditions go beyond their thermal comfort zone. Losses considered were: (1) decreased performance (growth, lactation), (2) increased mortality, and (3) decreased reproduction. The livelihood of locals in developing countries depends critically on local natural resource-based activities, such as crop and livestock production. As a result of negative weather impact on livestock rearing, the livelihood of shepherds/farmers whose size of income depends on the performance of these animals is endangered. Lack of balanced nutrition, improper housing system, inadequate livestock health care system, and poor management practises are factors that affect livelihood insecurities (Duff and Gaylean 2007). Housing and management technologies which can reduce climatic impacts on livestock are available, but the rational use of such technologies is crucial for the survival and profitability of the livestock enterprise (Hahn 1981; Gaughan et al. 2002). Figure 6.1 describes the direct and indirect effects of climate change on livestock production.

6.3 Significance of Multiple Stresses Experiments

Livestock raised in semi-arid environments are confronted with multiple stressors which tend to acutely impede the reproductive and productive capabilities of these animals (Naqvi and Sejian 2010). Often studies analyzing the effect of stress on farm animal productivity are usually focused on one stress at a time. Information about the effects of multiple stresses on farm animals is inadequate. As Moberg (2000) described in his hypothetical scheme, we now know that two stressors can synergistically influence normal functions in mammalian species and have greater impact than when one stressor is present. Despite the knowledge of seasonal variation in energy demands of thermally stressed animals, influence of thermal and nutritional stresses on sheep has not been adequately studied (Naqvi and Hooda 1991).

6.4 Concept of Multiple Stresses and Its Mechanism

Livestock researchers aiming at improving livestock production under a changing climatic scenario must aim at counteracting multiple stresses as these stresses are common occurrences in most ecological zones where livestock are reared. Under the changing climatic scenario, the concept of multiple stresses emerges as a potential threat to livestock production and its survival. Hence, research needs to be prioritized to tackle multiple stresses simultaneously to improve the profitability of livestock farms.

Generally, when animals are exposed to one stress at a time, they can effectively counter them based on their stored body reserves without altering the normal body functions. However, if they are exposed to more than one stress at a time, the summated effects of the different stressors might prove detrimental to these animals. This is because of their inability to cope with the combined effects of different stressors simultaneously. In such a case, the animal's body reserves are not sufficient to effectively counter such environmental extremes. As a result, their adaptive capability is hampered and the animals struggle to maintain normal homeothermy. Moberg (1999) explained that when animals are exposed to only one stress, they may not require the diversion of biologic resources needed for

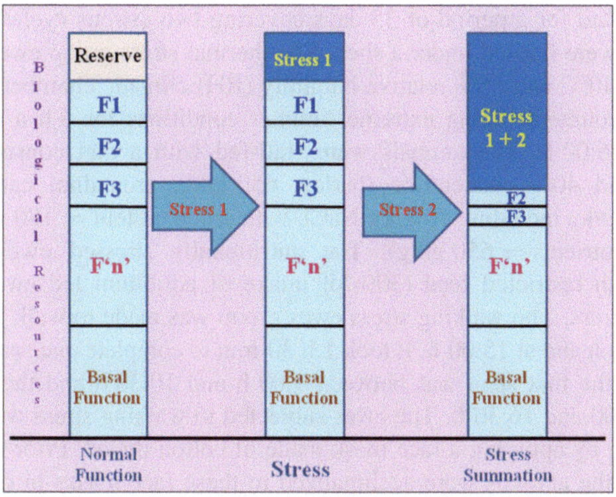

Fig. 6.2 Pictorial representation of summation effect of multiple stresses. When the animals are subjected to only one stress, they can cope easily with the reserve energy without compromising the productive performance. However, if the animals are subjected to two or more stresses simultaneously, the energy sources are deviated toward survival. In this case, the productive performances are severely hampered (*Source* Moberg 1999). F1, F2, F3 represents productive functions in an animal

other functions. If, however, two of these stressors occur simultaneously, the total cost may have a severe impact on other biologic functions. As a result, normal basal functions in these animals are drastically affected leading to decline in production. Figure 6.2 describes the schematic representation of the effect of multiple stresses on biologic functions in animals.

6.5 Experimental Findings on Multiple Stresses in Ewe

Livestock production in areas of hot and semi-arid environment faces numerous challenges in the form of compound/multiple stresses; decreased reproduction and production (Naqvi and Sejian 2010; Sejian et al. 2010a). Usually, studies focusing on the effect of stress on farm animal productivity have dealt with single stresses and studies on exploring effects of multiple stresses on farm animals are fewer. As a result of changing climatic conditions, multiple stresses have become a common occurrence in semi-arid tropical environment. Hence, it is vital to test the hypothesis that multiple stresses are more detrimental to livestock productivity than individual stresses. Several experiments were conducted in a climatic chamber and in a natural environment with the primary objective of establishing effects of multiple stresses on the adaptive capability of ewes, and parameters, such as changes in growth, physiologic response, blood biochemical response, endocrine responses, and reproductive performance, were evaluated. These studies were conducted for a period of 35 days covering two estrous cycles. The control group ewes were housed under a shed. The thermal stress group ewes were either housed in a 40°C and 55% relative humidity (RH) climatic chamber or in a 42°C natural environment during extreme summer conditions for 6 h a day, between 10:00 and 16:00 h. The animals were stall-fed with a diet consisting of 60% roughage and 40% concentrate (barley, 650 g/kg, groundnut cake, 320 g/kg, minerals 30 g/kg including 10 g/kg NaCl, with crude protein = 180 g/kg and total digestible nutrients = 650 g/kg). The nutritionally stressed ewes group was provided with restricted feed (30% of intake of adlibitum fed ewes) to induce nutritional stress. The walking stress ewes group was made to walk 14 km in two spans at 9:00 h and at 15:00 h. It took 1 h 30 min to complete one span (7 km) and accordingly the first span was between 9:00 h and 10:30 h and the second span between 15:00 and 16:30 h. The ewes subjected to walking stress were prevented from grazing by applying a face mask made of cotton thread. Prior to start of the experiment, the animals were acclimatized to these face masks in order to avoid any undue restraining stress. All ewes were synchronized for estrus using indigenously developed intravaginal sponges (Naqvi et al. 2001) to ensure all ewes were in estrus phase at the start of study. Blood samples were collected at weekly intervals. Thermal stress effects aggravated in sheep when coupled with feed restriction and exercise stress was evaluated. These series of experiments established the severity of multiple stresses on various biologic functions involved in adaptation in sheep. When both thermal and nutritional stresses were coupled, it had a severe impact on the biologic functions compared to when the animal was exposed to individual stress.

Physiologic adaptation is defined as a modification in an animal's behavioral or metabolic responses, which improves the ability of the animal to cope with a particular environmental challenge. The animals usually acquire this ability by their previous experience to such environmental extremes. Physiologic behavior of animals during exposure to environmental stress has been measured in terms of variations in the RR, body temperature, and heart rate. Table 6.1 depicts the differences in physiologic responses among individual, combined, and multiple stressed ewes. The experimental findings suggest that RR significantly increased in both the thermal and combined stress groups. Similar thermal stress-induced increase in RR after subjecting the sheep to climatic chamber has been reported by Al-Haidary et al. (2004). Marai et al. (2007) also reported that heat stress during summer is characterized by an increase in RR in sheep. The increase in the RR observed in these ewes may have more homeostatic relevance for the dissipation of excessive heat and the maintenance of a lower body temperature (Rahardja et al. 2011). The significant RR in the combined stress group compared to thermal stress group might be due to less energy available in combined stressed ewes (Naqvi 1987; Hooda and Naqvi 1990). Further, McManus et al. (2009) reported that respiratory mechanism is very important for thermolysis and maintenance of homeothermia in animals to avoid an increase in body temperature during thermal exposure. When walking stress was added to the combined stress, the RR was much lower than the combined stresses. Walking stress may have imposed severe

Table 6.1 Effect of thermal, nutritional, combined, and multiple stresses on physiologic responses of Malpura ewes

Parameter		Control	Heat stress	Nutritional stress	Combined stress	Multiple stress
RR (breaths/min)	Morning (M)	$24.4 \pm 0.4^{a,b,c}$	30.2 ± 0.8^{a}	20.6 ± 0.8^{c}	21.5 ± 2.1^{c}	21.97 ± 1.6^{c}
	Afternoon (A)	40.1 ± 2.4^{d}	130.8 ± 5.6^{a}	29.8 ± 1.2^{e}	107.5 ± 5.0^{b}	68.74 ± 3.8^{c}
	Variation (ΔA − M)	15.7 ± 2.1^{d}	100.6 ± 5.4^{a}	9.2 ± 1.0^{e}	86.0 ± 4.7^{b}	46.8 ± 2.1^{C}
Pulse rate (beats/Min)	Morning (M)	$57.5 \pm 1.2^{a,b}$	62.7 ± 2.0^{a}	$53.2 \pm 0.6^{a,b}$	50.2 ± 1.2^{b}	$52.81 \pm 1.7^{a,b}$
	Afternoon (A)	$74.6 \pm 2.8^{a,b}$	80.1 ± 2.5^{a}	67.5 ± 2.9^{b}	$69.4 \pm 3.6^{a,b,c}$	59.58 ± 2.1^{c}
	Variation (ΔA − M)	17.1 ± 2.1^{a}	17.4 ± 2.6^{a}	14.3 ± 2.6^{a}	19.1 ± 2.5^{a}	6.8 ± 0.6^{b}
Rectal temperature (°C)	Morning (M)	38.4 ± 0.1^{a}	38.5 ± 0.1^{a}	38.1 ± 0.1^{b}	37.8 ± 0.0^{b}	38.1 ± 0.1^{b}
	Afternoon (A)	38.9 ± 0.1^{b}	39.6 ± 0.1^{a}	38.7 ± 0.2^{c}	39.4 ± 0.1^{a}	39.2 ± 0.2^{a}
	Variation (ΔA − M)	0.5 ± 0.1^{c}	1.0 ± 0.1^{b}	0.6 ± 0.1^{c}	1.5 ± 0.1^{a}	1.1 ± 0.4^{b}

Combined stresses—thermal and nutritional stress; multiple stresses—thermal, nutritional, and walking stress
Means and SEM within a row having different superscripts differ significantly ($P < 0.05$)
(*Source* Sejian et al. 2010a, b)

stress to these ewes which ultimately culminated in a severe energy loss. The energy demand is more in these ewes as the locomotory activity requires substantial energy reserve to cope with walking as these ewes are additionally under nutritional stress. As a result, the energy level could be low in these ewes for higher respiratory muscular activity. The pulse rate reflects primarily the homeostasis of circulation along with the general metabolic status (Sejian et al. 2010b). The decreased pulse rate in the combined stress group might be due to decrease in the metabolic rate as a result of restricted feeding in this group of animals. This view was supported by the findings of several investigators who have reported that there is a correlation between heart rate and metabolic heat production (Yamamoto and Ogura 1985; Barkai et al. 2002). In parallel, Aharoni et al. (2003) have reported that heart rate decreases in livestock during thermal stress and this occurs as a general effort by the animal to decrease heat production. This reduction could be achieved by either reduced feed intake or metabolic activity or both (Al-Haidary et al. 2004). The rectal temperature in combined stress group, in the morning, is significantly lower than the control and individual stress groups. This could be due to thermolability mechanism which leads to passive lowering of rectal temperature (Slee et al. 1982). By this thermolability mechanism these ewes were able to vary their core body temperature on a daily basis. This could be the adaptive mechanism exhibited by these animals in order to cope with the water and feed scarcity (Wooden and Walsberg 2002). This provides evidence for the ability of these ewes to adjust their energy expenditure through body temperature regulation. This finding shows that the Malpura ewes have the capability to cope with feed and water scarcity, and daily changes in rectal temperature is a mechanism ewes use to adapt to semi-arid tropical environments.

Both hemoglobin (Hb) and packed cell volume (PCV) are significantly reduced in thermal, nutritional, and combined stress groups. But the magnitude of difference was more in combined stress groups than in individual stress groups (Table 6.2). Generally, PCV and Hb are considered to be the indices of the organic response to exercise stress (Garcia-Belenguer et al. 1996). Furthermore, McManus et al. (2009) reported a strong positive correlation between PCV and Hb concentration, indicating the significance of these parameters for heat tolerance in Brazilian sheep. Hb and PCV are considered to be good indicators of stress in farm animals (McManus et al. 2009). There are reports which suggest that during thermal stress both Hb and PCV decreased significantly (More and Sahni 1979; Yousef 1985; Maurya et al. 2007). This could be attributed to the hemodilution effect in which more water is transported in the circulatory system for evaporative cooling and increase in the blood volume of these animals (Al-Haidary 2004). In addition, Marai et al. (1991) reported that red blood cell destruction led to reduced Hb and PCV in thermal stressed animals. Further, ad libitum water availability could be the reason for reduced PCV due to more water intake in thermal stressed ewes even though the evaporation of water from the body increased. This view was supported by the findings of Sano et al. (1979). The reason for the highest magnitude of reduction in Hb and PCV in combined stress groups could be both reduction in synthesis of Hb due to nutritional restriction as well as hemodilution

Table 6.2 Effect of thermal, nutritional, combined, and multiple stresses on blood biochemical parameters of Malpura ewes

Parameters	Control	Thermal stress	Nutritional stress	Combined stress	Multiple stress
Hb (g/dl)	11.91 ± 0.35^{a}	9.48 ± 0.26^{c}	10.75 ± 0.32^{b}	8.59 ± 0.17^{d}	10.91 ± 0.12^{b}
PCV (%)	41.33 ± 1.46^{a}	31.31 ± 1.64^{b}	35.80 ± 1.73^{b}	26.54 ± 2.05^{c}	35.80 ± 0.75^{b}
Glucose (mg/dl)	52.08 ± 2.43^{a}	$47.80 \pm 1.61^{a,b}$	44.19 ± 1.91^{b}	42.99 ± 2.52^{b}	39.39 ± 0.53^{c}
Total protein (g/dl)	8.88 ± 0.33^{a}	$7.95 \pm 0.41^{a,b}$	$7.91 \pm 0.26^{a,b}$	7.08 ± 0.33^{b}	6.37 ± 0.08^{c}
Total Cholesterol (mg/dl)	52.31 ± 1.86^{a}	42.96 ± 2.29^{b}	42.65 ± 2.21^{b}	35.62 ± 2.33^{c}	30.58 ± 1.78^{d}
ACP (KA units)	2.14 ± 0.12^{a}	$1.48 \pm 0.14^{b,c}$	$1.78 \pm 0.09^{a,b}$	1.16 ± 0.18^{c}	–
ALP (KA units)	5.89 ± 0.25^{a}	5.10 ± 0.37^{b}	$5.57 \pm 0.25^{a,b}$	4.10 ± 0.33^{c}	–
T_3 (nmol/L)	1.71 ± 0.01^{a}	1.33 ± 0.00^{b}	1.42 ± 0.00^{b}	1.14 ± 0.01^{c}	1.39 ± 0.05^{b}
T_4 (nmol/L)	76.24 ± 4.17^{a}	58.90 ± 3.22^{b}	62.97 ± 3.64^{b}	45.94 ± 5.14^{c}	24.41 ± 0.10^{d}
Cortisol (nmol/L)	18.63 ± 1.30^{d}	76.98 ± 5.18^{a}	8.50 ± 1.06^{e}	46.44 ± 3.64^{b}	31.03 ± 1.08^{c}
Insulin (MicroIU/mL)	47.44 ± 2.92^{a}	$39.79 \pm 3.20^{a,b}$	$35.15 \pm 1.39^{b,c}$	26.41 ± 2.84^{c}	–

Combined stresses—thermal and nutritional stress; multiple stresses—thermal, nutritional, and walking stress

ACP acid phosphotase; *ALP* alkaline phosphatase; T_3 tri-iodo-thyronine; T_4 thyroxine

Means and SEM within a row having different superscripts differ significantly ($P < 0.05$)

(*Source* Sejian et al. 2010a, b)

and red cell destruction due to heat stress. However, both Hb and PCV increased significantly in multiple stresses groups. The reason for this could be severe hemoconcentration due to multiple stresses. The difference between combined stress and multiple stress groups with respect to Hb and PCV could be attributed to the fact that in the combined stress experiment the animals were allowed to get access to water ad libitum while in the multiple stress experiment the water availability was restricted.

The plasma glucose showed significant reduction in all stress groups. But the magnitude of reduction is more in both combined and multiple stress groups (Table 6.2). Several researchers have studied the effect of thermal stress on the blood glucose concentrations and conflicting results have been reported (More and Sahni 1980; Habeeb 1987; Sejian and Srivastava 2009). The decrease in glucose level during heat exposure is attributed to decrease in the concentration of insulin and thyroxine, which are closely associated to energy metabolism during heat exposure. Decrease in plasma glucose could also be due to the marked dilution of blood or increase in the plasma glucose utilization to produce more energy for greater muscular expenditure required for high muscular activity (Naqvi and Hooda 1991; Naqvi et al. 1991; Rasooli et al. 2004). Nutrient restriction combined with increased glucose utilization due to increased respiratory muscular activity after thermal exposure will further reduce glucose concentration in the combined stress group. The reduced plasma glucose in the multiple stress group as compared to single and two stresses could be attributed to increased utilization of glucose during locomotory activity. The total plasma protein was significantly reduced in individual stress groups, combined and multiple stress groups as compared to control group (Table 6.2). A negative correlation between plasma protein and elevated environmental temperature has been reported (Habeeb 1987; Sejian et al. 2008a). The biologic significance of thermal stress reducing protein is to support hepatic gluconeogenesis by glucocorticoids to increase the glucose level in an attempt to cope with thermal stress (Kamiya et al. 2006; Korde et al. 2007). The decrease in the plasma protein may be due to dilution of plasma proteins and decrease in protein synthesis as a result of depression of anabolic hormone secretion (El-Masry and Habeeb 1989). There is also increase in the glucocorticoids concentration during thermal stress and one of the principal functions of glucocorticoids is to favor protein catabolism to convert protein into amino acid to support gluconeogenesis (Sejian et al. 2008b; Sejian et al. 2009). The restricted feeding, of heat stressed animals in combined stress groups, also might have led to a decrease in the nitrogen concentration in the body, which leads to low protein level. The total cholesterol concentration also decreased significantly in individual, combined, and multiple stresses groups (Table 6.2). But the magnitude of reduction is more in multiple stresses group. This might be due to increase in utilization of fatty acids for energy production as a consequence of decrease in glucose concentration in the thermal stressed animals (Daader et al. 1989; Rasooli et al. 2004). It is an established fact that elevated cortisol concentration during thermal stress is to favor cholesterol catabolism to support gluconeogenesis to supply more energy in the form of glucose to heat stressed animals (Nazifi et al. 2003). The restricted feeding might have given additional stress load thereby limiting synthesis of cholesterol in combined stress groups, and significantly lowering total cholesterol concentration as compared to individual stress groups.

The association between thermal stress and increased secretion of cortisol, the principal glucocorticoid hormone in small ruminants, is well documented (Ali and Hayder 2008; Sejian et al. 2008a; Sejian and Srivastava 2009). The significant reduction in cortisol concentration in combined and multiple stresses groups as compared to thermal stress group indicates the difference in adaptive capability of Malpura ewes to these stresses. This difference could be attributed to the restricted feeding in combined and multiple stresses groups. Further, cortisol is thermogenic in nature and its action will contribute to additional heat load (Alvarez and Johnson 1973) and hence, the combined and multiple stressed animals showed the capability to adjust the cortisol level to minimal increase in an attempt to elicit heat stress relieving effects (Sejian et al. 2010b). Furthermore, when walking stress was combined with heat and nutritional stress, the level of cortisol is much lower than individual and combined stress groups (Table 6.2). This demonstrates that multiple stresses had imposed a much higher detrimental effect on these ewes than single and dual stresses.

Both thyroxine (T_4) and tri-iodo-thyronine (T_3) showed similar trends when ewes are exposed to dual and multiple stresses (Table 6.2). The level decreased significantly in all stress groups, But the magnitude of reduction in thyroid hormone levels is more in combined and multiple stresses groups than other groups. The major exogenous regulator of thyroid gland activity is the environmental temperature (Dickson 1993). Heat stresses associated with significant depression in thyroid gland activity resulting in lowering of thyroid hormone level (Nazifi et al. 2003; Rasooli et al. 2004). In addition, thyroid hormones in the blood are considered to be good indicators of the nutritional status of an animal (Riis and Madsen 1985). Following feed restriction or food deprivation, plasma thyroid hormone concentrations were reduced in sheep (Naqvi and Rai 1991; Abecia et al. 2001; Rae et al. 2002). These effects suggest that energy balance could play a major role in affecting the decrease in plasma thyroid hormone levels. Besides endogenous and environmental climatic factors, nutrition influences thyroid gland activity and blood thyroid hormone concentrations (Todini 2007). Hence, a combination of heat stress and nutritional stress could have resulted in a greater reduction in thyroid hormone concentrations in combined stress groups than in other groups. This could be attributed to the adaptive capability of combined stress ewes to avoid additional heat load due to metabolic activity. Furthermore, when walking stress was combined with thermal and nutrition stress, the level of thyroid hormone concentration was much lower.

Sheep that are exposed to high ambient temperatures reduce body heat by increasing the RR, body temperature, consumption of water, and by reducing feed intake (Marai et al. 2007). This finding of reduced feed intake after exposure to thermal stress agrees with later reports from Pereira et al. (2008) in which rate of feed intake by sheep was shown to decrease after exposure to heat stress. Padua et al. (1997) reported that feed conversion ratio decreased significantly in animals housed under hot conditions in a climatic chamber compared to animals housed under shelter during the spring. Marai et al. (2007) had postulated that when an animal is exposed to thermal stress, the peripheral thermal receptors which transmit suppressive nerve impulses to the appetite center in the hypothalamus are stimulated thereby causing a decrease in feed intake. This decrease in feed intake could be an adaptive mechanism in sheep to produce less body heat. The water intake of combined stress groups was significantly higher than ad libitum fed groups. This shows that the high water intake which is utilized to combat heat stress indicates that restrictedly fed animals are under more stress than unrestrictedly fed animals (Hooda and Naqvi 1990; Minka and Ayo 2009). Further, small ruminants that are deprived of food and minerals particularly potassium, impedes the sodium uptake from the reticulum-rumen causing hyponatraemia and hypo-osmolality (Dahlborn and Karlberg 1986; Holtenius and Dahlborn 1990). The average body weight of thermal, nutritional, combined, and multiple stresses groups was significantly lower than that of their respective control (Table 6.3). Growth, an increase in the live body mass or cell multiplication, is controlled genetically and environmentally (Marai et al. 2007). Elevated ambient temperature is one of the environmental factors that influence average daily weight gain

Table 6.3 Effect of thermal, nutritional, combined, and multiple stresses on growth and reproductive parameters of Malpura ewes

Parameters	Control	Thermal stress	Nutritional stress	Combined stress	Multiple stress
Initial body weight	33.75 ± 2.56^{a}	33.52 ± 1.85^{a}	34.68 ± 1.70^{a}	34.87 ± 1.46^{a}	32.63 ± 0.98^{a}
Final body weight	39.67 ± 2.65^{a}	35.19 ± 1.46^{a,b}	30.39 ± 1.50^{b}	30.04 ± 1.35^{b}	29.55 ± 1.22^{b}
ADG	169.14 ± 0.01^{a}	47.71 ± 0.07^{b}	–		
122.57 ± 0.06^{c}	–				
138.00 ± 0.07^{c}	-88.00				
Ewes in heat (%)	85.71	57.14	85.71	71.43	41.7
Estrus duration (hrs)	38.00 ± 2.41^{a}	23.40 ± 3.34^{b}	28.50 ± 5.68^{b,c}	18.75 ± 3.75^{b,d}	14.4 ± 2.78
Estrus cycle length (days)	18.17 ± 0.31^{b}	20.28 ± 0.74^{a,b}	18.00 ± 0.27^{b}	22.25 ± 1.67^{a}	23.56 ± 1.45
Conception rate (%)	71.43^{a}	42.86^{a,b}	57.14^{a,b}	28.57^{b}	–
Lambing rate (%)	71.43^{a}	42.86^{a,b}	57.14^{a,b}	28.57^{b}	–
Estradiol (pg/mL)	14.58 ± 0.96^{a}	12.06 ± 0.73^{b}	12.80 ± 0.91^{b}	10.04 ± 0.74^{c}	7.19 ± 0.23^{d}
Progesterone (ng/mL)	3.31 ± 0.56^{c}	4.48 ± 0.32^{a,b}	3.98 ± 0.26^{b,c}	5.19 ± 0.27^{a}	7.34 ± 0.28^{d}

Combined stresses—thermal and nutritional stress; multiple stresses—thermal, nutritional, and walking stress
ADG average daily gain
Means and SEM within a row having different superscripts differ significantly ($P < 0.05$)
(*Source* Sejian et al. 2011)

(Habeeb et al. 1992). Similar findings of impaired body weight and growth rate following exposure to elevated temperatures have been reported (Marai et al. 1991; Ismail et al. 1995). Decreased growth among animals exposed to elevated ambient temperatures has been attributed to associated adaptive mechanisms, such as decrease in anabolic activity and increase in tissue catabolism in these animals (Marai and Habeeb 1998; Marai et al. 1999). Increase in tissue catabolism, however, was due to increase in catecholamines and glucocorticoids in sheep following exposure to heat stress.

Endocrine response to stress results in suppression of productive functions, such as growth and reproduction, but favors maintenance and survival (Rivest and Rivier 1995; Stratakis et al. 1995; Lindsay 1996). Correlation between thermal stress and increased secretion of cortisol, the principal glucocorticoid hormone in small ruminants, is well documented (Ali and Hayder 2008; Sejian et al. 2008a). Reproduction processes are impacted during exposure of sheep to thermal stress (Tilbrook et al. 2002; Naqvi et al. 2004; Kornmatitsuk et al. 2008). Based on the above experimental results, we may thus conclude that glucocorticoids are paramount in mediating negative impacts of stress on reproduction in livestock. Supporting this notion, various studies have shown that administration of natural or synthetic glucocorticoids can inhibit the secretion of the gonadotrophins in sheep (Juniewicz et al. 1987; Tilbrook et al. 2000). Furthermore, glucocorticoids are capable of aggravating feedback effects of estradiol in livestock and reducing the stimulation of gonadotrophins releasing hormones (GnRH) receptor expression by estrogen (Adams et al. 1999; Daley et al. 1999). Glucocorticoids may also exert direct inhibitory effects on gonadal steroid secretion and sensitivity of target tissues to sex steroids (Magiakou et al. 1997). Inadequate nutrition, in addition, delays or prevents the onset of puberty, interferes with normal estrus cycles in female, and results in hypogonadism (Dunn and Moss 1992; Sejian et al. 2008b). This in turn affects all other reproductive processes.

Multiple stresses showed a highly significant influence on estrus cycle and duration (Table 6.3). This could be related to the high plasma progesterone concentration in multiple stressed ewes. Presumably, the longer estrus cycles were due to a slower rate of follicular maturation after corpus luteum (CL) regression. This statement was supported by the finding of Stewart and Oldham (1986), who reported that the nutritional effect on ovulation rate seems to be more due to mechanisms confined to final stages of folliculogenesis rather than change in secretion of GnRH, luteinizing hormone (LH) and follicle stimulating hormone (FSH). Nutritional influences on reproduction may be linked to variations in the interleukin growth factor (IGF-1) system (Roberts et al. 2001). Acute nutrient restriction to the point of decreasing IGF-1 secretion may affect the ability of developing follicles to respond to FSH through a reduction in FSH receptor expression (Kiyma et al. 2004).

Combined stress had relatively more detrimental effect on the conception rate than the individual stresses (Table 6.3). The probable reason for this could be reduced sex steroid receptor and altered sex steroids concentration. This view is supported by the fact that undernutrition affects a number of uterine sex steroid receptors, which subsequently affect the conception rate (Sosa et al. 2006). Sheep embryo is most susceptible to maternal heat stress (Thwaites 1971). In addition, variations in the physiologic range of peripheral progesterone concentration due to management factors such as nutrition may induce asynchrony between the embryo and uterus resulting in failure in the establishment of pregnancy after conception (Lawson and Cahill 1983). The probable reason for low conception rate in combined stress groups could be insufficient progesterone concentration to maintain pregnancy, because thermal and nutritional stress were withdrawn after mating which coincided with the end of the experimental period. Higher level of nutrition has been found to be associated with lower circulating progesterone concentrations in ewes (Mc Evoy et al. 1995; O'Callaghan et al. 2000). Ample evidence supports that undernutrition of ewes before and after mating increases embryonic mortality which consequently reduces the lambing rate (Rhind et al. 1989; Abecia et al. 2006).

Plasma estrogen and progesterone showed a reverse trend in individual, combined, and multiple stresses groups (Table 6.3). The effect of thermal stress is more severe than nutritional stress on plasma estradiol 17-β level. Kiyma et al. (2004) reported that serum concentrations of estradiol were lower in undernourished ewes. Similar results were reported in other species such as hamster and rats (Morin 1986; Otukonyong et al. 2000). Decreased concentration of estrogen may result from diminished ovarian follicular development caused by suppressed peripheral concentration of gonadotrophins (Gougeon 1996). The level of nutrition and peripheral progesterone concentrations are inversely related (Parr 1992;

Lozano et al. 1998) in ewes. This inverse relationship between level of feed intake and plasma progesterone concentration was attributed to the difference in the metabolic clearance rate of progesterone (Parr et al. 1993; Scaramuzzi et al. 2006). This view was corroborated by Forcada and Abecia (2006), who reported that the difference in the rate of clearance rather than differences in secretion levels explains the apparent inverse relationship between nutrition and peripheral progesterone concentrations in ewes. The increased plasma progesterone concentration in undernourished ewes in these studies might be due to the limited extravascular pool in such animals with low body fat content (Lamond et al. 1972).

6.6 Concluding Remarks

The series of studies conducted to establish the effect of more than one stress simultaneously in sheep clearly establishes the severe impact of these stresses on the biologic functions necessary to relieve stress. This is evident from the significant difference in all parameters studied in combined and multiple stresses groups. On an individual basis, ewes adapt better to nutritional stress than thermal stress. However, when both these stresses were combined, it had a severe impact on all the parameters studied in ewes. When compared to thermal stress, nutritional stress had less significant effect on the reproductive performance in ewes. Further, these studies reveal that ewes subjected to thermal and nutritional stress separately had less detrimental effects on the ewe's reproductive performance than when both stresses were combined. Hence, a pertinent conclusion may be drawn—when two or more stressors occur simultaneously or concurrently, the total impact on biologic functions in livestock may be severe. Furthermore, these studies reveal the importance of providing optimum nutrition to ewes to counteract thermal stress during summer. It is also evident from these experiments that when nutrition is not a limiting factor, then ewes coped with thermal stress better.

6.7 Future Perspective

Research in this area, in future, should address effects of multiple stresses in detail to have a clear understanding of environmental stress on livestock at the mechanistic level. This will help to fully exploit the potential of nutritionally manipulated reproduction in ewes under hot semi-arid environments. In addition, such experiments will be valuable in gaining a thorough understanding of adjusting the nutrient requirements to deal with the existing environmental condition. These will in turn aid in developing rational managerial decisions to optimize productivity in livestock under changing climatic conditions.

References

Abecia JA, Zuniga O, Forcada F (2001) Effect of melatonin treatment in spring and feed intake on wool growth and thyroxine secretion in rasa aragonesa ewes. Small Rumin Res 41:265–270

Abecia AJ, Sosa C, Forcada F, Meikle A (2006) The effect of undernutrition on the establishment of pregnancy in the ewe. Reprod Nutr Dev 46:367–378

Adams TE, Sakurai H, Adams BM (1999) Effect of stress like concentrations of cortisol on estradiol dependent expression of gonadotrophin-releasing hormone receptor in orchidectomized sheep. Biol Reprod 60:164–168

Aharoni Y, Brosh A, Kourilov P, Ariel A (2003) The variability of the ratio of oxygen consumption to heart rate in cattle and sheep at different hours of the day under different heat load conditions. Livest Prod Sci 79:107–117

Al-Haidary AA (2004) Physiologic responses of Niamey sheep to heat stress challenge under semi-arid environments. Int J Agric Biol 2:307–309

Ali A, Hayder M (2008) Seasonal variation of reproductive performance, foetal development and progesterone concentration of sheep in the subtropics. Reprod Domest Anim 43:730–734

Alvarez MB, Johnson HD (1973) Environmental heat exposure on cattle plasma catecholamines and glucocorticoids. J Dairy Sci 56:189–194

Amundson JL, Mader TL, Rasby RJ, Hu QS (2006) Environmental effects on pregnancy rate in beef cattle. J Anim Sci 84:3415–3420

Averos A, Martin S, Riu M, Serratosa J, Gosalvez LF (2008) Stress response of extensively reared young bulls being transported to growing-finishing farm under Spanish summer commercial conditions. Life Sci 119:174–182

Barkai D, Landau S, Brosh A, Baram H, Molle G (2002) Estimation of energy intake from heart rate and energy expenditure in sheep under confinement or grazing condition. Livest Prod Sci 73:237–246

Blanc F, Martin GB, Bocquier F (2001) Modelling reproduction in farm animals: a review. Reprod Nutr Dev 13:337–353

Boland MP, Lonergan P, O'Callaghan D (2001) Effect of nutrition on endocrine parameters, ovarian physiology, oocyte and embryo development. Theriogenology 55:1323–1340

Chadio SE, Kotsampasi B, Papadomichelakis G, Deligeorgis S, Kalogiannis D, Menegatos D, Zerwas G (2007) Impact of maternal undernutrition on the hypothalamic–pitutary-adrenal axis responsiveness in the sheep at different ages postnatal. J Endocrinol 192:495–503

Daader AH, Marai, IFM, Habeeb AA, Yousef HM (1989) Improvement of growth performance of Friesian calves under Egyptian sub-tropical conditions 1. Internal cooling technique using diuretics and drinking cool water. In: Proceedings of the 3rd Egyptian British conference on animal, fish and poultry production, vol 2. Alexandria, Egypt, pp 595–605

Dahlborn K, Karlberg BE (1986) Fluid balance during food deprivation and after intraruminal loads of water or isotonic saline in lactating and anoestral goats. Q J Exp Physiol 71:223–233

Daley CD, Sakurai H, Adams BM, Adams TE (1999) Effect of stress like concentrations of cortisol on gonadotroph function in orchidectomized sheep. Biol Reprod 60:158–163

Dickson WM (1993) Endocrine glands. In: Swenson MJ, Reece WO (eds) Dukes physiology of domestic animal, 11th edn. Comstock publishers association, Ithaca, London, pp 629–664

Duff GC, Galyean ML (2007) Recent advances in management of highly stressed, newly received feedlot cattle. J Anim Sci 85(3):823–840

Dunn TG, Moss GE (1992) Effects of nutrient deficiencies and excesses on reproductive efficiency of livestock. J Anim Sci 70:1580–1593

El-Masry KA, Habeeb AA (1989) Thyroid function in lactating friesian cows and water buffalos under winter and summer Egyptian conditions. In: Proceedings of 3rd Egyptian British conference on animal, fish and poultry production, vol 2. Alexandria, Egypt, pp 613–620

Forcada F, Abecia AJ (2006) The effect of nutrition on the seasonality of reproduction in ewes. Reprod Nutr Dev 46:355–365

Fuquay JW (1981) Heat stress as it affects animal production. J Anim Sci 52:164–174
Garcia-Belenguer S, Palacio J, Gascon M, Acena C, Revilla R, Mormede P (1996) Differences in the biologic stress responses of two cattle breeds to walking up to mountain pastures in the pyrenees. Vet Res 7:515–526
Gaughan JG, Goopy J, Spark J (2002) Excessive heat load index for feedlot cattle. In: Meat and livestock-Australia project report, FLOT. 316. MLA Ltd, Sydney, NSW
Gougeon A (1996) Regulation of ovarian follicular development in primates: facts and hypotheses. Endocr Rev 17:121–155
Habeeb AA (1987) The role of insulin in improving productivity of heat-stressed farm animals with different techniques. Ph.D. thesis, Faculty of agriculture, Zagazig University, Zagazig, Egypt
Habeeb AA, Marai IFM, Kamal TH (1992) Heat stress. In: Philips C, Piggens D (eds) Farm animals and the environment. C.A.B. International, pp 27–47
Hahn GL (1981) Housing and management to reduce climactic impacts on livestock. J Anim Sci 52(1):175–186
Hahn GL, Mader T, Spiers D, Gaughan J, Nienaber J, Eigenberg R, Brown-Brandl T, Hu Q, Griffin D, Hungerford L, Parkhurst A, Leonard M, Adams W, Adams L (2001) Heat wave impacts on feedlot cattle, considerations for improved environmental management. In: Stowell RR, Bucklin R, Bottcher RW (eds) Proceedings of the sixth international symposium—livestock environment VI. ASAE, St. Joseph, MI, pp 129–139
Hatfield J, Boote K, Fay P, Hahn L, Izaurralde C, Kimball BA, Mader T, Morgan J, Ort D, Polley W, Thomson A, Wolfe D (2008) Agriculture. In: The effects of climate change on agriculture, land resources, water resources, and biodiversity in the United States. A Report by the U.S. climate change science program and the subcommittee on global change research. Washington, USA, pp 362
Holtenius K, Dahlborn K (1990) Water and sodium movements across the ruminal epithelium in fed and food deprived sheep. Exp Physiol 75:57–67
Hooda OH, Naqvi SMK (1990) Physiologic responses of Malpura and avikalin sheep during thermal, nutritional and walking stresses. Indian J Anim Sci 60(8):958–961
Ismail E, Abdel-Latif H, Hassan GA, Salem MH (1995) Water metabolism and requirement of sheep as affected by breed and season. World Rev Anim Prod 30(1–2):95–105
Juniewicz PE, Johnson BH, Bolt DJ (1987) Effect of adrenal steroids on testosterone and LHsecretion in the ram. J Androl 81:90–196
Kamiya M, Kamiya Y, Tanaka M, Shioya S (2006) Milk protein production and plasma 3-methylhistidine concentration in lactating Holstein cows exposed to high ambient temperatures. Asian-Aust J Anim Sci 19(8):1159–1163
Kannan G, Terrill TH, Kouakou B, Gazal OS, Gelaye S, Amoah EA, Samake S (2000) Transportation of goats: effects on physiologic stress responses and live weight loss. J Anim Sci 78:1450–1457
Khalifa HH (2003) Bioclimatology and adaptation of farm animals in a changing climate. Interactions between climate and animal production. EAAP Tech Ser 7:15–29
Kiyma Z, Alexander BM, Van Kirk EA, Murdoch WJ, Hallford DM, Moss GE (2004) Effect of feed restriction on reproductive and metabolic hormones in ewes. J Anim Sci 82:2548–2557
Kleemann DO, Walker SK (2005) Fertility in south Australian commercial merino flocks: relationships between reproductive traits and environmental cues. Theriogenology 63: 2416–2433
Korde JP, Singh G, Varshney VP, Shukla DC (2007) Effects of long-term heat exposure on adaptive mechanism of blood acid-base in buffalo calves. Asian-Aust J Anim Sci 20(5):742–747
Kornmatitsuk B, Chantaraprateep P, Kornmatitsuk S, Kindahl H (2008) Different types of postpartum luteal activity affected by the exposure of heat stress and subsequent reproductive performance in Holstein lactating cows. Reprod Dom Anim 43:515–519
Koubková M, Knížková I, Kunc P, Härtlová H, Flusser J, Doležal O (2002) Influence of high environmental temperatures and evaporative cooling on some physiologic, hematological and biochemical parameters in high-yielding dairy cows. Czech J Anim Sci 47(8):309–318

Lamond RD, Gaddy GR, Kennedy WS (1972) Influence of season and nutrition on luteal plasma progesterone in rambuillet ewes. J Anim Sci 34:626–629
Lawson RAS, Cahill LP (1983) Modification of the embryo-maternal relationship in ewes by progesterone treatment early in the oestrous cycle. J Reprod Fertil 67:473–475
Lindsay DR (1996) Environment and reproductive behavior. Anim Reprod Sci 42:1–12
Lindsa, DR, Martin GB, Williams IH (1993) Nutrition and reproduction. In: King GJ (ed) Reproduction in domesticated animals. World animal science series. Elsevier Science Publishers, Amsterdam, pp 459–491
Lozano JM, Albecia JA, Forcada F, Zarazaga L, Alfaro B (1998) Effect of undernutrition on the distribution of progesterone in the uterus of ewes during the luteal phase of the estrous cycle. Theriogenology 49(3):536–546
Mader TL (2003) Environmental stress in confined beef cattle. J Anim Sci 81(14):E110–E119
Magiakou MA, Mastorakos G, Webster E, Chrousos GP (1997) The hypothalamic-pituitary-adrenal axis and the female reproductive system. Ann N Y Acad Sci 816:42–56
Marai IFM, El-Darawany AA, Fadiel A, Abdel-Hafez MAM (2007) Physiologic traits as affected by heat stress in sheep—a review. Small Rumin Res 71:1–12
Marai IFM, Habeeb AAM (1998) Adaptation of *Bos taurus* cattle under hot climate conditions. Ann Arid Zone 37(3):253–281
Marai IFM, Habeeb AAM, Farghaly HM (1999) Productive, physiologic and biochemical changes in imported and locally born friesian and holstein lactating cows under hot summer conditions of Egypt. Trop Anim Health Prod 31:233–243
Marai IFM, Abdel-Samee AM, El-gafaary MN (1991) Criteria of responses and adaptation to high temperature for reproductive and growth traits in rabbits. Options Mediterr Ser A 17:127–134
Marai IFM, Bhagat LB, Shalaby TH, Abdel-Hafez MA (2000) Fattening performances, some behavioral traits and physiologic reactions of male lambs fed concentrates mixture alone with or without natural clay, under hot summer of Egypt. Ann Arid Zone 39(4):449–460
Martin GB, Rogar J, Blache D (2004) Nutritional and environmental effects on reproduction in small ruminants. Reprod Fertil Dev 16:491–501
Maurya VP, Naqvi SMK, Joshi A, Mittal JP (2007) Effect of high temperature stress on physiologic responses of Malpura sheep. Indian J Anim Sci 77:1244–1247
Mc Manus C, Paludo GR, Louvandini H, Gugel R, Sasaki LCB, Paiva SR (2009) Heat tolerance in Brazilian sheep: physiologic and blood parameters. Trop Anim Health Pro 41:95–101
McEvoy TG, Robinson JJ, Aitken RP, Findlay PA, Palmer RM, Robertson IS (1995) Dietary-induced suppression of pre-ovulatory progesterone concentrations in superovulated ewes impairs the subsequent in vivo and in vitro development of their ova. Anim Reprod Sci 39:89–107
Minka NS, Ayo JO (2009) Physiologic responses of food animals to road transportation tress. Afr J Biotechnol 8(25):7415–7427
Moberg GP (1999) When does stress become distress? Lab Anim 28(4):422–426
Moberg GP (2000) Biologic responses to stress. Implications for animal welfare. In: Moberg GP, Mench JP (eds) Biology of animal stress. CAB International, Wallingford, UK
More T, Sahni KL (1979) Some haematological changes during three months after birth of Chokla lambs under semi-arid conditions. Indian Vet J 56:646–649
More T, Sahni KL (1980) Certain biochemical responses of sheep and goat exposed to thermal stress. Indian J Anim Sci 50:1012–1014
Morin LP (1986) Environment and hamster reproduction: responses to phase-specific starvation during estrous cycle. Am J Physiol 251:R663–R669
Morrison SR (1983) Ruminant heat stress: effect on production and means of alleviation. J Anim Sci 57:1594–1600
Naqvi SMK, Sejian V (2010) Physiologic adaptation of sheep to hot environmental conditions with special reference to climate change. In: Karim SA, Joshi A, Sankhyan SK, Shinde AK, Shakyawar DB, Naqvi SMK, Tripathi BN (eds) Climate change and stress management: sheep and goat production, 1st edn. Satish serial publishing house, Delhi, pp 259–282

Naqvi SMK, Hooda OK (1991) Influence of thermal, nutritional and exercise stress on some blood parameters of native and crossbred sheep. Indian J Anim Sci 61:660–662

Naqvi SMK, Joshi A, Das GK, Mittal JP (2001) Development and application of ovine reproductive technologies: an Indian experience. Small Rumin Res 39:199–208

Naqvi SMK, Rai AK (1991) Influence of dietary energy level on sheep for mutton during winter: effect on cardio respiratory responses, rectal temperature, some blood metabolites, enzymes and thyroidal hormones. Indian J Anim Sci 61:1126–1131

Naqvi SMK, Maurya VP, Gulyani R, Joshi A, Mittal JP (2004) The effect of thermal stress on superovulatory response and embryo production in bharat merino ewes. Small Rumin Res 55:57–63

Naqvi SMK (1987) Physiologic adaptation of sheep during energy crisis and thermal stress. Ph.D. thesis, Submitted to Kurukshetra University, Kurukshetra, India

Nazifi S, Saeb M, Rowghani E, Kaveh K (2003) The influence of thermal stress on serum biochemical parameters of Iranian fat-tailed sheep and their correlation with triiodothyronine (T_3), thyroxine (T_4) and cortisol concentrations. Comp Clin Pathol 12:135–139

National Research Council (NRC) (1981) Effect of environment on nutrient requirements of domestic animals. National Academy Press, Washington (Committee on animal nutrition. Subcommittee on environmental stress)

O'Callaghan D, Yaakub H, Hyttel P, Spicer LJ, Boland MP (2000) Effect of nutrition and superovulation on oocyte morphology, follicular fluid composition and systemic hormone concentrations in ewes. J Reprod Fertil 18:303–313

Otukonyong EE, Okutani F, Takahashi S, Murata T, Morioka N, Kaba H, Higuchi T (2000) Effect of food deprivation and leptin repletion on the plasma levels of estrogen (E2) and NADPH-d reactivity in the ventromedial and arcuate nuclei of hypothalamus in the female rats. Brain Res 887:70–79

Padua JT, Dasilva RG, Bottcher RW, Hoff SJ (1997) Effect of high environmental temperature on weight gain and food intake of Suffolk lambs reared in a tropical environment. In: Proceedings of 5th international symposium, Bloomington, Minnesota, USA, pp 809–815

Parr RA (1992) Nutrition-progesterone interactions during early pregnancy in sheep. Reprod Fertil Dev 4(3):297–300

Parr RA, Davis IF, Miles MA, Squires TJ (1993) Feed intake affects metabolic clearance rate of progesterone in sheep. Res Vet Sci 55:306–310

Pereira AMF, Baccari F Jr, Titto EAL, Almeida JAA (2008) Effect of thermal stress on physiologic parameters, feed intake and plasma TH concentration in alentejana, mertolenga, Frisian and limousine cattle breeds. Int J Biometeorol 52(3):199–208

Polkowska J (1996) Stress and nutritional influences on GnRH and somatostatin neuronal systems in the ewe. Acta Neurobiol Exp 56:797–806

Rae MT, Rhind SM, Miller DW, Brooks AN (2002) Maternal undernutrition alters triiodothyronine concentrations and pituitary response to GnRH in fetal sheep. J Endocrinol 173: 449–455

Rahardja DP, Toleng AL, Lestari VS (2011) Thermoregulation and water balance in fat-tailed sheep and kacang goat under sunlight exposure and water restriction in a hot and dry area. Animal. doi:10.1017/S1751731111000577

Rasooli A, Nouri M, Khadjeh GH, Rasekh A (2004) The influence of seasonal variations on thyroid activity and some biochemical parameters of cattle. Iran J Vet Res 5(2):1383–1391

Rhind SM, Mc Kelvey WAC, Mc Millen SR, Gunn RG, Elston DA (1989) Effect of restricted food intake, before and/or after mating, on the reproductive performance of Greyface ewes. Anim Prod 48:149–155

Riis PM, Madsen A (1985) Thyroxine concentration and secretion rates in relation to pregnancy, lactation and energy balance in goats. J Endocrinol 107:421–427

Rivest S, Rivier C (1995) The role of corticotrophin releasing factor and interleukin-1 in the regulation of neurons controlling reproductive functions. Endocr Rev 16:177–199

Rivington M, Matthews KB, Buchan K, Miller D, Russell G (2009) Investigating climate change impacts and adaptation options using integrated assessment methods. Aspects App Biol 93:85–92

Roberts AJ, Funston RN, Moss GE (2001) Insulin-like growth factor binding proteins in the bovine anterior pituitary. Endocrinol J 14:399–406

Robinson JJ (1996) Nutrition and reproduction. Anim Reprod Sci 42:25–34

Sano H, Takahashi K, Fujita M, Ambo K, Tsuda T (1979) Effect of environmental heat exposure on physiologic responses, blood constituents and parameters of blood glucose metabolism in sheep. Tohoku J Agric Res 30(2):76–86

Scaramuzzi RJ, Cambel BK, Downing JA, Kendall NR, Khalid M, Gutierrez M, Somchit A (2006) A review of the effect of supplementary nutrition in the ewe on the concentration of reproductive and metabolic hormones and the mechanism that regulate folliculogenesis and ovulation rate. Reprod Nutr Dev 46(4):339–354

Sejian V, Srivastava RS, Varshney VP (2008a) Pineal-adrenal relationship: modulating effects of glucocorticoids on pineal function to ameliorate thermal stress in Goats. Asian-Aust J Anim Sci 21:988–994

Sejian V, Sanyal S, Das PK, Ghosh PR, Sivakumar B, Pandiyan GDV (2008b) Effect of unilateral adrenalectomy on the blood biochemistry of black Bengal goats (*Capra hircus*). Turk J Vet Anim Sci 32(4):249–254

Sejian V, Srivastava RS (2009) Effects of melatonin on adrenal cortical functions of Indian goats under thermal stress. Vet Med Int. doi:10.4061/2010/348919

Sejian V, Srivastava RS, Varshney VP (2009) Effect of short-term thermal stress on biochemical profile in Marwari goats. Indian Vet J 87:503–504

Sejian V, Maurya VP, Naqvi SMK (2010a) Adaptability and growth of Malpura ewes subjected to thermal and nutritional stress. Trop Anim Health Prod 42:1763–1770

Sejian V, Maurya VP, Naqvi SMK (2010b) Adaptive capability as indicated by endocrine and biochemical responses of Malpura ewes subjected to combined stresses (thermal and nutritional) under semi-arid tropical environment. Int J Biometeorol 54:653–661

Sejian V, Maurya VP, Naqvi SMK (2011) Effect of thermal, nutritional and combined (thermal and nutritional) stresses on growth and reproductive performance of Malpura ewes under semi-arid tropical environment. J Anim Physiol Anim Nutr 95:252–258

Shelton M (2000) Reproductive perfoprmance of sheep exposed to hot environments. In: Malik RC, Razaqque MA, Al-Nasser AY (eds) Sheep production in hot and arid zones. Published by the Kuwait Institute for Scientific Research, pp 155–162

Slee J, Foster JE, Griffiths RG (1982) Thermolability in sheep induced by habituation to cold. J Therm Biol 7(3):131–181

Sosa C, Albecia JA, Forcada F, Vinoles C, Tasende C, Valares JA, Palacin I, Martin GB, Meikle A (2006) Effect of undernutrition on uterine progesterone and oestrogen receptors and on endocrine profiles during the ovine oestrous cycle. Reprod Fertil Dev 18:447–458

Sprott LR, Selk GE, Adams DC (2001) Review: factors affecting decisions on when to calve beef females. Prof Anim Sci 17:238–246

St-Pierre NR, Cobanov B, Schnitkey G (2003) Economic loss from heat stress by US livestock industries. J Dairy Sci 86(E Suppl):E52–E77

Stewart R, Oldham CM (1986) Feeding lupins for 4 days during the luteal phase can increase the ovulation rate. In: proceeding of the Australian society of animal production. Reviews of reproduction, vol 16, pp 367–370

Stratakis CA, Gold PW, Chrousos GP (1995) Neuroendocrinology of stress: implications for growth and development. Horm Res 43:162–167

Thwaites CJ (1971) Short term heat stress and embryo mortality in sheep. Aust J Exp Agri Anim Husb 11(50):265–267

Tilbrook AJ, Turne AI, Clark IJ (2000) Effect of stress on reproduction in non-rodent mammals: the role of glucocorticoids and sex differences. Rev Reprod 5:105–113

Tilbrook AJ, Turner AI, Clark IJ (2002) Stress and reproduction: central mechanisms and sex differences in non-rodent species. Stress 5(2):83–100

Todini I (2007) TH in small ruminants: effects of endogenous, environmental and nutritional factors. Animal 1:997–1008

Vinoles C, Forsberg M, Martin GB, Cajarville C, Repetto J, Meikle A (2005) Short-term nutritional supplementation of ewes in low condition affects follicle development due to an increase in glucose and metabolic hormones. Reproduction 129:299–309

Wingfield AS, Kitaysky JC (2002) Endocrine responses to unpredictable environmental events: stress or anti-stress hormone. Integr Comp Biol 42:600–609

Wooden KM, Walsberg GE (2002) Effect of environmental temperature on body temperature and metabolic heat production in a heterothermic rodent, *Spermophilus tereticaudus*. J Exp Biol 205:2099–2105

Yamamoto S, Ogura Y (1985) Variation in heart rate and relationship between heart rate and heat production of breeding Japanese black cattle. Jap J Livest Manage 3:109–118

Yousef HM (1985) Studies on some environmental factors affecting production and reproduction in some farm animals. M.Sc. thesis, Faculty of Agriculture, Zagazig University, Egypt

Part II
Management of Stresses

Chapter 7
Ameliorative Measures to Counteract Environmental Stresses

Veerasamy Sejian, Silvia Valtorta, Miriam Gallardo and Anoop Kumar Singh

Abstract The severity of heat stress on livestock production can be measured using certain weather parameters, which summate the intensity of heat stress exposure. Temperature-humidity-index (THI) is widely used to predict the severity of heat stress in domestic livestock. Reducing heat stress in livestock requires multidisciplinary approaches which emphasize animal nutrition, housing, and animal health. It is important to understand the livestock responses to environment, analyze them, in order to design modifications of nutritional and environmental management thereby improving animal comfort and performance. Researchers can attempt a variety of approaches to improve reproduction that involve modifying the environment (i.e., attempting to cool cows during reproduction), modifying the genetics of the animal (i.e., breeding of heat tolerant breeds), or intensifying reproductive management during periods of heat stress. Management alternatives, such as the strategic use of wind protection and bedding in the winter or sprinklers and shade in the summer, need to be considered to help livestock cope with adverse conditions. In addition to these changes, manipulation of diet energy density and intake may also be beneficial for livestock challenged by environmental conditions. Additionally, socio-economical status, technological tools, and financial infrastructure have instrumental roles in modifying environment stress. The ameliorative measures, to be incorporated, are therefore driven by socio-economical and environmental factors. This chapter addresses divergent sheltering methodologies that can be used to minimize the effects of multiple stresses, such as heat, nutritional, and health stress that an animal is exposed to.

V. Sejian (✉) · A. K. Singh
Adaptation Physiology Laboratory, Division of Physiology and Biochemistry,
Central Sheep and Wool Research Institute, Avikanagar, Jaipur,
Rajasthan 304501, India
e-mail: drsejian@gmail.com

S. Valtorta · M. Gallardo
Instituto de Patobiología INTA Castelar, 1712 Hurlingham, Buenos Aires, Argentina

V. Sejian et al. (eds.), *Environmental Stress and Amelioration in Livestock Production*, DOI: 10.1007/978-3-642-29205-7_7,

Keywords Cooling system • Fiber feeding • Shade • Shelter design • Ventilation

Contents

7.1 Introduction

The world's climate may be considered to consist of tropical, semitropical, temperate, and arctic zones/types. Each of these climatic zones represents complexes of physical or meteorological and biological factors, which constitute the natural environment. Each climate complex encompasses a multitude of plant and animal species that have evolved successfully by adaptive processes resulting from interactions with the meteorological environment (Kumar 2010). An environmental profile includes such factors as daily, seasonal, maximal and minimal temperatures, humidity, wind, radiation, length of season, and a biological assessment such as quality and quantity of protein and energy available. The physiological profiles are assessments of selected physiological processes, such as thermoregulation, hormonal balance, water balance, and energy balance. The fitting of the physiological or production profile to the environmental profile would be one basis for a 'bioclimatic' index. An animal residing in a region with a suitable bioclimatic index would be an individual with high physiological predictability for production of milk, growth, and reproduction under these environmental conditions. The application of meteorological and physiological principles to animal selection and management practices will increase availability of animal protein, especially in climatic-limiting zones of the world.

Heat stress is defined as the sum of forces external to a homeothermic animal that acts to displace body temperature from the resting state (Yousef 1984). Such stresses can disrupt the physiology and productive performance of an animal (West 2003). The increase in body temperature caused by heat stress has direct, adverse consequences on cellular function (Hansen and Arechiga 1999). Production losses in domestic animals are largely attributed to increases in maintenance requirements associated with sustaining constant body temperature, and altered feed intake (Mader et al. 2002; Davis et al. 2003; Mader and Davis 2004). Depending upon the intensity and duration of environmental stress, voluntary feed intake can average as much as 30% above normal under cold conditions, to as much as 50% below normal under hot conditions.

Increases in air temperature reduce livestock production during the summer seasons which may be partially offset during the winter season (Kadzere et al. 2002). Current management systems for ruminants do not usually provide as much shelter to buffer the effects of adverse weather conditions as for nonruminants. From that perspective, environmental management for ruminants exposed to global warming must consider: (1) general increases in temperature; (2) increases in night time temperatures; and (3) increases in the occurrence of extreme events (e.g., hotter daily maximum temperature and more/longer heat waves) (Nienaber and Hahn 2007).

Adaptation and mitigation strategies should contribute to reduced poverty and at the same time must benefit the most vulnerable communities without harming the environment (Das 2004). Information about climate change impacts, vulnerability patterns, coping, and adaptive capacity as well as facilitating location-specific adaptation and mitigation practices are of central concern.

7.2 Measurement of Severity of Heat Stress

The severity of heat stress on livestock production can be measured using certain weather parameters, which summate the intensity of heat stress exposure. The most noteworthy weather parameters included in such heat stress severity measurements are ambient temperature (including dry and wet bulb temperature) and relative humidity (Nienaber et al. 1999). Mathematical formulae have been developed by a variety of investigators to measure the severity of heat stress. Table 7.1 describes the different temperature-humidity-index (THI) used for evaluating heat stress effect on livestock.

All of these formulae calculate the heat stress severity in terms of scores on a 100 scale and compared on a standard THI chart to asses the stress imposed by the impending climatic condition on livestock over a period of time (Berman 2005). The THI has been used to represent thermal stress due to combined effects of air temperature and humidity and THI is used as a weather safety index to monitor and prevent heat stress-related losses (National Research Council 1971). Different livestock species have different sensitivity to ambient temperature and humidity. The capacity to tolerate heat stress is much higher in native breeds, particularly under higher temperatures at lower humidity than crossbred animals. This is mainly due to the fact that native breeds can dissipate excessive heat more effectively by sweating, whereas crossbreeds have reduced ability to sweat (Nienaber and Hahn 2007). The THI is used as a guide to measure heat stress by combining the effects of temperature and humidity into one value (Marai et al. 2001). There are three stress categories (temperatures given in Fahrenheit): Livestock alert is 75–78 degrees,

Table 7.1 Commonly used indices for evaluating heat stress effect on domestic animals

S.No.	THI indices	References
1	tdb + 0.36tdp + 41.5	Thom (1958)
2	(0.8 × Amb tem) + {[(RH/100) × (Amb tem − 14.4)} + 46.4	Thom (1959)
3	[0.4 × (Tdb + Twb)] × 1.8 + 32 + 15	Thom (1959)
4	(0.15 × Tdb + 0.85 × Twb) × 1.8 + 32	Bianca (1962)
5	(0.35 × Tdb + 0.65 × Twb) × 1.8 + 3.2	Bianca (1962)
8	0.55DB + 0.2DP + 17.5	Johnson (1965)
9	(Tdb + Twb) × 0.72 + 40.6	National Research Council (1971)
10	(0.55 × Tdb + 0.2 × Tdp) × 1.8 + 32 + 17.5	National Research Council (1971)
11	0.72(Td + Tdp) + 40.6	McDowell (1972)
13	Tdb + (0.36xTdp) + 41.2	Yousef (1985)
14	0.72(Td + Tw) + 40.6	World Mateorological Organization (1989)
15	db°F − {(0.55 − 0.55RH) (db°F − 58)}	LPHSI (1990)
16	(9/5 Temp°C + 32) − (11/2 − 11/2 × Humidity) × (9/5 Temp°C − 26)	Ravagnolo and Misztal (2000)
17	db°C − {(0.31 − 0.31RH) (db°C − 14.4)}	Marai et al. (2001)
18	t − {(0.31 − 0.31RH) (t − 14.4)}	Marai et al. (2002)
19	(0.85 × DBT + 0.15 × WBT) × $V^{-0.058}$	Tao and Xin (2003)
20	t air − [0.55 − (0.55 × RH/100)] × (t air − 58.8)	Oklahoma State University (2003)
21	Temp − (0.55 − (0.55 × (RH/100) × (Temp − 58)	Amundson et al. (2005)
22	BGHI = tbg + 0.36 tdp + 41.5	Buffington et al. (1981)
23	THVI = (0.85 × DBT + 0.15 × WBT) × $V^{-0.058}$	Tao and Xin (2003)
24	WBGTI = (0.7Twb) + (0.2Tbg) + (0.1Tdb)	Al-Tamimi (2005)
25	$HLI_{BG > 25}$ = 8.62 + (0.38 × RH) + (1.55 × BG temp) − (0.5 × wind speed) + ($e^{2.4 - \text{wind speed}}$)	Gaughan et al. (2008)
26	$HLI_{BG < 25}$ = 10.66 + (0.28 × RH) + (1.3 × BG) − wind speed	Gaughan et al. (2008)

Abbreviations: *THI* Temperature-humidity-index; *tdp* Dry bulb temperature; *twb* Wet bulb temperature; *tbg* Black globe temperature; *Tdp* Dew point temperature; *Tw* Wet bulb temperature; *Td* Dry bulb temperature; *RH* Relative humidity; *T air* Air temperature; *HLI* Heat Load index; *BG temp* Black globe temperature; *WBGTI* Wind black globe temperature index; *THVI* Temperature humidity and velocity index; *DBT* Dry bulb temperature; *WBT* Wet bulb temperature; *BGHI* Black globe humidity index

Livestock danger is 79–83 degrees, and Livestock emergency is 84+ degrees. Table 7.2 describes the different THI categories in *Bos taurus* cattle.

The higher the humidity the lower is the temperature for livestock alerts; for danger and emergency levels to occur. The two limitations in THI index are that it does not take into account the wind velocity and solar radiation. These two factors are vital meteorological variables that influence animal performance under semi-arid tropical environmental conditions. However, THI can be a reliable indicator for measurement of quanta of stress in livestock. Modifications to the THI have

Table 7.2 Different THI categories in *Bos taurus* cattle

THI category	Descriptive characteristics			
	Duration	THI[a] − h ≥ 79 (day)	THI[a] − h ≥ 84	Nighttime recovery (h # 72 THI[a])
1. Slight	Limited: 3–4 days	10–25	None	Good: 5–10 h/night
2. Mild	Limited: 3–4 days	18–40	#5/day	Some: 3–8 h/night
3. Moderate	More persistent (4–6 days usually)	25–50	#6/day	Reduced: 1–6 h/night
4. Strong	Increased Persistence (5–7 days)	33–65	#6/day	Limited: 0–4 h/night
5. Severe	Very Persistent (usually 6–8 days)	40–80	3–15/day on 3 or more successive days	Very limited: 0–2 h/night
6. Extreme	Very Persistent (usually 6–10 + days)	50–100	15–30/day on 3 or more successive days	Nil: #1 for 3 or more successive days

[a] Temperature-Humidity-Index (*THI*). Daily THI − h are the summation of the differences between the THI and the base level at each hr of the day. For example, if the THI value at 1,300 is 86.5 and the base level selected is 84, THI − h = 2.5. The accumu-lation for the day is obtained by summing all THI − h ≥ 84, and can exceed 24

Source Hahn et al. (1999)

been proposed to overcome shortcomings related to airflow and radiation heat loads. Based on recent research, Mader et al. (2006) and Eigenberg et al. (2005) have proposed corrections to the THI for use with feedlot cattle, based on measures of wind speed and solar radiation. While differences in the proposed adjustment factors are substantial, there are marked differences in the types and number of animals used in these two studies. Nevertheless, these approaches appear to merit further research to establish acceptable THI corrections, perhaps for a variety of animal parameters. Gaughan et al. (2002) developed a heat load index (HLI) as a guide to management of unshaded *Bos taurus* feedlot cattle during hot weather (>28°C). The HLI was developed following observation of behavioral responses (respiration rate and panting score) and changes in dry matter intake (DMI) during prevailing thermal conditions (Mader et al. 2006). The HLI is based on humidity, wind speed, and predicted black globe temperatures.

As a result of its broadly demonstrated success, the THI is currently the most widely accepted thermal index used for guidance of strategic and tactical decisions in animal management from moderate to hot conditions. Developing climatic indices of summer weather extremes (in particular, heat waves) for specific locations also provide livestock managers with information about how often those extremes (with possible associated death losses) might occur (Hahn et al. 2001). Panting score is one observation method used to monitor heat stress in cattle

(Mader et al. 2006). As temperature increases, cattle pant more to increase evaporative cooling. Respiratory dynamics change as ambient conditions change, and surrounding surfaces warm. This is a relatively easy method for assessing genotype differences and determining breed acclimatization rates to higher temperatures. In addition, shivering score or indices also have potential for use as thermal indicators of cold stress.

7.3 Measurement of Thermal Adaptability

The evaluation of adaptation to heat stress, based upon the lowest rectal temperature, respiratory frequency, and physiological variables as the main parameters under high temperatures, was found to be insufficient. The average relative deviations (ARD) from normal (regardless either positive or negative), due to exposure to hot climates, in thermal, water, and/or nitrogen balances of the animals (or in all traits measured), could be used in the estimation of parameters for detection of adaptability to hot climates (Habeeb et al. 1997; Marai and Habeeb 1998) as follows:

$$\text{Adaptability}\,(\%) = [100 - \text{ARD}] \times 100$$

In males, El-Darawany (1999) and Marai et al. (2006) used tunica dartos indices (TDI) to measure the ability of the male to tolerate increased ambient temperatures. The scrotum actively controls its own temperature through the function of the tunica dartos muscle which is interpreted as the distance between the testes and the abdominal wall. This muscle thus defines the magnitude of vascular heat exchange and is performed by the contraction of the tunica dartos muscle of the scrotum pulling the testes toward the body to increase its warmth, when the environmental temperature is low. During high ambient temperatures, the reverse process occurs in dissipating of the excess heat as much as possible from the testes (Taylor and Bogart 1988). The TDI can be used as an index to measure the ability of male to tolerate increased ambient temperatures, as it reflects the magnitude of vascular heat exchange.

7.4 Approaches for Alleviating Thermal Stress

Reducing heat stress on livestock requires multidisciplinary approaches which emphasize animal nutrition, housing, and animal health (Collier et al. 2003). Some of the biotechnological options may also be used to reduce thermal stress. It is important to understand the livestock responses to environment, analyze them, in order to design modifications of nutritional and environmental management thereby improving animal comfort and performance.

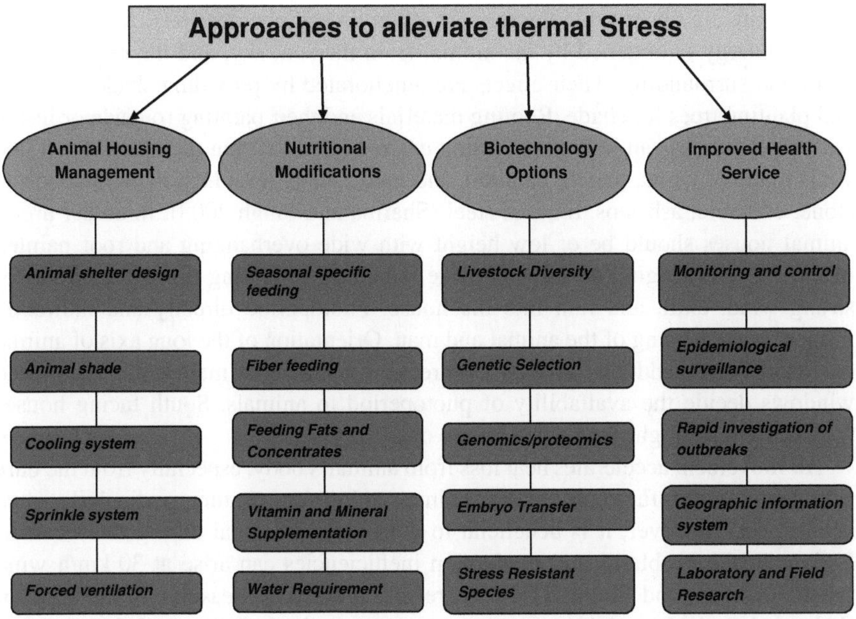

Fig. 7.1 Different approaches for alleviating thermal stress in livestock. The figure describes the various mitigation strategies to reduce the impact of heat stress in livestock. The strategies comprises of animal housing management, nutritional modifications, biotechnology options, and improved health services

So a range of technologies are needed, to match the different economic and other needs of small holders. Figure 7.1 describes the various approaches for ameliorating adverse effects of thermal stress on livestock.

7.4.1 Importance of Livestock Housing

Animal housing has been a matter of concern since livestock were domesticated. Its basic aim is to moderate the range of animal's microclimate and to optimize their production by protecting them from climate extremes. Consequently, animal housing should provide the 5 Freedoms (eating, resting, moving, voiding and respiring, and also comfort) to all types of stresses (climate, social, nutritional, and disease). House cannot be constructed frequently according to climatic needs. Hence, serious consideration of site selection, housing design, and future requirements for the livestock species to be reared at a particular locality are necessary (Sharma and Singh 2008). Loose housing with the provision of easy movement to and from sun in comparison to tie stall barn systems have been found appropriate to specific livestock species. Ambient temperature is the most important climatic variable affecting the design of the housing (Collier et al. 2006). The effects of temperature are observed in the form of radiation and photoperiods. About 50% of radiant energy is obtained by the animal from the sun, sky, and the rest from the earth and surrounding. Their effects are ameliorated by providing shade, shelters, and planting trees for shade. Roofing materials and their painting (outside or inside with black/white) can help in providing desired effects within the house. Roof and walls made of wood, thatch, bamboo, and mud radiate least heat in comparison to stone, concrete, asbestos, tin, and steel (Sharma and Singh 2008). In colder areas, animal houses should be of low height with wide overhanging and roof painted black. An overhanging of sufficient size is helpful in cutting down the effects of strong wind, chill, and rain into the house. Photoperiod directly and indirectly influences well-being of the animal and man. Orientation of the long axis of animal enclosures and paddocks as well as direction and size of main entry (gate) and windows decide the availability of photoperiod to animals. South facing houses receive more sunlight than north facing ones.

Air movement accelerates heat loss from animal's body, especially from the bare skin. Wind is harmful to unsheltered animals at low temperature, particularly when skin is wet. However, it is beneficial to nonsweating animal during hot weather. Serious health problems and production inefficiencies can arise at 30 km/h wind speeds (Sharma and Singh 2008). Therefore, protective measures in the form of shelter in hot, dry hot, and in temperate areas are required (Collier et al. 2006). Wind velocity is affected by local topography. In hill areas, the contours opposed to directional flow provides upward currents on the windward side. On steep slopes, reversed eddy currents over the crest and calm at the base are observed. An area on the base of leeward side is always better in areas of high wind velocity. In valley's, the forces of air are increased in core areas while on ridges the force is more at the peak. Therefore, selection of sites away from the core area and in depressions are useful in valley and ridge areas respectively (Sharma and Singh 2008).

The effects of humidity, both low (rapid evaporation, skin irritation, and general dehydration) and high (heat stagnation in confinement) are harmful to animals. In humid areas, animal housing should be open, having high roofs (spacious), airy and well-ventilated (von Borell 2001). Rainfall has direct and indirect effects on livestock production. Heavy rains increase humidity, reduces forage quality, grazing, and feed intake. Heavy rainfall areas are less suitable for animals possessing heavy coats. In such areas, the stilted, loose, airy, and ventilated and protective (from ecto-parasites) housing should be constructed as raised platforms.

7.4.2 Animal Shelter Design for Comfort

Physical modifications of the environment are based upon two basic concepts-protection of livestock from the factors contributing to heat stress, and enhancing evaporative heat loss by animals.

Differing environmental modification systems are classified according to their impacts on animal production and performance (Hahn 1981). One type are those systems that mitigate heat gain by radiation interception. Among the latter there is a range from forced air ventilation systems and those sprinkling or misting water, each separately, or some combination of these two systems, including evaporative cooling, and air conditioning.The various options are classified as follows:

1. Protective Methods
   - Natural shades
   - Artificial shade
2. Cooling methods

   2.1 Direct (cooling of the animal)

- Wetting the animal
- Forced ventilation
- Combination of ventilation and wetting

   2.2 Indirect (air cooling)

- Nebulizers
- Cross ventilation
- Wind tunnel
- Air conditioning

7.4.2.1 Protective Methods: Using Shades

The most obvious environmental modification to reduce the thermal stress of livestock during hot climate condition is the use of shade (Blackshaw and Blackshaw 1994). Shelter management is one of the key techniques for reduction of heat stress. Shade against direct solar radiation can be provided by trees like the banyan (*Ficus benghalensis*) or shelters made of straw and other locally available materials. Animals kept in or outside during summer are comfortable under tree shades that protect them from direct sunlight during peak hours of the day (Buffington et al. 1983). Tree shades provide an effective shelter to animals and moreover, plantation and forestation are beneficial to animals, humans, and the environment.

However, under grazing conditions the use of natural shades is not always appropriate, and is the reason that the use of artificial shades has come into use. Table 7.3 outlines comparative characteristics of natural and artificial shade systems.

Under grazing conditions, the artificial shades have proven effective in improving animal comfort and improving milk production (Valtorta et al. 1996, 1997). In this regard, at INTA Rafaela, Brondino et al. (2008) published a series of recommendations concerning design and construction of temporary housing facilities for animals, which are applicable to the conditions of the central dairy area of Argentina. In the case of permanent confinement systems, such as drylot or

Table 7.3 Some comparative differences between tree and artificial shade (*Adapted from* Valtorta and Gallardo 2004)

Characteristic	Tree shade	Artificial shade[a]
Shadow uniformity	Variable	High
Flooring	Natural soil	Consolidated
Resistance to ponding	Variable according to soil type	Good
Management of availability per animal	Complex	Simple
Availability after planning	Far	Immediate

[a] Considering a structure built following the proper design recommendations

free stall, earth movement to generate slopes is particularly important because of the volume of effluent to be handled routinely (Cook and Nordlund 2004).

Regardless of the management system, an appropriate measure to improve cow comfort is shades for holding pens. It should also be noted that milk production is a crucial period, during which there are a series of hormonal changes that lead to the ejection of milk. Any stressors during this period may alter this process. In the context of providing shade for holding pens, the results of the studies conducted in Argentina demonstrated (1) increased comfort as evidenced by the technical guidance note, which, at the time of maximum air temperature, reached 44°C in sunlight and under nets; (2) lower temperature of concrete floors, which at 1,500 h, reached an average of 52°C without shading and 27°C under shades. Shade is also important during evening milking due to marked increases in solar radiation, in addition to heat emitted by extremely hot floors and crowding of animals in confined areas (Schütz et al. 2010).

Properly designed shade structures provide adequate protection to livestock in the heat of summer and in winter (Armstrong 1994). For tropical climatic conditions, loose housing systems are considered most appropriate (Hahn 1981). The longer sides of the shelter should have an east–west orientation (Schütz et al. 2010). This orientation, reduces encroachment of direct sunlight shining on the side walls or when entering the shelter (Ugurlu and Uzal 2010). However, if mud is an issue, then north to south orientation will increase drying as shade moves across the ground during the day. In such situations animal shelters should be shielded from direct sunlight as much as possible by means of side covers of gunny bags or thatch (Schütz et al. 2010). In addition, the roof can be extended with additional shade material, and vertical shades moved to the outsides of the roof. Such devices result in improved protection from direct solar radiation and sun. The west side of the shed can also be protected similarly and fitted with side covers of gunny bags or curtains. The height of the shed structure should be greater than 2.4 m tall to allow sufficient air movement underneath the shade (Schütz et al. 2010). However, tall structures (more than 30.5 m high) are not economically viable.

Other ways to remove hot air trapped under the roof, inside the shed include outer coverings and shading, perhaps combined with a roof spray, are popular greenhouse technologies which can be applied in the structural design of shed to reduce the height of the shed. There are also a variety of roof systems giving improved natural ventilation by means of roof openings, enhanced solar chimney effect, etc. (Schütz et al. 2010). In addition, painting the roof white increases reflection of sunlight thus reducing the amount of solar energy absorbed. Increasing air flow is another important component for effective ventilation in animal shelters. In an animal shed air movement should be free in all areas of the shed (Bryant et al. 2007). To increase air flow, two main approaches may be used. One is the installation of fans so that air movement is increased, and the second is to provide open sides of the shed. In many cases walls of the shed may be made partially of concrete. In such situation, opening the lower level of the barn to increase air flow is not an option, so the addition of fans is essential to increase air circulation (Bryant et al. 2007). In sheds where sheet metal is used for walls, it may be practical to remove the side and install netting over these areas. The netting can be raised to increase air flow during the summer and lowered during the winter. Increasing the roof venting is yet another option that may be used for animal sheds. Another approach which can be used, similar to housing used for human, is the double wall approach. In this strategy, extra wall layers at either end of shed is constructed and the outer layer is kept 10 cm away from the inside wall, with vertical openings at both ends. The bottom opening allows cold air to enter, and the upper opening allows hot air to exit. The 10 cm of air between the two walls provides a thermal barrier to prevent conductive thermal energy from entering the animal house (Bryant et al. 2007). The double wall approach is an effective and proven technology in structural design for increasing or reducing ambient temperatures.

7.4.2.2 Cooling Methods

In addition to providing adequate shade, use of water as a cooling agent is an effective method for reducing heat stress of livestock particularly under dry heat/lower humidity and to keep their body temperature as cool as possible. Evaporative cooling can be accomplished by two approaches: (1) direct evaporation from the skin surface of the animal and (2) indirect evaporation involving cooling the microenvironment of the animals with cooling pads and fans in an enclosed shed.

Direct Methods

Wetting the Animal

Some studies report beneficial effects of wetting or misting animals through sprinklers during periods of high temperature. In Mexico, a region of sub-humid tropical climate, there was a 7% increase in milk production in cows sprinkled between 12:00 and 13:00 h under the shade. Also, positive results were reported in few studies conducted in Missouri when animals were sprinkled between 11:00 and 17:30 during moderate summer. In Israel, when cows were sprinkled for 1 h, four times a day, it relieved heat stress as evident from the increased milk production. In Australia when the cows were watered every time the temperature exceeded 26°C, similar effects were observed (Davison et al. 1996).

Sprinkling maximizes the amount of heat removed from the animal through evaporative cooling at reduced water costs (Gaughan et al. 2008). In addition, the ambient air temperature is lowered in the area immediately surrounding the animal, increasing the heat gradient and increasing the effectiveness of non-evaporative cooling mechanism (Morrison et al. 1981). To achieve adequate heat loss when sprinkling, droplets must wet the hides of the animal as accumulation of water in the hair may increase the humidity around the animal and reduce effective heat loss (Means et al. 1992). High pressure irrigation-type sprinklers can improve inexpensive wetting of animals, especially when coupled with fans, to increase air movement (Nienaber and Hahn 2007). However, cooled animals have limited ability to adapt to warm condition and may become reliant on sprinkling to keep cool even in milder conditions (Gaughan et al. 2008). Cessation of sprinkling during the day on hot days may increase the heat load in cattle, even though ambient temperature and humidity may decrease (Gaughan et al. 2008). Altering the microclimate of the sprinkled area helps in improving the well-being of feedlot cattle under extreme environment condition by reducing body temperature. Therefore, one point which should be kept in mind while using evaporative cooling system in hot and humid subtropical region is that cooling requires the use of forced ventilation.

Forced Ventilation

When there is no sufficient natural ventilation, the layer of air closest to the animal, warms. This reduces the rate of heat dissipation from the surface of the animal to the environment. The role of fans is to increase heat loss by convection. Therefore, ventilation is effective when the air temperature is lower than body temperature.

Fans should be located mainly in:

- The holding pen
- The milking parlor
- The resting pen (if any)
- The feeding area, the flow being directed toward the back of the animals.

Fans can reduce the body temperature by 0.3–0.4°C, provided that the temperature of the provided air is lower than the surface temperature of the animal. However, fans are not sufficient to reduce conditions of heat stress in high producing dairy cows in hot weather (Mader et al. 2007).

However, when working in high humidity environments, consideration should be given to any system introducing water into the environment as it may produce negative effects because water will evaporate, further increasing the humidity and saturating the air. It is for this reason that evaporative systems are designed to combine forced ventilation with wetting.

Combination of Forced Ventilation and Wetting

This system is based on the most important route of heat loss, which is evaporation from the surface of the skin under high temperature conditions.

It is important to consider how heat loss is enhanced on combining sprinkling and forced ventilation. Each gram of water evaporated from the skin of the animal represents a loss of 56 calories. However, there are large differences in the amount of water that evaporates through differing mechanisms:

- Passive diffusion, or perspiration, evaporating about 30 g/h, representing 16.8 kcal/h
- Active transpiration evaporates 170 g/h, equivalent to 95.2 kcal/h
- Wetting + forced ventilation evaporates 1,000 g/h, which means a loss of 560 kcal/h

This system is effective in all types of weather (dry and humid), since the forced high speed of the air allows drying of cows and prevents air saturation.

The combination of sprinklers and fans are suitable for both confined and grazing animals. In the latter case it can be implemented in holding pens.

Indirect Cooling Methods

Nebulizers

Nebulizers are a low-cost evaporative cooling system, characteristic of poultry houses. This process is based upon producing mists of fine droplets that must evaporate before reaching the ground, so as to cool the air in contact with animals. Some of these droplets may be deposited on the surface of the animal coat, which could alter insulative properties (Hahn 1985). However, if nebulization ensures substantial air movement, the system could lead to improvements in the environment for milk production in cows (Armstrong 1994).

Cross Ventilation

In this system, one of the side walls of the barn consists of a large refrigerated water panel, while on the opposite wall there are large pumps, as those used in poultry barns. These pumps expel the stale indoor air and force it to enter through the chilled panel, thus air-conditioning the barn.

Wind Tunnel

Wind tunnels are characterized by air inlets in one end of the barn and exhaust at the other end (Smith et al. 2006). This technology is based on the principle of increasing evaporative heat loss by removing excess heat and humidity of the air directly in contact with animals. It can provide adequate cooling in temperate climates (Stowell et al. 2001), but should be combined with other methods of cooling in warmer environments. Significant effects of the combination of evaporative cooling through the wind tunnel effect in swine and poultry facilities have been reported. Brouk et al. (2003) used the combination of wind tunnel with evaporative cells to cool dairy cows and reported significant reductions in rectal temperature and breathing rates during the afternoon and evening. Smith et al. (2006) also observed an increase of 2.4 kg milk/cow/day with this cooling system in the southeastern United States, where both temperature and humidity are high.

Air Conditioning

The use of this cooling system, for 24 h, produced a 10% increase in milk production in subtropical environments (Collier et al. 2006; West 2003). However, the costs associated with air-conditioning, together with the facilities necessary to provide a closed environment, or the conditioning ducts for zonal cooling, have made this technology a failure.

Order of Priorities for the Cooling of Cows

According to Catalá (2010), when different categories of animals are handled separately, the order of priority for cooling is the following:

1. Fresh cows (first 3 weeks postpartum)
2. Close pre-partum cows (3 weeks prepartum)
3. High producing cows (first 100 days)
4. Cows in mid lactation (100–200 days)
5. Dry cows (from dry to 3 weeks before delivery)
6. Late lactation cows (over 200 days)

An important aspect to consider is how long it takes to amortize the investment made in cooling systems. In intensive systems (free-stalls or dry-lot) this aspect has been very well-studied. In the U.S. and Israel, they believe that if the annual production increases 5–10%, the return on investment is between 2 and 3 years. If the increase is 20% the return on investment is 1 year (Catalá 2010). In pastoral or mixed systems, the analysis must consider that the cooling systems will not have permanent use. In many cases, this makes it difficult to determine the incidence of the costs involved in the cooling system.

A majority of livestock are kept by the smallholder farmers. Hence the above-mentioned protective measures may be unaffordable under most of the small and marginal farming conditions. Then alternative cost-effective systems can also be used like water application or sprinkling before milking, wallowing in case of buffaloes and low cost, renewable energy operated evaporative cooling systems can be used. So combinations of fans, wetting, shed, and well-designed housing can help alleviate the negative impact of high temperatures on animals in the tropics.

7.4.3 Nutritional Modification to Combat Heat Stress

7.4.3.1 Feed Requirement

During hot dry summer there is decrease in dietary feed intake which is responsible for reduced productivity. In these situations, the efficient practical approaches like frequent feeding, improved forage quality, use of palatable feeds, good nutritient balance, and greater nutrient density are required (Beede and Shearer. 1996). However, feeding excessive quantities of nutrients, like crude protein, can contribute to reduced efficiency of energy utilization, potentially adding to stress levels. Likewise if less forage is consumed, and the forage is high in quality, the cows' rumination activity may decrease. So, a through understating of dietary modification to minimize heat stress is necessary (NRC 2001).

Heat production from feed intake peaks 4–6 h after feeding. Therefore heat production in animal feed in the morning will peak in the middle of the day when environmental temperature is also elevated (Brosh et al. 1998). Consequently, it has been suggested that feeding animals later in the day prevents the coincidence of peak metabolic and environmental heat loads (Reinhardt and Brandt 1994; Brosh et al. 1998). Furthermore, limiting energy intake can effectively decrease basal metabolic heat production (Carstens et al. 1989) and therefore decrease total metabolic heat load of animals subjected to high environmental temperature.

7.4.3.2 Concept of Cold Diets

A cold diet is one that generates a high proportion of net nutrients for the synthesis and decreases heat generated during fermentation and metabolism. The salient features of a cold diet are:

1. Higher energy content per unit volume
2. More digestible fiber
3. Effective fiber (NDFef)
4. Lower protein degradability
5. More bypass nutrients.

In contrast, hot diets are characterized by marked imbalances between the basic nutrients: energy and protein. In general terms, hot diets may have a high proportion of undigestible fiber, accompanied in some cases with low protein concentration and/or energy. Also, they may be characterized by low NDFef with highly degradable protein, in relation to the amount of carbohydrates available in rumen.

In other cases, there are hot diets with a high proportion of rapidly degradable carbohydrates (starch and soluble sugars) in relation to the available nitrogen in rumen. These diets, in which lack of synchronization between nutrients, lead inevitably to lower conversion efficiency. Table 7.4 describes the differences between the two types of diet discussed.

During hot weather, declining DMI and high locational demand requires increased dietary mineral concentrations and further minerals are more easily depleted during hot summer months (Collier and Beede 1985). The increase in respiration will cause excessive water losses, thereby reducing mineral concentrations. As a result, mineral should be made available, 24 h a day during the summer. Potassium, sodium, magnesium, copper, selenium, zinc, and phosphorus levels should be supplied in the feed. Nutritional tools, such as antioxidant feeding (Vit-A, selenium, zinc etc.) and ruminant specific live yeast cultures can help in

Table 7.4 Characteristics of cold and hot diets

Diet characteristics	Cold	Hot
Digestibility	High	Low
Fiber	Low	High
Proteins	Low degradability	High Degradability
Minerals	High Na^+ and K^+	Low Na^+ and K^+
Examples	Tender pastures, high grain silages, high fat concentrates	Mature Pasture, Fibrous hays and silages, High fiber concentrates

protecting the animals against heat stress (Nayyar and Jindal 2010). Studies showing the addition of antioxidants in the diets of cows are able to reduce heat stress apart from limiting mastitis, optimizing feed intake, and reducing the negative impact of heat stress on milk production. Moreover, the use of antioxidants, such as Vit-E, Vit-A, selenium, and selenium enriched yeast help reduce the impact of heat stress on the redox balance, resulting in improved milk quality and cow health. A recent study in cattle showed that the supplementation of Vit-E helps in reducing the heat stress and improves the antioxidant status and lowers the incidence of mastitis, metritis, and retention of placenta (Sathya et al. 2007). During periods of heat stress, the incidence of rumen acidosis is increased particularly in high producing cows maintained on high concentrate diets (West 2003). Factors contributing to rumen acidosis problems in cow are related to DMI decreases, particularly lower forage intake and higher levels of fermentable carbohydrates (Patra 2007). Decreased rumination and decreased salivary activity reduces the buffering capacity of the rumen. Lowered rumen pH associated with subacute neous rumen acidosis (SARA) impairs fiber digestion efficiency due to pH effects on rumen fibrolytic bacteria (Krause et al. 2009). Table 7.5 describes the general feeding management practices to be followed for ruminants in hot climate.

7.4.3.3 Fiber Feeding

Because there is greater heat production associated with metabolism of acetate compared with propionate, there is a logical rationale for the practice of feeding low fiber rations during hot weather. Feeding more concentrate at the expense of fibrous ingredients increases ration energy density and reduces heat increment (Magdub et al. 1982). Altered proportions of ruminal volatile fatty acids (VFA) may explain a part of the differences in heat increment with fiber feeding during heat stress (Beatty 2005). VFA constitute a large proportion of the energy available to the cow, and declining intakes during heat stress reduces the quantity of VFA in the rumen because fermentable carbohydrate is reduced. Increased feeding of concentrates is a common practice during conditions conducive to heat stress, but maximal benefit from concentrates appears to be approximately 60–65% of the diet (Coppock 1985). Excessive concentrate feeding leads to acidosis and the

Table 7.5 General feeding management of ruminants in hot climate

Increase number of feeding
Feed will be fresh
Stimulate cows to come to manger
Select suitable time of feeding
Having fresh feed after milking
Feeding more feed during night time
Feeding a total maintenance ration (TMR)
Well-balanced combination for heat stress cows
Fiber is limited
Only good quality roughage is used
Provide good condition for fermentation
Under shade managers are preferable

associated production, health, and metabolic difficulties. The large amount of highly fermentable carbohydrate fed in typical high-concentrate diets should minimize the heat production observed in the very high fiber diets, which were used in research settings (Mader et al. 2002). Although high fiber diets contribute to heat stress, the level of intake is far more critical to the total amount of metabolic heat produced. Growing heifers fed pelleted rations containing 75% alfalfa or 25% alfalfa produced 48.8 and 45.5 MJ/d of heat (Reynolds et al. 1991). However, when the low and high intake (4.2 and 7.1 kg/d DMI) heifers were compared, heat production was 38.2 and 56.1 MJ/d. Therefore, intake effects have a substantial effect on heat production and must be considered in designing an effective nutritional and environmental management program. Research suggests that lower fiber, high grain diets may indeed reduce metabolic heat production and contribute to lower heat load in the animal (Holt et al. 2004). Further, low fiber, high grain diets provide more efficiently used end products, which contribute to lower dietary heat increments (Mader et al. 2002). However, low fiber, high grain diets must be balanced with the need for adequate fiber to promote chewing and rumination to maintain ruminal pH and cow health (Beatty 2005).

7.4.3.4 The Role of Effective Fiber

One of the main components of the dairy cow diet is fiber. The importance of fiber is that it is necessary for:

1. Adequate rumination activity (through the flow of adequate saliva)
2. Appropriate relationships between the main products of rumen fermentation, volatile fatty acids
3. Regulatory capacity of the ruminal acidity (ability to buffer or buffer capacity)
4. Modulation of the rate of passage and digestion of small particles in the ration.

The fiber in forage represents the plant cell wall and is determined in the laboratory as the component called neutral detergent fiber (NDF) (Holt et al. 2004).

Research pertaining to fiber requirement for animals must take into account not only its chemical function, such as nutrient precursor of VFA, but also its mechanical action (Mader et al. 2002). The NDFef is the fraction of NDF that affects chewing, rumination, insalivation, rumen pH, and movements (mixing cycle) (Holt et al. 2004). Because of the important functions of the effective fiber and the negative effects of heat on rumen function, cows are more likely to have subclinical ruminal acidosis during summer, especially when receiving rations with low forage: concentrate ratio. With less NDFef in the diet, reduced rumination activity and reduced saliva buffer capacity lead to lower rumen pH and often decrease the concentration of milk fat (West et al. 1999). Although forages are the main source of fiber, when they are ground and/or pelletized, NDFef can be seriously limiting, due to the small size of the particles (Beatty 2005). Furthermore, ground and pelletized feeds stimulate less salivation and cud chewing (Patra 2007).

On the other hand, if dietary fiber exceeds 40% NDF, consumption and rates of passage and digestion will be altered, because of the rumen fill effect, thus depressing appetite. This effect also depends on the nature of the fiber, being higher for those from mature forages or megatherm grasses (van Soest 1994). Exceptions are maize and sorghum due to their high grain content.

The NDFef can be measured indirectly by measuring the size and homogeneity of the particles. For the TMR diets, methods to make these measurements have been developed in the United States. One is the Penn State particles separation system, a shaker box, which has set of four screens, three of which have holes of different sizes and fourth collector or bottom pan (Heinrichs and Kononoff 2002).

When operating the system, the particles of the TMR are separated into four groups:

1. Top screen: retains particles larger than 2 cm
2. Screen 2: retains particles between 0.8 and 2 cm
3. Screen 3: retains particles between 0.2 and 0.8 cm
4. Bottom pan: retains particles smaller than 0.2 cm

The proportions of particles that are retained in each screen indirectly represent the mechanism of ruminal digestion. The particles retained in the top plate identify the coarse particles floating in the rumen, which contribute to chewing and insalivation. Middle screens include moderately digestible particles and the bottom pan collects the particles that are easily digestible or removable from the rumen.

According to the Penn State separation system guide (Heinrichs and Kononoff 2002), if a sample of an adequate corn silage-based TMR was separated, the proportion of the particles, on a moist basis, retained in each plate would be: upper screen 7–10%; second screen 40–50%; third screen less than 35% and bottom pan less than 20%. However, these recommendations are only guides and, these recommendations should be determined for different agro-ecological regions since the characteristics of the components of the diets vary from region to region.

Under grazing conditions, it is a common practice to provide small quantities (1.5–2 kg/cow/day) hay along with silage and concentrate mixes for high producing cows to substantially improve their performance. In summer, this management is particularly recommended because the pattern of grazing imposes a strong selectivity of the animal to consumption of leaves, which do not represent a source of NDFef (Heinrichs and Kononoff 2002).

7.4.3.5 Feeding Fats and Concentrates

The addition of fat to the diet of lactating dairy cows is common practice, and the higher energy density and the potential to reduce heat increment of high-fat diets may be particularly beneficial during hot weather (Beatty 2005). There are studies demonstrating that dietary fat can be added to the ration at up to 3–5% without any adverse effects to ruminal microflora (Collier et al. 2005). Improved efficiency and lower heat increments should make fat especially beneficial during hot weather. Ruminally protected fats allow the inclusion of a substantial quantity of fat in the diet, which could lower heat increment significantly.

7.4.4 Water Balance and Water Requirements

Water is the most essential element for the survival of animals. Water requirements for livestock can be met in three ways:

1. Metabolic water, derived from the oxidation of organic substrates and tissue
2. Water contained in food
3. Drinking water.

In any event the first route is the most important in quantitative terms and in summer it is by far the largest source. During the summer, any factor that limits access to water, directly affects the production of milk, which will fall sharply, mainly in high producing cows. Cows with water restrictions manifest higher body temperature, with a degree of heat stress higher than normal. Furthermore, water consumption and dry matter intake are closely related (NRC 2001). Under intense heat, ingestion of large volumes of water provides comfort to the animals by reducing the temperature of the rumen reticulum.

Dairy cows normally drink large amounts of water, but with intense heat they could take more than 120 L/day (NRC 2001). In a landmark study conducted in climatic chambers, it was recorded that water consumption of lactating cows increased by 29% when the temperature rose from 18 to 30°C. Concomitantly, fecal water loss decreased 33%, but losses via urine, skin, and respiratory tract increased by 15, 59, and 50%, respectively.

Regarding minerals, heat-stressed cows increase their need for Na^+ and K^+, due to the electrolyte imbalance generated at the cellular level. The higher needs of Na^+ are attributed to increased production of urine that, as explained above, reduces the plasma concentration of aldosterone (Sanchez et al. 1994). Instead, the increased demands for K^+ are attributable to an increased removal of this element with sweat.

In lactating cows fed a diet based on corn silage, hay, and concentrates, typical of many production models, it was found that the main factors that determined water intake were: dry matter consumed; the level of milk production, temperature and Na^+ intake. The following equation (NRC 2001) shows these relationships:

$$\mathrm{WI} = 16 + [(1.58 \pm 0.271) \times (\mathrm{DMI})] + [(0.9 \pm 0.157) \times (\mathrm{MP})] + [(0.05 \pm 0.023) \times (\mathrm{Na}^+)] + [(1.20 \pm 0.106) \times (\mathrm{T}_{\mathrm{md}})],$$

where

WI	Water intake (kg/day)
DMI	Dry matter intake (kg/day)
MP	Milk production (kg/day)
Na^+	sodium (g/day)
T_{md}	daily minimum temperature (°C)

The quality of drinking water is often one of the causes limiting its intake. Water quality is measured in chemical, bacteriological, and physical terms, through laboratory tests. To avoid significant production losses, each of these aspects must be carefully and regularly evaluated. Regarding chemical composition, the concentration of total dissolved solids (TDS) and the prevalent salts represent the quality factors that can seriously limit milk production in many regions. Ingestion of water with high levels of TDS is bad for dairy cattle, with a pronounced effect during hot weather (THI > 72); (NRC 2001). There is a controversy regarding the maximum levels of salts that affect the performance of dairy cows. For high producing cows (>35 l/day) water with TDS > 7,000 mg/l would not be suitable, but would have little effect on low producing animals (<25 l/day) (Bahman et al. 1993; NRC 2001). Experiments conducted in Israel (Solomon et al. 1998) showed that water with TDS above 4,000 mg/l produced negative effects on cows producing an average 35 l/day, when temperature was above 30°C.

The information available in Argentina (Taverna et al. 2001) indicates that under grazing conditions, water with 7,000–10,000 mg/l TDS, with 20–30% of sulfate, had little effect on productivity, for cows producing below 30 l/d.

All sulfate salts (Ca++, Na+, Mg++), when exceeding 1,500 mg/l, can decrease productivity because they have a laxative effect, the most potent being sodium sulfate. However, livestock, drinking water high in sulfates, (1,000–2,500 mg/l) initially suffer diarrhea, but then a process of habituation begins.

Moreover, ingestion of "light" water, i.e., very low in TDS, is also detrimental to productivity, especially when levels of sodium chloride are very low.

The temperature of drinking water could be another factor limiting intake. For example, in an experiment conducted in Texas, (Wilks et al. 1990) it was observed that cows drinking water cooled to 10°C presented lower respiration rates (70 vs. 81 rpm), lower rectal temperatures in the afternoon (39,8 vs. 40.2°C), and higher milk production (26.0 vs. 24.7 l/cow/day), as compared to animals drinking water at 27°C.

7.4.5 Biotechnology Options

The adverse effect of heat stress on livestock production had further increased in the recent years due to global warming. The desirable proposition in the present scenario is thus to develop thermotolerant animal breeds utilizing recent technology advances.

7.4.5.1 Livestock Diversity

In spite of livestock being reared in tropical environment, there are local indigenous breeds of livestock which can effectively perform countering environmental extremes. These local breeds can perform well in adverse climatic condition like high temperature, drought, and feed scarcity. Therefore, even under the changed climate scenario, the rich animal germplasm available may help to sustain the livestock productivity. In addition, there is a need to take up breeding programmes to develop climate change ready breed which performs better under stress caused due to climatic variability by using available rich germplasm.

7.4.5.2 Embryo Transfer

The reproductive efficiency of livestock is negatively influenced by high ambient temperatures resulting into silent; short estrus and hence low conception rate (Rutledge 2001). In this situation, we can use the embryo transfer technology in which in vitro produced embryo or embryo derived from donors, not exposed to high ambient temperature was used. With this technology, encouraging results have been obtained as a means to reduce adverse effect of heat stress on fertility (Al-Katanani et al. 2002). Caution should however be exercised as transfer of an embryo with non-compromised quality to a recipient subjected to the effects of the heat stress does not eliminate negative effect on endocrine axis and uterine environment. Moreover, embryo transfer is often not an economically or technically viable option for many countries in high temperature zones.

7.4.5.3 Genomics/Proteomics

Genomic and proteomic study play an important role to understand mechanism of thermoregulation and delineation of genes conferring superior thermotolerant capability in different livestock species. Earlier, attempts were made to evaluate histological response of skin to heat stress, association analysis of hair length and heat stress, embryonic resistance to heat stock, identification and characterization of heat stress response-related genes in cattle. Among the various proteins, the expression of heat shock protein (HSP) 70 is strictly stress inducible and can only be detected following a significant stress upon the cell and organisms (Satio et al. 2004; Khoei et al. 2004). The HSP70 helps in conferring the thermo adaptability and high level of thermotolerance. A recent study has shown that intracellular HSP70 expression in buffaloes is similar to the other livestock species. Higher intensity and duration of thermal exposure cause the higher HSP70 induction in buffalo lymphocytes to maintain cellular homeostasis with a threshold of thermal dose for maximum HSP70 expression (Patir and Upadhyay 2010). A few isolated studies have been carried out on heat stress-associated genes/transcripts (Lacetera et al. 2006; Moran et al. 2006; Collier et al. 2008).

The recent advancement in global expression technologies (whole genome arrays, RNA sequencing) is poised to be effectively utilized to identify those genes that are involved in key regulatory/metabolic pathway for thermal resistance and thermal sensitivity. Gene knockout technology will also allow better delineation of cellular metabolic mechanism required for acclimatization to thermal stress in dairy animals. By knowing the various genes responsible for thermotolerance we can change the genetic structure of animal and drift toward superior thermotolerant ability.

7.4.6 Improved Health Service

Increase in temperature and humidity due to climate change is strongly associated with emerging and re-emerging animal diseases by (1) increasing the numbers and geographic movement of insects (Culicoides imicola) that are major vectors of several arboviruses (e.g., bluetongue and African sickness); (2) increasing the survival of viruses from one year to the next; (3) improving conditions for new insect vectors that are now limited by colder temperatures (Mellor and Wittmann 2002; Colebrook and Wall 2004; Gould et al. 2006). These factors lead to production losses. Thus, improved disease control strategies and health services at larger scale are required.

As human, animal, and environmental health is interrelated therefore, strengthened communication and cooperation among professionals in these areas would be particularly valuable as we seek to predict, recognize, and mitigate the impact of global climate change on infectious diseases. For prevention, monitoring, and control of livestock diseases good data exchange mechanism are required at both state and national level. These should cover the distribution of animal diseases, ecological conditions including climate, and associated drugs and chemotherapeutants. In this contest, epidemiological surveillance is a critical component and it not only involves the early identification of emerging diseases and trends but also for resource planning and measuring the impact of control strategies. A global approach to epidemiological surveillance should be taken and should involve collaboration between professionals involved in human, animal, and environmental health. Of particular importance is the rapid investigation of unusual outbreaks. Such surveillance programmes are essential in allowing us to recognize and respond to emerging risk to climate change. It allows us to know what to expect and to be prepared with the right strategy in place and this might be a case of isolating the diseases and enforcing restriction zones.

We can also use the geographic information system (GIS) by which we can both monitor the level of stress and how our climate is changing and monitor the spread of diseases. We can use it to look for periods of heavy rainfall using a spatial analyst and illustrate it using GIS. This system tells us which pathogens will flourish, under their preferred condition. This tool can also help to pinpoint period of continually high minimum temperature. For example, in Israel 2000, the minimum temperatures were a key factor in the west Nile Virus outbreak and high night time temperatures were a feature of the 2003 heat wave in Europe. Likewise predictive modeling system can also be used to predict the probability of an outcome. It has potential to predict the probability of global climate change on ecological system and emerging hazards. Furthermore, laboratory and field research will also help in illuminating how climate changes influence pathogen characteristics, and models will help researchers and producers predict and plan for pathogen threats.

7.5 Conclusions

This chapter elaborates on ameliorative strategies that should be given consideration to prevent economic losses incurred due to environmental stresses on livestock productivity. Further, this chapter details the issues of imperfect information about the impact of climate and vulnerabilities, and the need for informed decisions on "resilient adaptation" by merging adaptation, mitigation, and prevention strategies. It offers new perspectives for policy-makers, institutions, societies, and individuals on improved ways of identifying most at risk communities and "best practices" of coping with current climate variability and extreme climate events.

References

Al-Katanani YM, Drost M, Monson RL, Rutledge JJ, Krininger CE III, Block J, Thatcher WW, Hansen PJ (2002) Pregnancy rates following timed embryo transfer with fresh or vitrified in vitro produced embryos in lactating dairy cows under heat stress conditions. Theriogenol 58:171–182

Al-Tamimi HJ (2005) Effect of solar radiation on thermophysiological and growth parameters of indigenous black bedwin goat kids in Southern Jordan. J Biol Sci 5(6):724–728

Amundson J, Rasby RJ, Mader TL, Hu QT (2005) The effects of temperature and temperature-humidity index on pregnancy rate in beef cows. Nebrasca Beef Cattle Report: 10–12

Armstrong DV (1994) Heat stress interaction with shade and cooling. J Dairy Sci 77:2044–2050

Bahman AM, Rooke JA, Topps JH (1993) The performance of dairy cows offered drinking water of low or high salinity in a hot arid climate. Anim Prod 57:23–28

Beatty DT (2005) Prolonged and continuous heat stress in cattle Physiology welfare and electrolyte and nutritional interventions. Murdoch Univ, Western Australia

Beede DK, Shearer JK (1996) Nutritional management of dairy cattle during hot weather Professional Dairy Management Seminar Dubuque IA, IA State University, p 15

Berman A (2005) Estimates of heat stress relief needs for Holstein dairy cows. J Anim Sci 83:1377–1384

Bianca W (1962) Relative importance of dry and wet bulb temperatures in causing heat stress in cattle. Nature 195:251–252

Blackshaw JK, Blackshaw AW (1994) Heat stress in cattle and the effect of shade on production and affects a review. Aust J Exp Agric 34:285–295

Brondino L, García K, Gastaldi L, Bulacio N, Ferreira M, Domínguez J, Sosa N, Massoni F, Walter E, Taverna M (2008) Instalaciones tipo "corral seco" para suministro de alimentos Ficha técnica n° 4. INTA, Proyecto Lechero

Brosh A, Aharoni Y, Degen AA, Wright D, Young BA (1998) Effects of solar radiation, dietary energy, and time of feeding on thermoregulatory responses and energy balance in cattle in a hot environment. J Anim Sci 76:2671–2677

Brouk MJ, Smith JF, Harner JP III (2003) Effect of utilizing evaporative cooling in tie-stall dairy barns equipped with tunnel ventilation on respiration rates and body temperatures of lactating dairy cattle. 5th International Dairy Housing Conference Am Soc Agric Eng Fort Worth TX, pp 312–318

Bryant JR, Lo′ pez-Villalobos N, Pryce JE, Holmes CW, Johnson DL, Garrick DJ (2007) Environmental sensitivity in New Zealand dairy cattle. J Dairy Sci 90:1538–1547

Buffington DE, Colazon-Arocho A, Canton GH, Pitt D (1981) Black globe humidity index (BGHI) as comfort equation for dairy cows. Trans Am Soc Agri Engi 24:711–714

Buffington DE, Collier RJ, Canton GH (1983) Shade management systems to reduce heat stress for dairy cows in hot, humid climates. Trans ASAE 26:1798–1802

Carstens GE, Johnson DE, Ellenberger MA (1989) Energy metabolism and composition of gain in beef steers exhibiting normal and compensatory growth, energy metabolism of farm animals. Eur Assoc Anim Prod Publ 43:131

Catalá M (2010) Instalaciones para el bienestar animal. 1er Simposio Regional Proleche de Bienestar Animal, 15–17de Sept Punta del Este, Uruguay

Colebrook E, Wall R (2004) Ectoparasites of livestock in Europe and Mediterranean region. Vet Parasitol 120:251–274

Collier RJ, Dahl GE, Van Baale MJ (2006) Major advances associated with environmental effects on dairy cattle. J Dairy Sci 89:1244–1253

Collier RJ, Beede DK (1985) Thermal stress as a factor associated with nutrient requirements and interrelationships, In: McDowell L (ed) Nutrition of Grazing Ruminants. Academic Press, New York, pp 59–71

Collier RJ, Collier JL, Rhoads RP, Baumgard LH (2008) Genes involved in the bovine heat stress response a invited review. J Dairy Sci 91:445–454

Collier RJ, Coppola C, Wolfgram A (2003) Novel approaches for the alleviation of climatic stress in farm animals, Interactions between climate and animal production. EAAP Tech Ser No7, Wageningen Academic Publ, Wageningen, pp 61–71

Collier RJ, Baumgard LH, Lock AL, Bauman DE (2005) Physiological limitations nutrient partitioning, Chapter 16, In: Wiseman J, Bradley R (eds) Yields of farmed Species constraints and opportunities in the 21st Century, Proceedings, 61st Easter School Nottingham, England. Nottingham Univ Press, Nottingham, pp 351–377

Cook NB, Nordlund K (2004) An update on dairy cow freestall design, Precconvention Seminar 7, Dairy herd problem investigation strategies. American Association of Biovine Practicioners 37th Annual Convention, Fort Worth, Texas, 20–22 Sept 2004

Coppock CE (1985) Energy nutrition and metabolism of the lactating dairy cow. J Dairy Sci 68:3403–3410

DAS HP (2004) Adaptation strategies required to reduce vulnerability in agriculture and forestry to climate change, climate variation and climate extremes. In: Management strategies in agriculture and forestry for mitigation of greenhouse gas emissions and adaptation to climate variability and climate change. Tech No 202, WMO-No 969, pp 41–92

Davis MS, Mader TL, Holt SM, Parkhurst AM (2003) Strategies to reduce feedlot cattle heat stress effects on tympanic temperature. J Anim Sci 81:649–661

Davison T, McGowan M, Mayer D, Young B, Jonsson N, Hall A, Matschoss A, Goodwin P, Gaughan J, Lake M (1996) Managing hot cows in Australia. Queensland Department of Primary Industry Brisbane, Australia

Eigenberg RA, Brown-Brandl TM, Nienaber JA, Hahn GL (2005) Dynamic response indicators of heat stress in shaded and non-shaded feedlot cattle, Part 2 Predictive relationships. Biosys Eng 91:111–118

El-Darawany AA (1999) Tunica dartos thermoregulatory index in bull and ram in Egypt. Ind J Anim Sci 69(8):560–563

Gaughan JB, Mader TL, Holt SM, Hahn GL, Young BA (2002) Review of current assessment of cattle and microclimate during periods of high heat load. Anim Prod Aus 24:77–80

Gaughan JB, Holt SM, Mader TL, Lisle A (2008) A new heat load index for feedlot cattle. J Anim Sci 86:226–234

Gould TW, Buss RR, Vinsant S, Prevette D, Sun W, Knudson CM, Milligan CE, Oppenheim RW (2006) Complete dissociation of motor neuron death from motor dysfunction by bax deletion in a mouse model of ALS. J Neurosci 26:8774–8786

Habeeb AA, Marai IFM, Kamal TH, Owen JB (1997) Genetic improvement of livestock for heat adaptation in hot climates. In: Proceedings of international conference on animal, poultry and rabbit production and health, dokki, Cairo, pp 11–16

Hahn GL (1985) Management and housing of farm animals in hot environments. In: Yousef M (ed) Stress physiology in livestock, vol 2 CRC Boca Raton Florida, pp 151–174

Hahn GL, Mader TL, Spiers D, Gaughan J, Nienaber JA, Eigenberg R, Brown-Brandl TM, Hu Q, Griffin D, Hungerford L, Parkhurst A, Leonard M, Adams W, Adams L (2001) Heat wave impacts on feedlot cattle, Considerations for improved environmental management. In: 6th International livestock. environment symposium, American Society of Agricultural Engineers St Joseph, MI, pp 129–130

Hahn GL (1981) Housing and management to reduce climatic impacts on livestock. J Anim Sci 52(1):175–186

Hahn GL, Mader TL, Gaughan JB, Hu Q, Nienaber JA (1999) Heat waves and their impacts on feedlot cattle. In: Proceedings, 15th Int Soci Biometeor Con, (ICB Paper No 11.1), Sydney, Sept 1999, pp 353–357

Hansen PJ, Arechiga CF (1999) Strategies for managing reproduction in the heat-stressed dairy Cow. J Anim Sci 77:36–50

Heinrichs J, Kononoff P (2002) Evaluating particle size of forages and TMRs using the New Penn State forage particle separator, DAS 02-42. Department of Dairy and Animal Science The Pennsylvania State University, http://www.das.psu.edu/teamdairy/

Holt SM, Gaughan JB, Mader TL (2004) Feeding strategies for grain-fed cattle in a hot environment. Aus Agri Res 55(7):719–725

Johnson HD (1965) Response of animals to heat. Agri Meteorol 28:109–114

Kadzere CT, Murphy MR, Silanikove N, Maltz E (2002) Heat stress in lactating cows, a review. Livest Prod Sci 77:59–91

Khoei S, Goliaei B, Neshasteh-Riz A, Deizadji A (2004) The role of heat shock protein 70 in the thermoresistance of prostate cancer cell line spheroids. FEBS Lett 561:144–148

Krause KM, Dhuyvetter DV, Oetzel GR (2009) Effect of a low-moisture buffer block on ruminal pH in lactating dairy cattle induced with subacute ruminal acidosis. J Dairy Sci 92:352–364

Kumar P (2010) Physiological adaptation of goats to climate change. In: Karim SA, Joshi A, Sankhyan SK, Shinde AK, Shakyawar DB, Naqvi SMK, Tripathi BN (eds) Climate change and management, Sheep and goat production, Satish serial publishing house, Azadpur, Delhi, pp 311–351

Lacetera N, Bernabucci U, Scalia D, Basirico L, Morera P, Nardone A (2006) Heat stress elicits different responses in peripheral blood mononuclear cells from Brown Swiss and Holstein Cows. J Dairy Sci 89:4606–4612

LPHSI (1990) Livestock and poultry indices, Agriculture engineering guide, Clemenson University Clemenson SC 29634, USA

Mader TL, Davis MS, Brown-Brandl T (2006) Environmental factors influencing heat stress in feedlot cattle. J Anim Sci 84:712–719

Mader TL, Holt SM, Hahn GL, Davis MS, Spiers DE (2002) Feeding strategies for managing heat load in feedlot cattle. J Anim Sci 80:2373–2382

Mader TL, Davis MS (2004) Effect of management strategies on reducing heat stress of feedlot cattle, feed and water intake. J Anim Sci 82:3077–3087

Magdub A, Johnson HD, Belyea RL (1982) Effect of environmental heat and dietary fiber on thyroid physiology of lactating Cows. J Dairy Sci 65(12):2323–2331

Marai IFM, Ayyat MS, Abd El-Monum UM (2001) Growth performance and reproductive traits at first parity of New Zealand white female rabbits as affected by heat stress and its alleviatin under Egyptian condition. Trop Anim Health Prod 33:457–462

Marai IFM, El-Darawany AA, Abou-Fandoud EI, Abdel-Hafez MAM (2006) Tonica dartos index as a parameter of adaptability of rams in sub-tropical conditions of Egypt. Anim Sci (Japan) 77:487–494

Marai IFM, Habeeb AAM, Gad AE (2002) Rabbits productive, reproductive and physiological performance traits as affected by heat stress a review. Livest Prod Sci 78:71–90

Marai IFM, Habeeb AAM (1998) Adaptation of Bos taurus cattle under hot climate conditions. Ann Arid Zone 37(3):253–281

McDowell RE (1972) Improvement of livestock production in warm climates. Freeman WH, Co, San Francisco

Means SL, Bucklin RA, Nordstedt RA, Beede DK, Bray DR, Wilcox CJ, Sanchez WK (1992) Water application rates for a sprinkler and fan dairy cooling system in hot humid climates. Appl Eng Agri 8(3):375–379

Mellor PS, Wittmann EJ (2002) Bluetongue virus in the Mediterranean Basin 1998–2001. Vet J 164:20–37

Moran DS, Eli-Berchoer L, Heled Y, Mendel L, Schocina M, Horowitz M (2006) Heat intolerance, does gene transcription contribute? J Appl Phy 100:1370–1376

Morrison SR, Prokop M, Lofgreen GP (1981) Sprinkling cattle for relief activation temperature, duration of sprinkling and pen area sprinkled. Trans ASAE 24:1299–1300

NRC (1971) A guide to environmental research on animals. National Acad Sci DC, Washington

Nayyar S, Jindal R (2010) Essentiality of antioxidant vitamins for ruminants in relation to stress and reproduction. Ira J Vet Res 11(1):1–9

Nienaber JA, Hahn GL, Eigenberg RA (1999) Quantifying livestock responses for heat stress management, a review. Int J Biomet 42:183–188

Nienaber JA, Hahn GL (2007) Livestock production systemmanagement responses to thermal challenges. Int J Biomet 52:149–157

NRC (2001) in Nutrient Requirements of Dairy Cattle, 7th edn National Acad Sci DC, Washington, pp 318–319

Oklahoma State University (2003) Livestock, Managing heat stress returns dividends. Extension Service-West Virginia University, pp 1–3

Patra AK (2007) Nutritional management in organic livestock farming for improved ruminant health and production, an overview. Livest Res Rural Dev 19(3), http://www.lrrd.org/lrrd19/3/patr19041.htm

Patir H, Upadhyay RC (2010) Purification, characterization and expression kinetics of heat shock protein 70 from *Bubalus bubalis*. Res Vet Sci 88(2):258–262

Ravagnolo O, Misztal I (2000) Genetic component of heat stress in dairy cattle, parameter estimation. J Dairy Sci 83:2126–2130

Reinhardt CD, Brandt RT (1994) Effect of morning vs evening feeding of limited-fed Holsteins during summer months. In: Proceedings Kansas State Univ Cattleman's Day Kansas State University, Manhattan, pp 38–39

Reynolds CK, Tyrrell HF, Reynolds PJ (1991) Effects of diet forage-to-concentrate ratio and intake on energy metabolism in growing beef heifers Whole body energy and nitrogen balance and visceral heat production. J Nutr 121:994–1003

Rutledge JJ (2001) Use of embryo transfer and IVF to bypass effects of heat stress. Theriogenology 55:106–111

Saito M, Tominaga L, Nanba E, Miyagawa I (2004) Expression of heat shock protein 70 and its mRNAs during ischemia-reperfusion in the rat prostate. Eur J Pharm 487:199–203

Sanchez WK, McGuire MA, Beede DK (1994) Macromineral nutrition by heat stress interactions in dairy cattle, review and original research. J Dairy Sci 77:2051–2079

Sathya A, Prabhakar S, Sangha SP, Ghuman SP (2007) Vitamin E and selenium supplementation reduces plasma cortisol and oxidative stress in dystocia-affected buffaloes. Vet Res Comm 31:809–818

Schütz KE, Rogers AR, Poulouin YA, Cox NR, Tucker CB (2010) The amount of shade influences the behavior and physiology of dairy cattle. J Dairy Sci 93(1):125–133

Sharma RJ, Singh DV (2008) Climate stress and its amelioration for improved production in dairy animals. In: 17th annual conference of society of animal physiologist of India and national symposium on current concepts in productivity management in livestock and poultry environment, nutrition and stress. College of Veterinary and animal Sciences, GB Pant University Agriculture and Techehnology Pantnagar Uttarakhand, pp 137–148

Smith TR, Chapa A, Willard S, Williams RJ, Crouch R, Riley T, Pogue D (2006) Evaporative tunnel cooling of dairy cows in the Southeast. In: Effect on body temperatures and respiration rates. J Dairy Sci 89:3904–3914

Solomon R, Miron J, Ben-Ghedalia D, Zomberg Z (1998) Performance of high producing dairy cows offered drinking water of high and low salinity in the Arava desert. J Dairy Sci (1995) 78: 620–624

Stowell RR, Gooch CA, Inglis S (2001) Performance of tunnel ventilation for freestall dairy facilities as compared to natural ventilation with supplemental cooling fans. 6th Int Symp Am Soc Agric Eng Louisville, KY pp 29–40

Tao X, Xin H (2003) Temperature-humidity-velocity index for market size broilers. Pro PAPER N 034037 Nevada

Taverna MA, Etcheverry F, Chávez MS, Quaino O (2001) Efecto de distintos tratamientos del agua de bebida de vacas sobre la producción y composición química de la leche abstract. Rev Arg Prod Anim 21(1):15–16

Taylor RE, Bogart R (1988) Scientific farm animal production. Macmillan Publishing Company, New York

Thom EC (1958) Cooling degree days. Air conditioning, heating and ventilating 55:65–69

Thom EC (1959) The discomfort index. Weatherwise 12:57–59

Ugurlu N, Uzal S (2010) The effect of new designed micro animal housing on the air speed distribution in the barn for providing of climatic comfort to the cattles. J Anim Vet Adv 9(1):169–172

Valtorta SE, Gallardo MR, Castro HC, Castelli ME (1996) Artficial shade and supplementation effects on grazing dairy cows in Argentina. Trans ASAE 39:233–237

Valtorta SE, Leva PE, Gallardo MR (1997) Evaluation of different shades to improve dairy cattle well-being in Argentina. Int J Biomet 41:65–67

Valtorta SE, Gallardo MR (2004) Evaporative cooling for Holstein dairy cows under grazing conditions. Int J Biomet 48:213–217

van Soest PJ (1994) Funtion of the ruminant forestomach. In: Nutritional ecology of the ruminant, 2nd edn. Cornell Univ Press, Ithaca and London

von Borell EH (2001) The biology of stress and its application to livestock housing and transportation assessment. J Anim Sci 79(E Suppl): E260–E267

West JW (1999) Nutritional strategies for managing the heat-stressed dairy cows. J Anim Sci 77:21–35
West JW (2003) Effects of heat stress on production in cattle. J Dairy Sci 86:2131–2144
Wilks DL, Coppock CE, Lanham JK, Brooks KN, Baker CC, Bryson WL (1990) Responses of lactating Holstein cows to chilled drinking water in high ambient temperatures. J Dairy Sci 73:1091–1099
WMO (1989) Animal health and production in extreme environments. WMO Tech Note 119:181
Yousef MK (1984) Stress physiology, definition and terminology. In: Yousef MK (ed) Stress physiology in livestock. CRC Press, Boca Raton, pp 3–7
Yousef MK (1985) Stress physiology in livestock. CRC press, Boca Raton

Chapter 8
Nutritional Manipulations to Optimize Productivity During Environmental Stresses in Livestock

Nira Manik Soren

Abstract Environmental stresses have huge impact on the production performance of livestock. Stress can be biotic or abiotic in nature. All animals perform better at thermoneutral zone, which are conducive for health and optimum performance. The upper and lower critical temperature is the point at which heat and cold stress begin to affect the animal, respectively. Apart from thermal stress, farm animals are also subjected to other types of stresses such as nutritional, walking, and transportation stress. The severity of the stress becomes pronounced when they occur simultaneously (multiple stresses), resulting in lowered performance and huge production losses. Farm animals try to cope up with stress to some extent by undergoing physiological and behavioral adjustment. Under these conditions livestock needs to be insulated against environmental stresses by providing optimum nutrition, proper managemental practices, and health care. Adverse environments can increase the nutritional requirements of animals directly, or they may reduce the supply of quality feed. Under these circumstances concerted effort must be taken to harmonize the welfare of animals by reducing environmental stress of food animals by nutritional manipulation and managemental practices. Further studies are required to have a clear understanding of these associations at a mechanistic level to fully exploit the potential of nutritionally manipulated production and reproduction in livestock. It is hoped that this approach will be valuable in gaining a thorough

N. M. Soren
Division of Animal Nutrition, Central Sheep and Wool Research Institute, Avikanagar, Jaipur, Rajasthan 304501, India

N. M. Soren (✉)
Division of Animal Nutrition, Central Sheep and Wool Research Institute, Avikanagar, Jaipur, Rajasthan 304501 India
e-mail: drmanik75@gmail.com

V. Sejian et al. (eds.), *Environmental Stress and Amelioration in Livestock Production*, DOI: 10.1007/978-3-642-29205-7_8,

understanding of adjusting the nutrient requirement to deal with existing environments and will therefore aid in developing rational managerial decisions to optimize productivity in livestock.

Keywords Environmental stress · Livestock production · Nutritional management

Contents

8.1 Introduction

Livestock plays a pivotal role in the agricultural and rural economies of developing countries such as India. Optimized livestock production depends on many factors such as genetic potentiality of the animals, environmental and climatic factors, health status, nutrients' availability, and other factors. Livestock production, under both intensive and extensive systems, is subjected to so many stresses, wherein livestock undergoes different types of physiological adjustments to cope with stressful conditions. Production is drastically reduced in the periods of stress and requires managemental interventions both in terms of optimum nutrition and health care. Stress can be defined as the cumulative detrimental effect of a variety of factors on the health and performance of animals. The stress in animals can be of abiotic and biotic origin. Abiotic stress is defined as the negative impact of non-living factors on the living organisms in a specific environment. The non-living variable must influence the environment beyond its normal range of variation to adversely affect the population performance or individual physiology of the organism in a significant way. Biotic stresses include living disturbances such as pathogenic microorganism, whereas abiotic stress factors, or stressors, are naturally occurring, often intangible, factors such as intense sunlight or wind that may cause harm to the plants and animals in the affected area. Abiotic stress is essentially unavoidable and affects both animals and plants. Abiotic stressors are most harmful when they occur together in combinations with other abiotic stress factors. Direct stresses result from temperature, solar radiation, humidity, rainfall, and wind velocity whereas indirect stresses comprise of those which affects the supply and availability of feed, and are determined by climatic and other factors which affect the growth of the plants that animals eat.

For livestock, the most stressful of all the abiotic stressors is heat, followed by cold. Many species are unable to regulate their internal body temperature. Even in those species that are able to regulate their own temperature, it is not always a completely accurate system. Temperature determines metabolic rates, heart rates, and other very important factors within the bodies of animals, so an extreme temperature change can easily distress the animal. Homeothermic animals are able to maintain a relatively constant core temperature by balancing the heat gained from metabolism against that gained from or given to the environment. This heat balance is achieved through the concerted effects of physiological, morphological, and behavioral thermoregulatory mechanisms. When the rate of heat loss from the body surface is very high it may lead to hypothermia and when it is very slow it leads to hyperthermia. These two conditions are not well tolerated by the body for an extended period. Under most conditions, there is a continual net loss of sensible heat from the body surface by conduction, convection, and radiation, and under all conditions there is a continual loss of insensible (evaporative) heat from the respiratory tract and skin surface. Thus homeostasis is maintained by the body.

High environmental temperature reduces productive and reproductive efficiency of livestock to a significant extent. There are reports which suggest that when thermal stress is coupled with nutritional stresses, it had severe impact on the productive and reproductive functions in contrast to subjecting to either thermal or nutritional stress separately (Sejian et al. 2010a, b). Most of the studies carried out in controlled condition (controlled-environmental chambers) have established upper critical temperatures for a number of production traits, which fall between 24 and 27°C for most traits and species of livestock. Upper critical temperatures in general will vary depending on several factors such as degree of acclimatization, rate of production (growth or lactation), pregnancy status, air movement around the animals, and relative humidity. Research conducted in controlled environment chambers has been valuable in establishing the basic parameters of stress. However, application of such information to field conditions is often difficult because of wide diurnal variations in ambient temperature and relative humidity, at one hand, and other aspect of the animals' environmental intrinsic compensatory mechanisms on the other hand. Under these conditions livestock needs to be protected against environmental stresses by providing optimum nutrition, proper managemental practices, and health care. Adverse environments can increase the nutritional requirements of animals directly, or they may reduce the supply and quality of the feed. Livestock

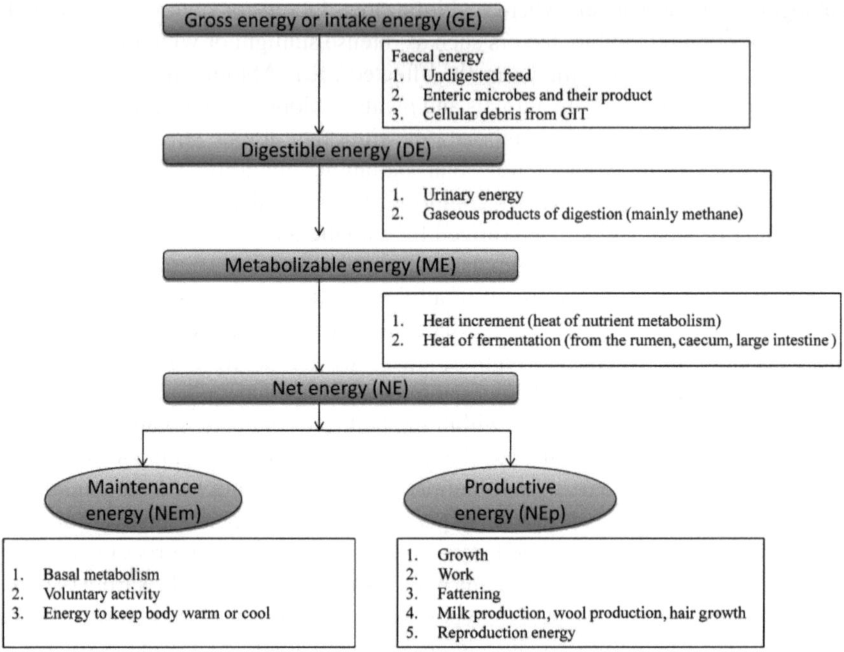

Fig. 8.1 Schematic representation of energy utilization by farm animals. This figure represents the pathways by which the gross energy or feed intake is converted to both maintenance energy and productive energy

production in adverse environments can be limited far more by these indirect effects than by an inability of the stock to maintain homeostatic equilibrium. Under these circumstances concerted effort must be taken to harmonize the welfare of animals by reducing environmental stress of food on animals through nutritional manipulation and managemental practices. A brief discussion of various stresses that are encountered in farm animals, their effect in production, and ameliorative measure are elaborated in this chapter.

8.2 Basic Concept of Thermoneutral Zone and Thermoregulation of Livestock

To have a better understanding of the different stressors that livestock species encounter during the course of production, it is imperative to become acquainted with the processes pertaining to the utilization of dietary energy, since the ambient temperature has a major impact on the energy metabolism of food producing farm animals. Some of the energy terminologies such as gross or intake energy, digestible energy, metabolizable energy, and net energy are commonly used when energy metabolism of farm animals is discussed (Fig. 8.1). Intake energy/gross energy is the combustible energy ingested per day and is determined from the combustible energy density of the feed, its opportunity for ingestion, and the appetite of the animal. Feed is not completely digested or absorbed; therefore, the gross energy (GE) of a food or other material provides no clue about the amount of energy available for livestock production. The amount of digestible energy is more useful for this purpose and can be defined as that portion of feed energy consumed which is not excreted in the feces. Similarly, metabolizable energy (ME) can be calculated by subtracting from the intake energy loss occurring in feces, urine, and gaseous product of digestion. Therefore, ME intake is that which is available to an animal for maintenance and productive function (Net energy). Further, the maintenance function involves the utilization and oxidation of ME for basal metabolism, voluntary activity, and for combating external stressors. The productive function is involved for growth, milk production, fattening, hair and wool production, reproduction, and storage.

Physiological processes in the body are associated with heat production, which is the sum total of non-productive energy utilized by the animal and of the energy "lost" in the course of converting the dietary nutrients. The non-productive energy is used for maintenance of essential physiological processes such as maintenance of the body temperature, the nervous system, organ functions, ion pumping, energy requirement for minimal activity, etc. The total amount of heat produced in the course of digestion, excretion, and metabolism of nutrients is called heat increment. Within a certain range of ambient temperature and besides unvarying feed and nutrient intake, the total heat production of the animal remains constant. This temperature range is called the thermoneutral zone (TNZ) and can be defined as the range of effective ambient temperature (Ames and Ray 1983) within which the heat from normal maintenance and productive functions of the animal in non-stress situations offsets the heat loss to the environment without requiring an increase in rate of metabolic heat production. In a thermoneutral environment, the heat production of the animal is at the minimum, and thus the dietary energy can be used for production (growth, egg and milk production) purpose efficiently (Johnson 1987). The lower critical ambient temperature range point and the upper critical ambient temperature range point define the limits of the TNZ (Robertshaw 1981). Generally, the TNZ ranges from lower critical temperature (LCT) to upper critical temperature (UCT) and depends on age, species, feed intake, diet composition, previous state of temperature acclimation or acclimatization, production, specific housing and pen conditions, tissue insulation (fat, skin), external insulation (coat), and behavior of an animal (Yousef 1985).

Unfavorable temperature (extreme cold or hot) often leads to an increased rate of heat production by the animal (more loss of energy), thereby less energy is available for production at the same level of energy intake, and the efficiency of energy utilization decreases. The upper and lower critical temperatures for different animal species and age groups are presented in Tables 8.1 and 8.2. The species, age, and body condition of the animals all have a significant influence on the critical temperature (Yousef 1985), but other environmental factors affecting their thermal sensation and heat dissipation, such as air velocity and air humidity,

Table 8.1 Lower critical temperature for different farm animals

Species	Body weight (kg)	Lower critical temperature (°C)	References
Dairy cows	600	−22	Charles (1994)
Beef cattle	400	−4	Charles (1994)
Veal calf	100	−14	Webster (1981)
Pig, fattening	20	21	Bruce (1981)
	40	19	Bruce (1981)
	60	17	Bruce (1981)
	80	16	Bruce (1981)
	100	16	Bruce (1981)
Sow	200	18	CIGR (1984)
Laying hen	2	12	CIGR (1984)
Broiler chicken	1	18	CIGR (1984)

Table 8.2 Upper critical temperature for different farm animals (FASS 2010)

Species	Body weight (kg)	Upper critical temperature (°C)
Dairy cow	–	24
Newborn dairy calf	–	−4
Lactating sow with piglets	–	26 for Sow
Prenursery	3–15	32
Nursery	15–35	26
Growing pigs	35–75	25
Finishing pigs	70–100	25
Sow, boar	>100	25
1-day old chicken	–	35
Finishing broiler	–	26
Laying hen	–	27–29

are also crucial. Increasing the airflow improves the efficiency of evaporative cooling, but higher humidity has the opposite effect. In cold and humid conditions, the heat conductivity of wet hair increases thus making the animal more sensitive to the lower ambient temperature.

Thermoregulation is the ability of the animals to maintain their body temperature in cold or hot environments, consisting of behavioral, physiological, and anatomical responses that affect energy metabolism. Moreover, in homoeotherms it is achieved by physiological and behavioral adjustments which involve the musculature, skin, sensory capacities, hypothalamus, and endocrine glands. Under thermal stress, animals exhibit anorexia, body extension, gasping, languor, lethargy, excessive drinking, bathing, decreased locomotor activities, group dispersion, and shade seeking (Hafez 1964). When animals are exposed to cold they show body flexure, huddling, hyperphagia, extra locomotor activities, depressed respiration, and nest building. Species and breed differences in the behavioral adjustments to unfavorable climates are related to habitat, morphological characteristics of body covering, degree of physiological adaptability, degree of physiological immaturity at birth or hatching, and the number of young. In a cold environment, the rate of oxidation increases, in other words, the body "burns" more nutrients, thus boosting its heat production, in order to compensate for the higher heat loss caused by the lower ambient temperature. Shivering is a tool aiding this process; since the energetic efficiency of muscle work is low, the resulting heat production is quite significant (Young et al. 1989). If heat loss exceeds heat production, the result will be hypothermia and ultimately death will occur. The thermoregulatory mechanisms of newborn and young animals (swine and poultry species) are poorly developed; therefore, in the cold environment there is every chance of increases in the number of mortalities (Mitchell and Kettlewell 1998).

The maintenance energy requirement of animals increases in a cold environment, which reduces the amount of energy available for production (Mader 2003). For example, a thin cow requires more energy for maintenance than fat cows. The more subcutaneous fat a cow has, the greater the ability to withstand colder temperatures. Cows that lose weight prior to calving can end up with weaker calves and poorer rebreeding rates. However, higher use of non-productive energy is not the only factor reducing the amount of energy available for production. Another contributing factor is the poorer digestibility of nutrients caused by the low ambient temperature (Kennedy et al. 1982). This means that in cold temperatures higher energy consumption is associated with a relatively lower energy supply. Higher feed or energy intake can help the animals to compensate for this lower energy supply to a certain extent (Galyean and Defoor 2003). From a practical perspective higher temperatures are much more hazardous for growing/finishing and breeding animals than a cold environment. Temperatures exceeding the higher critical level compromise animal performance not only by changing the energy and nutrient metabolism, but also by upsetting the body homeostasis, with detrimental consequences both for immune competence and for product quality. In general, livestock with high production potential are at greatest risk of heat stress, thereby requiring the most attention (Niaber and Hahn 2007). Temperature-humidity index (THI) could be used as an indicator of thermal climatic conditions. THI is determined by equation from the relative humidity and the air temperature and is calculated for a particular day according to the following formula (Kadzere et al. 2002):

$$\mathrm{THI} = 0.72(\mathrm{W} + \mathrm{D}) + 40.6$$

where

W—wet bulb temperature °C
D—dry bulb temperature °C

The principle of THI is that as the relative humidity at any temperature increases, it becomes progressively more difficult for the animal to cool itself. THI values of 70 or less are considered comfortable, 75–78 stressful, values greater than 78 cause extreme stress.

8.3 Different Stresses Encountered in Livestock Production

8.3.1 Cold Stress

Cold stress affects productivity of livestock to a significant extent. When animals are exposed to extreme cold stress, a number of biological processes are initiated to maintain body homeostasis. During extreme cold, the outside temperature falls rapidly and substantial amount of dietary energy may be diverted from productive functions to the generation of body heat. Failure to produce sufficient heat can result in death. A number of factors (environmental and animals) determine the severity of cold stress in animals. Both environmental and animal factors contribute to differences in heat loss from the animal to the environment. Environmental factors include air movement, precipitation, humidity, contact surfaces, and thermal radiation. Factors contributing to differences in animal heat loss from conduction, convection, and radiation are surface area (SA), which include surface or external insulation (EI), and internal or tissue insulation (TI). Evaporative losses are affected by respiration volume as well as SA, EI, and TI.

More often, cold stress leads to the development of secondary changes and possibly disease. With prolonged exposure to even mildly cold conditions, physiological adaptation occurs in animals resulting in an increase in thermal insulation, appetite, and basal metabolic intensity, as well as alterations in digestive functions. Much of the reduced productivity, and in particular the reduced nutritional efficiency, observed in ruminant production systems during the colder part of the year, can be accounted for by these adaptive changes.

Several experiments have demonstrated that during cold stress the ruminant appears to oxidize fat for heat production. The availability of such substrate (acetate, butyrate, and long-chain fatty acids) for oxidation to meet energy requirements for product synthesis and maintenance can, therefore, be influenced by the requirements for heat generation to maintain body temperature. Cold stressed animal will oxidize acetogenic substrate for heat production until 'surplus' acetogenic substrate is totally utilized, after which fat mobilization provides an extra source of metabolic fuel (Graham et al. 1959). Thus, the preferential oxidation of circulating acetate leaves a higher ratio of amino acids (and glucose) in the nutrients available for production than would be available to an animal in its zone of thermoneutrality.

Conversely, an animal that is not required to produce heat, above basal metabolism, will have more acetogenic substrates available for anabolic purposes. The environment can, thus, alter the partitioning of nutrients into productive functions and affect the efficiencies of feed utilization. The design of supplements, to balance diets for ruminants, needs to account for the varying demands for nutrients brought about by the thermal environment of the animal. It is recognized that cold stress in animals often increases voluntary feed intake and rumen turnover rate. In this way, it increases microbial cells moving to the lower tract, thus increasing the P:E ratio in the nutrients available for maintenance or production (Kennedy et al. 1986). The effects of cold stress on performance of livestock are summarized as under.

8.3.1.1 Feed Intake

The dry matter intake of the animals in general is increased during very cold weather. Experiment in sheep that was exposed to cold stress during grazing increased feed intake by 20–40% to compensate for heat loss (Graham et al. 1982). Baile and Forbes (1974) reported increased voluntary feed intake in cold stressed animals which was attributed to the activity of the thyroid gland (Gale 1973). The elevation of thyroid activity resulted in increased ruminoreticulum motility and higher rate of passage of digesta (Gonyou et al. 1979). In contrast to the above, cows consuming only range forage ate less at lower temperature when compared to those kept at warmer winter temperature (Adams et al. 1986). Kartchner (1996) also found that the intake of cows grazing on winter forage was below maintenance during harsh than during mild weather. If energy requirements for grazing and cold environment are considered simultaneously, the relationship between intake and requirement would be less favorable (Adams 1987). Decreased intake of forage results in lesser production of thermoneutral heat which in turn mobilize body fat to bridge energy gap to maintain body temperature (Webster 1974). During cold exposure elevated concentration of glucose and free fatty acids are found in the blood (Young 1975) and these elevated levels of free fatty acids are associated with reduced feed intake (Baile and Della-Fera 1993).

Effect of cold stress on voluntary feed intake (VFI) has also been reported from monogastric animals such as pigs, poultry, etc. When heavy pigs are exposed for an extended period of time to extremely low ambient temperatures (cold stress), feed intake can increase substantially, although feed conversion efficiency (FCE) and average daily gain (ADG) are reduced (Maenz et al. 1994). Cold temperatures also increase the pig's energy requirement, which means additional feed is required to maintain body temperature. Therefore, cold stress usually results in an increase in feed intake. In extreme conditions, feed intake can increase as much as 25%. Pigs exposed to cold stress have a higher metabolic rate (Dauncey and Ingram 1979; Herpin et al. 1987a, b; Herpin and Lefaucheur 1992) and therefore tend to eat more feed to supply the extra energy required for the increased metabolic heat production (Verstegen et al. 1982, 1984). The extra feed consumed for each °C below the lower critical temperature has been estimated at 25 and 39 g/d for growing and finishing pigs, respectively, while ADG is reduced by 10–22 g/d (Verstegen et al. 1982, 1984). It is important to note that these estimates are likely to be lower for pigs housed in larger groups, because of opportunities to huddle and thereby reduce cutaneous heat loss (NRC 1981). However, the extent to which these parameters are affected may depend on the genotype and the severity of cold stress. Younger pigs may be limited in their ability to increase feed intake to meet their nutrient intake requirements because of their limited gut size

(Quiniou et al. 2000). Addition of fibrous ingredients in the diet can also reduce the effects of cold temperatures. Oat and barley contain 7–12% fiber. Dietary fiber will increase the amount of heat produced by the pig and dilute the energy content of the feed. The result is more efficient use of the energy in the feed.

8.3.1.2 Digestibility of Nutrients

Cold stress also appears to affect feed efficiency by both reducing dry matter digestibility and diverting nutrients to heat generation. It has been reported that cold stress reduces dry matter digestibility by 1.8% for each 10°C reduction in temperature below 20°C. In cold stress, much of the reduction in digestibility is attributed to increased passage rate of feed through the digestive tract. An increase in ruminoreticulum motility during cold exposure enhances the rate of passage of small feed particles from the rumen by promoting their mixing, sorting, and fluid propulsion (Gonyou et al. 1979). The digestibility of long, chopped, and ground-pelleted form of hay was significantly reduced in sheep which might be due to increased rate of passage by cold exposure (Nicholson et al. 1980). Graham et al. (1980) indicated that cold temperature increased both fecal and urinary energy losses, resulting in decreased metabolizable energy availability. It is generally accepted that exposure of animals to severe cold decrease digestibility of feeds in confined ruminants (Kennedy et al. 1982) and lower forage digestibility during the periods of cold weather affects the performance of animals due to limited nutrients supply (Adams 1987). Bailey (1964) reported no decrease in digestibility during cold (−10°C) exposure of sheep but there was a slight increase after the animals were returned to a warm (20°C) environment. Horton (1978) did not observe a change in digestibility in sheep with exposure to temperatures of −12 to +10°C.

The average digestibilities of dry matter, energy, and crude protein was significantly lower in pigs when they were exposed to cold stress (6°C) than pigs exposed to 22°C (Phillips et al. 1982). The effect of ambient temperature on digestion of feedstuffs by growing hogs has also indicated a decrease in energy and nitrogen digestibility when the animals are exposed to cold (Fuller 1965).

8.3.1.3 Milk Production

Severe cold stress affects milk production in dairy cows. If cows are not fed additional feed or the quality does not allow them to eat enough to meet their additional energy requirements, body mass will be “burned” to produce metabolic heat. These cows lose weight as both feed energy and stored fat are diverted to maintain body temperature and vital functions. In this condition cows have lower milk production, increased neonatal mortality, and reduced growth rate in surviving calves. These cows usually have delayed return to estrus and poorer reproductive success. Cold exposure may directly limit the synthetic capacity of the mammary gland by reducing mammary gland temperature (Johnson 1976), or may act indirectly, by affecting the udder's blood supply (Thompson and Thomson 1977). The above adjustments along with changes in endocrine balance induced by cold exposure might significantly alter the metabolism of the mammary gland (Robertshaw 1981) thereby affecting milk production as a whole in cold-stressed ewe. Milk consists of about 90% water and flow of water from blood to milk is probably by osmosis. Lactose is the main osmolar component of milk, rate of milk secretion depends on lactose secretion. As blood glucose is the precursor of milk lactose, glucose uptake by the udder is important for milk secretion (Hardwick et al. 1963). Cold stress alters glucose metabolism in non-lactating animals, and blood glucose concentration, total turnover and oxidation of glucose are increased by cold exposure, thus output of glucose from liver increases (Depocas and Masironi 1960).

8.3.1.4 Body Condition

Dietary energy requirements of beef cattle are influenced by its size and breed, environment, and body condition (Avendaño-Reyes et al. 2010). Because of the insulation value of fat and its low maintenance requirements, fat cows may have lower winter energy requirements than thin cows (Pullar and Webster 1977). Hence, manipulation of body condition may be important economically to cow-calf producers and the ability to deposit external fat may be an important component of an animals' adaptability to cold climate.

If the winter is severe, the pregnant animals may spend whole of the winter using energy to generate heat. If nutrients are shunted to heat production, cattle begin to lose body condition. Cows, and particularly heifers, in poor body condition are at risk for calving-related problems. Therefore, it is imperative to monitor body condition scores (BCS) throughout the winter and to prepare supplements with increased energy, protein, vitamins, and minerals for cold stressed cattle. Cold stressed livestock need more feed just for maintenance than their unstressed counterparts.

8.3.2 Heat Stress

Heat stress can be defined as a group of conditions due to over exposure to or over exertion in excess environmental temperature. The condition includes heat cramp, heat exhaustion, and heat stroke. It is now widely accepted that heat stress is the major cause of lower production and economic loss in sheep and goat, poultry, swine, beef cattle, and dairy cattle. The thermal comfort zone for most animals is between 4 and 25°C. When temperature exceeds 25°C, animals suffer heat stress. In severe cases of heat stress, the deep body temperature (core temperature) rises, animal cells are affected, and production performance is reduced. The effect is more pronounced when the relative humidity is greater than 50%. Thresholds for animal heat stress are important for distinguishing between thermoneutral and thermally challenging environmental conditions. It is a well-recognized biometeorological problem that heat stress has a negative impact on productive performance and ultimately survival of livestock. Accurate estimates of heat stress thresholds are useful for evaluating heat dissipation capabilities, growth rate, feed conversion efficiency, and for comparing management techniques for overcoming the adverse effects of heat stress. For individual animals, the heat stress threshold can be used to characterize sensitivity to hot conditions and possibly identify sentinel animals to represent a group of animals. The current definition of the heat stress threshold emphasizes on the relationship between body and air temperature and the duration of exposure.

The biological mechanism by which heat stress affects production and reproduction in ruminants is partly explained by reduced feed intake, and also includes altered endocrine status, reduction in rumination and nutrient absorption, and increased maintenance requirements (Collier and Beede 1985; Collier et al. 2005) resulting in a net decrease in nutrient/energy availability for production. This decrease in energy results in a reduction in energy balance (EB), and partially explains why cows lose significant amounts of body weight when subjected to heat stress. A reduction in energy intake in lactating cows during heat stress result in a negative energy balance (NEB). Essentially, because of reduced feed and energy intake the dairy cow enters in a bioenergetic state, similar to the NEB observed in early lactation. The NEB associated with the early postpartum period is coupled with increased risk of metabolic disorders and health problems (Goff and Horst 1997; Drackley 1999), decreased milk yield and reduced reproductive performance (Lucy et al. 1992; Beam and Butler 1999; Baumgard et al. 2002, 2006). It is likely that many of the negative effects of heat stress on production, animal health, and reproduction indices are mediated by the reduction in energy balance. However, it is not clear how much of the reduction in performance (yield and reproduction) can be attributed or accounted for by the biological parameters effected by heat stress (i.e., reduced feed intake vs. increased maintenance costs).

Heat stress is one of the major concerns in pork production during summer because pigs do not have functional sweat glands like other livestock species to assist them in efficiently removing body heat. In swine, heat stress causes increased respiration rate leading to increased losses of carbon dioxide from the lungs, resulting in the reduction of partial pressure of carbon dioxide, and consequently the concentration of bicarbonate in the blood. The ensuing lowered concentration of hydrogen ions causes a rise in plasma pH, a condition widely known as alkalosis. Blood alkalosis is considered partially responsible for depressed feed intake and the consequent impaired performance in heat stressed animals. Animals respond to elevated temperature by reducing feed intake, increasing respiratory rate and water consumption, and decreasing activity in an attempt to improve heat loss and minimize the heat generation in the body. As a result of this compensatory mechanism, pigs exhibit poor growth rate and feed conversion, reduced milk production during lactation, impaired fertility, and increased mortality rates. Effects of heat stress in different farm animals are summarized as under.

8.3.2.1 Feed Intake

When cattle were exposed to heat stress a decrease in feed intake and reduction in metabolic rate was observed (Baccari et al. 1983). These responses help the animals in maintaining heat balance (Beede et al. 1983). However, in most temperate breeds of cattle, intake of good quality forages resulted in enhanced metabolic rate and increased requirements for water consumption for intermediary metabolism and thermoregulation (Springell 1968). Voluntary feed intake is affected in feedlot cattle exposed to temperature outside their thermoneutral zone. The reduction in dry matter intake from roughages-based diets becomes pronounced when environmental temperature increase is accompanied with high humidity (Warren et al. 1974; Bhattacharya and Hussain 1993). The feed intake in intensively managed livestock is less affected by heat stress, compared to grazing animals, wherein, during the period of heat stress grazing activity is significantly reduced to maintain thermal balance (Beede et al. 1983).

Moderate heat stress reduces feed intake and growth in young sheep consuming medium quality roughage diets (Dixon et al. 1999). However, exposure to high ambient temperature augments the body mechanism to dissipate body heat, in this situation an increase of respiration rate, body temperature, and enhanced water consumption and a decline in feed intake are evident (Marai et al. 2007). A higher heat increment is caused by the specific dynamic action that accompanies the metabolism of feed which is highest in the case of poor quality, fibrous feedstuffs (Marai et al. 2001). Factors such as water deprivation, nutritional imbalance and nutritional deficiency may exacerbate the impact of heat stress. Sheep, however, are less sensitive to heat stress, when compared to cattle, at a maintenance level of feeding. Research by different workers have reported that dry matter (DM) intake decrease significantly following exposure to heat stress in different breeds of sheep like Croix, Karakul, Rambouillet (Monty et al. 1991), Sardinian and Comisana (Nardone et al. 1991). Similarly, DM intake per kg body weight was reported to be lower and the maintenance requirements were higher at high ambient temperatures. The decrease in concentrate intake by rams was estimated to be approximately 13%, without altering the roughage consumption, when kept at 35°C in a climatic chamber (Nardone et al. 1991). In a recent study, Sejian et al. (2010a) subjected Malpura ewes (a native sheep breed of Rajasthan, India) to either heat stress (40°C; 55% RH) or nutritional stress (restricted feeding) or both in climatic chamber and reported significantly lower feed intake and higher water intake in both heat stress and combined stress group. The intensity of the above effect was very prominent when both the stressors were combined together.

Heat stress affects performance and pork production in swine especially during the summer months because pigs do not have functional sweat glands like other livestock species to assist them in efficiently dissipating body heat. Animals respond to elevated temperature by reducing feed intake, increasing respiratory rate and water consumption, and by decreasing activity in an attempt to improve heat loss and minimize the heat generation in the body. Environmental temperature has a marked effect on both VFI and feeding patterns (Quiniou et al. 2000). In growing pigs, VFI is reported to decrease curvilinearly with increasing environmental temperature while the size of meals is reduced (Collin et al. 2001). For each degree increase in temperature, a reduction in daily feed intake (462 g/d) was observed in multiparous sows that were reared in warm (23°C; RH: 93.8%) and hot (26.1°C; RH: 93.7%) seasons (Silva et al. 2009). The reduction in feed intake in hot seasons was attributed to increased relative humidity (RH). Between 25 and 27°C with a 50–60% RH, Quiniou and Noblet (1999) reported a reduction in feed intake equivalent to 254 g/d/°C. The greater daily feed intake reduction per degree Celsius found in their study (462 g/d°C) was related to the effect of the increased humidity observed during their study (85–98%). These results suggest that the negative effect of elevated ambient temperature may be accentuated by the increased RH in a tropical climate.

Marked changes are observed in poultry birds when subjected to heat stress. In poultry production, heat stress can be either acute or chronic. Acute heat stress refers to short and sudden periods of extremely high temperature, whereas chronic heat stress refers to extended periods of elevated temperature. Chronic stress has deleterious effects on birds that are reared on open-sided houses and causes a reduction in feed consumption and increasing water intake. Most of the reduction in feed consumption is due to reduced maintenance requirement. For every increase (1°C) in environmental temperature from 22 to 32°C, feed intake in broiler chicken was reduced by 3.6% (Ain Baziz et al. 1996). This reduction in intake was aimed to avoid increase of endogenous heat production since heat production is high when feed intake increases (Koh and Macleod 1999a, b).

8.3.2.2 Digestibility

Exposure of ruminants to heat stress in some studies has shown to increase digestibilities of dry matter, crude protein, cell solubles, and various fiber fractions, as a result of reduced rate of passage through the gastrointestinal tract and increased mean retention time (Lippke 1975; Warren et al. 1974). The increase in digestibility of feed associated with higher temperatures is probably due to depressed intake, which results in slower rate of passage (Davis and Merilan 1960). Weniger and Stein (1992) also reported significantly higher digestibility and nutrient degradation in the rumen at a high ambient temperature and low relative humidity (35°C and 50% humidity), than at a lower ambient temperature–high relative humidity (30°C and 60% humidity) in adult Merino rams.

While in some experiments the digestibility of DM, CP, ether extract, energy, and metabolizable energy of diets are reported to be depressed when the animals are exposed to high temperatures (Marai et al. 2001). The digestibility was lower when the proportion of roughage was 75% of the total diet in Awassi breed of sheep (Marai et al. 2007). In summer months, the nutrients content of the pasture declines with concomitant increase in higher fiber and lignin and there will be fall in the digestibility of nutrients because of an increase in the lignin or "wood" fiber content in the grasses. Other studies have shown the ability of ruminants to digest roughage-based diet increases in warmer temperatures while it decreases in colder environments (Conrad 1985).

Consistent decrease of protein and amino acid digestibilities of complete diets were observed in monogastrics animals like poultry subjected to heat stress (Wallis and Balnave 1984) and individual feed ingredients (Zuprizal et al. 1993). High ambient temperatures are associated with suppressed nutrient digestibility in poultry (Wallis and Balnave 1984; Sahin and Kucuk 2003). Larbier et al. (1993) observed that heat stress decreased protein digestibility. Bonnet et al. (1997) reported that the digestibility of proteins, fats, and starch decreased with exposure of broiler chickens to high temperatures. In addition, activities of trypsin, chymotrypsin, and amylase decrease significantly at a temperature of 32°C (Hai et al. 2000). The amylase and maltase activities in laying hens and broilers were reported to change under acute heat stress conditions, but not under chronic exposure to heat (Osman and Tanios 1983).

8.3.2.3 Growth Rate

Heat stress has appreciable effect on the growth performance of farm animals. Prolonged exposure of livestock to high temperatures negatively affects growth of livestock. For instance, in finisher pigs the optimum temperature range lies between 10 and 23.9°C (Myer and Bucklin 2001), and temperatures above 23.9°C decrease voluntary feed intake and pig growth (Kouba et al. 2001). Changes in growth performance with heat stress appear to be related to lower feed intake rather than changes in nutrient metabolism. Pigs exposed to heat stress (37°C) had lower (31%) ADG, lower (23%) feed intake, and 34% higher average daily water intake, compared with those in the non-heat stress conditions (Song et al. 2011). The ADG of crossbred wether lambs were significantly reduced when they were exposed to high temperature of 35°C. Reduced ADG during thermal stress occurs because maintenance requirement for energy increases during both cold and heat stress while energy intake declines during heat and does not increase as rapidly as maintenance costs during cold (Ames and Brink 1977).

Growth in animals is defined as increase in live body mass or cell multiplication and is controlled both genetically and environmentally. The available nutrients, hormones, and enzymes, as well as, elevated ambient temperatures are considered as some of the environmental factors that can influence ADG (Hafez 1987; Habeeb et al. 1992). The birth weight of lamb (temperate sheep breeds) showed a linear decrease when their dams were maintained for generations together in hot, semi-arid locations. Such a decline was shown from the sixth to the eleventh generation in Rambouillet lambs, while the reduction was apparent from the second generation in Suffolk lambs. In Merino and Dorset Horn, lamb birth weights showed no such a decline up to the sixth and third generations, respectively (Singh and Karim 1995).

The average daily gain of lambs are lower in summer than in winter, as well as in a psychrometric chamber (30–40°C) compared to a shelter (20–30°C), for Suffolk sheep (Marai et al. 1997; Padua et al. 1997). The effects of higher ambient temperature on growth performance can be explained in terms of a decrease in anabolic activity and the increase in tissue catabolism. This decrease in anabolism is essentially caused by a decrease in voluntary feed intake of essential nutrients. The decrease, especially ME for both body maintenance and weight gain, causes a loss in the production per unit of feed (Marai et al. 2007).

8.3.2.4 Milk Yield

High environmental temperature (heat stress) reduces milk yield in dairy cows especially in animals of high genetic merit. Johnson et al. (1962) demonstrated a linear reduction of dry matter intake (DMI) and milk yield when THI exceeded 70. The reductions were −0.23 and −0.26 kg/day per unit of THI for DMI and milk yield, respectively. Nardone et al. (1992) induced heat stress in climatic chambers and observed a decrease in milk yield to the tune of 35% in mid-lactating dairy cows and of 14% in early lactating dairy cows (Lacetera et al. 1996) kept under heat stress conditions. The optimal ambient temperature of dairy cows falls between 5 and 15°C. Over 15°C the animals start to sweat, although they are still able to maintain the equilibrium between heat production and heat dissipation. Heat dissipation by sweating gradually increases and although it becomes quite intense above the upper critical temperature (25°C) the cow is no more able to maintain the heat balance at such high temperatures. Kadzere et al. (2002) reported that on days of heat stress the amount of water lost through evaporation in dairy cows may be up to or even exceed the amount of water excreted in the milk. The high rate of water loss stresses the importance of water supply for dairy cows at high temperatures.

Milk production traits in ewes seem to have a higher negative correlation with the direct values of temperature or relative humidity than THI. The values of THI, above which ewes start to suffer from heat stress, seem to be quite different among breeds of sheep (Finocchiaro et al. 2005). Solar radiation seems to have a lesser effect on milk yield, but a greater effect on yield of casein, fat, and clot firmness in the milk of Comisana ewes (Sevi et al. 2001).

8.3.2.5 Egg Production and Meat Quality

During the period of extreme environmental temperature the production and quality of egg is affected in poultry. It has been observed that temperature and moisture of air are two major environmental factors controlling the heat stress of livestock (Bouraoui et al. 2002; St-Pierre et al. 2003). Heat loss in poultry is limited due to feathering and the absence of sweat glands. When the temperature and RH exceed the comfort level of a bird, it loses the ability to efficiently dissipate heat. This leads to physiological changes that are accompanied by a change in hormonal status and a reduction in feed intake to reduce metabolic heat production (Teeter et al. 1985) and lower growth rate as well as reduce feed efficiency (Geraert et al. 1996). Heat stress to different extents adversely affects egg size, laying percentage, mortality, body weight gain, and egg shell durability (Sterling et al. 2003; Lin et al. 2004; Franco-Jimenez and Beck 2007) in laying birds. The combination of high atmospheric temperature and relative humidity increase the severity of heat stress and results in the excessive generation of reactive oxygen species or free radicals due to increased metabolism in birds (Ramnath et al. 2008).

Heat stress has long been recognized as one of the major environmental factor influencing meat quality (Mckee and Sams 1997). Kadim et al. (2004) found strong negative effects of the hot season (34.3°C and 48.8% relative humidity) on the quality characteristics of beef meat, wherein, these authors reported higher pH, lower shear force value, and darker meat of *longissimus thoracis* muscle in heat stressed beef cattle when compared with muscle samples collected during the cool season. Sayre et al. (1963) reported stress-susceptible pigs that were subjected to heat stress (42–45°C for 20–60 min) prior to slaughter-developed pale, soft, and exudative (PSE) meat characteristics. Birds become vulnerable to heat stress at temperature above 30°C (Ensminger et al. 1990). Especially in the hot regions, heat stress is of major concern for the poultry industry because of the resulting in poor growth performances (lower body weight gain and carcass yield) and high mortality rates (Nardone et al. 2010). Selection of broiler chicken for rapid growth rate has been associated with increased susceptibility to heat stress (Berong and Washburn 1998; Cahaner et al. 1995). Northcutt et al. (1994) found that chickens that were subjected to heat (40–41°C for 1 h) and preconditioned (3 d at 35–36°C for 3 h) exhibited PSE meat characteristics. Environmental temperatures above 30°C can cause a reduction in feed intake, body weight, carcass weight, carcass protein, and muscle calorie content in broiler (Tankson et al. 2001). Feng et al. (2008) observed significant decrease of initial pH, drip loss, and shear force of breast muscle in heat stressed broilers. However, Northcutt et al. (1994) found that turkeys subjected to 30°C for 1 h had lower initial pH after slaughter than controls, but they did not differ in lightness or water-holding capacity from the controls. In addition, Froning et al. (1978) found that turkeys exposed to immediate pre-slaughter heat stress did not exhibit PSE meat characteristics. Stress is the factor which accelerates metabolism and quick exhaustion of muscle glycogen supplies.

8.3.2.6 Metabolism

During the heat stress a significant alteration in the metabolic activity takes place in the body of heat stressed animals. Vasodilatation of the blood vessels occurs, facilitating increased blood flow and thereby, helping in dissipation of excessive heat load from the body (Thatcher and Collier 1982). These changes cause reduction of blood supply to the internal organs, including the ruminant fore-stomach (von Engelhardt and Hales 1977), as the blood flow to the digestive system is greatly influenced by the level of feed intake (Lomax and Baird 1982), a factor which is influenced by heat stress (Attebery and Johnson 1969).

Another significant change encountered during high environmental temperature is the release of stress hormones. Glucocorticoids level increase many fold during the heat stress (Collier et al. 1995). Association between circulatory glucocorticoids and proteolytic activity in the digestive tract of animal results in increased urinary nitrogen and creatinine excretion (Tepperman 1980), which is indicative of greater protein catabolism and lower nitrogen retention in cattle that were maintained at high temperature of 48°C (Colditz 1972). Bhattacharya and Hussain (1993) concluded that heat stress resulted in a reduced metabolizable energy and nitrogen retention. A reduction in the level of plasma triiodothyronine, decrease in body weight gain, and poor feed conversion efficiency was reported in heat stress animals (Baccari et al. 1983).

Heat stress also results in the loss of large proportion of minerals such as sodium, potassium, magnesium, and chloride ions from the body (Jekinson and Mabon 1973). At high environmental temperature (40°C), Collier et al. (1995) reported a 28-fold increase in the urinary excretion of potassium in heat stressed cows compared to cows that were maintained at 15°C. Addition of dietary potassium (Mallonee et al. 1993) and sodium salts (Schneider et al. 1986) in the diet of heat stressed animals showed increase dry matter intake and milk production in the heat stressed cows.

8.3.3 Nutritional Stress

Nutrition plays an important role in the production performance of farm animals. Superior genetic potentiality of livestock is often veiled due to inadequate nutrition resulting in lower growth rate, milk yield, meat production, and other performances. Nutritional factors such as excessive or insufficient nutrition adversely affect the entire body function. Grazing animals in arid and semi-arid regions are generally subjected to periods of undernutrition during extreme hot environment due to nonavailability of feed and poor pasture conditions caused by lower availability of nutrients, which in turn results in low productivity. The animals suffer severe nutritional stress in the dry-season when the natural pasture is of low nutritional value and usually in scarce supply. During this time of the year, animals also waste a lot of energy, as they have to walk long distances in search of food and water. As a result of these dry season and adverse conditions, animals lose weight, body condition and have low milk yields, low conception rates, and increased calf mortalities, all of which culminate into heavy economic loses to the small holder farmers.

Sheep and goats due to their inherent grazing habit are often susceptible to nutritional stress. Extensive rearing of sheep and goats in most of the arid and semi-arid regions is characterized by grazing during day hours and housing of the animals during night time, with possible supplementation of concentrate mixtures and of straw or hay. In extensive production systems, the animals are free to move within a habitat that allows them to best perform their physiological and behavioral functions. However, grazing can sometimes also adversely affect the well-being of animal due to seasonal fluctuations of herbage quantity and quality, thus the grazing animals are usually subjected to a temporary nutritional stress. If the nutritional stress occurs during mating season, it can reduce sheep fertility (Rassu et al. 2004).

In the areas where sheep and goat breeding is more diffused, late spring and summer are characterized not only by poor grass availability and palatability but also by a marked reduction of its protein content (Negrave 1996). Therefore, grazing animals in extensive rearing can face nutritional imbalance/nutritional stress during this period of the year, with the alteration of rumen fermentation and protein synthesis, which affects their well-being and negatively influences milk, fat, and protein content. When goats graze in poor meadows with excessively fibrous vegetation, under bad weather conditions, and with limited time for herbage ingestion, they may show decreased milk production (Fedele et al. 1993). Pulina et al. (2006) found that short-term feed restriction strongly reduced milk yield and increased milk fat content in Sarda dairy ewes. Undernutrition significantly affected the milk fatty acid profile, as a consequence of body fat mobilization. Underfed ewes showed higher milk somatic cell count, indicating a metabolic stress of the animal and its mammary gland. Nutritional stress induced by restricted feeding (30% lower than control) and high environmental temperature (40°C) resulted in lower body weight gain and body condition score in Malpura ewes (Sejian et al. 2010a).

Young pigs can be susceptible to nutritional stress especially when they are weaned. This nutritional change is involuntarily added to the pig's sociological, environmental, and immunological stresses at weaning. In rapidly growing pigs, the rate of weight gain can exceed the development of the control and adaptation systems of the body. As a result, the body resistance to disease is decreased while the susceptibility of the animal to infection increases. Young animals are more sensitive to nutritional stress because their adaptive mechanisms are poorly developed. Over feeding and changing feeding patterns can cause stress and result in ulcers in pigs. Poor quality drinking water containing high levels of dissolved minerals, including sulfates, increases the incidence of diarrhea and worsens post-weaning growth in piglets.

8.3.4 Walking Stress

Livestock in tropical regions are subjected to walking stress, wherein, these animals have to travel long distances in search of grazing pasture. Small ruminants like sheep and goat spend a sizeable amount of energy in the walking activity, especially in hot summer. In contrast, temperate zone sheep do not suffer from heat stress, which is the main worry in the humid tropics; consequently, at least climatically, sheep adjust satisfactorily. Domestic livestock commonly spend around 7–12 h/day in grazing, which include time spent for searching as well as for consuming forages (Burns and Sollenberger 2002). As the grazing time increases, more energy is used for activity and less for production purpose, thus, the minimum time that results in adequate dry matter intake is considered optimum. Grazing time of livestock in pasture depends on the ease of ingesting, which varies with accessibility of plant parts, availability of total forages, and quality of the consumed diet (Burns and Sollenberger 2002). Grazing time is generally lowest when forage is abundant and of good quality and highest when forage is of low quality or availability is limited. Grazing time may fall when grazing animals are in a severe caloric deficit and forage availability is severely limited, thus contributing even further to decline in forage intake (Hodgson 1986).

The energy expenditure of locomotion contributes significantly to the energy requirement of animals in free-living conditions and must be included for accurate evaluation of the energy needs of the grazing animal. Most of the increase in energy expenditure of physical activity results from grazing and locomotion costs, whereas the contribution of other activities is usually considered to be negligible. Grazing animals have an extra daily maintenance requirement due to the demand of energy for the physical activities of forage intake and walking. However, when numbers of stressors such as heat stress, nutritional stress, and walking stress occur simultaneously, then the severity of the stress is increased and productive and reproductive performance of the animals are severely affected. Walking stress significantly altered the growth, physiological response, and hematobiochemical changes in the Malpura ewes that were subjected to walking for a distance of 14 km (Sejian et al. 2011) in semi-arid environment. In their study, there was a significant increase in respiration rate, rectal temperature, and plasma cortisol level, which was indicative of stress.

8.3.5 Transportation Stress

Stress associated with transport occurs in food animals essentially in commercial agriculture and to a lesser extent in the rural sector. Transportation stress may be more or less severe, depending on a number of different stress factors involved. The major stress factors in livestock transportation may be classified as noise, vibration (Dobson 1987), lack of exercise (Hutcheson and Cole 1986), prolonged standing (Lambooy and Engel 1991), and environmental temperature and humidity (Dantzer 1982). Animals may be transported for marketing, slaughter, restocking, from drought areas to better grazing areas and for a change of ownership. Typically, methods used to move animals are on hoof, by road (motor vehicle), by rail, on ship, and in some cases by air. Majority of livestock in developing countries, in general, are moved by trekking on the hoof for short distances and by road or rail, when transportation involves long distances. Transport of livestock is undoubtedly the most stressful (Maria et al. 2004) and injurious stage in the chain of operations between farm and slaughter house and contributes significantly to poor animal welfare, loss of production, and poor meat quality. Transportation of food animals is of great concern due to several reasons (Knowles et al. 1999a, b; Hartung 2003). These are:

1. Cause severe stress in animals, if welfare conditions are not provided.
2. Stressful transportation may adversely affect meat quality.
3. There is the risk of spread of infectious diseases over large distances.
4. Animal health can be impaired by various pre-transport and transport conditions.

The above conditions may cause injury, reduce performance, increase in morbidity and mortality rate, and consequently substantial economic losses due to loss of live weight and poor meat quality (Knowles et al. 1999a; Fazio and Ferlazzo 2003; Minka and Ayo 2007). Transportation exposes food animals to stress, resulting in increased morbidity and mortality. During transport the hypothalamo-pituitary-adrenal axis is activated by stressors (noise, vibration, prolonged standing, lack of food and water, and environmental temperature) resulting in increased concentrations of plasma catecholamines and cortisol (Sconberg et al. 1993). Catecholamines and cortisol are essential components of adaptation to stress. Deprivation of food and water for quite a long time during transportation further depresses the condition of the already stressed animal. The transported food animals are subjected to concomitant action of transportation and heat stress factors. These deleterious stress factors significantly weaken the body resistance and the affected animals become vulnerable to diseases.

8.4 Nutritional Management of Heat Stress in Domestic Animals

8.4.1 Dairy Cows

The primary impact of heat stress in dairy cows is a significant reduction in voluntary feed intake. Water intake is closely related to DMI and milk yield and its intake is increased by 1.2 kg/°C increase in minimum ambient temperature, but regardless of rate of increase it is obvious that abundant water must be available at all times under hot conditions. Thus water is the most important nutrient for dairy cow. Several nutritional strategies need to be implemented during this period and these include reformulations to account for reduced DMI, higher maintenance costs, and metabolic heat production from various feedstuffs (West 2003). The maintenance requirement of lactating dairy cows increases substantially as environmental temperature increases; therefore, the frequency of feeding may be increased so as to increase DMI. During the period of heat stress, the diurnal temperature rises to a significant level and animals are often reluctant to take feed in an effort to reduce heat production and feeding in the early morning or late evening hours can be practiced to reduce total heat load on the animal (Staples 2007). As the intake of DM generally declines with hot weather, effort should therefore be made to increase the nutrient density of the diet. Energy density can be increased by supplementing extra grain or fat in the diet and a reduction in forages. Feeding of dietary fat (rumen protected/rumen bypass) in the concentrate remains an effective strategy of providing extra energy during a time of negative energy balance. Compared to starch and fiber, fat has a much lower heat increment in the rumen and thus provides energy without a negative thermal side effect (Gaughan and Mader 2009).

Nutritionists often increase the energy or protein density of the ration during prolonged periods of heat stress. Caution should be exercised if increasing protein levels during hot weather since there is an energetic cost associated with feeding of excess protein. Feeding of excess N above requirements reduces ME by 7.2 kcal/g of N (Tyrrell et al. 1970). Studies have shown that the energetic cost associated with synthesizing and excreting urea can compromise milk production when feeding excess protein (West 2003). When diet containing 19 and 23% CP were fed to dairy cattle, milk yield was reduced by over 1.4 kg (Danfaer et al. 1980) and the energy cost associated with synthesizing and excreting urea accounted for the reduced milk yield (Oldham 1984). Heat stressed cows do have an extra requirement for dietary or rumen-produced glucose precursors like propionate. However, lower fiber/high fermentable carbohydrate rations may drive propionate, increase energy density, and lower heat increment, but these effects must be balanced with the potential for ruminal acidosis in animals already prone to acidosis due to intake variations and reduced rumen buffering capacity. Consideration should be given not only to the quantity of fermentable carbohydrate but also to the dynamic aspects of starch digestibility in fermented feeds such as corn silage or high moisture corn (Mahanna 2007).

During hot weather, declining DM intake and high lactation demand requires increased dietary mineral concentration. However, alterations in mineral metabolism also affect the electrolyte status of the cow during hot weather (West 2003). Important minerals such as sodium, potassium, and magnesium are lost from the body due to persisting perspiration and in this condition feeding diets with a high dietary cation–anion difference can improve intake in heat stressed cows (Tucker et al. 1988). Cattle utilize potassium (K+) as the primary osmotic regulator of water secretion from their sweat glands. As a consequence, K+ requirements are increased (1.4–1.6% of DM) during the summer, and this should be adjusted for in the diet. In addition, dietary levels of sodium (Na+) and magnesium (Mg+) should be increased, as they compete with potassium (K+) for intestinal absorption. Therefore, feeding elevated levels of potassium (1.5–1.6% of DM), sodium (0.45–0.60% of DM), and magnesium (0.35–0.40% of DM) is also recommended (Staples 2007) as they are the primary cations in bovine sweat. The role of vitamin nutrition in the management of heat stress in dairy cattle has been reported in some studies. Cattle supplemented with niacin (6 g/d) during summer months increased milk yield by 0.9 kg/d compared with controls (Muller et al. 1986). Recent research indicates that the addition of 12 g/day of encapsulated niacin may improve heat tolerance through elevation of cellular heat shock proteins and peripheral vasodilation (Burgos-Zimbelman et al. 2008).

8.4.2 Sheep and Goat

Nutrient requirements of sheep and goat increase during the period of heat stress. During heat exposure, energy requirement of domestic livestock increases because lot of energy is expended during respiratory muscular activity (Linn 1997), sweat gland activity (Shibasaki et al. 2006), and the calorigenic effect of hormones (Gudev et al. 2007). The adjustment of diets by reducing the roughage content, addition of dried beet pulp, and increasing fat content can be a practical approach toward minimizing the effect of heat during heat stress (Lofgreen 1974). Another practical approach can be to reduce the concentrate proportion in the diet, for instance, Moose et al. (1969) found that low-concentrate diets (35%) had lowered heat increment than higher concentrate diets (70%) when fed to lambs and reported that at temperatures above 25°C high heat increment can seriously impair the efficiency of diets containing higher percentages of roughage. Protein requirement during heat stress should be able to meet the need of animals to maintain nitrogen equilibrium (maintenance protein) and that needed for productive function. Ideally, dietary protein exceeding that needed for maintenance is used only for production (growth, wool, or milk); however, growth and other productive functions may be limited by available energy because of increased requirement of energy for maintenance during thermal stress. When energy is limiting, protein may then be catabolized and serve as an energy source (Crampton and Harris 1969).

8.4.3 Pigs

To counter heat stress the foremost mechanism adapted by growing pigs (Quiniou et al. 2000) and lactating sows (Quiniou and Noblet 1999; Renaudeau et al. 2001) is to reduce their feed intake, wherein, to lower internal heat production. Exposure to temperatures above 25°C can be considered as a heat stress situation, especially for lactating cows. The effects are accentuated when the high ambient temperature is combined with a high relative humidity (Renaudeau et al. 2003). In this situation, the best nutritional practice to minimize the negative effects of heat stress on feed intake would be to add supplemental fat to the diet because fat has a lower heat increment than either carbohydrate or protein, and to increase the concentration of other nutrients. Fat has a high caloric density that helps offset lowered caloric intake during heat exposure. The partial replacement of dietary CP by starch is associated with a reduced heat production in pig. Increased dietary fat content has similar effects on heat production and the effects of dietary CP reduction and fat addition are additive (Noblet et al. 2002). Based on these assumptions, it can be hypothesized that low CP diets (and/or fat supplemented diets) would be better tolerated by heat stressed pigs as the effect of heat stress on their energy intake would be attenuated.

Stahly et al. (1979) report an advantage of feeding synthetic lysine instead of natural protein, as it reduces heat increment of the diet. During heat stress, dietary crude protein may require adjustment. The daily feed intake decrease by about 40 g/°C of heat stress, and this is paralleled by a daily gain depression of 10–20 g/°C rise of temperature (NRC 1981). Some researchers have reported that dietary vitamin and mineral concentrations may need to be increased under heat stress conditions. But there is little evidence indicating that total daily requirements of these nutrients are affected by effective ambient temperature. Of course, as high temperatures reduce feed intake, it may be advisable to increase the concentration of certain vitamins and minerals in the diet to compensate. Peng and Heitman (1974) found that dietary thiamine requirement may be greater at 30 and 35°C compared with thermoneutral temperature. Holmes and Grace (1975) found more potassium in the urine of heat stressed pigs, but calcium retention was not affected.

8.4.4 Poultry

As poultry birds are very rapid growing with enhanced metabolic rate, thus they are susceptible to a number of stressors. High environmental temperature often deters optimum intake of feed. Energy intake is the most important nutrient limiting bird performance at high temperatures. Studies have shown that the energy requirement for maintenance decreases by about 30 kcal/day with an increase in the environmental temperature above 21°C. The maintenance energy is lower at higher temperature and most of this energy is wasted in heat dissipation, so the absolute energy requirement is not affected by heat stress. In general, the feed intake changes about 1.72% for every 1°C variation in ambient temperature between 18 and 32°C but when the rise of temperature is much higher (32–38°C), the rate of decline is much more higher (5% for each 1°C). Under these circumstances, corrective measures need to be taken to improve feed intake, which might include the addition of fat in the diet. When 5% fat was supplemented into the diet of heat stressed bird, feed consumption improved by 17% in heat stressed birds (Dale and Fuller 1979). The added fat provides an extra calorific value by decreasing the rate of passage of digesta, thereby increasing the utilization of nutrients (Latshaw 2008). Fats or oils with more saturated fatty acids are preferred in hot humid climates. The concentration of energy should be increased by 10% during heat stress, while the concentration of other nutrients should be increased by 25%. The requirements for protein and amino acids are independent of environmental temperature so heat stress does not affect bird performance as long as the protein requirement is met. As heat stress reduces feed intake, the levels of protein/amino acids therefore need to be increased with the environmental temperature up to 30°C. The supplementation of essential amino acids to a diet with a poor protein quality or amino acids imbalance helps to improve performance by reducing heat increment and the harmful effects of high temperature (Waldroup et al. 1976). In order to maximize nutrient intake, one must consider relatively high nutrient-dense diets, although these alone do not always ensure optimum growth. Relatively high protein (16–18% CP) with adequate methionine (2% of CP) and lysine (5% of CP) level together with high energy level (11.7–12.6 MJ/kg) are usually given to Leghorn hens, in hot weather situations (McNaughton et al. 1977).

Heat stress reduces calcium intake and the conversion of vitamin D_3 to its metabolically active form, 1, 25$(OH)_2$ D_3, which is essential for the absorption and utilization of calcium (Faria et al. 2001). The calcium requirement of layers, particularly older birds, is increased at high environmental temperatures. Therefore extra calcium should be provided at the rate of 1 g/bird in the summer months in the form of oyster shell grit or limestone. Supplementation should be made over the normal dietary calcium level (3.75 g/bird/d) recommended for layers. However, due care should be taken to prevent excessive levels of calcium, as it reduces feed intake. The phosphorus level in diet must not be forgotten as excessive phosphorus inhibits the release of bone calcium and the formation of calcium carbonate in shell gland, thereby reducing the shell quality. Supplementing the diet with 0.5% sodium bicarbonate or 0.3–1.0% ammonium chloride or sodium zeolites can alleviate the alkalosis caused by heat stress. Sodium bicarbonate stimulates feed and water intake and improves shell quality at high environmental temperature (Balnave and Muheereza 1997; Koelkebeck et al. 1993). The body weight gain can be increased up to 9% by addition of these chemicals in the feed of heat stressed broilers. The excretion of potassium through urine is significantly higher at 35°C than at 24°C. The potassium requirement increases from 0.4 to 0.6% with a rise in temperature from 25 to 38°C. A daily potassium intake of 1.8–2.3 g potassium is needed by each bird for maximum weight gain under hot conditions. To compensate for the reduced feed intake under heat stress, dietary allowances for electrolytes (sodium, potassium and chloride) may be increased by 1.5% for each 1°C rise in temperature above 20°C (Balnave and Zhang 1993; Smith 1994).

Additional allowances of ascorbic acid (vitamin C), vitamins A, E, and D_3 and thiamine can improve bird performance at higher temperatures (Ferket and Qureshi 1992). However, the loss of vitamin activity either in premix or in feed during storage particularly at elevated environmental temperature is a prime concern and probably explains the conflicting results on the effects of vitamin supplementation during heat stress. High temperature, moisture, rancid fats, trace minerals and choline speed up the denaturation of vitamins. Vitamin activity in feeds can be maintained by using feed antioxidants, gelatin encapsulated vitamins, appropriate storing conditions and adding choline and trace minerals separately from other vitamins. Ascorbic acid synthesis is decreased at elevated environmental temperature, making it an essential dietary supplement during the summer. The vitamin helps to control the increase in body temperature and plasma corticosterone concentration (Fenster 1989; Pardue and Thaxton 1986). It also improves eggshell quality via its role in the formation of the shell's organic matrix, Supplementation of ascorbic acid (200–600 mg/kg diet) improves growth, egg production, number of hatching eggs, feed efficiency, egg weight, shell quality, and livability during heat stress. The absorption of vitamin A declines at high temperatures. In broiler breeders, a threefold increase in supplementation has been found to be beneficial. Vitamin E protects the cell membrane and boosts the immune system, so additional dietary supplementation may be advantageous during hot weather.

8.5 Nutritional Management of Cold Stress

Many factors may influence the nutrient requirements of animals. Low environmental temperature increases the requirement of nutrients in farm animals. New born calf when subjected to low environmental temperature needs special care as the body energy reserve of new born animal is limited and the body energy stores in the form of fat and glycogen will not last for more than about one day under very cold conditions (Alexander et al. 1975; Okamoto et al. 1986; Rowan 1992). The thermo neutral zone in very young calves ranges from 15 to 25°C. Thus, when the environmental temperature drops below 15°C (LCT), the calf must expend energy to maintain its body temperature, thus the maintenance energy requirement is increased. However, for older calf the LCT may be as low as −5 to −10°C (Webster et al. 1978). Therefore, young calves should be fed extra energy during cold weather to satisfy the proportionate increase in maintenance energy requirements (NRC 2001). This can be done by increasing the amount of liquid diet with additional milk solids or by incorporating additional fat into the liquid diet (Schingoethe et al. 1986; Scibilia et al. 1987; Jaster et al. 1990). However, additional fat in milk replacer or starter decreases starter intake (Kuehn et al. 1994), which negates at least a portion of the increased energy density from fat supplementation.

The energy requirement of dairy cows in low ambient temperature is minimal because of high heat production due to consumption of large amounts of feed (NRC 2001). Even in barns with proper ventilation, it is unlikely that cows will require increased intake of energy to counteract cold environments if they are kept dry and are not exposed directly to wind. Young (1976) summarized experiments with ruminants exposed to cold temperature and reported that an average reduction in DM digestibility of 1.8% units for each 10°C reduction in ambient temperature below 20°C. Much of this lowered digestibility under cold stress was attributed to an increased rate of passage of feed through the digestive tract (Kennedy et al. 1976). Because of the effects of low temperature on digestibility, under extremely cold weather conditions, feed energy values could possibly be lower than expected. But for grazing ruminants lower environmental temperature increases the maintenance energy to a considerable extent because very cold temperature often affects active growth of plants and pasture grasses, a diet based on pasture and browse normally becomes inadequate for grazing ruminants. Though sheep and goats are able to meet their nutrient requirement form pasture but in case of large ruminants like cows, grazing with adequate supplementation of concentrate will meet nutritional requirement of stressed animal.

On the other hand, non-ruminant species, however, including pigs, and poultry require a more highly digestible diet to enable them to thrive and be highly productive. However, if they are kept in confinement, it becomes essential to provide all the important nutrients through diet. All species of livestock have a greater need for better quality feed when they are in full production (milk, eggs) or when they are growing fast. Poultry have increased energy requirements to maintain normal body temperature in cold ambient temperatures. The process of digestion of feed produces body heat and the amount of heat produced (heat increment) will vary according to the nutrient composition of the diet. In cold temperatures it may be desirable to formulate a diet with a higher heat increment and the opposite in hot temperatures.

Exposure of pig to environmental temperatures below its thermoneutral zone affects feed consumption. Conversely, feed consumption increases as environmental temperature is reduced within a moderate range. Finishing pigs in a cold environment eat more because their maintenance energy requirement is increased to maintain body temperature. Growth rate may not be affected, but poorer feed efficiency results in general. However, severely cold stressed pigs may not grow because they cannot consume sufficient amounts of energy above their maintenance requirement. There is a difference in the additional feed needed per °C of coldness between grouped pigs and those kept singly. Those in groups can huddle, thereby reducing body-surface exposure and heat loss to the cold environment. One can conclude that a pig weighing 20 kg should consume additional feed at the rate of 13 g/day/°C of coldness, and for pigs weighing 100 kg up to 35 g/day/°C. Extra feed intake is required during cold to compensate for reduced gain in restricted-fed pigs. Fat, protein, water, and ash gains, which together comprise body weight gains, may be reduced in several ways. Fat gain will be more reduced in the cold than protein gain, because fat is used primarily as fuel (Masoro 1966). In growing pigs, the same has been found by Hacker et al. (1973), Verstegen et al. (1973), and Brown et al. (1976). Close and Mount (1976) showed that reduction in protein gain in the cold is dependent on feeding level. Verstegen et al. (1978) concluded that average daily gain is depressed by 15 g/°C of coldness when feed intake rate remains constant.

8.6 Nutritional Management of Transportation Stress

Transportation of food animals for different purposes results in stress with substantial physiological changes in the body. Transportation can combine with a number of physical and psychological stressors, such as stress of loading and unloading, huddling of unfamiliar animals, loud noises, feed and water deprivation, extreme temperature, and new housing environment can be stressful. For that some basic managemental practices need to be followed. Provision should be made for care of animals during the journey and at the destination. Particular care should be taken with animals that are fatigued, old, young, infirm, pregnant, and/or nursing. Animals should be neither too loosely nor too tightly loaded so as to reduce the risk of excessive movement or overcrowding resulting in injury. The distance animals are transported, and the time taken, should be minimized. During transport animals should be protected from extremes of heat and cold and provided with adequate ventilation. Where livestock are transported over long distances, appropriate provision needs to be made for feeding and watering of the animals. The foremost essential nutrient is water, and has to be made available as temperature of the transport environment generally increases.

During transportation most of the animals react to the experience by becoming anorexic and adipsic. The stressful experiences of a novel environment, movement of the transportation vehicle, food and water sources that differ from those in the animal's previous environment for logistical reasons inhibit food and water consumption. The most appreciable change observed in farm animals that are being transported is loss of body weight, which is more rapid when transported than it would normally occur without feed and water. This consequence implies that transportation is stressful for reasons beyond the lack of feed and water. Provision of feed or water during transportation can be problematic because of food spoilage and water spillage; wetting of the floor by spilled water, which results in chilling, slipping, and injuries; animals' lack of ability to eat or drink while in motion; motion sickness; and lack of motivation to eat or drink during the trip. Thus, providing food or water may not be of any benefit during short trips because of lack of motivation to consume food and water.

Provision of feed and water during very long trips requires special attention, especially if the vehicle stops or has periods of stability during which animals may seek food and water. In cases where an animal may refuse food because it is presented in a novel form or source, animals should be adapted to the travel and post-travel diets and to feed and water dispensers before travel. Exposure to the food forms and water sources that will be used during travel before the trip may help to reduce dehydration and weight losses during transportation.

Water is the most important consideration for trips of intermediate length for most species. Small animals lose more heat, require more calories per unit of body mass, and become dehydrated more quickly than larger animals. Xin and Lee (1996) found that the provision of water (or a substitute) and feed were also important for sustaining day-old male chicks during long trips. Transportation stress in poultry impairs normal body functions, leading to increased morbidity and mortality, poor meat quality, and decreased productivity (Fazio and Ferlazzo 2003; Fallenberg and Speisky 2006; Franco-Jimenez and Beck 2007; Rozenboim et al. 2007). When poultry birds are huddled during transport, the combination of high ambient temperature and relative humidity provokes heat stress and the excessive generation of reactive oxygen species (ROS) or free radicals (FR) takes place as a result of increased metabolism in birds (Ramnath et al. 2008). Normally, non-enzymatic antioxidants, such as vitamin C, produced in the bird's kidneys, are involved in the elimination of excess FR or ROS from the body, but under praxis condition, they are either exhausted or overwhelmed; thus, exposing cells to their harmful effects (Maurice et al.

2002). Thus in such situation supplementation of vitamin C and E may prove beneficial in alleviating transportation stress in hot weather (Ajakaiye et al. 2010). Pigs are also quite susceptible to transportation stress and meat from such stressed pigs has bearing in meat quality (Tarrant 1989). In such situation sedation of stressed animals may be done, but such practices are not allowed in modern food industry due to residual effect in the food chain. Thus, alternative feed additives such as tryptophan, vitamin E, and herbal products can prove as an alternative to counter transportation stress in pigs (Peeters et al. 2004), thereby improving both their welfare and their meat quality without introducing a residue risk for food safety.

8.7 Conclusion

Stress is inevitable in livestock production system. In general it has adverse effect on livestock production and huge economic losses are sustained by farmers. Understanding of stress at mechanistic level and the different interaction of farm animals with environment will help in developing suitable strategies to overcome the adverse effect of stress in farm animals. Nutritional management of stress, however, will be helpful in overcoming the detrimental effect, provided sound knowledge of nutritional requirements of farm animals at different age, production level, and seasons are amalgamated in a holistic manner. Further, research works are needed to find the nutritional requirement of stressed animals in rapidly changing climate scenario.

References

Adams DC (1987) Influence of winter weather on range livestock. Proceedings of grazing livestock nutrition conference. Jackson, Yoming, July 23–24, pp 23–28

Adams DC, Nelsen TC, Reynolds WL et al (1986) Winter grazing activity and forage intake on range cows in the Northern Great Plains. J Anim Sci 62:1240–1246

Ain Baziz HA, Geraert PA, Padilha JCF et al (1996) Chronic heat exposure enhances fat deposition and modifies muscle and fat partition in broiler carcasses. Poult Sci 75:505–513

Ajakaiye JJ, Ayo JO, Ojo SA (2010) Effects of heat stress on some blood parameters and egg production of Shika Brown layer chickens transported by road. Biol Res 43:183–189

Alexander G, Bennett JW, Gammell RT (1975) Brown adipose tissue in the new born calf (*Bos taurus*). J Physiol 244:223–234

Ames DR, Brink DR (1977) Efficiency ratio effect of temperature on lamb performance and protein. J Anim Sci 44:136–144

Ames DR, Ray DE (1983) Environmental manipulation to improve animal productivity. J Anim Sci 57:209–220

Attebery JT, Johnson HD (1969) Effect of environmental temperature, controlled feeding and fasting on rumen motility. J Anim Sci 29:734–737

Avendaño-Reyes L, Fuquay JW, Moore RB et al (2010) Relationship between accumulated heat stress during the dry period, body condition score, and reproduction parameters of Holstein cows in tropical conditions. Trop Anim Health Prod 42:265–273

Baccari F Jr, Johnson HD, Hahn GL (1983) Environmental heat effects on growth, plasma T_3 and post-heat compensatory effects on Holstein calves. Proc Sco Exp Biol Med 173:312–318

Baile CA, Della-Fera MA (1993) Nature of hunger and satiety control systems in ruminants. J Dairy Sci 76:2899–2905

Baile CA, Forbes JM (1974) Control of feed intake and regulation of energy in ruminants. Physiol Rev 54:160–170

Bailey CB (1964) Effect of environmental temperature on feed digestion, water metabolism, body temperature and certain blood characteristics of sheep. Can J Anim Sci 44:68–75

Balnave D, Muheereza SK (1997) Improving eggshell quality at high temperatures with dietary sodium bicarbonate. Poult Sci 76:588–593

Balnave D, Zhang D (1993) Response of laying hens on saline drinking water to dietary supplementation with various Zn compounds. Poult Sci 72:603–609

Baumgard LH, Moore CE, Bauman DE (2002) Potential application of conjugated linoleic acids in nutrient partitioning. Proceedings southwest nutrition conference, pp 127–141

Baumgard LH, Odens LJ, Kay JK et al (2006) Does negative energy balance (NEBAL) limit milk synthesis in early lactation? Proceedings southwest nutrition conference, pp 181–187

Beam SW, Butler WR (1999) Effects of energy balance on follicular development and first ovulation in postpartum dairy cows. J Reprod Fertil 54:411–424

Beede DK, Mallonee PG, Schneider PL et al (1983) Potassium nutrition of heat-stressed lactating dairy cows. South Afr J Anim Sci 13:198–206

Berong SL, Washburn KW (1998) Effects of genetic variation on total plasma protein, body weight gains and body temperature responses to heat stress. Poult Sci 77:379–385

Bhattacharya AN, Hussain F (1993) Intake and utilization of nutrients in sheep fed different levels of roughage under heat stress. J Anim Sci 71:1761–1766

Bonnet S, Geraert PA, Lessire M et al (1997) Effect of high ambient temperature on feed digestibility in broilers. Poult Sci 76:857–863

Bouraoui R, Lahmar M, Majdoub A et al (2002) The relationship of temperature-humidity index with milk production of dairy cows in a Mediterranean climate. Anim Res 51:479–491

Brown DE, Hacker RR, King GJ (1976) Growth and ACTH responses to cold stress of young pigs fed ad libitum. Can J Anita Sci 56:365

Bruce JM (1981) Ventilation and temperature control criteria for pigs. In: Clark JA (ed) Enviromnental aspect of housoing for animal production. Butterworths, London

Burgos-Zimbelman RJ, Collier MB, Abdallah LH et al (2008) Improving resistance to thermal stress in dairy cows with protected niacin. Proceedings southwest nutrition conference, pp 114–126

Burns JC, Sollenberger LE (2002) Grazing behavior of ruminants and daily performance from warm-season grasses. Crop Sci 42:873–881

Cahaner A, Pinchasov Y, Nir I et al (1995) Effects of dietary protein under high ambient temperature on body weight gain, breast meat yield and abdominal fat deposition of broiler stocking differing in growth rate and fatness. Poult Sci 74:968–975

Charles DR (1994) Comparative climatic requirements. In: Wathes CM, Charles DR (eds) Livestock housing. CAB International, UK

CIGR (1984) Climatisation of animal house. Report of working group, Scottish Farm Building Investigation Unit, Craibstone, Aberdeen, Scotland

Close WH, Mount LE (1976) Gains in protein and fat in the growing pig in relation to environmental temperature. 7th Symposium on energy Metabolism. EAAP Publication No. 15, p 173, Vichy INRA

Colditz PJ (1972) The effect of diet and heat stress on feed intake, growth and nitrogen matabolosm in Friesian, F1 Braham × Friesian and Braham heifers. Aust J Agric Res 23: 717–723

Collier RJ, Beede DK (1985) Thermal stress as a factor associated with nutrient requirements and interrelationships. In: McDowell L (ed) Nutrition of grazing ruminants. Academic, New York
Collier RJ, Baumgard LH, Lock AL et al (2005) Physiological limitations: nutrient partitioning. In: Wiseman J, Bradley R (eds) Yields of farmed species: constraints and opportunities in the 21st century. Proceedings: 61st Easter School. Nottingham, England
Collier RJ, Beede DK, Thatcher WW et al (1995) Influences of environment and its modification on dairy animal health and production. J Dairy Sci 78:2221–2229
Collin A, Milgent J, Le Dividich J (2001) Modelling the effect of high, constant temperature on food intake in young growing pigs. Anim Sci 72:519–527
Conrad JH (1985) Feeding of farm animals in hot and cold environments. In: Yousef MK (ed) Stress physiology in livestock. CRC Press, Inc, Boca Raton
Crampton EW, Harris LE (1969) Applied animal nutrition, 2nd edn. W. H. Freeman and Co, San Francisco
Dale NM, Fuller HL (1979) Effects of diet composition on feed intake and growth of chicks under heat stress. Poultry Sci 58:1529–1534
Danfaer A, Thysen I, Ostergaard (1980) The effect of the level of dietary protein on milk production. 1. Milk yield, liveweight gain and health. Beret. Statens Husdyrbrugsfors, p 492
Dantzer R (1982) Research on farm animal transport in France: a survey. In: Moss R (ed) Transport of animals intended for breeding, production and slaughter. Curr Top Vet Med Anim Sci 18:218–231
Dauncey MJ, Ingram DL (1979) Effect of dietary composition and cold exposure on non-shivering thermogenesis in young pigs and its alteration by beta-blocker propranolol. Brit J Nutr 41:361–370
Davis AV, Merilan CP (1960) Effect of constant environmental temperature and relative humidity on feed digestion by lactating Holstein cows. J Dairy Sci 43:871 (Abstr)
Depocas F, Masironi R (1960) Body glucose as fuel for thermogenesis in the white rat exposed to cold. Am J Physiol 199:1051–1055
Dixon RM, Thomas R, Holmes JHG (1999) Interactions between heat stress and nutrition in sheep fed roughage diets. J Agric Sci 132:351–359
Dobson H (1987) Effect of transport stress on luteinizing hormone released by GnRH in dairy cows. Acta Endocrinol (Copenh) 115:63–66
Drackley JK (1999) Biology of dairy cows during the transition period: the final frontier? J Dairy Sci 82:2259–2273
Ensminger ME, Oldfield JE, Heinemann WW (1990) Feeds and nutrition. The Ensminger publishing company, Clovis
Fallenberg MA, Speisky H (2006) Antioxidants: their effects on broiler oxidative stress and its meat oxidative stability. World's Poult Sci J 62:53–70
Faria DE, Junqueira OM, Souza PA et al (2001) Performance, body temperature and egg quality of laying hens fed vitamins D and C under three environmental temperatures. Rev Bras Cienc Avic 3:49–56
FASS (2010) Federation of animal science societies. Guide for the care and use of agricultural animals in research and teaching, ISBN: 978-1884706-11-0, Champaign, Illinois, USA. www.fass.org/docs.agguide3rd/Ag_3rd_ed.pdf. Accessed January 2010
Fazio E, Ferlazzo A (2003) Evaluation of stress during transport. Vet Res Commun 27:519–524
Fedele D, Pizzillo M, Claps S et al (1993) Grazing behaviour and diet selection of goats on native pasture in southern Italy. Small Rumin Res 11:305–322
Feng J, Zhang M, Zheng S et al (2008) Effects of high temperature on multiple parameters of broilers in vitro and in vivo. Poult Sci 87:2133–2139
Fenster R (1989) Vitamin C and stress management in poultry production. Zootecnica Int 12: 16–22
Ferket PR, Qureshi MA (1992) Performance and immunity of heat-stressed broilers fed vitamin and electrolyte supplemented drinking water. Poultry Sci 71:88–97
Finocchiaro R, van Kaam JBCHM, Portolano B et al (2005) Effect of heat stress on production of Mediterranean dairy sheep. J Dairy Sci 88:1855–1864

Franco-Jimenez DJ, Beck MM (2007) Physiological changes to transient exposure to heat stress observed in laying hens. Poult Sci 86:538–544

Froning GW, Babji AS, Mather FB (1978) The effect of preslaughter temperature, stress struggle anesthetization on color and textural characteristics of turkey muscle. Poult Sci 57:630–633

Fuller MF (1965) The effect of environmental temperature on the nitrogen metabolism and growth of the young pig. Brit J Nutr 19:531–546

Gale CC (1973) Neuroendocrine aspects of thermoregulation. Ann Rev Physiol 35:391–399

Galyean ML, Defoor PJ (2003) Effects of roughage source and level on intake by feedlot cattle. J Anim Sci 81(14 suppl 2):E8–E16

Gaughan JB, Mader TL (2009) Effects of sodium chloride and fat supplementation on finishing steers exposed to hot and cold conditions. J Anim Sci 87:612–621

Geraert PA, Padilha JCF, Guillaumin S (1996) Metabolic and endocrine changes induced by chronic heat exposure in broiler chickens: Growth performance, body composition and energy retention. Brit J Nutr 75:195–204

Goff JP, Horst RL (1997) Physiological changes at parturition and their relationship to metabolic disorders. J Dairy Sci 80:1260–1268

Gonyou JW, Christopherson RJ, Young BA (1979) Effect of cold temperature and winter conditions on some aspects of behaviour of feedlot cattle. Appl Anim Ethol 5:113–119

Graham AD, Christopherson RJ, Thompson JR (1980) Endocrine and metabolic changes in sheep associated to acclimation to constant or intermittent cold exposure. Can J Anim Sci 61:81–90

Graham AD, Nicolson AM, Christopherson RJ (1982) Rumen motility responses to adrenaline and noradrenaline and organ weights of warm and cold acclimated sheep. Can J Anim Sci 62:777–786

Graham NMC, Wainman FW, Blaxter KL et al (1959) Environmental temperature, energy metabolism and heat regulation in sheep. I. Energy metabolism in closely clipped sheep. J Agric Sci 52:13–24

Gudev D, Popova-Ralcheva S, Moneva P et al (2007) Physiological indices in buffaloes exposed to sun. Archiva Zootechnica 10:127–133

Habeeb AA, Marai IFM, Kamal TH (1992) Heat stress. In: Philips C, Piggens D (eds) Farm animals and the environment. CAB International, UK

Hacker RR, Stefanovic MP, Batia TR (1973) Effects of cold exposure on growing pigs: growth, body composition and 17-ketosteroids. Can J Anita Sci 37:739

Hafez ESE (1987) Reproduction in farm animals, 5th edn. LEA and Febiger, Philadelphia

Hafez ESE (1964) Behavioral thermoregulation in mammals and birds. Int. J. Biometeor. 7: 231–240

Hai L, Rong D, Zhang ZY (2000) The effect of thermal environment on the digestion of broilers. J Anim Physiol Anim Nutr (Berl) 83:57–64

Hardwick DC, Linzell JL, Mepham TB (1963) The metabolism of acetate and glucose by the isolated perfused udder. 2. The contribution of acetate and glucose to carbon dioxide and milk constitutes. Biochem J 88:213–220

Hartung J (2003) Effects of transport on health of farm animals. Vet Res Commun 27:525–527

Herpin P, Lefaucheur L (1992) Adaptive changes in oxidative metabolism in skeletal muscle of cold-acclimated piglets. J Therm Biol 17:277–285

Herpin PR, McBride BW, Bayley HS (1987a) Effect of cold exposure on energy metabolism in the young pig. Can J Physiol Pharmacol 65:236–245

Herpin P, Bertin R, Le Devidich J et al (1987b) Some regulatory aspects of thermogenesis in cold-exposed piglets. Comp Biochem Physiol 87:1073–1081

Hodgson J (1986) Grazing behaviour and herbage intake. In: Frame J (ed) Grazing, British grassland society, occasional symposium No. 19, pp 51–64

Holmes CW, Grace ND (1975) A note on the metabolism of Ca, P, Mg, Na and K by pigs growing at a high ambient temperature. Anim Prod 21:341–343

Horton GMJ (1978) Manipulating energy metabolism in sheep exposed to cold. Am J Vet Res 42:146–148

Hutcheson DP, Cole NA (1986) Management of transit stress syndrome in cattle: Nutritional and environmental effects. J Anim Sci 62:555–560

Jaster EH, McCoy GC, Fernando RL (1990) Dietary fat in milk or milk replacers for young calves raised in hutches during winter. J Dairy Sci 73:1843–1850

Jekinson DM, Mabon RM (1973) The effect of temperature and humidity on skin surface pH and ionic composition of skin secretion in Ayshire cattle. Brit Vet J 129:282–290

Johnson HD (1976) Effects of temperature on lactation in cattle. In: Progress in animal biometeorology. The effect of weather and climate on animals, vol. 1. Part 1. Swets and Zeitlinger. B.V. Amsterdam

Johnson HD (1987) Bioclimate effects on growth, reproduction and milk production. In: Johnson HD (ed) Bioclimatology and the adaptation of livestock. Elsevier, Amsterdam

Johnson HD, Ragsdale AC, Berry IL et al. (1962) Effects of various temperature–humidity combinations on milk production of Holstein cattle. Univ Missouri Agric Exp Stn Res Bull (791)

Kadim T, Mahgoub O, Al-Ajmi DS et al (2004) The influence of season on quality characteristics of hot-boned beef *m. longissimus thoracis*. Meat Sci 66:831–836

Kadzere CT, Murphy MR, Silanikove N (2002) Heat stress in lactating dairy cows: a review. Liv Prod Sci 77:59–91

Kartchner RJ (1996) Effect of energy and protein supplementation of cows grazing native winter range forage on intake and digestibility. J Dairy Sci 79:639–646

Kennedy PM, Christopherson RJ, Milligan LP (1986) Digestive responses to cold. In: Milligan LP, Grovum WL, Dobson A (eds) Control of digestion and metabolism in ruminants. USA Prentice-Hall, New Jersey

Kennedy PM, Christopherson RJ, Milligan LP (1976) The effect of cold exposure of sheep on digestion, rumen turnover time, and efficiency of microbial synthesis. Br J Nutr 36:231

Kennedy PM, Christopherson RJ, Milligan LP (1982) Effect of cold exposure on feed protein degradation, microbial protein synthesis and transfer of plasma urea to the rumen of sheep. Brit J Nutr 47:521–535

Knowles TG, Brown SN, Edwards JE et al (1999a) Effect on young calves of a one-hour feeding stop during a 19-hour road journey. Vet Rec 144:687–692

Knowles TG, Warriss PD, Brown SN et al (1999b) Effects on cattle of transportation by road for up to 31-hours. Vet Rec 145:575–582

Koelkebeck KW, Harrison PC, Madindou TJ (1993) Effect of carbonated drinking water on production performance and bone characteristics of laying hens exposed to high environmental temperatures. Poult Sci 72:1800–1803

Koh K, Macleod MG (1999a) Circadian variation in heat production and respiratory quotient in growing broilers maintained at different food intakes and ambient temperatures. Brit Poult Sci 40:353–356

Koh K, Macleod MG (1999b) Effects of ambient temperature on heat increment of feeding and energy retention in growing broilers maintained at different food intakes. Brit Poult Sci 40:511–516

Kouba M, Hermier D, Le Dividich J (2001) Influence of a high ambient temperature on lipid metabolism in the growing pig. J Anim Sci 79:81–87

Kuehn CS, Otterby DE, Linn JG et al (1994) The effect of dietary energy concentration on calf performance. J Dairy Sci 77:2621–2629

Lacetera N, Bernabucci U, Ronchi B et al (1996) Body condition score, metabolic status and milk production of early lactating dairy cows exposed to warm environment. Riv Agric Subtrop Trop 90:43–55

Lambooy E, Engel B (1991) Transport of slaughter pigs by truck over a long distance: Some aspects of loading density and ventilation. Livest Prod Sci 28:163–174

Larbier ZM, Chagneau AM, Geraert PA (1993) Influence of ambient temperature on true digestibility of protein and amino acids of rapeseed and soybean meals in broilers. Poult Sci 72:289–295

Latshaw JD (2008) Daily energy intake of broiler chickens is altered by proximate nutrient content and form of the diet. Poult Sci 87:89–95

Lin H, Mertens K, Kemps B et al (2004) New approach of testing the effect of heat stress on egg quality: mechanical and material properties of eggshell and membrane. Br Poult Sci 45: 476–482

Linn JG (1997) Nutritional management of lactating dairy cows during periods of heat stress. Dairy update, Issue 125. http://www.ansci.umn.edu/dairy/dairyupdates/du125.htm

Lippke H (1975) Digestibility and volatile fatty acids in steers and wethers at 21 and 32°C ambient temperature. J Dairy Sci 58:1860–1869

Lofgreen GP (1974) Ration formulation for relief from heat stress. In: 13th California Feeders day, Imperial Valley. Exp. Stn., El Centro, California

Lomax MA, Baird GD (1982) Blood flow and nutrient exchange across liver gut of the dairy cows, effect of lactation and fasting. Brit J Nutr 49:481–491

Lucy MC, Staples CR, Thatcher WW et al (1992) Influence of diet composition, dry matter intake, milk production and energy balance on time of postpartum ovulation and fertility in dairy cows. Anim Prod 54:323–331

Mader TL (2003) Environmental stress in confined beef cattle. J Anim Sci 81:E110–E119

Maenz DD, Patience JF, Wolynetz MS (1994) The influence of the mineral level in drinking water and the thermal environment on the performance and intestinal fluid flux of newly weaned pigs. J Anim Sci 72:300–308

Mahanna WC (2007) Watch for changing starch digestibility. Feedstuffs, pp 12–13, June 11

Mallonee PG, Beede DK, Collier RJ et al (1993) Production and physiological responses of dairy cows to varying dietary potassium during heat stress. J Dairy Sci 76:819–825

Marai IFM, El-Darawany AA, Fadiel A et al (2007) Physiological traits as affected by heat stress in sheep-A review. Small Rumin Res 71:1–12

Marai IFM, Ayyat MS, Abd El-Monem UM (2001) Growth performance and reproductive traits at first parity of New Zealand White female rabbits as affected by heat stress and its alleviation, under Egyptian conditions. Trop Anim Health Prod 33:457–462

Marai IFM, Daader AM, Abdel-Samee AM et al (1997) Winter and summer effects and their ameleoration on lactating Friesian and Holstein cows maintained under Egyptian conditions. In: Proceedings of international conference on animal, poultry, rabbits and fish production and health, Cairo, Egypt

Maria GA, Villarroel M, Chacon G et al (2004) Scoring system for evaluating the stress to cattle of commercial loading and unloading. Vet Rec 154:818–821

Masoro EJ (1966) Effect of cold on metabolic use of lipids. Physiol Rev 46:67

Maurice DV, Lightsey SF, Abudabos A et al (2002) Factors affecting ascorbic acid biosynthesis in chickens: III. Effect of dietary fluoride on L-gulonolactone oxidase activity and tissue ascorbic acid (AsA) concentration. J Anim Physiol Anim Nutr 86:383–388

Mckee SR, Sams AR (1997) The effect of seasonal heat stress on rigor development and the incidence of pale, exudative turkey meat. Poultry Sci 76:1616–1620

McNaughton JL, Kubena LF, Deaton JW et al (1977) Influence of dietary protein and energy on the performance of commercial egg type pullets reared under summer conditions. Poult Sci 56:1391–1398

Minka NS, Ayo JO (2007) Effects of loading behaviour and road transport stress on traumatic injuries in cattle transported by road during the hot-dry season. Life Sci 107:91–95

Mitchell MA, Kettlewell PJ (1998) Physiological stress and welfare of broiler chickens in transit: solutions not problems! Poult Sci 77:1803–1814

Monty DE, Kelly LM, Rice WR (1991) Acclimatization of St. Croix, Karakul and Rambouillet sheep to intense and dry summer heat. Small Rumin Res 4:379–392

Moose MG, Ross CV, Pfander WH (1969) Nutritional and environmental relationships with lambs. J Anim Sci 29:619–627

Muller LD, Heinrichs AJ, Cooper JB et al (1986) Supplemental niacin for lactating cows during summer feeding. J Dairy Sci 69:1416–1420

Myer R, Bucklin R (2001) Influence of hot-humid environment on growth performance and reproduction of swine. Document AN107, extension, institute of food and agricultural sciences, University of Florida, http://edis.ifas.ufl.edu/AN107

Nardone A, Lacetera NG, Ronchi B et al (1992) Effetti del caldo ambientale Sulla produzione di latte e sui consumi alimentari di vacche Frisone. Prod Anim 5:1–15

Nardone A, Ronchi B, Lacetera NG et al (2010) Effects of climate changes on animal production and sustainability of livestock systems. Liv Sci 130:57–69

Nardone A, Ronchi B, Valentini A, (1991) Effects of solar radiation on water and food intake and weight gain in Sarda and Comisana female lambs. In: Animal husbandry in warm climates, vol 55. EAAP Publication

Negrave R (1996) Sheep grazing controls Calamagrostis Canadensis-dominated vegetation in the Boreal forest integrated forest vegetation management options and application, FRDA Report 251. p 2

Niaber JA, Hahn GL (2007) Livestock production system management responses to thermal challenges. Int J Biometeorol 52:149–157

Nicholson JWG, McQueen REB (1980) Effect of cold on digestibility of chopped or pelleted hay by sheep. Can J Anim Sci 60:571–578

Noblet J, Sève B, Jondreville C (2002) Valeurs nutritives pour les porcs. In Tables de composition et de valeur nutritive des matières premières destinées aux animaux d'élevage: porcs, volailles, bovins, ovins, caprins, lapins, chevaux, poissons, Coord. D. Sauvant, J.M. Perez and G. Tran, INRA Editions et AFZ, Paris

Northcutt JK, Foegeding EA, Edens FW (1994) Water-holding properties of thermally preconditioned chicken breast and leg meat. Poult Sci 73:308–316

NRC (1981) Effect of environment on nutrient requirements of domestic animals. National Academic Press, Washington

NRC (2001) Nutrient requirements of dairy cattle, 7th revised ed. National Academy of Sciences, Washington

Okamoto M, Robinson JB, Christopherson RJ et al (1986) Summit metabolism of new born calves with and without colostrum feeding. Can J AnimSci 66:937–944

Oldham JD (1984) Protein-energy interrelationships in dairy cows. J Dairy Sci 67:1090–1114

Osman AM, Tanios NI (1983) The effect of heat on the intestinal and pancreatic levels of amylase and maltase of laying hens and broilers. Comp Biochem Physiol 75A:563–567

Padua JT, Dasilva RG, Bottcher RW et al. (1997) Effect of high environmental temperature on weight gain and food intake of Suffolk lambs reared in a tropical environment. In: Proceedings of 5th international symposium, Bloomington, Minnesota, USA, pp 809–815

Pardue SL, Thaxton JP (1986) Ascorbic acid in poultry: a review. World's Poult Sci J 42: 107–123

Peeters E, Driessen B, Steegmans R et al (2004) Effect of supplemental tryptophan, vitamin E, and a herbal product on responses by pigs to vibration. J Anim Sci 82:2410–2420

Peng CL, Heitman H (1974) The effect of ambient temperature on the thiamin requirement of growing-finishing pigs. Br J Nutr 32:1–9

Phillips PA, Young BA, McQuitty JB (1982) Live weight, protein deposition and digestibility responses in growing pigs exposed to low temperature. Can J Anim Sci 62:95–108

Pulina G, Mazzette A, Battacone G et al (2006) Feed restriction alters milk production traits in Sarda dairy ewes. J Dairy Sci 89((Suppl.1)):59

Pullar JD, Webster AJF (1977) The energy cost of fat and protein deposition in the rat. Brit J Nutr 37:355–363

Quiniou N, Noblet J (1999) Influence of high ambient temperatures on performance of multiparous lactating sows. J Anim Sci 77:2124–2134

Quiniou N, Dubois S, Noblet J (2000) Voluntary feed intake and feeding behaviour of group-housed growing pigs are affected by ambient temperature and body weight. Liv Prod Sci 63:245–253

Ramnath V, Rekha PS, Sujatha KS (2008) Amelioration of heat stress induced disturbances of antioxidant defense system in chicken by Brahma Rasayana. Evid- Based Compl Alt Med 5:77–84

Rassu SPG, Enne G, Ligios S et al (2004) Nutrition and reproduction. In: Pulina G (ed) Dairy sheep nutrition. CABI Publishing, Wallingford

Renaudeau D, Ana C, Noblet J (2003) Effects of dietary fiber on performance of multiparous lactating sows in a tropical climate. J Anim Sci 81:717–725

Renaudeau D, Quiniou N, Noblet J (2001) Effect of high ambient temperature and dietary protein level on performance of multiparous lactating sows. J Anim Sci 79:1240–1249

Robertshaw D (1981) The environmental physiology of animal production. In: Clark JA (ed) Environmental aspects of housing for animal production. Butterworth, London

Rowan TG (1992) Thermoregulation in neonatal ruminants. In: Varley MA, Williams PEV, Lawrence TLJ (eds) Neonatal survival and growth, occasional Pulb. No.15, 13–24; Edinburgh, UK: Br Soc Anim Prod

Rozenboim I, Tako E, Gal-garber O et al (2007) The effect of heat stress on ovarian function of laying hens. Poult Sci 86:1760–1765

Sahin K, Kucuk O (2003) Heat stress and dietary vitamin supplementation of poultry diets. Nutr Abstr Rev Ser B Livest Feeds Feed 73:41R–50R

Sayre RN, Briskey EJ, Hoekstra WG (1963) Alteration of postmortem changes in porcine muscle by preslaughter heat treatment and diet modification. J Food Sci 28:292–297

Schingoethe DJ, Casper DP, Drackley JK et al (1986) Increased solids intake and feeding frequency for calves in hutches during cold weather. J Dairy Sci 69:1063–1069

Schneider PL, Beede DK, Wilcox CJ (1986) Response of lactation cows to dietary sodium source and quantity and potassium quantity during heat stress. J Dairy Sci 69:99–110

Scibilia LS, Muller LD, Kensinger RS et al (1987) Effect of environmental temperature and dietary fat on growth and physiological responses of new born calves. J Dairy Sci 70:1 426–1433

Sconberg S, Nockels CF, Bennet BW et al (1993) Effects of shipping, handling, adrenocorticotropic hormone, and epinephrine on α-tocopherol content of bovine blood. Am J Vet Res 54:1287–1293

Sejian V, Maurya VP, Naqvi SMK (2010a) Adaptability and growth of Malpura ewes subjected to thermal and nutritional stress. Trop Anim Health Prod 42:1763–1770

Sejian V, Maurya VP, Naqvi SMK (2011) Effect of walking stress on growth, physiological adaptability and endocrine responses in Malpura ewes in a semi-arid tropical environment. Int J Biometeorol. doi:10.1007/s00484-011-0420-y

Sejian V, Maurya VP, Naqvi SMK (2010b) Adaptive capability as indicated by endocrine and biochemical responses of Malpura ewes subjected to combined stresses (thermal and nutritional) in a semi-arid tropical environment. Int J Biometeorol 54:653–661

Sevi A, Annicchiarico G, Albenzio M et al (2001) Effects of solar radiation and feeding time on behavior, immune response and production of lactating ewes under high ambient temperature. J Dairy Sci 84:629–640

Shibasaki M, Wilson TE, Crandall CG (2006) Neural control and mechanisms of eccrine sweating during heat stress and exercise. J Appl Physiol 100:1692–1701

Silva BAN, Noblet J, Oliveira RFM et al (2009) Effects of dietary protein concentration and amino acid supplementation on the feeding behavior of multiparous lactating sows in a tropical humid climate. J Anim Sci 87:2104–2112

Singh G, Karim SA (1995) Changes in birth weight of temperate sheep breeds under tropical conditions. World Rev Anim Prod 30:55–60

Smith MO (1994) Effects of electrolyte and lighting regimen on growth of heat-distressed broilers. Poult Sci 73:350–353

Song R, Foster DN, Shurson GC (2011) Effects of feeding diets containing bacitracin methylene disalicylate (BMD) to heat-stressed finishing pigs. J Anim Sci doi:10.2527/jas.2010-3218

Springell PH (1968) Water content and water turnover in beef cattle. Aust J Agri Res 19:129–144

Stahly TS, Cromwell GL, Aviotti MP (1979) The effect of environmental temperature and dietary lysine source and level on the performance and carcass characteristics of growing swine. J Anim Sci 49:1242
Staples CR (2007) Nutrient and feeding strategies to enable cows to cope with heat stress conditions. 22nd annual southwest nutrition management conference, University of Arizona, Tempe, pp 93–108
Sterling KG, Bell DD, Pesti GM, Aggrey SE (2003) Relationships among strain, performance, and environmental, temperature in commercial laying hens. J Appl Poult Res 12:85–91
St-pierre NR, Cobanov B, Schnitkey G (2003) Economic losses from heat stress by US livestock industries. J Dairy Sci 86:52–77
Tankson JD, Vizzier-Thaxton Y, Thaxton JP et al (2001) Stress and nutritional quality of broilers. Poult Sci 80:1384–1389
Tarrant PV (1989) The effect of handling, transport, slaughter and chilling on meat quality and yield in pigs: A review. Ir J Food Sci Technol 13:79–107
Teeter RG, Smith MO, Owens FN et al (1985) Chronic heat stress and respiratory alkalosis: Occurrence and treatment in broiler chicks. Poult Sci 64:1060–1064
Tepperman J (1980) Metabolic and endocrine physiology. Year Book Medical, Inc, Chicago
Thatcher WW, Collier RJ (1982) Effect of heat on animal productivity. In: Rechcigl M Jr (ed) Handbook of agricultural productivity. vol. II. CRC press, Inc, Boca Raton
Thompson GE, Thomson EM (1977) Effect of cold exposure on mammary circulation, oxygen consumption and milk secretion in the goat. J Physiol 272:187–196
Tucker WB, Harrison GA, Hemken RW (1988) Influence of dietary cation-anion balance on milk, blood, urine, and rumen fluid in lactating dairy cattle. J Dairy Sci 71:346–354
Tyrrell HF, Moe PW, Flatt WP (1970) Influence of excess protein intake on energy metabolism of the dairy cow. Fifth symposium energy metabolism. Farm Anim, pp 69–71
Verstegen MWA, Brandsma HA, Mateman G (1982) Feed requirement of growing pigs at low environmental temperatures. J Anim Sci 55:88–94
Verstegen MWA, Brandsma HA, Mateman G et al (1984) Effect of cold thermal environment on feed requirements, growth rate and slaughter quality in pigs. Archiv Exp Vet 38:431–438
Verstegen MWA, Brascamp EW, van der Hel W (1978) Growing and fattening of pigs in relation to temperature of housing and feeding level. Can J Anim Sci 58:1
Verstegen MWA, Close WH, Start IB et al (1973) The effects of environmental temperature and plane of nutrition on heat loss, energy retention and deposition of protein and fat in groups of growing pigs. Br J Nutr 30:21
Von Engelhardt W, Hales JRS (1977) Partition of capillary blood flow in the rumen, reticulum and omasum of sheep. Am J Physiol 232:453–459
Waldroup PW, Mitchell RJ, Payne JR et al (1976) Performance of chicks fed diets formulated to minimize excess levels of essential amino acids. Poult Sci 55:243–253
Wallis IR, Balnave D (1984) The influence of environmental temperature, age and sex on the digestibility of amino acids in growing broiler chickens. Br Poult Sci 25:401–407
Warren WP, Martz FA, Asay KH et al (1974) Digestibility and reate of passage by steers fed fescue, alfalfa and orchard grass hay in 18 and 38°C ambient temperatures. J Anim Sci 39:93–96
Webster AJF, Gordon JG, McGregor R (1978) The cold tolerance of beef and dairy type calves in the first weeks of life. Anim Prod 26:85–92
Webster AJF (1974) Prediction of heat losses from cattle exposed to cold outdoor environments. J Appl Physiol 30:634–643
Webster AJF (1981) Optimal housing criteria for ruminants. In: Clark JA (ed) Enviromnental aspect of housing for animal production. Butterworths, London
Weniger JH, Stein M (1992) Influence of environmental temperature and humidity on nutrient digestibility of sheep. 1. Aims, experimental procedure and digestibility. Zuchtungskunde 64:148–155
West JW (2003) Effects of heat-stress on production in dairy cattle. J Dairy Sci 86:2131–2144
Xin H, Lee K (1996) Use of Aqua-Jel® and feed for nutrient supply during long journey air transport of baby chicks. Trans ASAE 39:1123–1126

Young BA, Walker B, Dixon AE et al (1989) Physiological adaptation to the environment. J Anim Sci 67:2426–2432

Young BA (1975) Temperature induced changes in metabolism in body weight of cattle (*Bos taurus*). Can J Physiol Pharmacol 53:947–953

Young BW (1976) Effects of cold environments on nutrient requirements of ruminants. In: Fonnesbeck PV, Harris LF, Kearl LC (eds) Proceedings 1st international symposium on feed composition, animal nutrient requirements, and computerization of diets, Logan, Utah State University, pp 491–496

Yousef MK (1985) Basic principles. Stress physiology in livestock, vol 1. CRC Press, Boca Raton

Zuprizal M, Larbiel A, Chagneau M et al (1993) Influence of ambient temperature of true digestibility of protein and amino acids of rapeseed and soybean meal in broilers. Poult Sci 72:289–295

Chapter 9
Role of Pineal Gland in Relieving Environmental Stress

Veerasamy Sejian, Saumya Bahadur, Vijay Kumar Bharti and Rajendra Swaroop Srivastava

Abstract Pineal gland is considered as a neuroendocrine transducer of cyclic photic input, which is responsible for the seasonal changes in the reproductive capability of various animal species. Considerable evidence has now been accumulated to indicate its participation in a wide range of reproductive processes and associated organs, among which pineal-adrenal, pineal-thyroid, and pineal-immune system relationships are the thrust areas of research investigations. Pineal gland is known to have an anti-stressogenic effect in mammals and birds. It is also known to have a tranquilizing effect on animals. Melatonin is included in the feed of pigs raised in commercial piggeries to protect them against occurrence of peptic ulcers. Although the antistress properties of melatonin are established, yet such reports are very meager in domestic livestock species. Several studies conducted on goats suggest that there is a strong interrelationship between the pineal gland and adrenal cortex in relieving thermal stress. The significant effect of melatonin on various adrenal cortex secretions and functions during thermal stress establishes such relationship between the two endocrine glands. These studies established the anti-stress properties of melatonin in goats. Several recent studies conducted in goats had established that apart from melatonin, there are several other peptides produced from the pineal gland, which have anti-thermal stress properties.

V. Sejian (✉) · S. Bahadur
Adaptation Physiology Laboratory, Division of Physiology and Biochemistry,
Central Sheep and Wool Research Institute, Avikanagar, Via-Jaipur,
Rajasthan 304501, India
e-mail: drsejian@gmail.com

V. K. Bharti
Environmental Physiology and Toxicology Laboratory, DIHAR,
Defense Research and Development Organization, Leh Ladakh 901205, India

R. S. Srivastava
Division of Physiology and Climatology, Indian Veterinary Research Institute,
Izatnagar, Bareilly, Uttar Pradesh 243122, India

V. Sejian et al. (eds.), *Environmental Stress and Amelioration in Livestock Production*, DOI: 10.1007/978-3-642-29205-7_9,

The data generated from these studies help us to understand the functional relationship between pineal and adrenal glands, and how these influence each other for the well-being of the domestic and farm animals during thermal stress. Given the importance of thermal stress in hampering animal productivity to a greater extent in tropical countries, these findings have greater significance in terms of improving the economy of farm households as well as poor farmers.

Keywords Pineal · Adrenal · Immune system · Stress · Melatonin · Cortisol

Contents

9.1 Introduction

Environment is a complex system, comprising multiple elements networking among each other. Out of the wide network of components, climate happens to be of paramount importance. Climate is constituted of photoperiod, temperature, humidity, solar radiation, etc., which strongly governs the behavioral patterns of animals. The climatic factors induce neuroendocrinal responses in animals, thereby affecting their health and production.

Environmental stress consists of physical 'abiotic' characters, such as climatic factors, toxins, and radiation or 'biotic' character, such as parasites and competitors (Kristensen 2004). The physiology of stress involves three systems directly; the nervous, the endocrine, and the immune, all of which can be prompted by perceived threats (Everly 2002). Stress is viewed as a general biologic and usually functional response to environmental and bodily demands. Adaptations to demands of physical environment through behavioral; physiologic; neuroendocrine; and metabolic responses, so as to maintain homeostasis, are a general biologic phenomenon in animals. Cannon (1915) was the pioneer in bringing the concept of adaptation linking the brain, behavior, and endocrine system to light. Seyle (1946) emphasized on the role of adrenal cortex as the chief coordinator of nonspecific bodily response to stress. This concept led to the idea that the level of corticosteroids in blood reflects the state of stress in an organism. General adaptation syndrome (Seyle 1950), hypothesizes an alarm, a resistance, and an exhaustion phase of stress response of adrenal gland.

In the tropical and sub tropical environment, heat stress is the most vital climatic stress that adversely affects livestock. Heat stress with high humidity, tends to further disturb animal physiology. Animals employ some response mechanisms to adapt to the hot climate and although these mechanisms are helpful for survival, they are detrimental to animal productivity. Exposure to different environmental stressors elicits various physiologic changes. These changes are either recognized as emergency, activating adrenomedullary system and releasing catecholamines or general adaptation response in which pituitary-adrenal axis is activated to release corticosteroids.

The adrenal gland has been known to be influenced by the environment, more than any other endocrine glands of the body. Changes in the external environmental stimuli (photoperiod, sound, temperature, etc.) and emotional or traumatic stress appear to increase the activity of the hypothalamo-hypophyseal-adrenal axis (HHAA), resulting in increased adrenocortical/medullary activity. In the area of endocrine research, thrust investigations are presently being carried out to establish relationship between Pineal-Adrenal and Pineal-Adrenal-Immune system.

Pineal gland is recognized as a neuroendocrine organ transducing circadian rhythm/cyclic photic input, which affects seasonal changes in reproductive capability of various species (Reiter 1991). Substantial data have been gathered to illustrate the participation of pineal gland in a variety of extra-reproductive processes (Johnson 1982) wherein, pineal-adrenal, pineal-thyroid, and pineal-immune (PI) system interplay are the most attractive areas of scientific exploration. Pineal-adrenal liaison research dates back to the discovery of adrenoglomerulotropin in pineal extracts of rats (Farrell 1960). Ever since then, researchers have made attempts to set up the possible role of the pineal gland in influencing the (HHAA), (Mess 1983). Early on, proposition of a bilateral relationship between the pineal and adrenal glands or melatonin and corticosteroids was reinforced by experimental and clinical findings suggesting that melatonin may ably protect organisms against stress-induced pathologies (Khan et al. 1990).

It has been suggested that melatonin may act as an important regulator of adrenal functions (Touitou et al. 1989; Vaughan 1984). Glucocorticoids are one amid other endogenous compounds that have been shown to influence melatonin production in various vertebrates (Bauer et al. 1989; Demisch et al. 1988; Zawilska and Sadowska 2002). In vivo and in vitro experiments revealed a suppressive action of melatonin on glucocorticoids production and release (Ogle and Kitay 1978; Vaughan et al. 1972). They also aid termination of stress response by acting at extrahypothalamic regulatory centers, the hypothalamus and the pituitary. The negative feedback mechanism of glucocorticoid on adrenocorticotrophic hormone (ACTH) secretion tends to limit the duration of total tissue exposure to glucocorticoid, thereby playing down its catabolic, lipogenic, antireproductive and immunosuppressive effect (Tsigos and Chrousos 2002).

Extensive research has been conducted to establish the PI system interactions. It has been indicated by in vivo and in vitro experiments that pineal gland, via its hormone melatonin, boosts immune responses (Guerrero and Reiter 2002; Nelson 2004). In fact, the PI axis involving melatonin and immunocompetent cells, is an integral part of innate immune response (Markus et al. 2007). Melatonin possesses the ability to upregulate the immunosuppression although not directly on immunocompetent cells, but mediated via the endogenous opioid system upon antigen activation of T cells (Pierpaoli and Maestroni 1987). Furthermore, melatonin is required for photoperiodic regulation of circulating leukocytes and neural immune interactions that mediate several aspects of immune functions (Wen et al. 2007). Melatonin plays significant immunomodulatory role in immunocompromised animals and also stimulates production of cytokines (Guerrero and Reiter 2002). Corticosteroids possess potent immuno-suppressive properties in lymphocytes both in vitro and in vivo (Cupps and Fauci 1982). Melatonin reverses the depression of antibody response induced by corticosterone (Maestroni et al. 1988).

9.2 Pineal Gland and its Significance

Pineal gland is a small endocrine gland, situated in the caudal portion of the third ventricle. It is white in color and pinecone shaped. It is also called pineal body, epiphysis cerebri, epiphysis or the "third eye" because of its light-transducing ability. The importance of the pineal gland was recognized long ago in the 16th century and the gland was thought to be the seat of the soul. However, it is only in the past three decades that remarkable advances in the knowledge of the functional significance of the epiphysis have been made (Vincenzo et al. 1996). René Descartes, who dedicated much time to the study of the pineal gland, called it the "seat of the soul." He believed that it was the point of connection between the intellect and the body.

Histologically, the pineal is composed of pinealocytes and glial cells. Pinealocytes produce and secrete serotonin derivative of melatonin, a hormone that plays a major role in sexual development, hibernation in animals, metabolism, and seasonal breeding. The pineal regulates circadian rhythms or biologic rhythms through the secretion of melatonin. Seasonal changes, which bring forth changes in day length have profound effects on reproduction in many species, and melatonin is the key player in controlling such events. The reproductive function shows both qualitative and quantitative differences especially in the seasonal cycle in both females and males. This feature is especially evident in animals that breed seasonally. Observed variations in the reproductive cycle reveal that breeding periods are timed by changes in the photoperiod. The pineal regulates seasonal changes in the reproductive function of these animal species through its endocrine activity (Reiter 1991; Weaver et al. 1993). Long-day breeders, such as the hamster, are reproductively active during the summer (Silman 1993). Because reproductive regression was associated with the extended period of melatonin elevation, the terms antigonadal or antigonadotrophic were commonly applied to melatonin. Antigonadal action is exerted by inhibition of gonadotrophin-releasing hormone (GnRH) (Buchanan and Yellon 1991). In contrast, in short-day breeders, such as the sheep, are engaged in reproductive activity (breeding season), which is associated with autumn and winter seasons with decreased day length. Melatonin exerts a stimulatory effect on the reproductive axis in these species (Karsch et al. 1984) and can be referred to as being progonadotrophic (Coelho et al. 2006; Wagner et al. 2008; Chemineau et al. 2008). In addition, melatonin receptors have been identified in hypothalamic neurons governing the release of pituitary gonadotrophs (Roy et al. 2001; Dubocovich and Markowska 2005), in gonadotropins of the anterior pituitary (Johnston et al. 2003; Balik et al. 2004), and in both female and male gonads (Woo et al. 2001; Frungieri et al. 2005).

Melatonin may play a protective, anti-stress role in the gastric mucosa via a mechanism involving the central nervous system and may inhibit the induction of gastric ulcers by restraining stress or centrally administered thyrotropin releasing hormone (TRH). Both in vitro and in vivo experiments show that the pineal gland, via its hormone melatonin, enhances immune function. Mechanism involved in the immunostimulatory effect is not well understood, but some evidences suggest the existence of specific binding sites for melatonin on immune cells. Additionally, in both in vitro and in vivo experiments, melatonin has been found to protect cells, tissues, and organs against oxidative damage induced by a variety of free radical generating agents and processes. Melatonin's function as a free radical scavenger (Tan et al. 1993) and antioxidants is likely assisted by the ease with which it crosses morphophysiologic barriers, e.g., the blood–brain barrier, and enters cells and subcellular compartments (Hardeland 2005). Melatonin has been reported to stimulate/alter the activities of enzymes (Rodriguez et al. 2004), which improve the total antioxidative defense capacity of the organism, i.e., superoxide dismutase (SOD), glutathione peroxidase (GPX), glutathione reductase (GR), glucose-6-phosphate dehydrogenase, and nitric oxide synthase (NOS). Melatonin, ubiquitously acting as antioxidant has implications for the optimal function of cells and organs, including those of the reproductive system (Tamura et al. 2008). Although the direct free radical scavenging actions of melatonin are accomplished without an interaction with specific receptors, the stimulatory effects of melatonin on antioxidative enzymes are likely mediated either by membrane receptors or by nuclear- or cytosol-binding sites (CBS) (Tomas-Zapio and Coto-Montes 2005).

Melatonin participates in homeostasis by controlling the level of proliferation and differentiation of cells and thereby preventing the growth of malignant cells (Sanchez Barcelo et al. 2003). Melatonin is highly lipophilic and its effects on tumor cells are independent of melatonin receptor mediated pathway. Experimental studies revealed that melatonin controls the cell proliferation by regulating both the intracellular growth stimulating hormone (GSH) and nitric oxide (NO) levels, both of which are essential for the maintenance of homeostasis (Stephanie et al. 2002). Melatonin prevents cancerous growth by potentiating the immunocompetent mechanism (Vijayalaxmi et al. 2002). Melatonin has also been shown to influence the growth of natural killer (NK) cells through interleukin-2 (IL-2) production and this action explains melatonin's antiproliferative effect. NK cells play an important role in immunosurveillance against the development of tumors and metastases (Srinivasan 2000; Vijayalaxmi et al. 2002).

9.3 Pineal Gland Secretions

The pineal gland contains a number of peptides, including GnRH, TRH, oxytocin, and vasotocin, along with a number of important neurotransmitters, such as somatostatin, norepinephrine, serotonin, and histamine. Apart from these hormones, there are other peptides, such as tryptamine, 5-methoxytryptamine, epithalamine and epithalon, that are secreted from the pineal gland (Sibarov et al. 2002), The major pineal hormone, however, is melatonin, a derivative of the amino acid tryptophan. Figure 9.1 represents the schematic description of synthesis of melatonin.

9.4 Pineal-Adrenal Relationship

Stress is expressed as a response of an organism to external stimuli or change. The most often used nomenclature defines environmental stimuli that lead to an imbalance of homeostasis as "stressor" and the corresponding defense reaction of an animal as "stress response" (Mostl and Palme 2002). Different types of stressors may elicit varying degrees of responses in animals. The cellular and molecular mechanisms underlying these specific responses tailored for individual circumstances present a major challenge for the future. During stress, various endocrine responses are involved to soothe the individual. The front-line hormones that overcome stressful situations are the glucocorticoids and catecholamine. The secretion of glucocorticoids is a classic endocrine response to stress (Kannan et al. 2000). Currently, it appears that glucocorticosteroids provide an initial integrating signal, which in conjunction with other hormones and paracrine secretions may determine specific behavioral and physiologic responses that help the animal adapt to different environmental conditions (Wingfield and Kitaysky 2002). There are recent reports, which suggest the involvement of melatonin in heat stress relief or tolerance in farm animals (Collier et al. 2008; Sejian et al. 2008a). Furthermore, there is evidence in support of the role of melatonin and prolactin in the up-regulation of heat shock protein 70 (HSP 70) expression during heat stress (Collier et al. 2008).

During thermal stress, glucocorticoids can modify the level of melatonin in the body so that the anti-stress action of melatonin could be enhanced. The action of glucocorticoids on the level of pineal melatonin shows that a

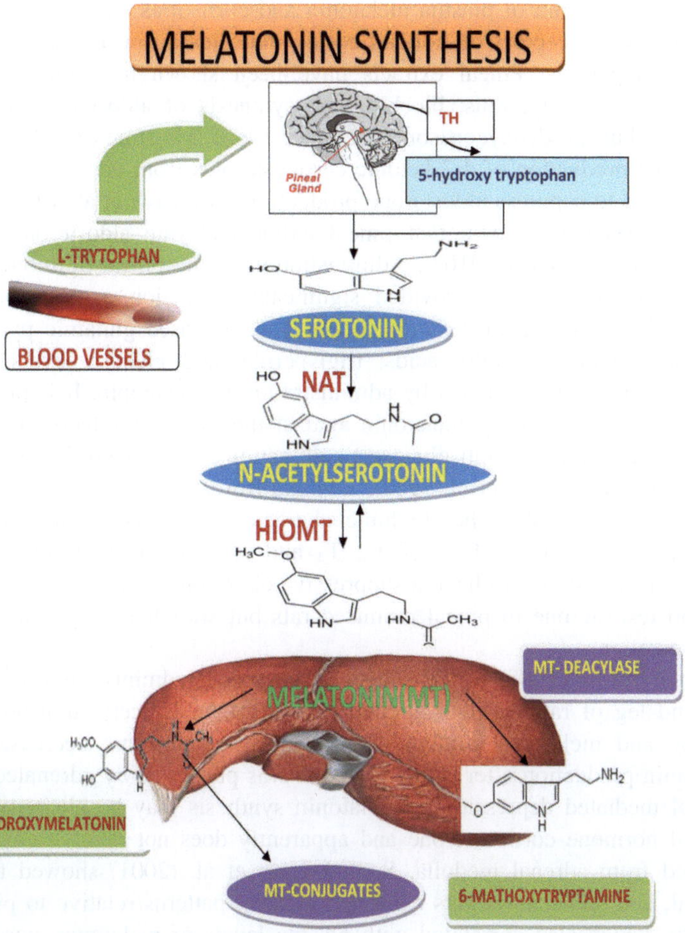

Fig. 9.1 Schematic description of synthesis of melatonin. The synthesis of melatonin takes place in the pinealocytes, the cellular units of the pineal gland. Biosynthesis of melatonin starts with the active absorption of amino acid tryptophan from the blood. At first, 5-hydroxy-tryptophan is formed through tryptophanehydroxylase (*TH*). Subsequently, the transformation into serotonin (5-hydroxy-tryptamine) takes place by the enzyme aromatic-amino-acid-decarboxylase (*AAMD*). The next step is the transformation of serotonin into N-acetylserotonin through N-acetyl-transferase (*NAT*) as the limitating step for melatonin synthesis. Finally, the synthesis of melatonin (N-acetyl-5-methoxy-tryiptamine) through the enzmye hydroxyindole-O-methyltransferase (*HIOMT*) takes place

mechanism exists for glucocorticoids to act on the pineal to control level of melatonin (Sejian et al. 2008b). In addition, melatonin may act as an important regulator of adrenal function (Sejian et al. 2010a; Sejian et al. 2010b; Sejian and Srivastava 2010a). In vitro and in vivo experiments revealed a suppressive action of melatonin on adrenal glucocorticoid production and release. The relationship between the rhythm of plasma melatonin and cortisol as well as the presence of the melatonin receptors on adrenal cortex indicates a direct effect of melatonin on steroidogenesis. Pineal extracts have been shown to inhibit adrenal beta hydroxylase activity, thus, blocking the synthesis of aldosterone, cortisol and cortisone, but not deoxycortisone (Lommer 1966). Melatonin has been shown to reduce the production of 4-3-ketonic corticosteroids from endogenous precursors in vitro and to suppress aldosterone production (Gromava et al. 1967). However, some studies have shown that pineal extracts elevate aldosterone production (Sejian and Srivastava 2010a). Administration of the pineal hormone melatonin to growing female rats provided significant protection against the injurious effects of glucocorticoid dexamethasone. The increased glutamic pyruvic transaminase (GPT), free fatty acids, triglyceride, and glucose levels caused by glucocorticoid were reversed by administration of melatonin. It is proposed that the protection offered by melatonin against the injurious effects of dexamethasone is due to a direct anti-glucocorticoid action and does not involve any other endocrine organ (Mori et al. 1984). Zwirska-Korczala et al. (1991) reported that pinealectomy abolishes the rhythmic character of corticosterone secretion and alters the circadian rhythms of T_3, T_4 and testosterone secretion. Exogenous melatonin was shown to have a suppressive effect on the diurnal secretion of T_3, T_4, and testosterone in pinealectomised rats but stimulated rhythmic corticosterone secretion.

Troiani et al. (1988) observed that subcutaneous administration of saline into the hind leg of rats at night elicited a decrease in N-acetyl transferase (NAT) activity and melatonin content of the pineal gland. This decrease in pineal melatonin production after saline injection was prevented by adrenalectomy. The adrenal mediated depression in melatonin synthesis may be elicited by adrenal cortical hormone corticosterone and apparently does not involve any chemicals released from adrenal medulla. Sudhakumari et al. (2001) showed that pineal, adrenal, and gonadal weights exhibited cyclical patterns relative to pineal gland activity, which also correlated with plasma levels of melatonin, corticosterone, and gonadal steroids in Jungle bush quail. They also observed that increased photoperiod, ambient temperature, and rainfall were positively correlated with adrenal and gonadal functions and inversely related to pineal gland activity. Pineal gland is known to have anti-stressogenic effect in mammals and birds. It is also known to have a tranquilizing effect on animals. Melatonin is included in the feed of pigs raised in commercial piggeries to protect them against occurrence of peptic ulcers. There are also report indicating role of melatonin to upregulate HSP 70 gene expression during heat stress (Collier et al. 2008). Experimental data indicate that melatonin has anti-stress properties in goats. The significant effect of synthetic glucocorticoid on melatonin level during thermal stress establishes the relationship between these two endocrine glands. Given the importance of thermal stress in hampering animal productivity in tropical countries, this finding has potential impact on the revenue of commercial and subsistence farmers.

9.5 Pineal-Adrenal-Immune System Relationship

In recent years, much attention has been devoted to the possible interaction between pineal gland, adrenal gland, and the immune system (Esquifino et al. 2004; Maestroni 2001; Guerrero and Reiter 2002). Maestroni et al. (1986) first showed that inhibition of melatonin synthesis causes inhibition of cellular and humoral responses in mice. The diurnal and seasonal changes in the immune system have been shown to correlate with melatonin synthesis and secretion (Skwarlo-Sonta 2002). Melatonin is synthesized by human lymphocytes and this finding adds further support to the hypothesis that melatonin plays a role in the regulation of the human immune system (Carrillo-Vico et al. 2004). Melatonin also plays a significant role in guarding the non-specific responses of laboratory animals and birds (Paredes et al. 2007; Sanchez et al. 2004). Melatonin provides a functional link between the neuroendocrine and immuno-hematopoietic systems (Claustrat et al. 2005).

During the last 20 years, the number of published papers unfolding bidirectional interrelationships between the pineal gland, melatonin, and the immune system has increased several folds (Conti and Maestroni 1994), most of them aimed to elucidate the mechanism(s) involved. In humans, there is evidence showing an inverse relationship between plasma melatonin and cortisol circadian rhythms. The connection between rhythms of plasma melatonin and cortisol as well as the occurrence of melatonin receptor on adrenal cortex can be taken as a validation for direct effects of melatonin on steroidogenesis (Torres-Farfan et al. 2003). Mechanisms which melatonin uses to influence immune system function are complex, but include participation of mediators (endogenous opioids, cytokines, hormones, zinc pool) as well as specific binding sites on the immune cells. Melatonin, being a highly lypophilic compound, may also penetrate immune cells without mediation of specific receptors. By synthesizing and secreting soluble factors, cytokines, the immune system may influence pineal gland function and thereby closing the information loop that sustains homeostasis during harmful environmental conditions.

Deciphering the melatonin message within the body is critical to adapt to the physiologic functions of an animal to environmental conditions and needs, and this adaptation would increase the probability of its survival. Previously, immune system was considered to function autonomously, but now ample experimental evidences support bidirectional interactions with the nervous and endocrine systems (Goetzl and Sreedharan 1992; Fabris 1994; Homo-Delarche and Dardenne 1993). Adrenal corticosteroids were the first hormonal factors considered to be regulators of the diurnal rhythm of the immune system. There is substantial evidence, which shows that particular subtypes of immune cells and other immune parameters fluctuate differentially over a 24 h period and exhibit different phase relationships with circulating corticosteroid levels (Angeli et al. 1990; McNulty et al. 1990; Levi et al. 1992). Thus, a three-way coordination between the pineal

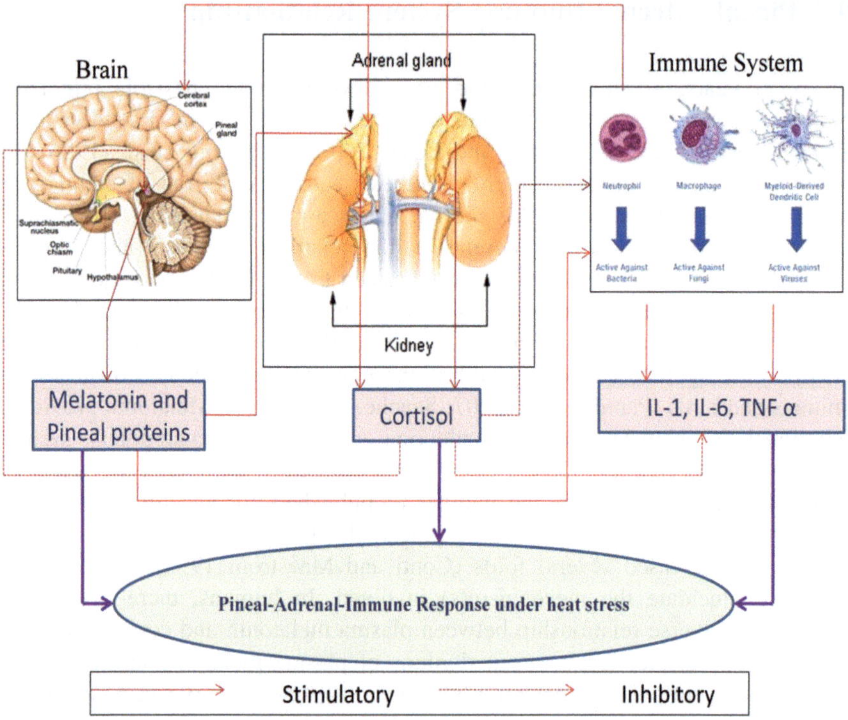

Fig. 9.2 Pictorial representation of pineal-adrenal-immune system relationship. The adrenal cortex hormone cortisol inhibits both melatonin as well as immune functions. However, pineal hormone melatonin stimulates both adrenocortical secretions as well as immune functions. Immune system has positive feedback stimulation on pineal melatonin secretions

gland, adrenal gland, and immune system is well explored through scientific tools. Figure 9.2 describes the pictorial representation of pineal-adrenal-immune system relationship.

9.6 Experimental Findings for Pineal-Adrenal-Immune Relationship Under Thermal Stress

One approach to study the functional involvement of pineal-adrenal-immune system in alleviating stress in animals is to investigate changes, which occur in these animals after suppressing the production of major hormonal products from the pineal-adrenal-immune system (Sejian 2007). Schematic representations of three experimental technical details that depict pineal-adrenal-immune system relation hypothesis have been described in Fig. 9.3.

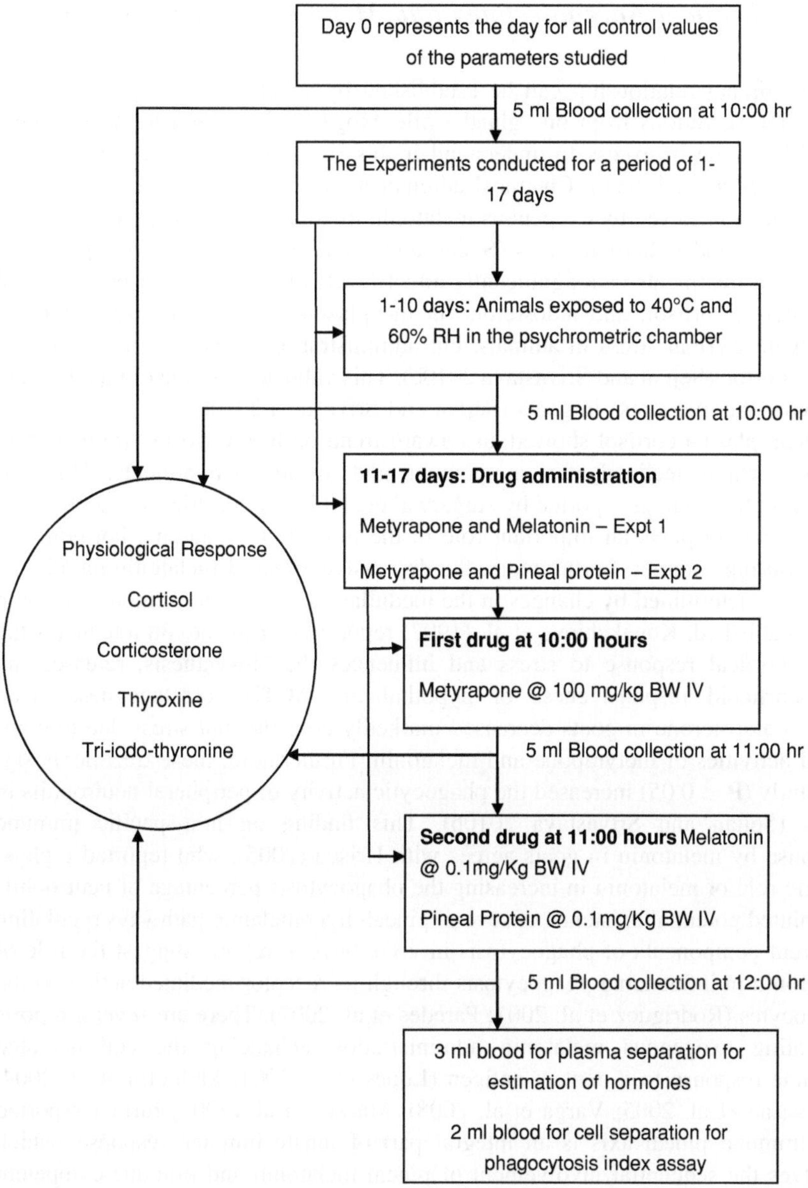

Fig. 9.3 Schematic representations of experimental technical details to fulfill pineal-adrenal-immune system relationship. This relationship is established in three different ways (1) Suppressing adrenal cortex secretion by metyrapone and administering melatonin, (2) Suppressing adrenal cortex secretion by metyrapone and administering pineal proteins and (3) Suppressing pineal secretions by propranolol and administering hydrocortisone

9.6.1 Chemical Adrenalectomy and Melatonin Administration

Pineal-adrenal relationship can be established by suppressing the production of endocrine secretions from one gland while exogenously administering the other gland hormone in excess to understand if this resumed the normal secretion of already suppressed gland. Chemical adrenalectomy can be achieved by injecting metyrapone followed by exogenous melatonin treatment to support pineal-adrenal association under thermal stress (Sejian and Srivastava 2010b). Metyrapone and melatonin treatments can significantly affect levels of glucose, total protein, total cholesterol, cortisol, and aldosterone in the plasma. Metyrapone treatment can aggravate thermal stress in animals, but administration of melatonin ameliorates the condition (Sejian and Srivastava 2010c). This validates the role of melatonin in relieving thermal stress in goats (Sejian and Srivastava 2010b).

Mean plasma cortisol showed an upward trend on heat exposure (Sejian et al. 2010a), which declined after metyrapone and melatonin treatments. This was similar to the findings reported by Aggarwal et al. (2005) in cattle. It was observed that melatonin plays an important role in the regulation of adrenal hormones in cattle during summer. Furthermore, the depressive effect of melatonin on adrenal glands as determined by changes in the medullary and cortical hormones in cattle was established. Konakchieva et al. (1997) reported that melatonin attenuates the adrenocortical response to stress and influences the biosynthesis, release, and glucocorticoid responsiveness of hypothalamic ACTH secretagogues. Mean plasma aldosterone in goats decreased markedly after thermal stress due to combined activities of metyrapone and melatonin. Furthermore, these treatments significantly ($P < 0.05$) increased the phagocytic activity of peripheral neutrophils in goats (Sejian and Srivastava 2010b). This finding on non-specific immune response by melatonin in goats agrees with Hriscu (2005), who reported a physiologic role of melatonin in increasing the phagocytosis percentage of neutrophils and hinted potential existence of several pineal–hypothalamic pathways regulating different components of phagocytosis in vivo. Several reports suggest the role of melatonin in influencing phagocytosis through its receptor mediated action on the phagocytes (Rodriguez et al. 2001; Paredes et al. 2007). There are several reports indicating exogenous melatonin administration enhancing the cell-mediated immune response to a contact antigen (Lopes et al. 2001; Malhotra et al. 2004; Srinivasan et al. 2005; Varga et al. 2008). Markus et al. (2007) further reported that immune–pineal axis is an integral part of innate immune response, which involves the sequential involvement of pineal melatonin and immune-competent cells.

There was remarkable decrease in cortisol level after metyrapone treatment, which was further reduced by melatonin administration, indicating the attenuating effect of melatonin on adrenal corticoids (Sejian and Srivastava 2010c). In response to metyrapone and melatonin treatments, an increase in aldosterone level was observed. These results reiterate the stress relieving properties of melatonin and the role of pineal gland in adrenal cortical functions. Further studies on this

Table 9.1 Effect of heat stress, metyrapone and melatonin treatment on endocrine parameters

Hypothesis	Experiment	Treatment	Parameters				
			Cortisol (nmol/L)	T_3 (nmol/L)	T_4 (nmol/L)	Insulin μIU/ml	Aldosterone pg/ml
Melatonin acts as an anti-stress agent	Chemical adrenalectomy + Melatonin administration	Control	18.76 ± 2.44^{a}	1.28 ± 0.04^{a}	87.04 ± 6.3^{a}	9.16 ± 0.75^{a}	3.60 ± 0.89^{a}
		Heat stress	82.74 ± 4.33^{b}	1.43 ± 0.12^{b}	61.59 ± 3.15^{b}	7.22 ± 1.12^{a}	1.46 ± 0.27^{b}
		Metyrapone	76.62 ± 4.93^{c}	1.10 ± 0.16^{c}	53.83 ± 2.46^{c}	13.21 ± 1.53^{b}	1.17 ± 0.19^{b}
		Melatonin	62.98 ± 3.65^{d}	1.20 ± 0.15^{c}	51.01 ± 3.83^{C}	23.40 ± 2.45^{c}	0.82 ± 0.05^{C}

Values bearing different superscript within a column differs significantly @ $P < 0.05$

will help us understand the functional relationship between pineal, adrenal, and immune system, and how this relationship modulates the non-specific immune response for the well-being of goats during thermal stress. Table 9.1 describes the effect of heat stress, metyrapone and melatonin treatment on endocrine parameters.

9.6.2 Chemical Adrenalectomy and Pineal Protein Administration

Sejian and Srivastava (2010d) conducted a study to establish the probable anti-stress effect of total precipitated pineal proteins (PP) in goats. Biochemical parameters, enzyme profile, and non-specific immune responses were used to study animals housed in psychrometric chamber with temperature and relative humidity (RH) conditions of 40°C and 60%, respectively for a period of 17 days. Chemical adrenalectomy was achieved by injecting animals with appropriate concentrations of metyrapone followed by exogenous total precipitated PI treatment. Thermal stress and chemical adrenalectomy significantly affected glucose, total protein, total cholesterol, and ALP contents of plasma. In addition, thermal stress, chemical adrenalectomy, and PP affected phagocytosis index (PI) (Garcia-Perganeda et al. 1999). While thermal stress decreased the PI, metyrapone and pineal protein treatments increased the levels of PI. Administration of total precipitated PP and metyrapone to stressed induced animals successfully relieved adverse affects of heat stress on the animals (Sejian and Srivastava 2010d). Table 9.2 describes the effect of heat stress, metyrapone and PP treatment on endocrine parameters.

9.6.3 Pinealectomy and Hydrocortisone Administration

Sejian et al. (2008a) also established the pineal-adrenal-thyroid-immune interaction of goats by suppressing the pineal secretions chemically and administering hydrocortisone exogenously under heat stress. Chemical pinealectomy was achieved in goats by injecting propranolol intravenously followed by exogenous hydrocortisone treatment upon exposure to thermal stress. On subjecting the animals to thermal stress, the level of melatonin increased significantly. Perhaps, higher concentration of melatonin is required to combat the stressful condition in order to maintain the homeothermy. This shows the protective role of melatonin during thermal stress. Further, melatonin plays an important role in thermoregulation (John et al. 1978). Barriga et al. (2002) reported direct effect of corticosterone on pinealocytes in reducing melatonin level. From this, it is evident that, glucocorticoids have direct action on pinealocytes to alter melatonin level. Hence, during adverse pinealectomy condition in these goats, glucocorticoids could have stimulated the pinealocytes to release melatonin in these animals in order to relieve

Table 9.2 Effect of heat stress, metyrapone, and pineal protein treatment on endocrine profile in goats

Hypothesis	Experiment	Treatment	Parameters				
			Cortisol (nmol/L)	T_3 (nmol/L)	T_4 (nmol/L)	Insulin µIU/ml	Aldosterone pg/ml
Pineal protein acts as an anti-stress agent	Chemical adrenalectomy + Pineal protein administration	Control	26.80 ± 3.77^a	1.27 ± 0.20^a	86.49 ± 10.82^a	8.32 ± 0.16^a	2.08 ± 0.75^a
		Heat stress	80.20 ± 5.75^b	1.78 ± 0.28^b	116.74 ± 10.60^b	6.32 ± 1.16^b	2.70 ± 1.09^b
		Metyrapone	71.78 ± 7.78^c	0.82 ± 0.12^c	62.74 ± 6.22^c	20.64 ± 1.85^c	1.08 ± 0.02^c
		Pineal Protein	69.00 ± 4.88^c	1.12 ± 0.15^c	60.20 ± 2.87^c	30.63 ± 0.32^d	0.41 ± 0.02^d

Values bearing different superscript within a column differs significantly @ $P < 0.05$

Table 9.3 Effect of heat stress, propranolol and hydrocortisone treatment on endocrine parameters

Hypothesis	Experiment	Treatment	Parameters				
			Cortisol (nmol/L)	T_3 (nmol/L)	T_4 (nmol/L)	Insulin µIU/ml	Aldosterone pg/ml
Melatonin acts as an anti-stress agent	Chemical Pinealectomy + Hydrocortisone administration	Control	15.73 ± 0.84^a	1.41 ± 0.11^a	69.00 ± 2.93^a	10.67 ± 0.69^a	2.87 ± 0.58^a
		Heat Stress	76.74 ± 2.43^b	2.05 ± 0.13^b	111.82 ± 9.14^b	9.88 ± 0.50^a	1.50 ± 0.87^b
		Propranolol	205.24 ± 9.37^c	1.70 ± 0.12^c	103.08 ± 3.41^b	12.09 ± 1.88^b	0.58 ± 0.08^c
		Hydrocortisone	531.53 ± 3.65^d	1.75 ± 0.05^c	99.02 ± 6.77^b	15.21 ± 1.33^c	0.88 ± 0.05^c

Values bearing different superscript within a column differs significantly @ $P < 0.05$

thermal stress (Sejian and Srivastava 2010e). Propranolol and hydrocortisone treatments also influenced plasma T_3, and T_4 concentrations indicating the role of melatonin in controlling thyroid gland hormones (Sakamoto et al. 2000; Abecia et al. 2005). Table 9.3 describes the effect of heat stress, propranolol, and hydrocortisone treatment on endocrine parameters.

Chemical pinealectomy significantly affected plasma levels of plasma glucose, total protein, total cholesterol, cortisol, insulin, aldosterone, melatonin, and corticosterone and these could be significantly counteracted by administration of hydrocortisone (Sejian et al. 2008a). Furthermore, propranolol and hydrocortisone treatments also significantly affected levels of plasma calcium, ACP, ALP, and PI (Sejian et al. 2008b; Sejian et al. 2010c). The decline of PI after pinealectomy could be attributed to the reduction in pineal secretions, such as melatonin, which modulates phagocytosis in macrophages (Kanchev et al. 2006; Roy et al. 2001). This finding establishes profound influence of pineal gland on the non-specific immune response. Hydrocortisone counteracted the effect of propranolol treatment by increasing the phagocytosis index, showing immunopotentiative nature of glucocorticoids in the absence of pineal endocrine secretions. The immunostimulative capacity of glucocorticoids with non-specific immune response was reported by Forner et al. (1995). Furthermore, it has been shown by several researchers that pineal gland and its hormone, melatonin, play central role in the control of the circadian organization of hypophysial-adrenal and immune system (Reiter 1995; Karasek 2004). Propranolol treatment aggravated thermal stress; although administration of hydrocortisone could ameliorate the condition. This indicates the role of pineal in support of thermoregulation. These results establish the modulating effects of glucocorticoids on pineal activity to relieve thermal stress in goats.

9.7 Conclusions

Pineal gland and its secretions, melatonin and PP, play the central role in the control of the circadian organization of hypophysial–adrenal and immune system. Experimental data generated from various studies have helped researchers in the field to understand the functional relationship among pineal, adrenal, and immune system, and how these relationships modulate stress and non-specific immune response for the well-being of animals (goats) under thermal stress. Notably, melatonin and PP play vital roles in mitigating negative impacts of thermal stress in animals. The effect of glucocorticoids on melatonin levels during thermal stress establishes the relationship between these two endocrine glands under heat stress. As thermal stress is the most significant factor influencing livestock production in tropical countries, understanding the endocrine relationship that plays a major role in relieving such stress could pave the way for improved livestock production to improve the economy of farm households and underprivileged farmers.

9.8 Future Scope of Research

It has become evident that there exists an elaborate interplay between the pineal-adrenal-immune system axis, the details of which need to be experimentally proved. Further detailed studies are required to fully understand the mechanisms of interactions between these glands (pineal-adrenal, pineal-adrenal-immune system etc.). It is still not apparent whether melatonin controls thermal stress by acting directly on adrenal cortex or by acting on hypothalamus or anterior pituitary. In addition, there are some unanswered questions on whether there is a direct link between glucocorticoids and melatonin production and secretion. If one attempts to establish this multi-gland relationship, research and experiments should be conceived in such a way to unveil the precise mechanism of the interactions among pineal, adrenal, and immune system.

Acknowledgments The authors are highly thankful to the Senior Research Fellows Miss Saumya Bahadur, Miss Rajni Chhetri, Miss. Indu Shekhawat, Mr. Anoop Kumar, and Mr. Kamal Kumar for their valuable help in preparing this manuscript.

References

Abecia JA, Valares JA, Forcada F (2005) The effect of melatonin treatment on wool growth and thyroxine secretion in sheep. Small Rumin Res 56:265–270

Aggarwal A, Upadhyay RC, Singh SV, Kumar P (2005) Adrenal–thyroid pineal interaction and effect of exokgenous melatonin during summer in crossbred cattle. Indian J Anim Sci 75:915–921

Angeli A, Gatti G, Masera R, Sartori ML, Carignola R (1990) Chronobiologic aspects of neuroenodcrine-immune interactions. Int J Neurosci 51:341–343

Balik A, Kretschmannova K, Mazna P, Svobodova I, Zemkova H (2004) Melatonin action on neonatal gonadotrophs. Physiol Res 53(suppl 1):S153–S166

Barriga C, Martin MI, Ortega E, Rodriguez AB (2002) Physiologic concentration of melatonin and corticosterone in stress and their relationship with phagocytic activity. J Neuroendocrinol 14:691–696

Bauer MS, Poland RE, Whybow PC, Frazer A (1989) Pituitary, adrenal and thyroid effects on melatonin content of the rat. Psychoneuroendocrinology 14:165–175

Buchanan KL, Yellon SM (1991) Delayed puberty in the male Djungarian hamster: effect of short photoperiod or melatonin treatment on the Gn-RH neuronal system. Neuroendocrinology 54:96–102

Cannon WB (1915) Bodily changes in pain, hunger, fear, and rage. Appleton, New York

Carrillo-Vico A, Calvo JR, Abreu P, Lardone PJ, Garcia-Maurino S, Reiter RJ, Guerrero JM (2004) Evidence of melatonin synthesis by human lymphocytes and its physiologic significance: possible role as intracrine, autocrine, and/or paracrine substance. FASEB J 18:537–539

Chemineau P, Guillaume D, Migaud M, Thiery JC, Pellicer-Rubio MT, Malpaux B (2008) Seasonality of reproduction in mammals: intimate mechanisms and practical applications. Reprod Domest Anim 43(suppl 2):40–47

Claustrat B, Brun J, Chazot G (2005) The basic physiology and pathophysiology of melatonin. Sleep Med Rev 9:11–24

Coelho LA, Rodrigues PA, Nonaka KO, Sasa A, Balieiro JC, Vincente WR, Cipolla-Neto J (2006) Annual pattern of plasma melatonin and progesterone concentrations in hair and wool ewe lambs kept under natural photoperiod at lower latitudes in the southern hemisphere. J Pineal Res 41:101–107

Collier RJ, Collier JL, Rhoads RP, Baumgard LH (2008) Invited review: genes involved in the bovine heat stress response. J Dairy Sci 91(2):445–454

Conti A, Maestroni GJM (1994) Melatonin-induced immuno-opioids: role in lymphoproliferative and autoimmune diseases. Adv Pineal Res 7:83–100

Cupps T, Fauci AS (1982) Corticosteroid mediated immunoregulation in man. Immunol Rev 65:133–155

Demisch L, Demisch K, Nickelsen T (1988) Influence of dexamethasone on nocturnal melatonin production in healthy adult subjects. J Pineal Res 5:317–321

Dubocovich ML, Markowska M (2005) Functional MT1 and MT2 melatonin receptors in mammals. Endocrine 27:101–110

Esquifino AI, Pandi-Perumal SR, Cardinali DP (2004) Circadian organization of the immune response: a role for melatonin. Clin Appl Immunol Rev 4:423–433

Everly G (2002) A clinical guide to the treatment of the human stress response. Kluwer Academic Press, New York

Fabris N (1994) Neuroendocrine regulation of immunity. Adv Pineal Res 7:41–56

Farrell G (1960) Adrenoglomerulotropin. Circulation 21:1009–1015

Forner MA, Barriga C, Rodriguez AB, Ortega E (1995) A study of the role of corticosterone as a mediator in exerciseinduced stimulation of murine macrophage phagocytosis. J Physiol 488:789–794

Frungieri MB, Mayerhofer A, Zitta K, Pignataro OP, Calandra RS, Gonzalez-Calvar SI (2005) Direct action of melatonin on Syrian hamster testes: melatonin subtype 1a receptors, inhibition of androgen production, and interaction with local corticotropin-releasing hormone system. Endocrinology 146:1541–1552

Garcia-Perganeda A, Guerrero JM, Rafil-EI-Idrissi M, Paz Romero M, Polo d Calvo JR (1999) Characterization of membrane melatonin receptors in mouse peritoneal macrophage: inhibition of adenylyl cyclase by a pertussis toxin-sensitive G-protein. J Neuroimmunol 1:85–94

Goetzl EJ, Sreedharan SP (1992) Mediators of communication and adaptation in the neuroendocrine and immune system. FASEB J 6:2646–2652

Gromova EA, Kraus M, Krecek J (1967) Effect of melatonin and 5-hydroxytryptamine on aldosterone and corticosterone production by adrenal glands of normal and hypophysectomized rats. J Endocrinol 39(3):345–350

Guerrero JM, Reiter RJ (2002) Melatonin-immune system relationships. Curr Top Med Chem 2:167–179

Hardeland R (2005) Antioxidative protection by melatonin: multiplicity of mechanisms from radical detoxification to radical avoidance. Endocrine 27:119–130

Homo-Delarche F, Dardenne M (1993) The neuroendocrineimmune axis. Springer Semin Immunopathol 14:221–238

Hriscu ML (2005) Modulatory factors of circadian phagocytic activity. Ann N Y Acad Sci 1057:403–430

John TM, Itoh S, George JC (1978) On the role of pineal in thermoregulation in the pigeon. Horm Res 9(1):41–56

Johnson LY (1982) The pineal as a modulator of the adrenal and thyroid axes. In: Reiter RJ (ed) The Pineal gland. Extra reproductive effects, vol 3. CRC Press, Boca Ratan, pp. 107–112

Johnston JD, Messager S, Ebling FJ, Williams LM, Barrett P, Hazlerigg DG (2003) Gonadotrophin-releasing hormone drives melatonin down-regulation in the developing pituitary gland. Proc Natl Acad Sci U S A 100:2831–2835

Kanchev LN, Baichev J, Kamenov I, Baikov B, Hallak AK (2006) Melatonin, corticosterone, stress and phagocytic activity. Bul J Vet Med 9:257–264

Kannan G, Terrill TH, Kouakou B, Gazal OS, Gelaye S, Amoah EA, Samake S (2000) Transportation of goats: effects on physiologic stress responses and live weight loss. J Anim Sci 78:1450–1457

Karasek M (2004) Melatonin, human aging, and age-related diseases. Exp Gerontol 39:1723–1729

Karsch FJ, Bittman EL, Foster DL, Goodman RL, Legan SJ, Robinson JE (1984) Neuroendocrine basis of seasonal reproduction. Recent Prog Horm Res 40:185–232

Khan R, Daya S, Potgieter B (1990) Evidence for a modulation of the stress response by the pineal gland. J Exp Edu 46(8):860–862

Konakchieva R, Mitev Y, Almeida OF, Patchev VK (1997) Chronic melatonin treatment and the hypothalamo–pituitary–adrenal axis in the rat: attenuation of the secretory response to stress and effects on hypothalamic neuropeptides content and release. Biol Cell 89(9):587–596

Kristensen TN (2004) Environmental and genetic stress—integrating studies on Drosophila and farm animals. PhD Thesis. Department of Genetics and Ecology, Institute of Biologic Sciences, University of Aarhus, Denmark

Lcvi F, Canon C, Depres-Brummer P, Adam R, Bourin P, Pati A, Florentin I, Misset JL, Bismuth H (1992) The rhythmic organization of the immune network: implications for the chronopharmacologic delivery of interferons, interleukins and cyclosporin. Adv Drug Deliv Rev 9:85–112

Lommer D (1966) Inhibition of corticosteroid 11-beta-hydroxylation by an extract of pineal body. Experientia 22(2):122–123

Lopes C, Mariano M, Markus RP (2001) Interaction between the adrenal and the pineal gland in chronic experimental inflammation induced by BCG in mice. Inflamm Res 50:006–011

Maestroni GJ (2001) The immunotherapeutic potential of melatonin. Expert Opin Investig Drugs 10:467–476

Maestroni GJ, Conti A, Pierpaoli W (1986) Role of the pineal gland in immunity. circadian synthesis and release of melatonin modulates the antibody response and antagonizes the immunosuppressive effect of corticosterone. J Neuroimmunol 13:19–30

Maestroni GJM, Conti A, Pierpaoli W (1988) Role of the pineal gland in immunity III. melatonin antagonizes the immunosuppressive effects of acute stress opiatergic mechanism. Immunology 63:465–469

Malhotra S, Sawhney G, Pandhi P (2004) The therapeutic potential of melatonin: a review of the science. Medsc Gen Med 6:46–62

Markus RP, Ferreira ZS, Fernandes PACM, Cecon E (2007) The immune-pineal axis: a shuttle between endocrine and paracrine melatonin sources. NeuroImmunomodulation 14:126–133

McNulty JA, Relfson M, Fox LM, Fox LM, Kus L, Handa RJ, Schneider GB (1990) Circadian analysis of mononuclear cells in the rat following pinealectomy and superior cervical ganglionectomy. Brain Behav Immun 4:292–307

Mess B (1983) The pineal gland and its endocrine role. Plenum press, New York, pp 477–508

Mori W, Aoyama H, Mori N (1984) Melatonin protects rats from injurious effects of a glucocorticoid, dexamethasone. Jpn J Exp Med 54(6):255–261

Mostl E, Palme R (2002) Hormones as indicators of stress. Dom Anim Endocrinol 23(1–2):67–74

Nelson RJ (2004) Seasonal immune function and sickness responses. Trends Immunol 25:187–192

Ogle TF, Kitay JI (1978) In vitro effects of melatonin and serotonin on adrenal steroidogenesis. Proc Soc Exp Biol Med 157:103–109

Paredes SD, Terron MP, Marchena AM, Barriga C, Pariente JA, Reiter RJ, Rodriguez AB (2007) Effect of exogenous melatonin on viability, ingestion capacity, and free radical scavenging in heterophils from young and old ringdoves (Streptopelia risoria). Mol Cell Biochem 304:305–314

Pierpaoli W, Maestroni GJ (1987) Melatonin: a principal neuroimmunoregulatory and anti-stress hormone: its anti-aging effects. Immunol Lett 16(3–4):355–361

Reiter RJ (1991) Neuroendocrine effects of light. Int J Biometereol 35:169–175

Reiter RJ (1995) The pineal gland and melatonin in relation to aging: the summary of theories and of the data. Exp Gerontol 30:199–212

Rodriguez AB, Terron MP, Duran J, Ortega E, Barriga C (2001) Physiologic concentrations of melatonin and corticosterone affect phagocytosis and oxidative metabolism of ringdove heterophils. J Pineal Res 31:31–38

Rodriguez C, Mayo JC, Sainz RM, Antolin I, Herrera F, Martin V, Reiter RJ (2004) Regulation of antioxidant enzymes: a significant role for melatonin. J Pineal Res 36:1–9

Roy D, Angelini NL, Fujeda H, Brown GM, Belsham DD (2001) Cyclical regulation of GnRH gene expression in GTI-7 GnRH-secreting neurons by melatonin. Endocrinology 142:4711–4720

Sakamoto S, Nakamura K, Inoue K, Sakai T (2000) Melatonin stimulates thyroid-stimulating hormone accumulation in the thyrotropes of the rat pars tuberalis. Histochem Cell Biol 114:213–218

Sanchez S, Paredes SD, Martin MI, Barriga C, Rodriguez AB (2004) Effect of tryptophan on circulating levels of melatonin and phagocytic activity. J Appl Biomed 2:169–177

Sanchez-Barcelo EJ, Cos S, Fernandez R, Mediavilla MD (2003) Melatonin and mammary cancer: a short review. Endocr Relat Cancer 10:153–159

Sejian V (2007) Role of pineal gland and adrenal cortex to combat thermal stress in farm animals. In the short course manual on "Animal behavior and welfare" under centre of advanced studies in veterinary physiology, Division of Physiology and Climatology, Published by Director, IVRI, Izatnagar, pp 147–149

Sejian V, Srivastava RS (2010a) Pineal-adrenal relationship under thermal stress with emphasis on effect of pineal proteins on endocrine profiles in chemically adrenalectomized does. Glob Vet 4(3):249–254

Sejian V, Srivastava RS (2010b) Effects of melatonin on adrenal cortical functions of Indian goats under thermal stress. Vet Med Int 2010:6 (Article ID 348919) doi:10.4061/2010/348919

Sejian V, Srivastava RS (2010c) Pineal-adrenal-immune system relationship under thermal stress: effect on physiologic, endocrine and non-specific immune response in goats. J Physiol Biochem 66(4):339–349. doi:10.1007/s13105-010-0040-8

Sejian V, Srivastava RS (2010d) Effects of pineal proteins on biochemical profile, enzyme profile and non-specific immune response of Indian goats under thermal stress. Anim Prod Res Adv 6(1):1–6

Sejian V, Srivastava RS (2010e) Interrelationship of endocrine glands under thermal stress: effect of exogenous glucocorticoids on mineral, enzyme, thyroid hormone profiles and phagocytosis index of Indian goats. Endocr Regul 44:101–107

Sejian V, Srivastava RS, Varshney VP (2008a) Pineal-adrenal relationship: modulating effects of glucocorticoids on pineal function to ameliorate thermal stress in goats. Asian Australas J Anim Sci 21(7):988–994

Sejian V, Srivastava RS, Varshney VP (2008b) Pineal-adrenal-thyroid relationships under thermal stress: effect on circulating levels of thyroid hormones in goats. Indian J Anim Sci 78:1263–1264

Sejian V, Srivastava RS, Varshney VP (2010a) Effect of thermal stress on endocrine profile and phagocytosis index in Marwari goats. Indian Vet J 87:1156–1157

Sejian V, Srivastava RS, Varshney VP (2010b) Effect of exogenous melatonin on different organ weights of adrenalectomized rats. Indian Vet J 87:15–16

Sejian V, Srivastava RS, Varshney VP (2010c) Effect of short-term thermal stress on biochemical profile in Marwari goats. Indian Vet J 87:503–504

Selye H (1946) General adaptation syndrome and the diseases of adaptation. J Clin Endocrinol 6:117

Selye H (1950) The physiology and pathology of exposure to stress. Acta Montreal, Canada

Sibarov DA, Kovalenko RI, Nozdrachev AD, Malinin VV, Khavinson VK (2002) Effects of pineal-gland peptides on the electric activity of pinealocytes in rats. Dokl Biol Sci 385: 331–333

Silman RE (1993) Melatonin: a contraceptive for the nineties. Eur J Obstet Gynecol Reprod Biol 49:3–9

Skwarlo-Sonta K (2002) Melatonin in immunity: comparative aspects. Neuro Endocrinol Lett 23(1):61–66

Srinivasan V (2000) Physiologic role of pineal in homeostasis. In the short course manual on "Physiology of pineal gland" under centre of advanced studies in veterinary physiology, Division of Physiology and Climatology, Published by Director, IVRI, Izatnagar, pp 69–72

Srinivasan V, Maestroni GJM, Cardinali DP, Esquifino AI, Perumal SRP, Miller SC (2005) Melatonin, immune function and aging. Immun Ageing 2:17

Stephanie WF, Siu SWF, Lau KW, Tam PC, Shiu SYW (2002) Melatonin and prostate cancer cell proliferation: interplay with castration, epidermal growth factor, and androgen sensitivity. Prostate 52(2):106–122

Sudhakumari CC, Haldar C, Senthilkumaran B (2001) Seasonal changes in adrenal and gonadal activity in the quail, Perdicula asiatica: involvement of the pineal gland. Comp Biochem Physiol B 128:793–804

Tamura H, Nakamura Y, Korkmaz A, Manchester LC, Tan DX, Sugino N, Reiter RJ (2008) Melatonin and the ovary: physiologic and pathophysiologic implications. Fertil Steril (in press). published online ahead of print 18 Sept 2008. doi:10.1016/j.fertnstert..05.016

Tan DX, Chen LD, Poeggeler B, Manchester LC, Melatonin Reiter RJ (1993) A potent, endogenous hydroxyl radical scavenger. Endocrine J 1:57–60

Tomas-Zapio C, Coto-Montes A (2005) A proposed mechanism to explain the stimulatory effect of melatonin on antioxidative enzymes. J Pineal Res 39:99–104

Torres-Farfan H, Richter G, Rojas-Garcia P (2003) Mt1 melatonin receptor in the primate adrenal gland: inhibition of adrenocorticotropin-stimulated cortisol production by melatonin. J Clin Endocrinol Metab 88(1):1450–1458

Touitou Y, Bogdan A, Auzeby A, Touitou C (1989) Activity of melatonin and other pineal indoles on the in vitro synthesis of cortisol, cortisone, and adrenal androgens. J Pineal Res 6(4):341–350

Troiani ME, Reiter RJ, Vaughan MK, Gonzalez-Brito A, Herbert DC (1988) The depression in rat pineal melatonin production after saline injection at night may be elicited by corticosterone. Brain Res 450(1–2):18–24

Tsigos C, Chrousos GP (2002) Hypothalamic-pituitary-adrenal axis, neuroendocrine factors and stress. J Psych Res 53:865–871

Varga G, Ehrchen J, Tsianakas A, Tenbrock K, Rattenholl A, Seeliger S, Mack M, Roth J, Sunderkoetter C (2008) Glucocorticoids induce an activated, anti-inflammatory monocyte subset in mice that resembles myeloid-derived suppressor cells. J Leukoc Biol 84:644–650

Vaughan GM (1984) Melatonin in humans. Pineal Res Rev 2:141–201

Vaughan MK, Vaughan GM, Reiter RJ, Benson B (1972) Effect of melatonin and other pineal indoles on adrenal enlargement produced in female mice by pinealectomy, unilateral adrenalectomy, castration and stress. Neuroendocrinology 10:139–154

Vijayalaxmi, Thomas CR Jr, Reiter RJ, Herman TS (2002) Melatonin: from basic research to cancer treatment clinics. J Clin Oncol 20(10):2575–2601

Vincenzo A, Vincenzo S, Alberto M (1996) The pineal gland and reproduction. Hum Reprod Update 2:225–235

Wagner GC, Johnston JD, Clarke IJ, Lincoln GA, Hazlerigg DG (2008) Redefining the limits of daylength responsiveness in a seasonal mammal. Endocrinology 149:32–39

Weaver DR, Stehle JH, Stopa EG, Reppert SM (1993) Melatonin receptors in human hypothalamus and pituitary: implications for circadian and reproductive responses to melatonin. J Clin Endocrinol Metab 76:295–301

Wen JC, Dhabhar FS, Prendergast BJ (2007) Pineal dependent and -independent effects of photoperiod on immune function in Siberian hamsters (Phodopus sungorus). Horm Behav 51(1):31–39

Wingfield JC, Kitaysky AS (2002) Endocrine responses to unpredictable environmental events: stress or anti-stress hormones? Integr Comp Biol 42(3):600–609

Woo MM, Tai CJ, Kang SK, Nathwani SP, Pang SF, Leung PC (2001) Direct action of melatonin in human granulosa-luteal cells. J Clin Endocrinol Metab 86:4789–4797

Zawilska JB, Sadowska M (2002) Prolonged treatment with glucocorticoid dexamethasone suppresses melatonin production by the chick pineal gland and retina. Pol J Pharmacol 54:61–66

Zwirska-Korczala K, Kniazewski B, Ostrowska Z, Buntner B (1991) Influence of melatonin on rat thyroid, adrenals and testis secretion during the day. Folia Histochem Cytobiol 29:19–24

[illegible]

Part III
Livestock Adaptation

Chapter 10
Basic Principles Involved in Adaption of Livestock to Climate Change

John B. Gaughan

Abstract Animal agriculture accounts for approximately 70% of all agricultural land use and accounts for approximately 40% of the world's agriculture gross domestic product (GDP), with the livestock sector contributing to the livelihood of over one billion people. Potentially, climate change will have large negative impacts on the livestock production in many countries, especially when the animals used are not adapted to the changed environmental conditions. Determining animal responses to climate change is a challenge. The impacts may be direct, e.g. effects of heat load on animals or indirect, e.g. prolonged droughts. There is a need to select animals (and species) that are suited to the current climatic conditions, as well as the predicted future conditions. This chapter will discuss the effect of changing climate on animal performance, animal response to environmental stressors and the processes of adaptation.

Keywords Thermal adaptation · Livestock · Climate change

Contents

J. B. Gaughan (✉)
School of Agriculture and Food Sciences, The University of Queensland,
Gatton 4343, Australia
e-mail: j.gaughan@uq.edu.au

V. Sejian et al. (eds.), *Environmental Stress and Amelioration in Livestock Production*, DOI: 10.1007/978-3-642-29205-7_10,

10.1 Introduction

Worldwide livestock production accounts for approximately 70% of all agricultural land use (Steinfeld et al. 2006). Potentially, climate change will have large negative impacts on the livestock sectors in many countries. The ability to manage animals within a potentially harsh environment is as much about the people as it is the animals. A lack of knowledge on how to ameliorate the impact of changing climate on livestock production is one problem, but the major problem in many regions is the lack of the financial resources to manage change. Housing animals is an obvious method of ameliorating the effects of climate. However, housing is expensive to build and maintain and if housing is not correctly designed it could exacerbate many problems. In many regions housing of animals is not a suitable option. Therefore, the focus should be on the adaptive capacity of the animal population in question. There is a growing need to select animals (and species) that are suited to the current climatic conditions, as well as the predicted future conditions. This is not an easy task. For example, there are considerable breed variations, within and between breeds, for thermal tolerance and overall stress tolerance. The ability of livestock breeders to identify phenotypes, which carry specific genes, is difficult partly because phenotypic variance is due to the combined effects of genetic and environmental components. Therefore, there is a reliance on selecting animals from within the environment or from a similar environment in which they are expected to live. Livestock need to have 'adequate' performance in three key areas: survivability (to reproductive age), growth and fertility, in addition to their specific production purpose, e.g. milk, wool, egg production. Thus reliance on 'natural' selection is fundamentally the correct approach. However, animals with adequate performance in the three key areas, may achieve this at the expense of their production purpose. Furthermore, the extent to which the location is likely to be favourable or unfavourable to the species or breed concerned in the future needs to be considered.

10.2 Defining Climate Change: With Livestock in Mind

When discussing livestock adaptation to climate change, we need to define what we mean by climate change. The web-based dictionary—Dictionary.com defines climate change as "Any long-term significant change in the weather patterns of an area" (http://dictionary.reference.com/browse/climate+change). The Australian Bureau of Meteorology defines climate and weather as follows: *Climate is the sum or synthesis of all the weather recorded over a long period of time. It tells us the average or most common conditions, or extremes, or counts of events, or frequencies. Weather is a description of conditions over a short period of time—a "snap shot" of the atmosphere at a particular time.* Climate change models and their predictions are often reported in relation to global climate change. In terms of livestock adaptation, what is happening to the climate at a regional or local level is probably more important (at least in the short to medium term) in regards to animals' ability to cope. Within the predicted changes in climate there are also indications of an increase in extreme events. Animals, generally, have a better ability to adapt to longer (generational) change than they do to an acute event such as a heat wave. However, the effects of climate change on a regional and local basis are not as easy to model or predict (Angilletta 2009). Indeed, Jenouvrier and Visser (2011) stated that "current predictive models linking climate to populations largely ignore phenological changes and other ecological responses to climate change, which may radically change their (climate change) predictions". Livestock breeders are generally more concerned with local climatic conditions than regional or global change because the local changes have the biggest immediate impact on animal performance and it is this current performance that biases selection.

10.3 Effect of Changing Climate on Animal Performance

Determining animal responses to climate change is a challenge. Direct effects of higher temperature on growth rate or reproductive performance are relatively easy to predict based on current knowledge. The indirect effects are more difficult to predict. These effects will come about due to changes in water and feed availability, and an increased risk of disease and parasitism. Environmental stressors such as high ambient temperature, high humidity, drought, and lack of food resources have profound effects on livestock. The economic impact of climate extremes and climate change on livestock production has been reviewed by a number of authors (e.g. St-Pierre et al. 2003; Thornton et al. 2009). In general, animals are able to cope with environmental stressors in a number of ways, for example.

(i) By migration (limited options in livestock production)—i.e. they move to a more favourable area,

(ii) changing behaviour, e.g. grazing at night, shade seeking, wallowing and
(iii) changing metabolism, e.g. reduced production and fertility.

Climate change may manifest itself as rapid changes in climate in the short term (a couple of years) or more subtle changes over decades. The ability of animals to adapt to environmental change is dependent on a number of factors. Acute challenges are very different for chronic long-term challenges, and in addition animal responses to acute or chronic stress are also very different. The extents to which animals are able to adapt are primarily limited by physiological and genetic constraints (Devendra 1987; Parsons 1994). Additionally, climatic conditions do not affect an animal equally at all stages of its life cycle (Angilletta 2009). For example, young animals may be more susceptible to temperature extremes than mature animals. The ability of animals to cope with climatic extremes also depends on the level of production. For example, high production dairy cows (>30 kg/milk/day; 13.7% reduction) are more susceptible to high heat loads than low production cows (<19 kg/milk/day; 4.1% reduction) (Gaughan and Lees 2010).

10.4 Animal Response to Environmental Stressors

When environmental conditions change, an animal's ability to cope (or adapt) to the new conditions is determined by its ability to maintain performance and oxidative metabolism (Pörtner and Knust 2007). The stress response is influenced by a number of factors including: species, breed, previous exposure to the stressor, health status, performance, body condition, mental state and age. Insufficient acclimatisation or adaptation would determine what an animal experiences as stressful. Subsequent acclimatisation or adaptation may alleviate the stress response (Kassahn et al. 2009), but performance may not return to pre-stress levels. Animal response to temperature stress (which could be considered as part of the adaptation process) usually results in a loss of performance (e.g. growth or reproduction) before cellular and molecular stress responses are activated (Kassahn et al. 2009). This suggests that the use of biological stress markers as an aid in selection may be limited. And indeed little progress has been made in this area to date. Typically the first noticeable livestock response to a heat challenge is a decrease in feed intake. A reduction in feed intake has a direct impact on performance, for example reductions in milk yield, milk quality, meat and egg production. In addition, the cellular and molecular stress responses will have an effect on mammary gland metabolism, energy partitioning, immune status and reproductive success. The effects on reproduction efficiency may be long term, e.g. months rather than days (Santolaria et al. 2010). All of these changes lead to economic losses in livestock systems (St Pierre et al. 2003). Biological effects on livestock due to global warming are not predictable simply in terms of a response to increased heat load (Marcogliese 2001). Among other things, increasing

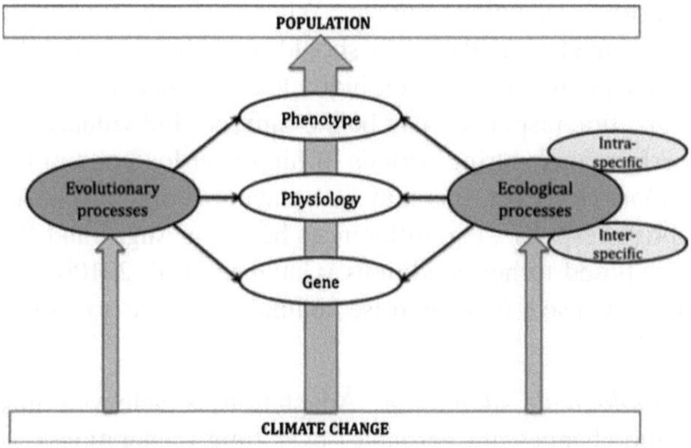

Fig. 10.1 A model for animal adaptation to climate change (Jenouvrier and Visser 2011)

temperature may also increase exposure, and susceptibility of animals to parasites and disease (Marcogliese 2001; Sutherst 2001), especially vector-borne diseases (Tabachnick 2010). However, little effort has been dedicated to understand the potential impact of climate change on parasite populations and subsequent effects on animal production (Marcogliese 2001; Tabachnick 2010).

10.5 Defining Adaptation

In the context of animal science, adaptation is often confused with the related concepts of: acclimatization, acclimation and habituation, and at times it's difficult to separate these (Loeschcke and Sørensen 2005). Adaptation is defined in the Oxford Dictionary as; *to make suitable* (*to or for a purpose*): *modify*, *alter*. The Glossary of Terms for Thermal Physiology (2001) defines adaptation as "a change which reduces the physiological strain produced by stressful components of the total environment". Adaptation can be further defined in terms of genetic or phenotypic (physiological) changes which occur in an animal in response to internal and external stimuli (Hafez 1968a; Khalifa 2003; Willmer et al. 2006).

It is likely that no single process occurs, which leads to an end point of animal adaptation to a given stressor. It is more likely that animals will go through a series of 'steps' which may or may not lead to permanent genetic change. The ability of an animal to successfully respond to an environmental stressor is a function of both its genetic capacity for change, and the duration × intensity of the stressor. The capacity to adapt can be determined by the ability of an animal to adjust to both average environmental conditions as well as climate extremes (McManus et al. 2009).

A useful integrative model for adaptation (Fig. 10.1) was developed by Jenouvrier and Visser (2011). Although the model was developed with wild animal populations in mind, the approach is useful for determining the process by which livestock adapt to a changing climate. When we are thinking about animal adaptation to climate change, the focus should be on how a particular species or a population will adapt and not focus on individuals or phenotype *per se*. Of course, some of the adaptation responses may be the same for individuals and phenotypes, e.g. reduced performance during periods of high heat load or similar responses to drought conditions. However, this may be an acute coping response rather than a long-term adaptive response, e.g. differences between Angus and Brahman cattle, when they are exposed to hot conditions (Gaughan et al. 2010b).

The terminology used to categorise animal response to climate change is defined as follows:

- *Genetic or Biological Adaptation*. Adaptation is achieved through genetic change (i.e. the changes are heritable) over time (generations), and are rarely reversible (Willmer et al. 2006). This type of adaptation involves changes brought about by interactions between evolutionary process, changes induced by environmental stimulation and experiences during an animal's lifetime (Hafez 1968a; Price 1984). This type of adaptation can come about via natural selection and also by selection of favoured animals by humans (Hafez 1968a). The latter indirectly includes natural selection because animals selected by humans in a particular location are usually selected on the basis of their performance within that location and although there may be specific selection criteria it is usually the animals that perform in a particular environment that are selected (but not always). The biological properties of animals are a result of interactions between stress intensity, magnitude of environmental fluctuations and the energy available from resources (Parsons 1994).
- *Phenotypic or Physiological Adaptation*. Animals have the ability to respond to acute or sudden environmental change, e.g. shivering when exposed to cold or vasodilatation when exposed to heat (Folk 1974; Hafez 1968a; Langlois 1994). In addition, there may be some adaptation to long-term exposure to climate change (e.g. altering thyroid and adrenal cortex activities during the hot season) although this type of adaptation may be somewhat limited.
- *Acclimatisation*. Acclimatisation is the functional compensation (i.e. physiological or phenotypic change) over a period of days to weeks in response to a complex of environmental factors, as in seasonal climate change or the longer term subtle changes in climate.
- *Acclimation*. Acclimation is the functional compensation (i.e. physiological or phenotypic change) over a period of days to weeks in response to a single environmental factor (often ambient temperature), as in controlled experiments. Controlled chamber studies, including those which this author has conducted are often used as models to predict animal response to climate change. Unfortunately these studies are short term and do not truly reflect long-term adaptation.
- *Habituation*. Habituation can be described as specific or general. (i) Specific—specific to a particular repeated stimulus and specific to the part of the body which has been repeatedly stimulated, and (ii) General—a change in the

physiological set of the organism relevant to the repeated stimulus and the conditions incidental to its application.

- *Learning*. Learning is the acquisition of a new response, or a qualitative change of an existing response, or an inhibition or facilitation of an existing response by a new stimulus (Hafez 1968a).
- *Conditioning*. Conditioning is the transfer of an existing response to a new stimulus (Hafez 1968a).
- *Hardening*. Transitory adaptation to an extreme temperature that followed brief exposure at a sub-lethal temperature (Bowler 2005).

However it should be noted that it is difficult to make definitions that will apply universally to all animals or situations (Loeschcke and Sørensen 2005).

10.6 Processes of Thermal Adaptation

Adaptation is a progressive reduction in physiological strain in an animal when the animal is repeatedly exposed to a non-lethal stress (Taylor 2006). Adaptation is conducive to animal survival because it increases the ability of the animal to tolerate stress (Taylor 2006). Within a limited range of body temperatures, animals have some capacity to modify their behavioural, physiological or morphological characteristics in response to a thermal challenge (Angilletta 2009).The range in body temperature is due to optimised structural and kinetic coordination of molecular, cellular and systemic processes (Pörtner and Farrell 2008). An animal's ability to survive once ambient temperature approaches the animals thermal limits, depends on its capacity to maintain or restore proteins following a thermal challenge (Hofmann and Todgham 2009) (see *Biochemical adaptations* below). Thermotolerance is a cellular adaptation that is a result of a single non-lethal heat exposure that allows an animal to survive subsequent insults that would otherwise be lethal (Moseley 1997). What we tend to see in the short term is phenotypic adaptation, what happens with long-term chronic stress is difficult to predict largely because scientific studies have not been undertaken. Indeed, the precise physiological and biochemical mechanisms that define the upper and lower thermal tolerance limits in animals are largely unknown (Hofmann and Todgham 2009). Furthermore, the outcome of climatic change within any particular environment is unknown. Climate change is one of the several intriguing factors making animal production difficult in many areas of world. In reality animals can, will and do adapt to climate change, if they have sufficient time. If they are not able to adapt they face regional or global extinction.

The response mechanisms employed by an animal when exposed to stressors are helpful for survival but are usually detrimental to productive performance which means that animals that are actually adapting to specific changes may not be selected because they are doing so at the expense of production. The potential climate change impact factors on animals include an overall increase in prolonged chronic heat stress and an increase in the number of extreme climatic events. Stress

Table 10.1 Types of selection on animals[a]

	R—selection	*K*—selection	*A*—selection
Environment			
Stability	Low	High	High
Abiotic stress	High	Low	High
Energy	Low	High	Low
Individuals			
Body size	Small	Large	Small or Large
Lifespan	Short	Long	Long
Maturity	Early	Late	Late
Reproduction			
Pattern	Semelparous[b]	Iteroparous[c]	Either
Generation time	Short	Long	Either
Fecundity	High	Low	Low
Offspring	Many, small	Few, large	Either
Parental care	Absent	Common	Possible
Populations			
Density	Fluctuating	High	Low, or fluctuating
Stability	Fluctuating	Steady	Fluctuating
Range	High	Low	Either
Competition	Low	High	Low
Biotic interactions	Few, simple	Many, complex	Few, simple
Overview	Small	Large	Very varied
	Rapid reproductive output	Slow reproductive output	Usually slow
	Colonists	Complex communities	Simple climax
	Generalist	Specialists	Specialists

[a] Adapted from Willmer et al. (2006)
[b] Only produces offspring once in its life
[c] Has many reproductive cycles throughout its life

can be looked upon as being either abiotic (physical and chemical factors) or biotic (direct and indirect effects of other organisms including competition for food) (Willmer et al. 2006). The extents to which animals are able to adapt to stressors are largely limited by physiological (genetic) constraints (Devendra 1987; Parsons 1994). Selection criteria of livestock and poultry (and possibly domestic animals in general) needs to be considered in the context of climate change, and whether the location (habitat) is likely to be favourable or unfavourable to the species of concern.

Interactions between environmental stressors, the magnitude of short-term and long-term climate change and the available feed resources interact to determine the type of animals found in a particular environment, and the type of selection or adaptation that occurs within the environment (Willmer et al. 2006). Three types of selection are recognised and these are based on the type of habitat where the animals are expected to live (Greenslade 1983 cited by Parsons 1994; Willmer et al. 2006) (Table 10.1). The classifications used are: *R* selection—(unpredictable habitat; animals are small in size, rapidly reproducing, early maturing, produce large number of offspring and are short lived, e.g. many rodent species), *r* refers to the rate of population increase (Willmer et al. 2006); *A* selection—adversity selection (predictably unfavourable environments which have high environmental stress but low magnitude of fluctuation, e.g. deserts, mountain habitats; animals have high stress resistance, low fecundity, late maturing, long lives and low levels of biotic interaction (Willmer et al. 2006), e.g. camels, desert sheep and goats, Zebu cattle); and *K* selection—(predictably favourable low stress environments; animals are relatively large, slowly reproducing, long lived and only produce a few young, e.g. European cattle and sheep, and kangaroos): *k* represents the carrying capacity of the environment. Within these selection classifications animals have developed a number of adaptive mechanisms.

10.7 Mechanisms of Adaptation

Mechanisms of adaptation include morphological, behavioural, biochemical and physiological change. These mechanisms interact to produce a change in the animal which improves an animal's fitness in a particular environment. For example, Sudanese Desert sheep tolerate thermal stress by panting, cutaneous evaporation and by concentrating urine (Gaughan et al. 2009). Fundamentally, all mechanisms of adaptation are controlled by genes. Therefore, the changes which occur in an animal are a result of gene × environment interaction.

Hansen (2004) has written an excellent review on adaptation of zebu cattle to heat stress and this review is a good resource for further information.

10.7.1 Morphological Adaptations

Livestock have adapted, either naturally or via human selection, to a variety of environmental conditions. These conditions include deserts, alpine regions, wet and dry tropics, arctic and temperate zones. Within a species there is considerable variation within breeds and between breeds in regard to their capacity to cope with an environmental challenge (Gaughan et al. 2010b). The structural adaptations of animals to thermal environments are the most pronounced (Hafez 1968a). For example, the long angular ears and large dewlap of a Brahman (*Bos indicus*) cow compared with the short round ears and small dewlap of an Angus (*Bos taurus*) cow; the former enhancing heat loss and the latter reducing heat loss. Properties of the hair coat in zebu cattle enhance conductive and convective heat loss and reduce absorption of solar radiation (Hansen 2004). Heat tolerant cattle (Sanga breeds) have slightly different skeletal structures compared to temperate breeds. These changes influence the angle of the rib cage which leads to an improvement in respiratory dynamics so that the animal can lose heat without excessive panting (Ramsey 2010). Tropically adapted *Bos taurus* cattle breeds such as those defined as Criollos in the Americas

Table 10.2 The main morphological adaptations in farm animals

Environmental stress	Morphological adaptation	Animal
Solar radiation	Long limbs	Camel
	Long open coats	Desert sheep and goats
High temperature	Hair shedding in summer	
	Increased surface area (skin folds)	Brahman cattle
	Small body long ears	Rabbit
	Loose coarse wool	Awassi sheep
	Fine dense wool	Merino sheep
Low temperature	Long and fine hair	Temperate cattle e.g. Angus
	Thick subcutaneous fat	Arctic species
	Abundant brown fat	Many species—usually neonates. e.g. piglets
	Thick heavy coat	Rabbit, sheep, cattle, horse
High humidity	Dark pigmentation, sparsely haired	Water buffalo
Seasonality in food	Adipose tissue reserves	Awassi sheep
	Hump	Camel
	Fat-tail	Sheep
	Fat-rump	Sheep (Somalia)
Desert	Thick skin, hard tissue around mouth; thick mouth, lined with long papillae, increased drinking capacity, hump, conservation of metabolic water, ability to survive dehydration	Camel
High altitude	Increased O_2 carrying capacity in blood, increased concentration of RBCs, ability to transfer O_2 from capillary blood to tissue cells, high efficiency in nutrient extraction	Llama, alpaca

Adapted from Hafez (1968b) and Khalifa (2003)

have developed a series of distinctive characteristics which enhances heat loss (De Alba 1987). These morphological adaptations include:

- Short hair
- Sparse hair
- Unmedullated hair
- Pigmented skin
- Very short ears with little hair
- Very mobile and slender tail.

The coat and skin characteristics of sheep and goats that have evolved in tropical and desert areas are different from those that evolved in temperate climates. For example, the loose open fleece of hair and wool of Awassi sheep enhances heat loss via convection. Sudanese desert goats have long legs and long ears, and relatively large size (33 kg) which present a mechanism for evaporative heat loss (Devendra 1987). Black Bedouin goats are able to store high volumes of body water, and have considerable sweating capacity which allows them to cope in a hot environment even with the additional heat load resulting from their dark colouration (Shkolnik and Choshniak 1985).

The main morphological adaptations of livestock are external insulation (coat and fur depth, hair type, hair density and subcutaneous fat), fat storage in hump or tail especially under desert conditions, skin colour and body size (Khalifa 2003) (Table 10.2).

10.7.2 Behavioural Adaptations

Behaviour is one of the most effective adaptive mechanisms (Hafez 1968b), at least in the short term. Behavioural change allows animals to avoid or reduce the impact of stressors. However, not all behavioural changes have a positive impact. For example, there may be increased aggressive behaviour when cattle, with limited shade or water access, are exposed to high heat load, i.e. the dominant animals will push less dominant ones from shade, or they may block water trough access. However, most behavioural changes would be considered to be attempts to reduce the impact of a given stressor. These changes include: seeking shelter (e.g. shade or wind breaks), changing posture (e.g. standing or altering orientation to the sun to reduce gain of heat from solar radiation), wallowing, reducing feed intake (when exposed to hot conditions) or increasing feed intake (when exposed to very cold conditions) and increasing or decreasing water intake again based on ambient conditions. Other behavioural changes such as bunching, by heat stressed cattle, may be a psychological response to the stress and is not considered to be a positive behaviour. It is largely thought that bunching will decrease heat loss from already heat stressed cattle. On the other hand, huddling in pigs and puppies is a positive behavioural adaptation to cold stress. Behavioural changes are an innate component of the animal. These behaviours may be enhanced by human selection within a breed or species. However, as with many other adaptive responses, there may be subsequent production losses.

Understanding the 'normal' behaviours of animals under non-stressful conditions is paramount for our understanding of behavioural changes when animals are stressed.

10.7.3 Biochemical Adaptations

Biochemical adaptations to thermal stress are complex and varied. The responses may be changes in proteins, membrane lipids and metabolic rate (Bowler 2005). "Critical biological processes, including development, growth, and fitness, are all temperature dependent" (Spees et al. 2002). The responses may be due to a direct effect of high temperature on metabolic function or may be a result of heat impacting gene expression. Three specific examples will follow:

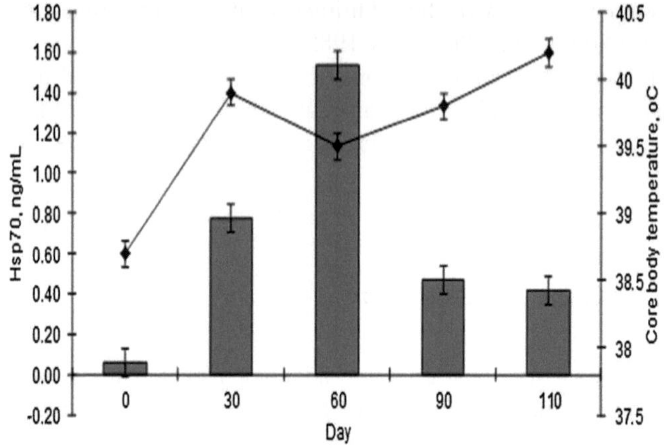

Fig. 10.2 Relationship between Hsp70 (*bars*) and body temperature (*line*) of angus steers housed in a feedlot over summer on days 0, 30, 60, 90 and 110 of the feeding period (Gaughan and Bonner 2011)

Production of heat shock proteins (HSP) is a cellular and tissue defence response to thermal stress (Iwaki et al. 1993; Spees et al. 2002; Marruchella et al. 2004). Heat shock proteins are members of highly conserved families of molecular chaperons that have multiple roles in animals (Hecker and McGarvey 2011), and are named according to their molecular weight expressed in kilodaltons (kDa) (e.g. Hsp70, Hsp90) (Marruchella et al. 2004). Heat shock proteins protect against thermal stress by refolding proteins that have been damaged (denatured) by heat and by preventing protein aggregation (protein aggregation is harmful to animals) (Hofmann and Todgham 2009). However, intense or prolonged exposure may lead to permanent protein damage even in the presence of HSP. Among the heat shock protein families, Hsp70 has a significant role in the thermotolerance of cells (Beckham et al. 2004), and in animal survival (King et al. 2002). However, heat shock protein expression is not limited to heat stress. Other stressors such as changes in nutrition, housing and mixing animals may induce a heat shock response. In a feedlot study, Gaughan and Bonner (2011) demonstrated the combined effects of multiple stressors (including temperature) on HSP expression in Angus steers (Fig. 10.2).

It is well accepted that there exist in cattle, genes which regulate body temperature and enhance cellular resistance to elevated temperature. Although largely unidentified, the existence of these genes offers the possibility for their incorporation into non-tolerant breeds through crossbreeding or on an individual gene basis (Hansen 2007).

There is evidence that cattle that evolved in hot climates have acquired genes that protect cells from the effects of elevated temperature (Kamwanja et al. 1994; Hansen 2007). Genetic resistance to cellular effects of elevated temperature are seen in both *Bos indicus* as well as in tropically evolved *Bos taurus* breeds such as Senepol and Romosinuano (Hansen 2007). Little is currently known regarding the molecular basis for the improved cellular resistance to elevated temperature in thermotolerant cattle. Survivability of heat stressed bovine lymphocytes was used by Kamwanja et al. (1994) to determine the differences in heat tolerance of Brahman, Senepol and Angus cattle. There were no significant differences between Brahman, Senepol and Angus in the induction of Hsp70 in heat shocked lymphocytes. Hansen (2007) suggested that the tendency for lower amounts of Hsp70 in Brahman and Senepol may indicate that protein denaturation in response to elevated temperature is reduced in the Brahman and Senepol breeds. Across the three breeds the viability of lymphocytes was reduced by high temperature (42°C). However, lymphocyte viability was significantly better in Brahman (56.8%) and Senepol (54.2%) than in Angus (44.3%) (Kamwanja et al. 1994).

It is not clear at this time if heat shock protein expression will be a useful biomarker for selection of animals for heat tolerance.

10.7.4 Physiological Adaptations to Heat

Physiological adaptation can be looked at in terms of short-term (hours to a few days) or long-term changes (weeks or permanent).

Short-term: Short-term changes are often caused by acute stressors, e.g. a heat wave. These changes will be manifested as an increased sweating rate, body temperature, respiration rate, heart rate and a reduction in metabolic heat production. The physiological responses of livestock to acute heat stress have been well documented (Nienaber et al. 1999; Mayer et al. 1999; Kadzere et al. 2002; Huynh et al. 2005; Hamadeh et al. 2006; Mader et al. 2006; Gaughan et al. 2010a; Sullivan et al. 2011; Alhidary et al. 2012).

Long-term change: Long-term change may be due to chronic stress whereby some physiological aspect of the animal establishes itself at a new adjusted level. This adjusted level is often at the expense of production. Long-term changes may also be due to the effect of acute stressors on gene expression (where the gene expression is a permanent change).

Epigenetic temperature adaptation: Environmental changes during gestation of mammals or during incubation of birds can have an effect on gene expression in embryos (Nichelmann 2004). These effects may have positive or negative effects during later development. Low birth weight of calves induced by malnutrition of the cow has long-term effects on growth and development of the calf. Epigenetic temperature adaptation has been observed in some avian species (May et al. 1987; Yahav and Plavnik 1999; Nichelmann 2004) and is considered to be a positive response. Epigenetic adaptation is the response of an animal to an environment that will be encountered in its future development (Nichelmann 2004). Epigenetic adaptation is the result of environmental effects on gene expression and is not genetically fixed (Nichelmann 2004). The epigenetic effect was demonstrated by exposing 5-day-old broiler chickens to high temperatures (38°C) for a 24 h period (Yahav and Plavnik 1999). This short-term exposure increased the birds' heat

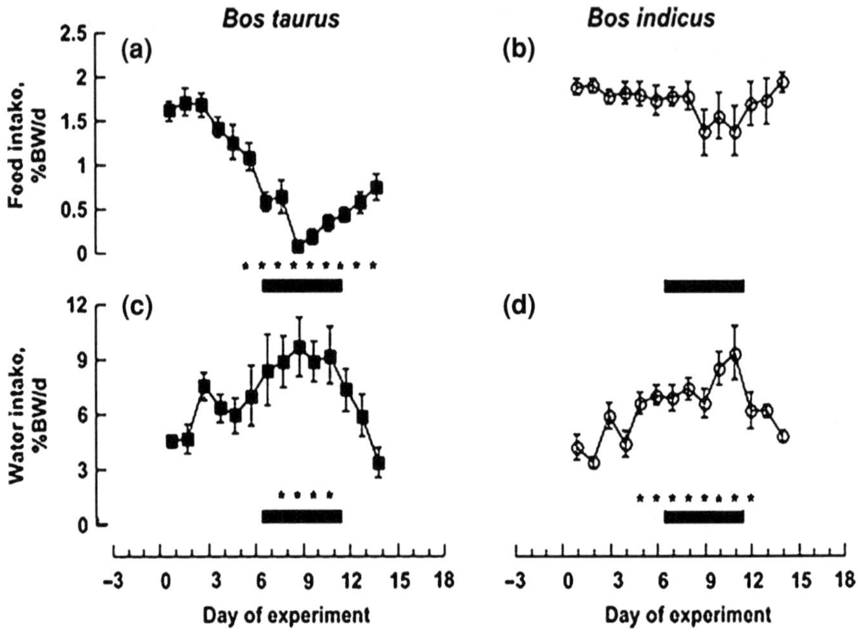

Fig. 10.3 Mean daily feed intake (**a** and **b**) and water intake (**c** and **d**) for *Bos taurus* and *Bos indicus* heifers. *Points* show the mean ± SEM for each of six animals. The *horizontal bar* under each figure indicates the hottest 5 day of the experiment. *Asterisks* under the data denote $P < 0.05$ for the day marked versus the control days (day 1 and 2) (Beatty et al. 2006)

tolerance to latter heat events. Nichelmann (2004) reported that incubation temperatures at the end of the incubation period improved the thermal tolerance of ducks but not turkeys.

Reduced metabolic rate: Low metabolic rates resulting from reduced growth rates and milk yields of many zebu breeds is a major contributing factor to their thermotolerance. There is also evidence that the basal metabolic rate of *Bos indicus* is lower than for *Bos taurus* (Hansen 2004). When *Bos taurus* cattle adapt to prolonged heat stress they will also reduce metabolic rate by reducing feed intake (Fig. 10.3). The reduction in feed intake may lead to reductions in growth of feedlot cattle (Sullivan et al. 2011) and a decline in milk production from dairy cows (Bohmanova et al. 2007). Similar reductions in performance are seen for many species that are not heat tolerant.

10.8 Conclusions

There is little doubt that climate change will have an impact on livestock performance in many regions. The capacity of animals to adapt in the short to medium term will be limited primarily by their genetics. However, financial resources and management capacity will have a major role. Adaptation to prolonged stressors will most likely be accompanied by a production loss, and input costs may also increase. Increasing or maintaining current production levels in an increasingly hostile environment is not sustainable. It may make better sense to look at using adapted animals, albeit with lower production levels (and also lower input costs) rather than try to infuse 'heat tolerance' genes into a non-adapted breed.

References

Alhidary IA, Shini S, Al Jassim RAM, Gaughan JB (2012) Physiological responses of Australian Merino wethers exposed to high heat load. J Anim Sci 90:212–220

Angilletta MJ (2009) Thermal adaptation—a theoretical and empirical synthesis. Oxford University Press Inc, New York

Beatty DT, Barnes A, Taylor E, Pethick D, McCarthy M, Maloney SK (2006) Physiological responses of Bos taurus and Bos indicus cattle to prolonged, continuous heat and humidity. J Anim Sci 84:972–985

Beckham JT, Mackanos MA, Crooke C, Takahashi T, O'Connell-Rodwell C, Contag CH, Jansen ED (2004) Assessment of cellular response to thermal laser injury through bioluminescence imaging of heat shock protein 70. Photochem Photobiol 79:76–85

Bohmanova J, Misztal I, Cole JB (2007) Temperature-humidity indices as indicators of milk production losses due to heat stress. J Dairy Sci 90:1947–1956

Bowler K (2005) Acclimation, heat shock and hardening. J Therm Biol 30:125–130

De Alba J (1987) Special cattle groups in the tropics. In: Johnson HD (ed) Bioclimatology and the adaptation of livestock. Elsevier, Amsterdam

Devendra C (1987) Goats. In: Johnson HD (ed) Bioclimatology and the adaptation of livestock. Elsevier, Amsterdam

Folk GE (1974) Textbook of environmental physiology. Lea & Febiger, Philadelphia

Gaughan JB, Bonner SL (2011) Heat shock protein 70 expression as an indicator of chronic stress in cattle. Proceedings of the 5th international congress on stress response in biology and medicine, Quebec City, 21–25 August 2011, p 188

Gaughan JB, Bonner S, Loxton I, Mader TL, Lisle A, Lawerence R (2010a) Effect of shade on body temperature and performance of feedlot steers. J Anim Sci 88:4056–4067

Gaughan J, Lacetera N, Valtorta SE, Khalifa HH, Hahn L, Mader T (2009) Response of domestic animals to climate change. In: Ebi KL, Burton I, McGregor GR (eds) Biometeorology for adaptation to climate variability and change. Springer, Germany

Gaughan JB, Lees JC (2010) Categorising heat load on dairy cows. Proc Aust Soc Anim Prod 28:105

Gaughan JB, Mader TL, Holt SM, Sullivan ML, Hahn GL (2010b) Assessing heat tolerance of 17 beef cattle genotypes. Int J Biometeorol 54:629–635

Glossary of Terms for Thermal Physiology (2001) IUPS thermal commission. Japan J Physiol 51:245–280

Hafez ESE (1968a) Principles of animal adaptation. In: Hafez ESE (ed) Adaptation of domestic animals. Lea & Febiger, Philadelphia

Hafez ESE (1968b) Behavioral adaptation. In: Hafez ESE (ed) Adaptation of domestic animals. Lea & Febiger, Philadelphia

Hamadeh SK, Rawda N, Jaber JS, Habre A, Abi Said M, Barbour EK (2006) Physiological responses to water restriction in dry and lactating Awassi ewes. Livest Sci 101:101–109

Hansen PJ (2004) Physiological and cellular adaptations of zebu cattle to thermal stress. Anim Reprod Sci 82–83:349–360

Hansen PJ (2007) Exploitation of genetic and physiological determinants of embryonic resistance to elevated temperature to improve embryonic survival in dairy cattle during heat stress. Theriogenology 68S:S242–S249
Hecker JG, McGarvey M (2011) Heat shock proteins as biomarkers for the rapid detection of brain and spinal cord ischemia: a review and comparison to other methods of detection in thoracic aneurysm repair. Cell Stress Chaperon 16:119–131
Hofmann GE, Todgham AE (2009) Living in the now: physiological mechanisms to tolerate a rapidly changing environment. Annu Rev Physiol 72:22.1–22.19
Huynh TTT, Aarnink AJA, Versegen MWA, Gerrits WJJ, Heetkamp B, Canh TT (2005) Effcets of increasing temperature on physiological changes in pigs at different relative humidities. J Anim Sci 83:1385–1396
Iwaki K, Chi SH, Dillmann WH, Mestril R (1993) Induction of HSP70 in cultured rat neonatal cardiomyocytes by hypoxia and metabolic stress. Circulation 97:2023–2032
Jenouvrier S, Visser ME (2011) Climate change, phonological shifts, eco-evolutionary responses and population viability: toward a unifying predictive approach. Int J Biometeorol 55:905–919
Kadzere CT, Murphy MR, Silanikove N, Maltz E (2002) Heat stress in lactating dairy cows: a review. Livest Prod Sci 77:59–91
Kamwanja LA, Chase CC Jr, Gutierrez JA, Guerriero V Jr, Olson TA, Hammond AC, Hansen PJ (1994) Responses of bovine lymphocytes to heat shock as modified by breed and antioxidant status. J Anim Sci 72:438–444
Kassahn KS, Crozier RH, Pörtner HO, Caley MJ (2009) Animal performance and stress: responses and tolerance limits at different levels of biological organisation. Biol Rev 84: 277–292
Khalifa HH (2003) Bioclimatology and adaptation of farm animals in a changing climate. In: Lacetera N, Bernabucci U, Khalifa HH, Ronchi B, Nordone A (eds) Interactions between climate and animal production. Wageningen Academic Publishers, The Netherlands
King YT, Lin CS, Lin JH, Lee WC (2002) Whole-body hyperthermia-induced thermo tolerance is associated with the induction of heat shock protein 70 in mice. J Exp Biol 205:273–278
Langlois B (1994) Inter-breed variation in the horse in regard to cold adaptation: a review. Livest Prod Sci 40:1–7
Loeschcke V, Sørensen JG (2005) Acclimation, heat shock and hardening—a response from evolutionary biology. J Therm Biol 30:255–257
Mader TL, Davis MS, Brown-Brandl T (2006) Environmental factors influencing heat stress in feedlot cattle: feed and water intake. J Anim Sci 82:712–719
Marcogliese DJ (2001) Implications of climate change for parasitism of animals in the aquatic environment. Can J Zool 79:1331–1352
Marruchella G, Di Leonardo M, Di Guardo G, Romanucci M, Mara M, Tiscar PG, Mosca F, Della Salda L (2004) Heat shock proteins (HSPs) 27, 72 and 73 in normal and pre-ulcerative mucosa of the gastric pars oesophagea in swine. J Comp Path 131:10–17
May JD, Deaton JW, Branton SL (1987) Body temperature of acclimated broilers during exposure to high temperature. Poult Sci 66:378–380
Mayer DG, Davison TM, McGowan MR, Young BA, Matshoss AL, Hall AB, Goodwin PJ, Gaughan JB (1999) Extent and economic impact effect of heat loads on dairy cattle production in Australia. Aust Vet J 77:804–808
McManus C, Prescott E, Paludo GR, Bianchini E, Louvandini H, Mariante AS (2009) Heat tolerance in naturalized Brazilian cattle breeds. Livest Sci 120:256–264
Moseley PL (1997) Heat shock proteins and heat adaptation of the whole organism. J App Physiol 83:1413–1417
Nichelmann M (2004) Perinatal epigenetic temperature adaptation in avian species: comparison of turkey and Muscovy duck. J Therm Biol 29:613–619
Nienaber JA, Hahn GL, Eigenberg RA (1999) Quantifying livestock responses for heat stress management: a review. Int J Biometeotol 42:183–188
Parsons PA (1994) Habitats, stress, and evolutionary rates. J Evolution Biol 7:387–397
Pörtner HO, Farrell AP (2008) Physiology and climate change. Science 322:690–692

Pörtner HO, Knust R (2007) Climate change affects marine fish through the oxygen limitation of thermal tolerance. Science 315:95–97
Price EO (1984) Behavioral aspects of animal domestication. Q Rev Biol 59:1–32
Ramsey K (2010) Adaptive traits of sanga cattle: their importance in meeting the challenges associated with climate change in the tropics and sub tropics. Adv Anim Biosci 1:381–382
Santolaria P, Lopez-Gatius F, Garcia-Ispierto I, Bech-Sabat G, Angulo E, Carretero T, Sanchez JA, Yaniz J (2010) Effects of cumulative stressful and acute variation of farm climate conditions on late embryo/early feta loss in high producing dairy cows. Int J Biometeorol 54:93–98
Shkolnik A, Choshniak I (1985) Physiological responses and productivity of goats. In: Yousef MK (ed) Stress physiology in livestock, volume II, Ungulates. CRC Press, Inc, Boca Raton
Spees JL, Chang SA, Snyder MJ, Chang ES (2002) Thermal acclimation and stress in the American lobster, *Homarusamericanus*: equivalent temperature shifts elicit unique gene expression patters for molecular chaperones and polyubiquitin. Cell Stress Chaperones 7: 97–106
St-Pierre NR, Cobanov B, Schnitkey G (2003) Economic losses from heat stress by US livestock industries. J Dairy Sci 86(E Suppl):E52–E77
Steinfeld H, Gerber P, Wasenaar T, Castel V, Rosales M, de Haan C (2006) Livestocks long shadow: environmental issues and options. FAO, Rome
Sullivan ML, Cawdell-Smith AJ, Mader TL, Gaughan JB (2011) Effect of shade area on performance and welfare of short-fed feedlot cattle. J Anim Sci 89:2911–2925
Sutherst RW (2001) The vulnerability of animal and human health to parasites under global change. Int J Parasitol 31:933–948
Tabachnick WJ (2010) Challenges in predicting climate and environmental effects on vector-borne disease episystems in a changing world. J Exp Biol 213:946–954
Taylor NAS (2006) Ethnic differences in thermoregulation: genotypic versus phenotypic heat adaptation. J Therm Biol 31:90–104
Thornton PK, van de Steeg J, Notenbaert A, Herrero M (2009) The impacts of climate change on livestock and livestock systems in developing countries: a review of what we know and what we need to know. Agric Sys 101:113–127
Willmer P, Stone G, Johnston I (2006) Environmental physiology of animals, 2nd edn. Blackwell Publishing Ltd, Oxford
Yahav S, Plavnik I (1999) Effect of early-stage thermal conditioning and food restrictions on performance and thermo tolerance of male broiler chickens. Br Poult Sci 40:120–126

Pörtner HO, Knust R (2007) Climate change affects marine fishes through the oxygen limitation of thermal tolerance. Science 315:95–97
Price EO (1984) Behavioral aspects of animal domestication. Q Rev Biol 59:1–32
[illegible]

Chapter 11
Neuroendocrine Regulation of Adaptive Mechanisms in Livestock

Sheba M. J. MohanKumar, Priya Balasubramanian, Meenambigai Dharmaraj and Puliyur S. MohanKumar

Abstract Neuroendocrine responses to stress play an integral role in the maintenance of homeostasis in livestock. In general, activation of the stress circuitry inhibits functions such as growth and reproduction. Substantial evidence suggests that neuroendocrine responses varies with the type of stressor and are specific and graded, rather than 'all or none'. While acute responses have important adaptive functions and are vital to coping and survival, chronic stressors elicit endocrine responses that may actually contribute to morbidity and mortality. Integration of these responses is possible through the network of mutual interactions that exist between the immune system, the central nervous system, and the endocrine system. A crucial component of this network is the stress axis. Activation of the stress axis is accomplished through the release of several neurotransmitters and hormones. The stress axis or the hypothalamo-pituitary-adrenal (HPA) axis consists of 3 components: corticotrophin releasing hormone (CRH) neurons in the hypothalamus, corticotrophs in the anterior pituitary, and the adrenal cortex. A variety of molecular mediators have been implicated in the stimulation of CRH neurons ranging from neurotransmitters such as catecholamines to proinflammatory cytokines. CRH is an obligatory and primary stimulus for

S. M. J. MohanKumar (✉) · P. Balasubramanian
Department of Pharmacology and Toxicology, College of Veterinary Medicine,
Michigan State University, East Lansing, MI 48824, USA
e-mail: mohankumrs@cvm.msu.edu

M. Dharmaraj
Department of Animal Biotechnology,
Tamil Nadu Veterinary and Animal Sciences University, Chennai, 600 007, India

P. S. MohanKumar
Department of Pathobiology and Diagnostic Investigation, College of Veterinary Medicine,
Michigan State University, East Lansing, MI 48824, USA

V. Sejian et al. (eds.), *Environmental Stress and Amelioration in Livestock Production*, DOI: 10.1007/978-3-642-29205-7_11,

adrenocorticotropin hormone (ACTH) secretion by the pituitary gland. Subsequently, ACTH stimulates glucocorticoid synthesis from the cells of adrenal cortex. The secretion of ACTH, which is very crucial in this neuroendocrine response, seems to be regulated by a variety of peptides, but principally by CRH and vasopressin (VP; arginine vasopressin in most mammalian farm animal species; lysine vasopressin in pigs). CRH seems to be active mainly in the acute phase of stress while VP is proposed to maintain HPA axis activity after repeated stimulation. In addition to the more traditional regulation of pituitary corticotrope function by hypothalamic peptides, proinflammatory cytokines are now recognized to play an important regulatory role in the HPA axis. Currently, interleukin-1, interleukin-6, and tumor necrosis factor are implicated strongly as stimulators of the stress axis. Cytokines are also capable of stimulating the secretion of the hormone leptin from adipocytes. Leptin is now recognized as an inhibitor of stress axis activity. Therefore, both leptin and glucocorticoids complete the negative feedback circuit to suppress stress axis activity and maintain homeostasis. We will explore the relationship between these hormonal mediators, immune cytokines, brain neurochemicals, and stress axis activation and the implications for livestock production.

Keywords Stress · HPA axis · CRH · Vasopressin · Leptin · Cytokines · Reproduction · Growth · Thyroid function · Appetite regulation

Contents

11.1 Introduction

Stress is a necessary response in all living beings to maintain homeostasis. In mammals, the stress response is a highly integrated, delicately balanced phenomenon (Chrousos and Gold 1992). Animals respond to a variety of social and environmental stressors through a coordinated effort of the brain, the pituitary, the adrenal, and other organ systems. Livestock in particular are subject to several stressors that are not encountered by other animals. These include thermal stress, production stress, transportation stress, nutritional stress, immune stress, stress due to crowding, drought, and other environmental conditions (Kelley 1980). They respond to these stressors in specific ways and develop specific adaptive mechanisms ranging from simple behavioral alterations to more involved physiological adaptations. The response exhibited by animals to these stressors varies depending on the duration of stress and when the stress occurs relative to their lifespan.

The adaptations that animals have developed to different stressors vary with the type of stressor and with the individual animal. While one stressor can elicit a marked reaction in one animal, the same stressor may leave another animal untouched. Individual differences aside, the response or reaction initiated by different types of stressors varies. These responses often result in altered utilization of biological resources at the disposal of the individual animal. The biological costs are low for acute stressors but quite devastating for chronic stresses leading to reduced growth and production in livestock (Moberg 2000). Nevertheless, the ultimate goal of stress axis activation is to effectively manage stressful situations and enhance the probability of survival. The neuroendocrine system plays a central role in integrating the stress response and the regulation of a variety of body functions, such as growth, reproduction, feeding, energy homeostasis, etc. Therefore, it is important to understand the neuroendocrine regulation of these functions and how they are influenced by stress.

11.2 Evaluating Stress and its Impact on Livestock

Stress responses in livestock have been classified into four types (Curtis 2011):

1. Understress: When an environment lacks certain features such as social companions or objects of interest, it causes the animal to enter boredom and develop tendencies towards stereotypic behavior that may be annoying, distracting, or a nuisance.

2. Eustress or good stress: makes the animal produce an extraordinary response that is tolerated very well by the animal. This stressful experience may even be enjoyable in some instances.
3. Overstress is when environmental conditions activate the stress axis mildly. Fitness and performance may decline but they rebound once the stressor is removed.
4. Distress or bad stress provokes major stress responses leading to the generation of negative emotions, such as fear, anxiety, frustration, and pain. Distress can cause the animals to undergo suffering and has a profound effect on production and performance. The animal may or may not regain normalcy after removal of the stressor.

While raising livestock, farmers engage in several management practices that may not be perceived as stressful by the farmer, but leaves a marked impact on livestock production. For example, routine procedures, such as vaccinations, milking, truck loading, etc. can be particularly stressful for livestock if they are not handled properly (Siegel 2007). Loud noises used to control animals can be very stressful to livestock that is reflected in an increase in heart rate (Waynert 1999; Marchant et al. 2003). Stressed animals need at least 30 min to calm down and have their heart rate return to normal (Stermer et al. 1981). On the other hand, calm animals are less stressed. Negative handling techniques are known to directly impact livestock production. Animals raised in a calm environment have better weight gain and better meat quality (Petherick et al. 2002; King et al. 2006). Egg production dropped in layer hens (Barnet et al. 1992) and milk yield decreased in dairy cattle as a result of stressful episodes (Hemsworth et al. 2000). Stressful handling can also decrease conception rates after breeding in sheep, cattle, and swine (Grandin 2010a). Cattle prodded with an electric goad 15 min prior to slaughter had tough meat (Warner et al. 2007). Pigs are particularly sensitive to this kind of treatment. Prodding increases lactic acid levels, decreasing the pH of meat making it pale, soft, and exudative, which is a severe quality defect (Grandin 2010b). Therefore, it is clear that stress has a marked effect on livestock production affecting it at various levels from growth and breeding, to production. These effects involve several neuroendocrine systems that regulate growth, lactation, and the stress response itself. In this chapter, we will focus on the neuroendocrine control of the stress response and how it integrates with other neuroendocrine functions to help the animal adapt to acute and chronic stress.

11.3 The Stress Axis and its Organization

The stress axis or the hypothalamo-pituitary-adrenal (HPA) axis is activated when the animal perceives a threat to its existence. Activation of the HPA axis is a consistent neuroendocrine response to stress. It may be a conscious response to stressors that are apparent to livestock, such as the sight of a predator, overcrowding, excessive environmental temperatures, presence of guard dogs, the sight of people, or poor stockmanship (Grandin 2010a). This involves sensory input from various sense organs and integration by the limbic system. The stress axis may also be stimulated as an involuntary response to immune stressors, such as infections, increased internal parasitic load, and poor nutrition due to famine or drought. This response is coordinated by other brain centers that receive sensory stimuli from the viscera or stimuli from the circulation.

Activation of the stress axis involves three main components. Stress that is perceived by the brain results in the alteration of several neurotransmitters and neuropeptides in specific brain regions. Some of the main neurotransmitters include catecholamines, such as norepinephrine (NE), dopamine (DA), epinephrine (EPI), indoleamines, such as serotonin (5-HT), other neurotransmitters, such as acetylcholine (Ach), gamma amino butyric acid (GABA), and neuropeptide Y (NPY). These neurotransmitters are released rapidly from nerve terminals since they are stored in vesicles. They affect the activity of corticotrophin-releasing hormone (CRH) neurons (Mueller and Nistico 1989). CRH neurons are distributed in distinct populations all over the brain; from the hindbrain to the hypothalamus. In the hypothalamus, they are concentrated in the paraventricular nucleus (PVN) and their terminals extend to the median eminence. Stimulation of CRH neurons results in the release of CRH at the median eminence where they enter the portal circulation to reach the anterior pituitary. In the anterior pituitary, CRH acts on corticotrophs to stimulate the secretion of adrenocorticotrophic hormone (ACTH) and other peptides such as β-endorphin. Besides CRH, vasopressin that is released from the anterior pituitary is also a potent stimulator of ACTH. ACTH that enters the general circulation acts on the adrenal cortex to stimulate the secretion of glucocorticoids (Mueller and Nistico 1989).

Glucocorticoids have a number of physiological effects to enhance the ability of an animal to 'fight or fly'. It rapidly mobilizes glycogen from the liver to provide an immediate supply of energy in the form of glucose (Rose et al. 2010). This is quickly used by the brain and muscles. It also enhances glucose utilization that is essential for maintaining physiological stability. Cardiac output and respiration are increased to provide more oxygen and glucose to tissues, while gastrointestinal motility and activity is suppressed (Salak-Johnson 2011). Cell growth and reproduction are temporarily suspended under the influence of glucocorticoids. Ultimately, glucocorticoids act on the brain through a negative feedback mechanism to terminate the stress response. Besides acting through ACTH to stimulate glucocorticoid secretion from the adrenal, the PVN also contributes to the stress response by increasing sympathetic nervous system (SNS) activity (Pacak et al. 1995). There are direct neuronal connections between the PVN and the intermediolateral cellular column of the spinal cord. These connections are important for the stress-induced increase in cardiac output and respiration. They are also crucial for the stimulation of catecholamine secretion by the adrenal medulla that causes vasoconstriction in the visceral region forcing blood to the periphery where it is most needed for the animal to fight or fly. This would be a normal sequence of events if the stress were to be consciously perceived by the animal through sensory cues.

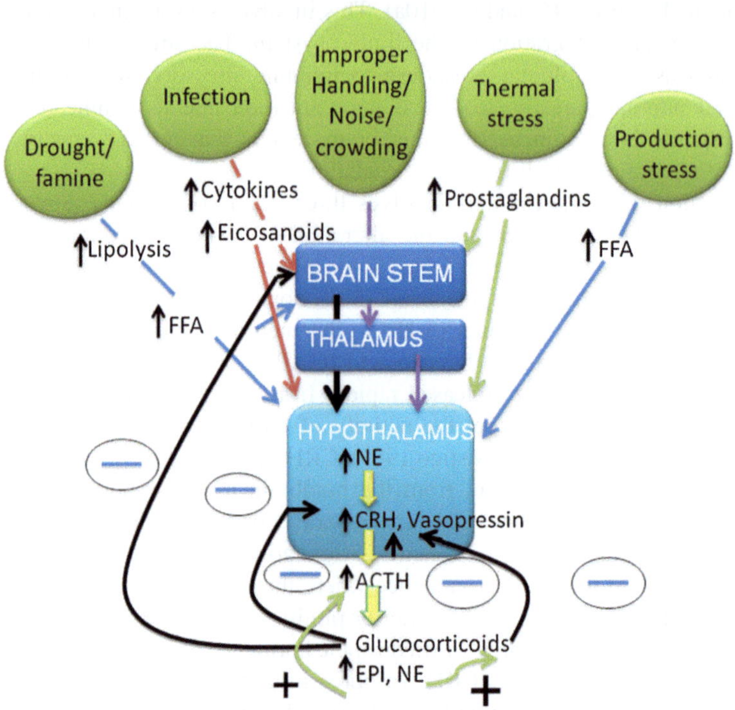

Fig. 11.1 Effect of various stressors on the stress axis. This figure demonstrates how various stress generators impact the CNS to stimulate stress hormone production and how glucocorticoids and adrenal steroids produce feedback effects on the brainstem, hypothalamus, and the pituitary to suppress stress axis functions. Intermediary molecules, such as cytokines, free fatty acids (FFA), and eicosanoids are specific to the stress that the animals are exposed to. The net result of the various stressors is increase in neurotransmitter levels such as norepinephrine (NE) in the hypothalamus which stimulates the secretion of corticotrophin releasing hormone (CRH), that results in increased adrenocorticotrophic hormone (ACTH) secretion by the pituitary which culminates in elevated levels of glucocorticoids, and adrenal steroids, epinephrine (EPI), and NE

In the case of autonomic responses caused by changes in the internal milieu, the mechanism by which stress axis is activated could be slightly altered, but once the CRH neurons are stimulated, it culminates in the secretion of glucocorticoids. For example, with a bacterial infection, increases in lipopolysaccharides in the circulation would cause elevations in proinflammatory cytokines. These could bind to their receptors on para-abdominal ganglia of the vagus (Watkins et al. 1995). Information that is transmitted through the vagal trunk reaches noradrenergic nuclei in the brain stem such as the *nucleus tractus solitarius*, which ultimately causes activation of the PVN through its ascending projections (Sehic and Blatteis 1996; MohanKumar et al. 2000). On the other hand, environmental hardships such as drought and famine which are particularly severe in certain parts of the world can result in an increase in serum triglycerides, free fatty acids (FFA) as a result of fat mobilization for energy purposes. FFA can cross the blood brain barrier and access specific neuronal populations. There is evidence that FFA such as oleic acid are capable of stimulating NE, DA, and 5-HT release from the hypothalamus directly (Jagannathan et al. 2009) and this could contribute to the activation of the stress axis that is observed with oleic acid administration (Widmaier et al. 1992). The complex interaction between various stressors and how they act through factors, such as CRH, vasopressin, ACTH, glucocorticoids, cytokines, FFA, and other mediators in influencing stress axis function is depicted in Fig. 11.1.

11.3.1 The Stress Axis: A Historical Perspective

The fact that stress causes activation of the adrenal cortex was first recognized by Hans Selye in 1939. Harris (1948) first suggested that the phenomenon could be mediated by neurons in the hypothalamus that regulate hormone secretion by the anterior pituitary gland. Other investigators provided evidence that indeed hypothalamic neurons were capable of secreting certain factors that controlled the secretion of ACTH from the pituitary (Guillemin and Rosenberg 1955; Saffran et al. 1955; Porter and Jones 1956). Saffran and colleagues (Saffran et al. 1955) first identified the hypothalamic factor that stimulates the HPA axis and termed it "Corticotropin releasing factor (CRF)". It was determined to be a 41 aminoacid peptide by Vale and colleagues in 1981 who renamed the molecule as "CRH" (Spiess et al. 1981). During the early years, when the field of neuroendocrinology was still in its infancy, other investigators were exploring the role of other factors and hormones that could potentially stimulate ACTH secretion. Martini and colleagues (Martini 1955) and Ono et al. (1985) were examining the in vivo effects of vasopressin on ACTH secretion as Saffran and colleagues identified CRF. Since vasopressin was capable of increasing ACTH levels, they hypothesized that it was another potential CRF. Then came the controversy of which of the two molecules was more important for HPA activation: CRF or vasopressin. Crediting its molecular characterization by Vale et al. (1981) and detailed studies on its physiological role (Rivier and Plotsky 1986), it had been concluded that although CRH and vasopressin both stimulate ACTH secretion, they do so independently or in concert with each other. Vasopressin has a potentiating effect on CRH-induced ACTH secretion in bovine (Carroll et al. 2007) and pigs (Minton and Parsons 1993). CRH can also increase adrenal blood flow and directly act on the bovine adrenal gland to stimulate glucocorticoid secretion (Jones and Edwards 1990; Carroll et al. 1996). Vasopressin has a similar effect on bovine adrenocortical cells (Carroll et al. 1996). CRH can also act on adrenal medullary chromaffin cells to synthesize ACTH (Mazzocchi et al. 1994). Thus, CRH and vasopressin act at the level of both the pituitary and the adrenal gland to cause activation of the HPA axis.

Besides CRH and vasopressin, several other factors are known to influence the HPA axis. EPI and oxytocin also stimulate ACTH secretion. Besides, adrenergic and oxytocinergic receptors have been identified in the pituitary gland in rats

(Petrovic et al. 1983; Antoni 1986). Several other factors, such as cytokines, growth factors, etc. can also stimulate the HPA axis.

Besides secreting glucocorticoids, the adrenal gland also secretes NE and EPI from the medulla. EPI appears to be associated with fear and anxiety responses (Fell et al. 1985; Abelson et al. 1996) while NE is the primary catecholamine in most livestock species. It is capable of stimulating CRH release, ACTH secretion from the pituitary (Mezey et al. 1984; Dinan 1996), or cortisol from the adrenal gland. Therefore, the increase in NE levels in the circulation, results in further activation of the stress axis, perhaps in an attempt to prolong the stress response. Selye proposed that all stressors would result in activation of the HPA axis and that there is an "all or none" response to stress. But the stress response is more complex than that, involving specific responses for specific stressors and also relating those stressors to past stressful events in order to mount a quicker and stronger response when exposed to the same stressor subsequently. The involvement of particular neurohumoral mechanisms probably influences the intensity and duration of the stress response. Some investigators believe that the magnitude of the stress response depends on the intensity and power of the stressors themselves (Natelson et al. 1981).

The need for different regulatory systems (Fig. 11.1) to influence HPA activation is not clear. It could be a fail-safe mechanism to have a functional stress response in the absence of a single activating pathway. This emphasizes the importance of the stress axis for survival. The other possibility is that each of these pathways or combination of pathways may be activated in response to specific stressors. For example, a physiological stress response to altered internal milieu may involve neuronal pathways that are very different from those that are activated in response to the sight of a predator.

11.3.2 Short-Term vs Long-Term Activation of the HPA Axis

Short-term increase in glucocorticoids can be protective and facilitates normal physiological and behavioral adaptive processes. On the contrary, high levels of glucocorticoids or prolonged increases in glucocorticoid levels can adversely affect various regulatory processes (Salak-Johnson 2011). Glucocorticoid receptors are present in both CRH neurons and noradrenergic neurons in the brain that regulate its secretion. Therefore, glucocorticoids have the capacity to inhibit these neurons through negative feedback mechanisms or stimulate them when the stress axis has to be activated. With short-term increases in glucocorticoid levels, physiological processes such as growth and reproduction are temporarily suppressed. This enables the animal to make better use of biological resources for reflex actions such as flight (Salak-Johnson 2011). However, long-term stress can interfere with the functioning of the endocrine system, which in turn affects growth and reproduction and affects an animal's productivity and even lower its chances of survival (Salak-Johnson 2011). Food intake and appetite are also affected after exposure to long-term stress (Sapolsky et al. 2000). This combined with an increase

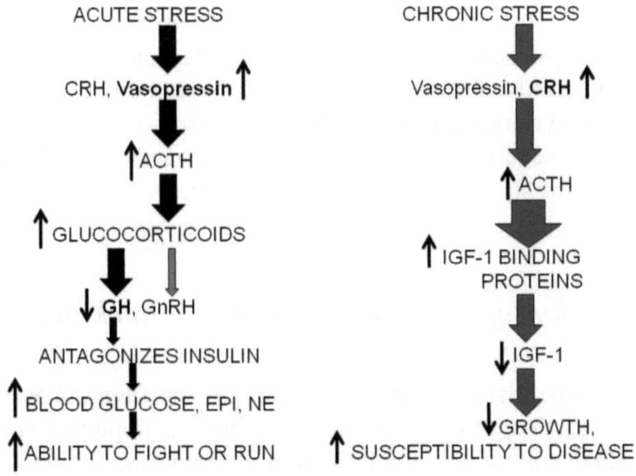

Fig. 11.2 Differences between acute and chronic stress responses. This figure depicts the differences in hormone secretion in response to acute and chronic stressors. While acute stressors increase Vasopressin more than CRH in the hypothalamus, chronic stressors have the opposite effect. The resultant effect on growth hormone and blood glucose levels enable the animal to rapidly respond to acute stressors by increasing their ability to run or fight. Chronic stress on the other hand has long-term effects such as reduced growth and increased susceptibility to diseases

in catabolic activity and a reduction in anabolism leads to a net loss in weight and productivity. Ultimately, repeated chronic stress can lead to pathological consequences (Salak-Johnson 2011). Although CRH and vasopressin are implicated in stress axis activation, they may play specific roles during short-term vs long-term stimulation of the stress axis. Both hormones are secreted in a pulsatile manner and stimulate ACTH secretion by acting through G-protein coupled receptors. Studies in sheep have found that repeated or prolonged stimulation with CRH or vasopressin can cause a reduction in ACTH secretion suggesting desensitization of these receptors. However, there is a subtle difference. Desensitization to vasopressin occurs much rapidly than CRH. This involves phosphorylation of the V1b receptor and internalization. Resensitization of this receptor could be blocked with a protein phosphatase 2B inhibitor suggesting that the rapid desensitization of the ACTH response to vasopressin is involved in acute stress rather than chronic stress (Mason et al. 2002). These adaptations are outlined in Fig. 11.2.

11.3.3 Negative Feedback Inhibition of the Stress Axis and its Impact

The homeostatic mechanisms that come into play to restore the animal to normal function involve suppression of the stress axis through negative feedback mechanisms. These could involve negative feedback inhibition by glucocorticoids on brainstem noradrenergic neurons, CRH neurons, or corticotrophs in the pituitary.

This would lead to reduction in NE levels in the hypothalamus, and a reduction in CRH, ACTH, and glucocorticoid levels. These feedback mechanisms may be more apparent during periods of acute stress. However, chronic stress on the other hand can develop adaptive mechanisms. In this case, stress axis activity may remain elevated in spite of negative feedback mechanisms and can only dissipate when the underlying cause for stress axis activation is removed.

11.3.4 Adaptation of the Stress Axis in Livestock to Stress

A 'fit' animal is one that is capable of adapting to its environment and not only surviving but is also capable of thriving. Adaptation is important not only for the individual animal but for the whole system to propagate it. In instances when an animal fails to adapt to its environment, it becomes evident in its way of being. An animal that is unable to adapt is experiencing stress. Failure to adapt has negative consequences for the animal's state of being (Curtis 2011). It directly imparts homokinetic mechanisms in the animal's body, altering its internal environment and causes it to be less productive and in serious circumstances even cause reproductive failure (Curtis 2011). The animal has a range of adaptive responses to stressors that are described by Broom (1993).

The animal's response ranges from minor to extreme responses that result in variable loss of fitness and well-being from which the animal may or may not recover once the stressor is removed. What should be noted is that the animal may mount an extraordinary response to a stressor without showing any sign of adaptation. The impact of environmental stressors can be estimated by measuring the effects on the animal. This will be reflected in the animal's survival and production or performance (Salak-Johnson et al. 2007).

Adaptation to stress is believed to be a 2-phase process with the time taken for adaptation ranging from weeks to months. There is not only a change in the rate of secretion of hormones but there is also modulation of their receptors at target sites (Collier et al. 2006). Adaptive mechanisms to stress are covered by allostasis. Allostasis is a process similar to homeostasis. It involves the whole body and helps the animal determine what kind and intensity of response is needed to effectively manage a particular stressor (Sterling and Eyer 1988; McEwen and Wingfield 2003). The neuroendocrine system plays a key role in this phenomenon and is responsible for the specific responses to a stressor that results from coordinated interaction between specific brain regions. An animal that has a wide range of regulatory allostatic mechanism is always at an advantage because it can cope with the stressor much better than an animal that does not. Sometimes allostatic mechanisms may fail making the animal unable to mount a suitable response or preventing it from turning off the response even in the absence of the stressor. This will cause a deviation of the regulatory systems from its normal operating plateau to a new plateau predisposing the animal to pathology. The effect of allostasis on an animal is termed 'allostatic load'. Allostatic load can be low when the animal is not exposed to any stressors and can be exceedingly high when there are unpredictable environmental events in addition to adaptive responses to seasonal and other demands (Salak-Johnson 2011). Normal allostatic load is essential for the good health of the animal (Korte et al. 2007). A normal allostatic load will enable the animal to use a wide range of physiological and behavioral adaptations to match the environmental demand. On the other hand, high or low allostatic loads can threaten the well-being of the animal making it vulnerable to disease states.

Adaptive mechanisms in livestock to stress leave a marked impression on not only the stress axis itself but on a variety of systems that impact livestock production. The reproductive system, growth axis, and the thyroid axis are some of the neuroendocrine systems that are affected by stress. Neural circuits that regulate the immune system and the appetite regulatory system are also involved. In the next few sections let us examine the effect of stress on these individual systems and how they adapt after different types of stress that livestock are exposed to such as thermal stress, undernutrition, etc.

11.4 Reproductive Axis and its Adaptations to Stress

11.4.1 Organization of the Reproductive Axis in Animals

The reproductive axis is organized along similar lines across livestock species. The reproductive axis or the hypothalamo-pituitary-gonadal (HPG) axis is comprised of the hypothalamus that contains gonadotropin releasing hormone (GnRH) neurons that secrete GnRH in pulses, the pituitary which releases the gonadotropins, Luteinizing hormone (LH) and follicle stimulating hormone (FSH), and the gonads (ovary/testis) which secrete the gonadal hormones estradiol and progesterone in females or testosterone in males. GnRH is a decapeptide and it binds to type I GnRH receptor, a G-protein coupled receptor to produce its effects (Sealfon et al. 1997). GnRH neurons are distributed in distinct regions of the hypothalamus. There may be minor variations between species, but generally speaking, GnRH neurons are concentrated in the organum vasculosum lamina terminalis (OVLT), the diagonal band of Broca (DBB), the medial preoptic area (MPA), the suprachiasmatic nucleus, and the arcuate nucleus. In the cow, they are predominantly located in the DBB, MPA, and anterior hypothalamus (Weesner et al. 1993). Terminals of these neurons are located in the median eminence where they release GnRH that has immediate access to the hypothalamo-pituitary portal circulation that transports this hormone to the anterior pituitary. Here, GnRH neurons act on gonadotrophs to cause the secretion of LH and FSH. GnRH secretion occurs in pulses and therefore, LH and FSH release is pulsatile as well (Mueller and Nistico 1989).

There are cyclical changes in LH and FSH secretion that help the reproductive tract go through the motions of preparing itself for receiving sperms, fertilization, and

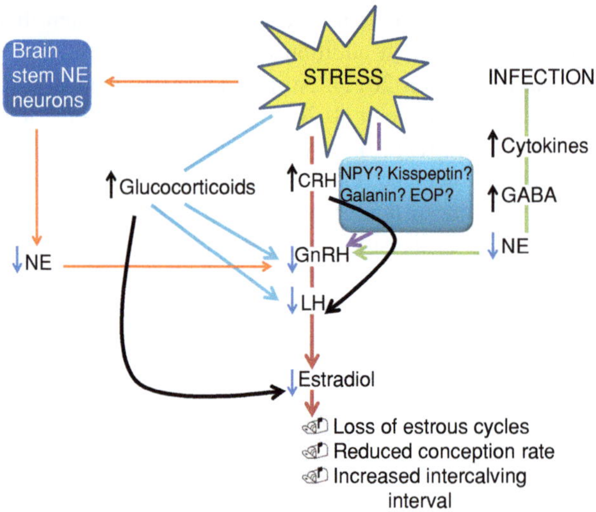

Fig. 11.3 Effect of stress on reproductive functions. Stress inhibits reproductive functions. This is probably mediated through a reduction in norepinephrine (NE) levels in the hypothalamus, an increase in corticotrophin releasing hormone (CRH), or through an increase in cytokines that increases gamma amino butyric acid (GABA) to suppress NE levels. The net result is a reduction in gonadotropin releasing hormone (GnRH), luteinizing hormone (LH), and estradiol levels that functionally translates to loss of estrous cyclicity, reduced ovulation and conception rates and increasing intercalving intervals

implantation of the fertilized ovum/zygote/embryo. This cyclical change that occurs in the female reproductive system of the animals is called the estrous cycle. The length of estrous cycle and the time of ovulation differ among livestock species. In the cow, sow and doe, estrous cycles last for 21 days while in the ewe, it is 17 days long. The cow and sow cycle throughout the year while sheep and goat are seasonal breeders. This is timed in such a way as to provide the best chances of survival of the young one. Although the length of the estrous cycle and the frequency of occurrence vary among species, the hormonal patterns and their functions are the same. Basically, FSH from the pituitary acts on the ovaries to stimulate follicular growth. As the follicles mature, they produce estradiol that gradually increases during the follicular phase, and acts on the brainstem and the hypothalamus through a positive feedback mechanism to secrete GnRH that in turn, stimulates the LH surge. The surge in LH is critical for ovulation and also promotes the growth and development of the corpus luteum in females (Herbison 1997).

11.4.2 Stress and Reproductive Functions

During stressful conditions like high ambient temperature, poor nutritional status, infection, etc., reproductive functions in livestock are held in temporary suspension until the conditions become favorable for successful development of the progeny. This adaptive mechanism compromises fertility in livestock through complex interactions between the stress and reproductive axes. Stress could impair fertility by affecting GnRH–LH pulsatility, estradiol profiles, LH surge timing and levels, follicular recruitment and development, and ovulation (Dobson et al. 2003). This has been depicted in Fig. 11.3. Acute psychological/thermal stress produces very short term increases in LH but FSH is unaffected. Certain stressful conditions such as lameness, is observed in livestock on a regular basis and interferes with livestock reproduction in a major way. Lameness is a painful, long-term stressful condition that is associated with poor reproductive fitness. Lame cows require more number of inseminations per pregnancy, have a lower pregnancy rate to first insemination, and have longer intercalving intervals, all leading to decreased production (Collick et al. 1989; Whay et al. 1997; Melendez et al. 2003; Hernandez et al. 2005a, b). Lame animals are also less prone to overt estrus expression that is required for appropriately timed artificial insemination (Walker et al. 2008).

In sheep, glucocorticoids, vasopressin, ACTH, CRH, and opioids have an inhibitory effect on LH secretion (Daley et al. 1999a, b). Glucocorticoids can inhibit gonadal steroid secretion and alter the sensitivity of target tissues to these hormones. Studies in rats suggest that CRH has an inhibitory effect on GnRH neurons. In one study, CRH deficient mice still had suppressed LH secretion after restraint stress suggesting that mediators other than CRH are most likely involved in stress-induced reduction in LH levels (Jeong et al. 1999). Vasopressin is a possible alternate candidate to CRH, as it is also released from the PVN of the hypothalamus along with CRH during stress and vasopressin receptor antagonists were able to suppress hypoglycemia–induced inhibition of LH secretion (Heisler et al. 1994). Also, destruction of PVN does not restore LH pulsatility during stress (Rivier and Rivest 1991) suggesting the involvement of other brain regions like amygdala or bed nucleus of the stria terminalis (BNST) in stress-induced decrease in LH levels. In sheep, CRH administration stimulates LH secretion (Caraty et al. 1997). It is likely that the effect of stress on reproductive functions varies based on the type, intensity of stressor and the species affected (Matteri 2000).

Recently, studies have suggested the involvement of a new group of peptides called kisspeptins in the MPA and arcuate nucleus of the hypothalamus in the initiation of puberty and positive feedback regulation of the HPG axis. In ewes, kisspeptin, and its receptor kiss1r (a G protein coupled receptor) expression has been documented in the arcuate nucleus of the hypothalamus (Smith 2008) and intracerebroventricular administration of kisspeptin produced a dramatic increase in GnRH levels in the cerebrospinal fluid with a concomitant increase in serum LH levels (Messager et al. 2005), suggesting a pivotal role for kisspeptins in GnRH regulation. Also, stress-induced suppression of LH levels in rats was accompanied by downregulation of components of the kisspeptin-kiss1r signaling pathway in the MPA and arcuate nucleus (Kinsey-Jones et al. 2009). Since kisspeptin and related molecules play an important role in GnRH regulation, it is not surprising that their levels are modulated as a result of stress. However, they do not play a direct role in mediating the effects of stress on GnRH neurons.

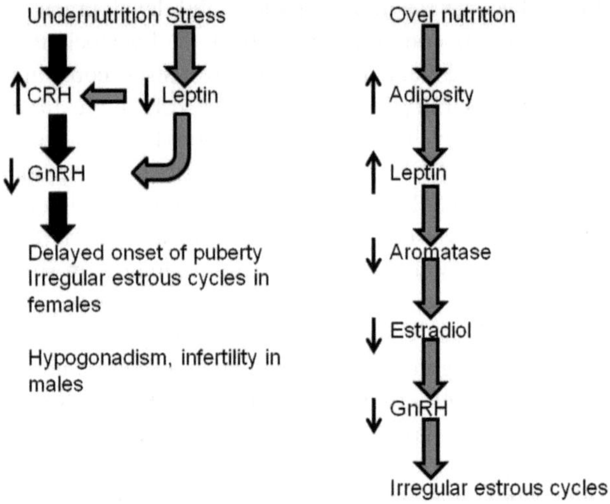

Fig. 11.4 Effect of under or overnutrition on reproductive functions: Stress caused by undernutrition or fasting can result in increased corticotrophin releasing hormone (CRH). Depletion of adipose tissue stores can result in lower leptin levels. Changes in both these hormones directly impact gonadotrophin releasing hormone (GnRH) resulting in impaired reproductive functions. Overnutrition on the other hand, results in elevated leptin levels that can decrease estradiol synthesis by the ovaries, that results in failure of gonadotrophin hormone releasing hormone (GnRH) neurons to be stimulated culminating in reduced luteinizing hormone and irregular estrous cycles

11.4.3 Undernutrition and its Effects on Reproduction

Nutrition is critical for reproductive functions, pregnancy, and the generation of healthy progeny (Schillo 1992; Chilliard et al. 1998). Females that are subjected to underfeeding or caloric restriction experience delayed onset of puberty and altered estrous cycles. In males, undernutrition has been shown to cause hypogonadism and infertility. This is mainly attributed to a decrease in GnRH release from the hypothalamus (Schillo 1992). CRH is also believed to play a role in this phenomenon since inhibition of CRH abolishes the suppressive effect of hypoglycemia on LH secretion in rhesus monkeys (Van Vugt et al. 1997) and ewes (Smith et al. 2003).

In the context of nutritional stress and reproduction, it is important to understand the role of leptin, a cytokine produced by adipose tissue on the HPG axis. Leptin is a 17 kDa protein that is produced by adipocytes and acts as a signaling molecule in the brain relaying information about the nutritional status of the body (Zhang et al. 1994). Undernutrition and fasting results in low leptin levels in the circulation which probably signals the brain to suspend reproduction. This is supported by studies demonstrating that chronic diet restriction in ovariectomized ewes reduces LH levels and intracerebroventricular administration of leptin restores LH levels (Henry et al. 2001). Diet restriction also causes differential expression of long-form leptin receptor in the anterior pituitary and hypothalamus of ewes (Dyer et al. 1997). On the flip side, overnutrition could lead to increased levels of leptin and hyperleptinemia which is not conducive for reproduction either. Studies in rats suggest that higher leptin levels can directly inhibit LH release from the gonadotrophs of the pituitary (Smith et al. 2002). Leptin could also act through the leptin receptors in the ovary to inhibit aromatase enzyme resulting in reduced estradiol synthesis and impaired positive feedback on the brain and pituitary resulting in poor reproductive functions (Ghizzoni et al. 2001). Heavier ewes have been noted to develop cystic follicles and reduced preovulatory gonadotropin surge (Christman et al. 2000). This mechanism is depicted in Fig. 11.4.

11.4.4 Thermal Stress and its Effects on Neuroendocrine Systems Governing Reproduction

Livestock have a range of temperatures that are conducive to their health and well-being. This range of ambient temperatures is termed the thermoneutral zone. Animals are affected by heat stress at the upper critical temperature. Some of the factors that contribute to heat stress are radiant energy from the sun, humidity, and ambient temperature. The animal is affected by thermal stress when it cannot dissipate an adequate amount of heat to maintain normal body temperatures (Gaughan et al. 2008). The temperature-humidity-index (THI) can be used to determine the level of stress that an animal is exposed to. THI = (dry bulb temperature °C + 0.36 × dewpoint temperature °C + 41.2). Animals begin to experience heat stress when THI is greater than 72. Milk production and reproduction are affected during moderate heat stress when THI is between 80 and 89. When THI is 90–98, animals experience severe stress and their milk yield and reproductive capacity are severely compromised. At THI greater than 98, probability of death increases. The loss in milk yield due to thermal stress poses a significant economic problem (St-Pierre et al. 2003).

Besides undernutrtion, heat stress can also negatively impact reproduction. Heat stressed cows have reduced duration and manifestations of estrus, reduced LH pulse and amplitude and increased incidence of anestrus or silent ovulation. Gonadotropin secretion is markedly reduced by high ambient temperatures (Donoghue et al. 1989; Barb et al. 1991; Gilad et al. 1993). Besides producing less GnRH, the sensitivity of the pituitary to GnRH is also reduced by heat stress (Gilad et al. 1993). In vitro studies using isolated follicles from Holstein heifers showed that heat stress decreases gonadotropin-induced estradiol synthesis and increases progesterone synthesis (Bridges et al. 2005). Similar findings have been observed in vivo where acute heat stress reduces estradiol production by granulosa cells (Wolfenson et al. 1997). This altered balance between estradiol and progesterone causes premature luteinization of follicles and decreases ovulation and fertility (Bridges et al. 2005). In males, the process of spermatogenesis is highly temperature sensitive. Normal spermatogenesis requires 4–5°C less than the core body temperature. Hence in heat stressed bulls, the percentage of viable and motile spermatozoa is reduced and the number of sperm abnormalities is increased (Rahman et al. 2011). Thus, heat stress reduces conception rate because of poor quality semen.

Furthermore, heat stress alters the uterine environment making it unfavorable for successful implantation. Heat stress during early gestation reduces uterine blood flow and embryonic attachment to the uterine wall thus favoring embryonic loss (Roman-Ponce et al. 1978). Also, chronic heat stress decreases oxygen consumption by the fetal placenta with retardation in placental growth resulting in fetal growth restriction (Early et al. 1991). Similar to high environmental temperatures, long-term exposure to low temperatures can decrease gonadotropins as well (Wada 1993; Matteri and Becker 1996). This could be attributed to shifting of resources to energy conservation, increased thermogenesis at low temperatures, decreased fat deposition, reduced leptin secretion, and decreased reproductive functions (Matteri 2000). In poultry, the critical THI for causing heat stress is 78 and above. In birds, heat stress reduces reproduction as well (Etches et al. 1995). This has been associated with a reduction in preovulatory LH and hypothalamic GnRH (Donoghue et al. 1989; Novero et al. 1991). In another study, no difference was observed in LH levels; however, a direct effect on ovarian function was observed (Rozenboim et al. 2007). The loss of ovarian function can be attributed to the reduction in blood flow to the ovaries (Wolfenson et al. 1981) since most of the blood flow is directed to the skin in an effort to lose heat through conduction, convection, and radiation. In poultry, increases in prolactin levels have been observed in response to heat stress (El Halawani et al. 1984; Donoghue et al. 1989). The increase in prolactin is believed to cause a reduction in gonadotropins and ovarian regression in these birds (Youngren et al. 1991; Rozenboim et al. 1993).

11.4.5 Immune Stress and How it Impacts the HPG Axis

Activation of the immune system in response to various stressors, ranging from immobilization; bacterial and viral infections, produces an increase in the circulating levels of cytokines, such as interleukin-1 (IL-1), tumor necrosis factor alpha (TNF-α), etc. Proinflammatory cytokines, especially IL-1, produce marked changes in reproductive functions by affecting specific brain circuits. IL-1β has been shown to decrease LH surges during afternoon of proestrous day in rats to inhibit ovulation (Sirivelu et al. 2009a). This effect is believed to be mediated through an increase in CRH levels in the hypothalamus, or by increasing GABA, an inhibitory neurotransmitter in the brain (Sirivelu et al. 2009b). Other neuropeptides such as α-melanocyte stimulating hormone (α-MSH) and endogenous opioids can also participate in this response by directly inhibiting GnRH neurons as a response to stress. Other GnRH stimulating neurotransmitters such as NE and NPY may decrease as a consequence ultimately leading to reduced GnRH output and decreased LH surges resulting in anovulation. These mechanisms help to conserve energy by placing reproductive functions on hold and favoring more important life processes. Offspring that are born in times of stress may be susceptible to epigenetic changes that may manifest as diseases during adulthood. It may also result in lower rate of survival of the offspring. It is therefore beneficial to the species as a whole that reproduction is not favored during times of stress.

Taken together, it is clear that stress-induced alterations in the reproductive axis are highly complicated, involving a variety of molecular mediators as depicted in Fig. 11.3. These mediators influence how the animal responds to these stressors to produce a temporary suspension of reproductive functions and how these functions rebound once the animal emerges from the influence of stressful situations. Although all the mechanisms described above have not been exclusively studied in livestock, it is reasonable to assume it could be similar in these species as well.

11.5 Stress Axis and Development/Developmental Programming

Exposure of animals/livestock to chronic stressors during gestation can leave a profound impact on the developing offspring. Although animals have an inherent mechanism for protecting the developing foetus from glucocorticoid exposure, chronic elevations in glucocorticoids do leave their mark. Glucocorticoids effects at this level are modified by an enzyme 11-hydoxysteroid dehydrogenase β (11-HSDβ), that converts active glucocorticoids to the inactive form (Cottrell and Seckl 2009). This enzyme is present at relatively high levels in the placenta to protect the developing offspring from being exposed to glucocorticoids. However, during chronic stress, increase in glucocorticoid levels impacts the growth of the foetus resulting in a condition called intrauterine growth retardation (IUGR) (Cottrell and Seckl 2009). This has been reported in sheep (Darp et al. 2010). This is a potential problem in livestock that are exposed to malnutrition as a result of drought and famine. Undernutrition can also contribute to IUGR. Besides affecting the growth of the offspring, it also 'programs' neural circuits so that their behavior in adulthood is affected (Markham and Koenig 2011).

11.6 Stress and Milk Production

Besides affecting ovulation and cyclicity, stress can also impact milk production. The mechanism by which heat stress affects reproduction is by decreasing food intake. Stress causes dairy cattle to eat less roughage, and decreases the rate of rumination. This results in reduced volatile free fatty acid production and alterations in the ratio of ketone bodies in circulation. Rumen pH is reduced and there is electrolyte imbalance especially when cattle are subjected to heat stress (Collier et al. 1982). Reduced nutrient absorption is commonly observed in heat stress and more energy is spent on maintenance (Collier et al. 2005). Heat stressed animals produce more insulin in response to a glucose challenge, and are unable to oxidize fatty acids and ketones for energy. They are entirely dependent on glucose as the energy source, which is then rerouted from the mammary gland and this probably contributes to the reduction in milk yield (Baumgard 2007). This could be a local adaptive mechanism that does not involve neural pathways. However, these mechanisms have not been explored in the heat-stressed cow.

Higher levels of stress hormones in sheep are known to decrease milk production and increase IL-1 concentrations in the milk (Caroprese et al. 2010). In sows, higher ambient temperatures decreased milk yield and decreased piglet growth rate (Renaudeau and Noblet 2001). The reduction in milk yield is most likely related to a reduction in prolactin levels. However, prolactin levels increase concurrently with glucocorticoids during stress (Yayou et al. 2010; Collier et al. 1982), and are most likely not related to the reduction in milk yield. The effect of stress on prolactin secretion is profound and uniform across species. Unlike other anterior pituitary hormones, prolactin does not have a prolactin releasing factor, but it is negatively regulated by hypothalamic dopamine (DA). Dopamine, secreted by the tuberoinfundibular dopaminergic neurons, directly inhibits prolactin secretion. However, it can also be stimulated by other factors such as thyrotropin releasing hormone (TRH), neurophysins, and substance P (Henriksen et al. 1995; Shin et al. 1995; Watanobe and Sasaki 1995).

In livestock, prolactin plays an important role in the synthesis and secretion of milk. It is also involved in immune functions, osmotic balance, metabolism, etc. (Nicoll 1980). The reason for the increase in prolactin levels during stress is not clear. It is believed to protect against the effects of chronic stress (Drago et al. 1989). Higher environmental temperatures increase prolactin levels (Mueller et al. 1974). This effect is believed to be mediated through increases in DA levels in the ventromedial hypothalamus (VMH) (Colthorpe et al. 1998). Specific D1 receptor antagonists can inhibit the prolactin response in sheep to high ambient temperatures but not to psychological stress. Milk yield drops dramatically in dairy cattle in response to acute stress. It is not clear how this happens in the context of high prolactin levels. A reduction in prolactin receptors in mammary tissue in response to stress could explain the reduced milk yield in spite of the higher circulating levels of prolactin. In contrast to acute stress, chronic stress decreases prolactin levels (Van den Berghe and de Zegher 1996). Fasting and undernutrition cause prolactin levels to drop in pigs (Steele et al. 1985). Similarly, anorexia observed in chronic infections and debilitating conditions can contribute to the decrease in prolactin resulting in low milk yield.

Immunosuppression caused by glucocorticoids can predispose adult animals and neonates to infections to which they easily succumb. Cattle are prone to E.coli-induced mastitis when they are immunosuppressed causing a severe drop in milk yield and at times the infection can be so severe, resulting in death (Burvenich et al. 2003). Endotoxin-induced mastitis further increases cortisol levels in dairy cows (Shuster et al. 1993). In other studies, stress caused by chronic lameness decreases the animal's fertility (Collick et al. 1989) and increases the duration to postpartum estrus (Petersson et al. 2006). Even when these animals begin to show signs of estrus, the intensity of the heat signs are reduced (Walker et al. 2008). This makes selecting these animals for artificial insemination difficult. Lame animals probably experience greater difficulty during the later stages of gestation as well.

11.7 Effects of Stress on Growth Hormone

Besides affecting the HPA axis to mount a glucocorticoid response, stressors, especially those that are capable of affecting the animal on a chronic basis, have profound effects on the growth of livestock.

11.7.1 Organization of the Somatotrophic Axis

The growth axis or somatotrophic axis controls the secretion of growth hormone. The somatotrophic axis is made-up of growth hormone releasing hormone (GHRH) neurons in the hypothalamus, and somatotrophs of the anterior pituitary that secrete growth hormone. The distribution of GHRH neurons varies greatly between species. The cell bodies range in distribution from the arcuate nucleus, medial perifornical region of the lateral hypothalamus, the ventromedial nucleus, the paraventricular and supraoptic nuclei, and the dorsomedial nucleus of the hypothalamus (Mueller and Nistico 1989). Their terminals extend to the median eminence where GHRH is released and moves into the anterior pituitary through the hypophyseal portal system. Growth hormone secretion is inhibited by somatostatin, a tetradecapeptide that is produced in the hypothalamus. Neurons that secrete this hormone are localized in the preoptic area, mediobasal hypothalamus, and the amygdala. Terminals of these neurons also extend to the median eminence where they directly regulate the secretion of GHRH. While GHRH increases growth hormone secretion, somatostatin inhibits it. Growth hormone that is released from the pituitary acts on the liver to produce and release insulin-like growth factor-1 (IGF-1). Growth and development of tissues such as muscle and bone are dependent on IGF-1. Growth hormone can directly affect these tissues as well.

11.7.2 Effect of Thermal and Undernutrition Stress on Growth Hormone Secretion

Stress is known to decrease growth hormone secretion and IGF-1 secretion in rats (Peisen et al. 1995). However, in livestock, growth hormone levels increase, but

IGF-1 levels decrease in response to stress (Bruggeman et al. 1997; McCusker 1998; Carroll et al. 1999). This response has been noticed with restraint stress (Farmer et al. 1991; Cataldi et al. 1994; Rushen et al. 1995). An increase in GHRH levels is believed to be responsible for this effect. Other studies suggest that the simultaneous activation of the HPA axis facilitates the increase in GHRH (Burguera et al. 1990). This could be mediated through NE levels in the hypothalamus since this neurotransmitter is stimulatory to both GHRH and CRH neurons (Mueller and Nistico 1989). The resultant increase in glucocorticoids can suppress IGF-1 levels.

Fasting and undernutrition also increase growth hormone and decrease IGF-1 levels in livestock (Ketelslegers et al. 1995). In this study, receptors for growth hormone appeared to be down-regulated in the liver. This could be another mechanism by which IGF-1 levels are reduced after stress. A similar effect has been observed after weaning in piglets (Carroll et al. 1998). The increases in growth hormone in response to stress oppose the action of insulin and decrease glucose utilization by tissues thus reserving them for the acute stress response. The reduction in IGF-1 levels result in reduced growth, and the conservation of energy and nutrients for survival. This is believed to be an important adaptive response, shifting energy resources from growth to survival (Haussmann et al. 2000). This is depicted in Fig. 11.5.

At higher ambient temperatures, growth hormone levels increase, but they decrease at colder temperatures (Barb et al. 1991; Ozawa et al. 1994). In this case, more energy is spent on thermoregulation rather than growth. The effect of higher ambient temperatures on the growth axis is also dependent on the developmental stage of the animal. Heat stress during late gestation decreases fetal growth and alters the endocrine status of the dam (Collier et al. 1982). Lower temperatures do not decrease growth hormone secretion or IGF-1 levels in piglets but decrease growth hormone receptor mRNA and IGF-1 mRNA in the liver (Carroll et al. 1999).

11.8 Stress and the Thyroid Axis

Thyroid hormones are important for basal metabolism, thermogenesis, mental acuity, and for good body conformation (Danforth and Burger 1984). The thyroid axis comprises of TRH neurons in the hypothalamus, thyrotrophs in the pituitary and the thyroid gland. TRH neurons are distributed throughout the hypothalamus especially in the preoptic-suprachiasmatic region, the PVN, periventricular region, lateral hypothalamus, ventromedial, and dorsomedial hypothalamic nuclei. Although a large number of TRH terminals have been identified in the median eminence, they are also present in the amygdala, lateral septal nucleus, lower brain stem, and spinal cord. TRH is under the stimulatory control of a variety of neurotransmitters and neuropeptides (Mueller and Nistico 1989). Acute stress is known to increase thyroid hormone levels within 10 min (Farmer et al. 1991).

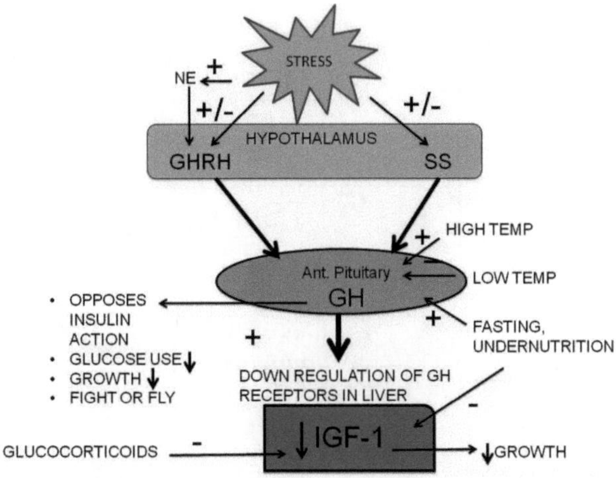

Fig. 11.5 Pathways by which stress influences growth hormone secretion. Various stress paradigms affect the growth axis differentially. High temperature stimulates growth hormone (GH) secretion, whereas low temperatures inhibit GH secretion. GH is also differentially affected by acute or chronic stress. While acute stressors decrease glucose utilization and prepare the animal to fight or fly, chronic stress decreases insulin-like growth factor (IGF)-1 production and inhibit growth

The benefit of this short-term increase in thyroid hormones is not clear. It probably plays a role in rapid increase in metabolism resulting in readily available energy sources for the fight or flight response (Fig. 11.6). Long-term exposures to cool temperatures stimulate the thyroid axis (Arancibia et al. 1996) mainly by increasing TSH levels. The resulting increase in basal metabolic rate and increased body temperature can inhibit the thyroid axis. In the context of survival, reduced thyroid activity would decrease metabolism and corresponding energy usage. This is especially useful when food is scarce (Chilliard et al. 1998). However, activation of the thyroid axis becomes beneficial during cold stress. The resultant increase in Endogenous opioids and glucocorticoid levels can also cause suppression of the thyroid axis (Harvey et al. 1987) (Fig. 11.6).

11.9 Stress and Feeding: Implications for Livestock Production

General or perceived stress results in decreased feed intake in livestock and this is believed to be mediated through a variety of neurotransmitters and neuropeptides. A number of stressors affect neural circuits that regulate appetite and are therefore capable of affecting food intake in livestock. Appetite is a function that is controlled by feeding centers in the hypothalamus. Several hormones, neurotransmitters, and

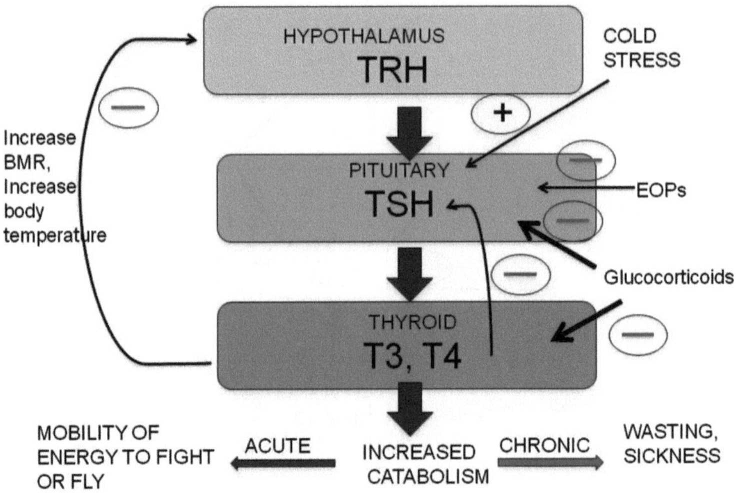

Fig. 11.6 Influence of various stressors on the thyroid axis: thyroid hormone (T3, T4) secretion is mediated through thyrotropin releasing hormone (TRH) and thyroid stimulating hormone (TSH). Increases in thyroid hormones can negatively regulate TRH secretion. Thyroid hormone secretion is also influenced at the level of the pituitary and the thyroid gland by endogenous opioid peptides (EOPs), glucocorticoids and a variety of stressors. While acute increases in thyroid hormone levels can promote the fight or flight response in animals, chronic increases can lead to wasting and sickness

neuropeptides are known to influence feeding. NPY is a 36 aminoacid peptide that is synthesized and released in the PVN. Underfeeding increases NPY levels and NPY mRNA in sheep hypothalamus (McShane et al. 1993). Leptin, an adipocyte-derived hormone can also act on the hypothalamus to decrease NPY levels and shut down feeding (Dyer et al. 1997). Two other important mediators are CRH and melanotropin concentrating hormone (MCH). They do not just decrease feeding behavior but may also inhibit visceromotor pathways that help identify food (Carr et al. 2002). Intensely painful, psychologically threatening stressors directly inhibit food intake (Shimizu et al. 1989). Mild stressors on the other hand can stimulate feeding (Morley et al. 1983). The hypothalamus especially nuclei such as the AN, PVN, LH, and VMH play an important role in feeding behavior (Kalra et al. 1999). Neuropeptides such as CRH and MCH play a critical role in maintaining normal body weight and energy balance.

Leptin, a hormone secreted by adipocytes, inhibits feeding and this effect is believed to be mediated through the melanocortin (MC-4) receptors (Fan et al. 1997; Seeley et al. 1997) located in Pro-opiomelanocortin (POMC) neurons of the arcuate nucleus. The PVN also plays an important role in feeding behavior. CRH is a potent inhibitor of feeding and another CRH-like peptide. Urocortin also plays a role in suppressing feeding activities (Spina et al. 1996). PVN receives input from not only the arcuate nucleus but is affected by orexigenic signals from different

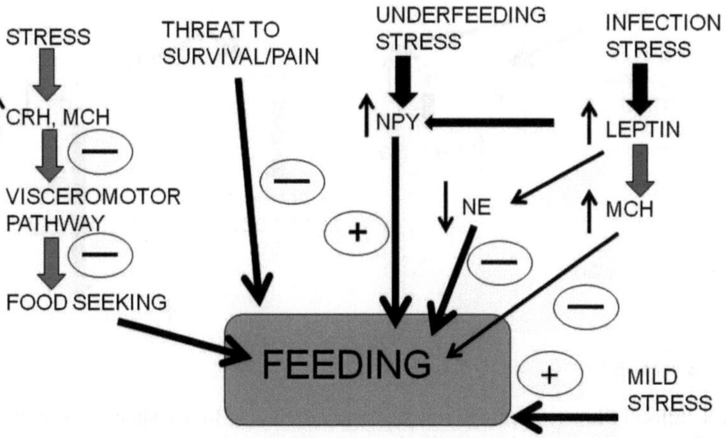

Fig. 11.7 Effect of various stressors on feeding behavior. Different stressors act through various intermediary molecules specific to each stressor to stimulate or inhibit feeding. Increases in corticotrophin releasing hormone (CRH) and melanin concentrating hormone (MCH) can inhibit feeding behavior by acting through the visceromotor pathway. On the other hand, when energy consumption becomes vital to certain stress responses, feeding is stimulated through neuropeptide Y (NPY)

parts of the brain and therefore acts as a control center synthesizing and modulating feeding-related activities (Cowley et al. 1999). Since stress increases CRH to stimulate the HPA axis, CRH becomes an effective mediator for stress-induced suppression of feeding. For example, restraint stress induces anorexia in rats and this effect is completely blocked by immunoneutralization of CRH (Shibasaki et al. 1988).

The reduction in appetite that is seen after infection is probably mediated through several mechanisms. Activation of the HPA axis and increase in CRH levels as a result of elevated cytokines, could contribute to reduced appetite. Proinflammatory cytokines themselves are capable of causing anorexia (MohanKumar et al. 2003). Other than these, cytokines also stimulate the secretion of leptin (Francis et al. 1999, 2000). Increases in circulating leptin levels can decrease food intake by acting at the level of the hypothalamus (Clark et al. 2006). It can also reduce NE levels in the PVN that plays an important role in food intake (Clark 2008; Fig. 11.7). Other anorexigenic and orexigenic peptides could also be involved, but have not been studied in livestock.

11.9.1 Feeding Responses to Cold Stress

Under normal conditions, leptin and NPY act in concert to maintain energy homeostasis. Under stressful conditions, depending on how severe the stress is,

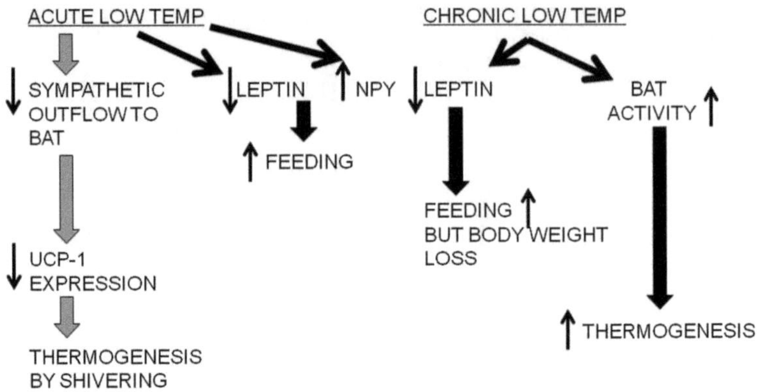

Fig. 11.8 Effect of low temperature stress on thermogenesis: Brown adipose tissue (BAT) plays a vital role in thermogenesis in animals. Acute exposures to low temperatures are managed by thermogenesis induced by shivering and increased feeding in response to changes in leptin and neuropeptide Y (NPY) levels. However, chronic exposures to low temperatures increase BAT activity and result in thermogenesis associated with higher energy consumption

food intake is affected. Exposure to low temperatures causes a reduction in leptin levels (Trayhurn et al. 1995) that would result in increased food intake. In contrast to leptin, NPY levels increase in the hypothalamus in response to cold exposure resulting in increased appetite. Administration of NPY also decreases sympathetic outflow to brown adipose tissue and reduces the expression of uncoupling protein-1 (UCP-1) (Egawa et al. 1991; Giraudo et al. 1994). UCP-1 uncouples mitochondrial respiration, increases energy utilization and helps animals cope with lower temperatures by using non-shivering thermogenesis. Exposure to cold temperatures for short periods of time therefore induces heat generation by increasing thermogenesis that involves shivering. In contrast to acute cold exposures, chronic exposures of 21 days or more results in decreased leptin, but no increase in NPY (Bing et al. 1998). Animals tend to lose weight in spite of eating more. The main reason for this could be the increase in thermogenesis by brown adipose tissue (Fig. 11.8). However, in animals that lack brown adipose tissue such as the pig, the adaptive mechanism is quite different. Piglets respond to cooler temperatures by drinking more milk (Forbes and Kyriazakis 1995). They also gain more weight (Jensen et al. 1969; Holmes and Mount 1966).

11.10 Stress Axis and its Influence on the Immune System

Besides the endocrine system that is commonly affected with chronic stress, another important system that is affected is the immune system. The relationship between stress axis and the immune system has been described time and again.

Briefly, the immune system is a powerful line of defense for the organism especially against infectious agents. Acute stress may inhibit, activate, or have no effect on the immune system (Salak-Johnson 2011). Chronic stress on the other hand is known to suppress the immune system making the animal more susceptible to infectious agents. Glucocorticoid-induced immune suppression may have a positive role especially in autoimmune disorders.

The mechanism by which glucocorticoids affect the immune system is mainly by decreasing the levels of proinflammatory cytokines like IL-1 and IL-6. Glucocorticoids are capable of affecting these cytokines at the transcriptional level (Liberman et al. 2007). In the meantime, glucocorticoids also increase levels of antiinflammatory cytokines such as IL-10 (Gayo et al. 1998). Altogether it appears that glucocorticoids have an antiinflammatory role which may serve the animal well especially to mitigate the effects of acute inflammation. However, chronic suppression of the immune system by glucocorticoids may be less than desirable. As glucocorticoid levels drop with time, the immune system rebounds and regains its original capacity (Mazzocchi et al. 1994). It has also been observed that subsequent exposure to stressors could result in a more robust glucocorticoid response leading to a more profound suppression of the immune system (Ragavan and Frantz 1981).

Although the duration of the stressor plays an important role in determining the immune response of an animal to stress, there are other important factors that can modulate the stress response. We discussed earlier that the response mounted by individual animals to the same stressor is different. This difference is clearly linked to the social status of the animal (Wingfield and Ramenofsky 1999). Social status has an impact on the animals' coping ability, changing the animals' behaviour in response to a stressor. Most common behavioral changes observed in animals subjected to psychological or physical stress is the exhibition of fear and anxiety. Stressed livestock refrain from exploratory activity and social interaction, exhibit vocalization, and sometimes also have stereotypic behavior (Grandin 2010b). It is believed that changes in behavior help the animal to cope with the stress without producing a physiological response (McGlone 1993). For example, during mild thermal stress, seeking water or shade could help the animal solve the problem than mounting a stress response. In fact, the animals have to experience the thermal stress for a prolonged period of time or the temperature has to climb beyond the manageable level for the animal to mount a physiological response to the heat stress. Even in this case, there are physiological adaptive mechanisms such as increased sweating, respiration, vasodilation in the skin, etc. that can help an animal remain close to homeostasis. The behavioral response observed in livestock in response to stress is a reflection of central mechanisms that prepare the animal and help it adapt better to environmental events (Schulkin 2003). Neuropeptides are believed to play a central role in this phenomenon. One of the main neuropeptides is CRH (Koob and Bloom 1985; Liang and Lee 1988; De Souza 1995; Heinrichs and Richard 1999). Besides CRH, other neuropeptides, such as NPY, vasopressin, and oxytocin can also influence behavioral responses to stress to cause anxiety, fear, and decrease in social interactions (Thorsell 2010; Litvin et al. 2011).

11.11 Molecular Adaptations to Stress, Learning, and Memory

Stress is believed to facilitate learning and memory only when 1. The stress is experienced in the context and around the time of the event that needs to be remembered (pain at the time of branding is associated with the place in which branding takes place etc.) and 2. When hormones and neurotransmitters are released in response to the stressor, act through the same circuits as those activated by the situation (pain pathways activated at the time of branding must converge with pathways activated by crowding, noise, etc. prior to branding—limbic pathways). This helps the animal focus its attention and improves memory of information relevant to the stressful episode (Joels et al. 2006).

Stress hormones also modulated function within the brain by changing the structure of neurons and neuronal synaptic connections. One of the most sensitive areas of the brain that are prone to such modifications is the hippocampus. There is a ramification of neuronal connections between the granule neurons in the dentate gyrus and Contractors Association (CA3) neurons in the hippocampus. There is a 600-fold amplification of excitatory impulses and a 300-fold amplification of inhibitory impulses as a result of these connections (McEwen 1999). New neurons are continuously produced in the dentate gyrus, even through adult life. Dendrites of CA3 cells undergo reversible remodeling during chronic stress especially when food restriction is combined with increased physical activity (Popov et al. 1992; Lambert et al. 1998; Magarinos et al. 2006). This may be particularly relevant for grazing livestock in periods of drought or famine when they are constantly on the move in search of food. The purpose of this type of adaptation may be to protect against permanent damage and enhance vulnerability to damage.

One type of change that occurs in the hippocampus is the replacement of neurons. The dentate gyrus has cells in the subgranular layer that gives rise to granule neurons (Kempermann and Gage 1999; Seri et al. 2001). These have a short half-life of 28 days (Cameron, Hazel et al. 1998) and are influenced by a variety of hormonal factors (Aberg et al. 2000; Czeh et al. 2001; Trejo et al. 2001). A variety of neurochemicals, such as GABA, serotonin, NE, endogenous opioid peptides, brain-derived neurotrophic factor (BDNF), IGF-1, etc. and glucocorticoids influence this neurogenesis. Endogenous opioid peptides and excitatory aminoacids through N-methyl D-aspartate (NMDA) receptors are particularly capable of suppressing neurogenesis after certain acute stress and chronic stress paradigms (Gould et al. 1997).

Besides neurogenesis, stress also influences structural plasticity in the hippocampus, Amygdala and prefrontal cortex. There is a reduction in the number of giant spines on dendrites in response to stress (Stewart et al. 2005). Endosome like structures in these dendritic spines are reduced. Active synaptic zones between giant spines and mossy fiber terminals constantly change and adapt. For example, chronic restraint stress for 21 days can cause retraction and simplification of dendrites. All these changes are believed to be mediated through changes in the cytoskeleton (McEwen 1999; Radley et al. 2008). Adrenal steroids promote this remodeling by interacting with several neurochemical systems, such as serotonin, endogenous opioid peptides, calcium currents, GABA, benzodiazepine receptors, and Glutamate-NMDA receptors. Other molecules such as neural cell adhesion molecules and tissue plasminogen activator and BDNF also play a role in synaptic plasticity (McEwen 1999). Remodeling of dendrites during chronic stress helps to decrease excitatory input and prevent permanent damage to cells.

11.12 Conclusions

In general, the neuroendocrine responses to different stressors vary depending on the type, intensity, and duration of the stressful event. Nevertheless, the neuroendocrine system responds to these stressors by producing a marked increase in stress hormone levels. With chronic exposures or even intense acute stressors, survival mechanisms are chosen over growth, production, and reproduction. The stress response initiated by a particular stressor integrates a variety of homeostatic mechanisms ranging from physiological, immune to behavioral adaptations. Allostasis helps the animal to maintain these adaptations in the face of adverse, constantly fluctuating environmental demands. Both phenomena are tightly coordinated by the brain and involve the whole body. Neuroendocrine adaptations help conserve energy, promote thermogenesis, and help use energy sources intelligently and effectively promoting long-term survival of the animal. As a result, effective stress management by the animal leaves no impact on livestock productivity. On the other hand, inability of the animal to meet the demands posed by chronic stressors produces a major biological consequence to the animal impairing growth, productivity, reproduction, and survivability. It is therefore essential to use good management practices, promote the welfare of livestock to help them maintain optimal responses to stressors.

References

Abelson JL, Weg JG, Nesse RM, Curtis GC (1996) Neuroendocrine responses to laboratory panic: cognitive intervention in the doxapram model. Psychoneuroendocrinology 21:375–390

Aberg MA, Aberg ND, Hedbacker H, Oscarsson J, Eriksson PS (2000) Peripheral infusion of IGF-I selectively induces neurogenesis in the adult rat hippocampus. J Neurosci 20: 2896–2903

Antoni FA (1986) Oxytocin receptors in rat adenohypophysis: evidence from radioligand binding studies. Endocrinology 119:2393–2395

Arancibia S, Rage F, Astier H, Tapia-Arancibia L (1996) Neuroendocrine and autonomous mechanisms underlying thermoregulation in cold environment. Neuroendocrinology 64:257–267

Barb CR, Estienne MJ, Kraeling RR, Marple DN, Rampacek GB, Rahe CH, Sartin JL (1991) Endocrine changes in sows exposed to elevated ambient temperature during lactation. Domest Anim Endocrinol 8:117–127

Barnett JL, Hemsworth PH, Newman EA (1992) Fear of humans and its relationships with productivity in laying hens at commercial farms. Br Poult Sci 33:699–710

Baumgard LH, Rhoads RP (2007) The effects of hyperthermia on nutrient partitioning. In: Proceedings of Cornell nutrition conference, pp 93–104

Bing C, Frankish HM, Pickavance L, Wang Q, Hopkins DF, Stock MJ, Williams G (1998) Hyperphagia in cold-exposed rats is accompanied by decreased plasma leptin but unchanged hypothalamic NPY. Am J Physiol 274:62–68

Bridges PJ, Brusie MA, Fortune JE (2005) Elevated temperature (heat stress) in vitro reduces androstenedione and estradiol and increases progesterone secretion by follicular cells from bovine dominant follicles. Domest Anim Endocrinol 29:508–522

Broom DM, Johnson KG (1993) Stress and animal welfare, 1st edn. Kluwer Academic Publishers, Norwell, pp 35–53

Bruggeman V, Vanmontfort D, Renaville R, Portetelle D, Decuypere E (1997) The effect of food intake from two weeks of age to sexual maturity on plasma growth hormone, insulin-like growth factor-I, insulin-like growth factor-binding proteins, and thyroid hormones in female broiler breeder chickens. Gen Comp Endocrinol 107:212–220

Burguera B, Muruais C, Penalva A, Dieguez C, Casanueva FF (1990) Dual and selective actions of glucocorticoids upon basal and stimulated growth hormone release in man. Neuroendocrinology 51:51–58

Burvenich C, Van Merris V, Mehrzad J, Diez-Fraile A, Duchateau L (2003) Severity of E. coli mastitis is mainly determined by cow factors. Vet Res 34:521–564

Cameron HA, Hazel TG, McKay RD (1998) Regulation of neurogenesis by growth factors and neurotransmitters. J Neurobiol 36:287–306

Caraty A, Miller DW, Delaleu B, Martin GB (1997) Stimulation of LH secretion in sheep by central administration of corticotrophin-releasing hormone. J Reprod Fertil 111:249–257

Caroprese M, Albenzio M, Marzano A, Schena L, Annicchiarico G, Sevi A (2010) Relationship between cortisol response to stress and behavior, immune profile, and production performance of dairy ewes. J Dairy Sci 93:2395–2403

Carr JA, Brown CL, Mansouri R, Venkatesan S (2002) Neuropeptides and amphibian prey-catching behavior. Comp Biochem Physiol B: Biochem Mol Biol 132:151–162

Carroll JA, Willard ST, Bruner BL, McArthur NH, Welsh TH Jr (1996) Mifepristone modulation of ACTH and CRH regulation of bovine adrenocorticosteroidogenesis in vitro. Domest Anim Endocrinol 13:339–349

Carroll JA, Veum TL, Matteri RL (1998) Endocrine responses to weaning and changes in post-weaning diet in the young pig. Domest Anim Endocrinol 15:183–194

Carroll JA, Buonomo FC, Becker BA, Matteri RL (1999) Interactions between environmental temperature and porcine growth hormone (pGH) treatment in neonatal pigs. Domest Anim Endocrinol 16:103–113

Carroll JA, McArthur NH, Welsh TH Jr (2007) In vitro and in vivo temporal aspects of ACTH secretion: stimulatory actions of corticotropin-releasing hormone and vasopressin in cattle. J Vet Med A Physiol Pathol Clin Med 54:7–14

Cataldi M, Magnan E, Guillaume V, Dutour A, Sauze N, Mazzocchi L, Conte-Devolx B, Oliver C (1994) Acute stress stimulates secretion of GHRH and somatostatin into hypophysial portal blood of conscious sheep. Neurosci Lett 178:103–106

Chilliard Y, Bocquier F, Doreau M (1998) Digestive and metabolic adaptations of ruminants to undernutrition, and consequences on reproduction. Reprod Nutr Dev 38:131–152

Christman SA, Bailey MT, Head WA, Wheaton JE (2000) Induction of ovarian cystic follicles in sheep. Domest Anim Endocrinol 19:133–146
Chrousos GP, Gold PW (1992) The concepts of stress and stress system disorders. Overview of physical and behavioral homeostasis. JAMA 267:1244–1252
Clark KA, MohanKumar SM, Kasturi BS, MohanKumar PS (2006) Effects of central and systemic administration of leptin on neurotransmitter concentrations in specific areas of the hypothalamus. Am J Physiol Integr Comp Physiol 290:R306–R312
Clark KA, Shin AC, Sirivelu MP, MohanKumar SM, MohanKumar PS (2008) Systemic administration of leptin decrease plasma corticosterone: role of hypothalamic norepinephrine. Brain Res 1195:89–95
Collick DW, Ward WR, Dobson H (1989) Associations between types of lameness and fertility. Vet Rec 125:103–106
Collier RJ, Beede DK, Thatcher WW, Israel LA, Wilcox CJ (1982) Influences of environment and its modification on dairy animal health and production. J Dairy Sci 65:2213–2227
Collier RJ, Baumgard, LH, Lock AL, Bauman DE (2005) Physiological limitations: nutrition partitioning. Chapter 16. In: Wiseman J, Bradley R (eds) Yields of farmed species: constraints and opportunities in the 21st century proceedings: 61st Easter school, Nottingham, England, pp 351–377
Collier RJ, Dahl GE, VanBaale MJ (2006) Major advances associated with environmental effects on dairy cattle. J Dairy Sci 89:1244–1253
Colthorpe KL, Anderson ST, Martin GB, Curlewis JD (1998) Hypothalamic dopamine D1 receptors are involved in the stimulation of prolactin secretion by high environmental temperature in the female sheep. J Neuroendocrinol 10:503–509
Cottrell EC, Seckl JR (2009) Prenatal stress, glucocorticoids and the programming of adult disease. Front Behav Neurosci 3:19
Cowley MA, Pronchuk N, Fan W, Dinulescu DM, Colmers WF, Cone RD (1999) Integration of NPY, AGRP, and melanocortin signals in the hypothalamic paraventricular nucleus: evidence of a cellular basis for the adipostat. Neuron 24:155–163
Curtis SE (2011) Adaptation and stress: state of being. In: Ullrey DE, Baer CK, Pond WG (eds) Encyclopedia of animal science, vol 1, 2nd edn. CRC press, Boca Raton, pp 6–9
Czeh B, Michaelis T, Watanabe T, Frahm J, de Biurrun G, van Kampen M, Bartolomucci A, Fuchs E (2001) Stress-induced changes in cerebral metabolites, hippocampal volume, and cell proliferation are prevented by antidepressant treatment with tianeptine. Proc Natl Acad Sci U S A 98:12796–12801
Daley CA, Macfarlane MS, Sakurai H, Adams TE (1999a) Effect of stress-like concentrations of cortisol on follicular development and the preovulatory surge of LH in sheep. J Reprod Fertil 117:11–16
Daley CA, Sakurai H, Adams BM, Adams TE (1999b) Effect of stress-like concentrations of cortisol on gonadotroph function in orchidectomized sheep. Biol Reprod 60:158–163
Danforth E Jr, Burger A (1984) The role of thyroid hormones in the control of energy expenditure. Clin Endocrinol Metab 13:581–595
Darp RA, de Boo HA, Phua HH, Oliver MH, Derraik JG, Harding JE, Bloomfield FH (2010) Differential regulation of igf1 and igf1r mRNA levels in the two hepatic lobes following intrauterine growth restriction and its treatment with intra-amniotic insulin-like growth factor-1 in ovine fetuses. Reprod Fertil Dev 22:1188–1197
De Souza EB (1995) Corticotropin-releasing factor receptors: physiology, pharmacology, biochemistry and role in central nervous system and immune disorders. Psychoneuroendocrinology 20:789–819
Dinan TG (1996) Noradrenergic and serotonergic abnormalities in depression: stress-induced dysfunction? J Clin Psychiatry 57(4):14–18
Dobson H, Ghuman S, Prabhakar S, Smith R (2003) A conceptual model of the influence of stress on female reproduction. Reproduction 125:151–163

Donoghue DJ, Krueger BF, Hargis BM, Miller AM, el Halawani M (1989) Thermal stress reduces serum luteinizing hormone and bioassayable hypothalamic content of luteinizing hormone-releasing hormone in hens. Biol Reprod 41:419–424

Drago F, D'Agata V, Iacona T, Spadaro F, Grassi M, Valerio C, Raffaele R, Astuto C, Lauria N, Vitetta M (1989) Prolactin as a protective factor in stress-induced biological changes. J Clin Lab Anal 3:340–344

Dyer CJ, Simmons JM, Matteri RL, Keisler DH (1997) Leptin receptor mRNA is expressed in ewe anterior pituitary and adipose tissues and is differentially expressed in hypothalamic regions of well-fed and feed-restricted ewes. Domest Anim Endocrinol 14:119–128

Early RJ, McBride BW, Vatnick I, Bell AW (1991) Chronic heat stress and prenatal development in sheep: II. Placental cellularity and metabolism. J Anim Sci 69:3610–3616

Egawa M, Yoshimatsu H, Bray GA (1991) Neuropeptide Y suppresses sympathetic activity to interscapular brown adipose tissue in rats. Am J Physiol 260:328–334

El Halawani ME, Silsby JL, Behnke EJ, Fehrer SC (1984) Effect of ambient temperature on serum prolactin and luteinizing hormone levels during the reproductive life cycle of the female turkey (meleagris gallopavo). Biol Reprod 30:809–815

Etches RJ, John TM, Verrinder-Gibbins GAM (1995) Behavioural, physiological, neuroendocrine and molecular responses to heat stress. In: Daghir NJ (ed) Poultry production in hot climates. CAB International, Cambridge, pp 31–66

Fan W, Boston BA, Kesterson RA, Hruby VJ, Cone RD (1997) Role of melanocortinergic neurons in feeding and the agouti obesity syndrome. Nature 385:165–168

Farmer C, Dubreuil P, Couture Y, Brazeau P, Petitclerc D (1991) Hormonal changes following an acute stress in control and somatostatin-immunized pigs. Domest Anim Endocrinol 8:527–536

Fell D, Gough MB, Northan AA, Henderson CU (1985) Diazepam premedication in children. Plasma levels and clinical effects. Anaesthesia 40:12–17

Forbes JM, Kyriazakis I (1995) Food preferences in farm animals: why don't they always choose wisely? Proc Nutr Soc 54:429–440

Francis J, MohanKumar PS, MohanKumar SM, Quadri SK (1999) Systemic administration of lipopolysaccharide increases plasma leptin levels: blockade by soluble interleukin-1 receptor. Endocrine 10:291–295

Francis J, MohanKumar SM, MohanKumar PS (2000) Correlations of norepinephrine release in the paraventricular nucleus with plasma corticosterone and leptin after systemic lipopolysaccharide: blockade by soluble IL-1 receptor. Brain Res 867:180–187

Gaughan JB, Mader TL, Holt SM, Lisle A (2008) A new heat load index for feedlot cattle. J Anim Sci 86:226–234

Gayo A, Mozo L, Suarez A, Tunon A, Lahoz C, Gutierrez C (1998) Glucocorticoids increase IL-10 expression in multiple sclerosis patients with acute relapse. J Neuroimmunol 85:122–130

Ghizzoni L, Barreca A, Mastorakos G, Furlini M, Vottero A, Ferrari B, Chrousos GP, Bernasconi S (2001) Leptin inhibits steroid biosynthesis by human granulosa-lutein cells. Horm Metab Res 33:323–328

Gilad E, Meidan R, Berman A, Graber Y, Wolfenson D (1993) Effect of heat stress on tonic and GnRH-induced gonadotrophin secretion in relation to concentration of oestradiol in plasma of cyclic cows. J Reprod Fertil 99:315–321

Giraudo SQ, Kotz CM, Grace MK, Levine AS, Billington CJ (1994) Rat hypothalamic NPY mRNA and brown fat uncoupling protein mRNA after high-carbohydrate or high-fat diets. Am J Physiol 266:1578–1583

Gould E, McEwen BS, Tanapat P, Galea LA, Fuchs E (1997) Neurogenesis in the dentate gyrus of the adult tree shrew is regulated by psychosocial stress and NMDA receptor activation. J Neurosci 17:2492–2498

Grandin T (2010a) How to improve livestock handling and reduce stress. Improving animal welfare: a practical approach. CABI Publishing, NY, pp 64–87

Grandin T (2010b) The importance of measurement to improve the welfare of livestock, poultry and fish. In: Grandin T (ed) Improving animal welfare: a practical approach. CABI publishing, NY, pp 1–20

Guillemin R, Rosenberg B (1955) Humoral hypothalamic control of anterior pituitary: a study with combined tissue cultures. Endocrinology 57:599–607

Harris GW (1948) Neural control of the pituitary gland. Physiol Rev 28:139–179

Harvey S, Lam SK, Scanes CG (1987) Opiate participation in thyroid hormone regulation in domestic fowl. Gen Pharmacol 18:245–247

Haussmann MF, Carroll JA, Weesner GD, Daniels MJ, Matteri RL, Lay DC Jr (2000) Administration of ACTH to restrained, pregnant sows alters their pigs' hypothalamic-pituitary-adrenal (HPA) axis. J Anim Sci 78:2399–2411

Heinrichs SC, Richard D (1999) The role of corticotropin-releasing factor and urocortin in the modulation of ingestive behavior. Neuropeptides 33:350–359

Heisler LE, Tumber AJ, Reid RL, Van Vugt DA (1994) Vasopressin mediates hypoglycemia-induced inhibition of luteinizing hormone secretion in the ovariectomized rhesus monkey. Neuroendocrinology 60:297–304

Hemsworth PH, Coleman GJ, Barnett JL, Borg S (2000) Relationships between human-animal interactions and productivity of commercial dairy cows. J Anim Sci 78:2821–2831

Henriksen JS, Saermark T, Vilhardt H, Mau SE (1995) Tachykinins induce secretion of prolactin from perifused rat anterior pituitary cells by interactions with two different binding sites. J Recept Signal Transduct Res 15:529–541

Henry BA, Goding JW, Tilbrook AJ, Dunshea FR, Clarke IJ (2001) Intracerebroventricular infusion of leptin elevates the secretion of luteinising hormone without affecting food intake in long-term food-restricted sheep, but increases growth hormone irrespective of bodyweight. J Endocrinol 168:67–77

Herbison AE (1997) Noradrenergic regulation of cyclic GnRH secretion. Rev Reprod 2:1–6

Hernandez JA, Garbarino EJ, Shearer JK, Risco CA, Thatcher WW (2005a) Comparison of milk yield in dairy cows with different degrees of lameness. J Am Vet Med Assoc 227:1292–1296

Hernandez JA, Garbarino EJ, Shearer JK, Risco CA, Thatcher WW (2005b) Comparison of the calving-to-conception interval in dairy cows with different degrees of lameness during the prebreeding postpartum period. J Am Vet Med Assoc 227:1284–1291

Holmes CW, Mount LE (1966) Heat loss from young pigs, individually and in groups, at ambient temperatures of 9, 20 and 30 degrees C. J Physiol 186:75–77

Jagannathan L, MohanKumar SM, MohanKumar PS (2009) Direct effects of oleic acid on neurotransmitter release from the rat hypothalamus in vitro program no 86419/BB33 neuroscience meeting planner, Chicago, IL

Jensen AH, Kuhlman DE, Becker DE, Harmon BG (1969) Response of growing-finishing swine to different housing environments during winter seasons. J Anim Sci 29:451–456

Jeong KH, Jacobson L, Widmaier EP, Majzoub JA (1999) Normal suppression of the reproductive axis following stress in corticotropin-releasing hormone-deficient mice. Endocrinology 140:1702–1708

Joels M, Pu Z, Wiegert O, Oitzl MS, Krugers HJ (2006) Learning under stress: how does it work? Trends Cogn Sci 10:152–158

Jones CT, Edwards AV (1990) Adrenal responses to corticotrophin-releasing factor in conscious hypophysectomized calves. J Physiol 430:25–36

Kalra SP, Dube MG, Pu S, Xu B, Horvath TL, Kalra PS (1999) Interacting appetite-regulating pathways in the hypothalamic regulation of body weight. Endocr Rev 20:68–100

Kelley KW (1980) Stress and immune function: a bibliographic review. Ann Rech Vet 11:445–478

Kempermann G, Gage FH (1999) New nerve cells for the adult brain. Sci Am 280:48–53

Ketelslegers JM, Maiter D, Maes M, Underwood LE, Thissen JP (1995) Nutritional regulation of insulin-like growth factor-I. Metabolism 44:50–57

King DA, Schuehle Pfeiffer CE, Randel RD, Welsh TH Jr, Oliphint RA, Baird BE, Curley KO Jr, Vann RC, OHale DS, Savell JW (2006) Influence of animal temperament and stress responsiveness on the carcass quality and beef tenderness of feedlot cattle. Meat Sci 74:546–556

Kinsey-Jones JS, Li XF, Knox AM, Wilkinson ES, Zhu XL, Chaudhary AA, Milligan SR, Lightman SL, O'Byrne KT (2009) Down-regulation of hypothalamic kisspeptin and its

receptor, Kiss1r, mRNA expression is associated with stress-induced suppression of luteinising hormone secretion in the female rat. J Neuroendocrinol 21:20–29
Koob GF, Bloom FE (1985) Corticotropin-releasing factor and behavior. Fed Proc 44:259–263
Korte SM, Olivier B, Koolhaas JM (2007) A new animal welfare concept based on allostasis. Physiol Behav 92:422–428
Lambert KG, Buckelew SK, Staffiso-Sandoz G, Gaffga S, Carpenter W, Fisher J, Kinsley CH (1998) Activity-stress induces atrophy of apical dendrites of hippocampal pyramidal neurons in male rats. Physiol Behav 65:43–49
Liang KC, Lee EH (1988) Intra-amygdala injections of corticotropin releasing factor facilitate inhibitory avoidance learning and reduce exploratory behavior in rats. Psychopharmacology (Berl) 96:232–236
Liberman AC, Druker J, Perone MJ, Arzt E (2007) Glucocorticoids in the regulation of transcription factors that control cytokine synthesis. Cytokine Growth Factor Rev 18:45–56
Litvin Y, Murakami G, Pfaff DW (2011) Effects of chronic social defeat on behavioral and neural correlates of sociality: vasopressin, oxytocin and the vasopressinergic V1b receptor. Physiol Behav 103:393–403
Magarinos AM, McEwen BS, Saboureau M, Pevet P (2006) Rapid and reversible changes in intrahippocampal connectivity during the course of hibernation in European hamsters. Proc Natl Acad Sci U S A 103:18775–18780
Marchant-Forde JN, Lay DC Jr, Pajor EA, Richert BT, Schinckel AP (2003) The effects of ractopamine on the behavior and physiology of finishing pigs. J Anim Sci 81:416–422
Markham JA, Koenig JI (2011) Prenatal stress: role in psychotic and depressive diseases. Psychopharmacology (Berl) 214:89–106
Martini L (1955) Hypothalamic control of secretion of adrenocorticotropic hormone. Ann Endocrinol (Paris) 16:670–675
Mason D, Hassan A, Chacko S, Thompson P (2002) Acute and chronic regulation of pituitary receptors for vasopressin and corticotropin releasing hormone. Arch Physiol Biochem 110:74–89
Matteri RL, Becker BA (1996) Gonadotroph function in 3-week-old gilts reared in a warm or cool thermal environment. Life Sci 59:27–32
Matteri RL, Carroll JA, Dyer CJ (2000) Neuroendocrine response to stress. In: Moberg GP, Mench JA (eds) The biology of animal stress. CABI Publishing, NY, pp 43–76
Mazzocchi G, Malendowicz LK, Markowska A, Nussdorfer GG (1994) Effect of hypophysectomy on corticotropin-releasing hormone and adrenocorticotropin immunoreactivities in the rat. Mol Cell Neurosci 5:345–349
McCusker RH (1998) Controlling insulin-like growth factor activity and the modulation of insulin-like growth factor binding protein and receptor binding. J Dairy Sci 81:1790–1800
McEwen BS (1999) Stress and hippocampal plasticity. Annu Rev Neurosci 22:105–122
McEwen BS, Wingfield JC (2003) The concept of allostasis in biology and biomedicine. Horm Behav 43:2–15
McGlone JJ (1993) What is animal welfare? J Agric Environ Ethics 6:26–36
McShane TM, Petersen SL, McCrone S, Keisler DH (1993) Influence of food restriction on neuropeptide-Y, proopiomelanocortin, and luteinizing hormone-releasing hormone gene expression in sheep hypothalami. Biol Reprod 49:831–839
Melendez P, Bartolome J, Archbald LF, Donovan A (2003) The association between lameness, ovarian cysts and fertility in lactating dairy cows. Theriogenology 59:927–937
Messager S, Chatzidaki EE, Ma D, Hendrick AG, Zahn D, Dixon J, Thresher RR, Malinge I, Lomet D, Carlton MB, Colledge WH, Caraty A, Aparicio SA (2005) Kisspeptin directly stimulates gonadotropin-releasing hormone release via G protein-coupled receptor 54. Proc Natl Acad Sci U S A 102:1761–1766
Mezey E, Reisine TD, Brownstein MJ, Palkovits M, Axelrod J (1984) Beta-adrenergic mechanism of insulin-induced adrenocorticotropin release from the anterior pituitary. Science 226:1085–1087
Mueller E, Nistico G (1989) Brain messengers and the pituitary. Academic, New York

Minton JE, Parsons KM (1993) Adrenocorticotropic hormone and cortisol response to corticotropin-releasing factor and lysine vasopressin in pigs. J Anim Sci 71:724–729
Moberg GP (2000) Biological response to stress: implications for animal welfare. The biology of animal stress, vol 1. CABI Publishing, NY pp 1–21
MohanKumar SM, Smith CL, MohanKumar PS (2003) Central adaptation to chronic administration of interleukin-1 beta in rats. Brain res Bull 62:71–76
MohanKumar SM, MohanKumar PS, Quadri SK (2000) Effects of bacterial lipopolysaccharide on central monoamines and fever in the rat: involvement of the vagus. Neurosci Lett 284: 159–162
Morley JE, Levine AS, Rowland NE (1983) Minireview. Stress induced eating. Life Sci 32: 2169–2182
Mueller GP, Chen HT, Dibbet JA, Chen HJ, Meites J (1974) Effects of warm and cold temperatures on release of TSH, GH, and prolactin in rats. Proc Soc Exp Biol Med 147: 698–700
Natelson BH, Tapp WN, Adamus JE, Mittler JC, Levin BE (1981) Humoral indices of stress in rats. Physiol Behav 26:1049–1054
Nicoll CS (1980) Ontogeny and evolution of prolactin's functions. Fed Proc 39:2563–2566
Novero RP, Beck MM, Gleaves EW, Johnson AL, Deshazer JA (1991) Plasma progesterone, luteinizing hormone concentrations, and granulosa cell responsiveness in heat-stressed hens. Poult Sci 70:2335–2339
Ono N, Bedran de Castro J, Khorram O, McCann SM (1985) Role of arginine vasopressin in control of ACTH and LH release during stress. Life Sci 36:1779–1786
Ozawa A, Johke T, Hodate K (1994) Plasma insulin-like growth factor-I response to cold exposure in barrows. Endocr J 41:725–730
Pacak K, Palkovits M, Kopin IJ, Goldstein DS (1995) Stress-induced norepinephrine release in the hypothalamic paraventricular nucleus and pituitary-adrenocortical and sympathoadrenal activity: in vivo microdialysis studies. Front Neuroendocrinol 16:89–150
Peisen JN, McDonnell KJ, Mulroney SE, Lumpkin MD (1995) Endotoxin-induced suppression of the somatotropic axis is mediated by interleukin-1 beta and corticotropin-releasing factor in the juvenile rat. Endocrinology 136:3378–3390
Petersson KJ, Strandberg E, Gustafsson H, Berglund B (2006) Environmental effects on progesterone profile measures of dairy cow fertility. Anim Reprod Sci 91:201–214
Petherick JC, Holroyd RG, Doogan VJ, Venus BK (2002) Productivity, carcass and meat quality of lot-fed bos indicus cross steers grouped according to temperament. Aust J Exp Agric 42:389–398
Petrovic SL, McDonald JK, Snyder GD, McCann SM (1983) Characterization of beta-adrenergic receptors in rat brain and pituitary using a new high-affinity ligand, [125I]iodocyanopindolol. Brain Res 261:249–259
Popov VI, Bocharova LS, Bragin AG (1992) Repeated changes of dendritic morphology in the hippocampus of ground squirrels in the course of hibernation. Neuroscience 48:45–51
Porter JC, Jones JC (1956) Effect of plasma from hypophyseal-portal vessel blood on adrenal ascorbic acid. Endocrinology 58:62–67
Radley JJ, Rocher AB, Rodriguez A, Ehlenberger DB, Dammann M, McEwen BS, Morrison JH, Wearne SL, Hof PR (2008) Repeated stress alters dendritic spine morphology in the rat medial prefrontal cortex. J Comp Neurol 507:1141–1150
Ragavan VV, Frantz AG (1981) Opioid regulation of prolactin secretion: evidence for a specific role of beta-endorphin. Endocrinology 109:1769–1771
Rahman MB, Vandaele L, Rijsselaere T, Maes D, Hoogewijs M, Frijters A, Noordman J, Granados A, Dernelle E, Shamsuddin M, Parrish JJ, Van Soom A (2011) Scrotal insulation and its relationship to abnormal morphology, chromatin protamination and nuclear shape of spermatozoa in holstein-friesian and belgian blue bulls. Theriogenology 76(7):1246–1257
Renaudeau D, Noblet J (2001) Effects of exposure to high ambient temperature and dietary protein level on sow milk production and performance of piglets. J Anim Sci 79:1540–1548

Rivier CL, Plotsky PM (1986) Mediation by corticotropin releasing factor (CRF) of adenohypophysial hormone secretion. Annu Rev Physiol 48:475–494

Rivier C, Rivest S (1991) Effect of stress on the activity of the hypothalamic-pituitary-gonadal axis: peripheral and central mechanisms. Biol Reprod 45:523–532

Roman-Ponce H, Thatcher WW, Caton D, Barron DH, Wilcox CJ (1978) Thermal stress effects on uterine blood flow in dairy cows. J Anim Sci 46:175–180

Rose AJ, Vegiopoulos A, Herzig S (2010) Role of glucocorticoids and the glucocorticoid receptor in metabolism: insights from genetic manipulations. J Steroid Biochem Mol Biol 122:10–20

Rozenboim I, Tabibzadeh C, Silsby JL, el Halawani ME (1993) Effect of ovine prolactin administration on hypothalamic vasoactive intestinal peptide (VIP), gonadotropin releasing hormone I and II content, and anterior pituitary VIP receptors in laying turkey hens. Biol Reprod 48:1246–1250

Rozenboim I, Tako E, Gal-Garber O, Proudman JA, Uni Z (2007) The effect of heat stress on ovarian function of laying hens. Poult Sci 86:1760–1765

Rushen J, Nay TS, Wright LR, Payne DC, Foxcroft GR (1995) Stress and nursing in the pig: role of HPA axis and endogenous opioid peptides. Physiol Behav 58:43–48

Saffran M, Schally AV, Benfey BG (1955) Stimulation of the release of corticotropin from the adenohypophysis by a neurohypophysial factor. Endocrinology 57:439–444

Salak-Johnson JL (2011) Adaptation and stress: neuroendocrine, physiological and behavioral responses. In: Ullrey DE, Baer CK, Pond WG (eds) Encyclopedia of animal science, vol 1, 2nd edn. CRC Press, Boca Raton, pp 1–5

Salak-Johnson JL, Niekamp SR, Rodriguez-Zas SL, Ellis M, Curtis SE (2007) Space allowance for dry, pregnant sows in pens: body condition, skin lesions, and performance. J Anim Sci 85:1758–1769

Sapolsky RM, Romero LM, Munck AU (2000) How do glucocorticoids influence stress responses? Integrating permissive, suppressive, stimulatory, and preparative actions. Endocr Rev 21:55–89

Schillo KK (1992) Effects of dietary energy on control of luteinizing hormone secretion in cattle and sheep. J Anim Sci 70:1271–1282

Schulkin J (2003) Allostasis: a neural behavioral perspective. Horm Behav 43:21–7, discussion 28–30

Sealfon SC, Weinstein H, Millar RP (1997) Molecular mechanisms of ligand interaction with the gonadotropin-releasing hormone receptor. Endocr Rev 18:180–205

Seeley RJ, Yagaloff KA, Fisher SL, Burn P, Thiele TE, van Dijk G, Baskin DG, Schwartz MW (1997) Melanocortin receptors in leptin effects. Nature 390:349

Sehic E, Blatteis CM (1996) Blockade of lipopolysaccharide-induced fever by subdiaphragmatic vagotomy in guinea pigs. Brain Res 726:160–166

Seri B, Garcia-Verdugo JM, McEwen BS, Alvarez-Buylla A (2001) Astrocytes give rise to new neurons in the adult mammalian hippocampus. J Neurosci 21:7153–7160

Shibasaki T, Yamauchi N, Kato Y, Masuda A, Imaki T, Hotta M, Demura H, Oono H, Ling N, Shizume K (1988) Involvement of corticotropin-releasing factor in restraint stress-induced anorexia and reversion of the anorexia by somatostatin in the rat. Life Sci 43:1103–1110

Shimizu N, Oomura Y, Kai Y (1989) Stress-induced anorexia in rats mediated by serotonergic mechanisms in the hypothalamus. Physiol Behav 46:835–841

Shin SH, Heisler RL, Lee CS (1995) Neurophysin stimulates prolactin release from primary cultured rat pituitary cells. J Endocrinol 144:225–231

Shuster DE, Kehrli ME Jr, Stevens MG (1993) Cytokine production during endotoxin-induced mastitis in lactating dairy cows. Am J Vet Res 54:80–85

Siegel PB, Gross WB (2007) General principles of stress and well-being. In: Grandin T (ed) Livestock handling and transport, 3rd edn. CABI publishing, Wallingford

Sirivelu MP, Burnett R, Shin AC, Kim C, MohanKumar PS, MohanKumar SM (2009a) Interaction between GABA and norepinephrine in interleukin-1beta-induced suppression of the luteinizing hormone surge. Brain Res 1248:107–114

Sirivelu MP, Shin AC, Perez GI, MohanKumar PS, MohanKumar SM (2009b) Effect of L-dopa on interleukin-1 beta-induced suppression of luteinizing hormone secretion in intact female rats. Hum Reprod 24:718–725

Smith JT (2008) Kisspeptin signalling in the brain: steroid regulation in the rodent and ewe. Brain Res Rev 57:288–298

Smith GD, Jackson LM, Foster DL (2002) Leptin regulation of reproductive function and fertility. Theriogenology 57:73–86

Smith RF, Ghuman SP, Evans NP, Karsch FJ, Dobson H (2003) Stress and the control of LH secretion in the ewe. Reprod Suppl 61:267–282

Spiess J, Rivier J, Rivier C, Vale W (1981) Primary structure of corticotropin-releasing factor from ovine hypothalamus. Proc Natl Acad Sci USA 78:6517–6521

Spina M, Merlo-Pich E, Chan RK, Basso AM, Rivier J, Vale W, Koob GF (1996) Appetite-suppressing effects of urocortin, a CRF-related neuropeptide. Science 273:1561–1564

Steele NC, McMurtry JP, Rosebrough RW (1985) Endocrine adaptations of periparturient swine to alteration of dietary energy source. J Anim Sci 60:1260–1271

Stermer R, Camp TH, Stevens DG (1981) Feeder cattle stress during transportation. American society for agricultural engineers, St Joseph, Michigan paper no. 81-6001

Sterling P, Eyer J (1988) Allostasis: a new paradigm to explain arousal pathology. In: Fisher S, Reason J (eds) Handbook of life stress, cognition and health. Wiley, NY, p631

Stewart MG, Davies HA, Sandi C, Kraev IV, Rogachevsky VV, Peddie CJ, Rodriguez JJ, Cordero MI, Donohue HS, Gabbott PL, Popov VI (2005) Stress suppresses and learning induces plasticity in CA3 of rat hippocampus: a three-dimensional ultrastructural study of thorny excrescences and their postsynaptic densities. Neuroscience 131:43–54

St-Pierre NR, Cobanov B, Schnitkey G (2003) Economic losses from heat-stress by US livestock industries. J Dairy Sci 86:52–77

Thorsell A (2010) Brain neuropeptide Y and corticotropin-releasing hormone in mediating stress and anxiety. Exp Biol Med (Maywood) 235:1163–1167

Trayhurn P, Duncan JS, Rayner DV (1995) Acute cold-induced suppression of ob (obese) gene expression in white adipose tissue of mice: mediation by the sympathetic system. Biochem J 311(3):729–733

Trejo JL, Carro E, Torres-Aleman I (2001) Circulating insulin-like growth factor I mediates exercise-induced increases in the number of new neurons in the adult hippocampus. J Neurosci 21:1628–1634

Van den Berghe G, de Zegher F (1996) Anterior pituitary function during critical illness and dopamine treatment. Crit Care Med 24:1580–1590

Van Vugt DA, Washburn DL, Farley AE, Reid RL (1997) Hypoglycemia-induced inhibition of LH and stimulation of ACTH secretion in the rhesus monkey is blocked by alprazolam. Neuroendocrinology 65:344–352

Wada M (1993) Low temperature and short days together induce thyroid activation and suppression of LH release in Japanese quail. Gen Comp Endocrinol 90:355–363

Walker SL, Smith RF, Jones DN, Routly JE, Dobson H (2008) Chronic stress, hormone profiles and estrus intensity in dairy cattle. Horm Behav 53:493–501

Warner RD, Ferguson DM, Cottrell JJ, Knee BW (2007) Acute stress induced by preslaughter use of electric prodders causes tougher meat. Aust J Exp Agric 47:782–788

Watanobe H, Sasaki S (1995) Effect of thyroid status on the prolactin-releasing action of vasoactive intestinal peptide in humans: comparison with the action of thyrotropin-releasing hormone. Neuroendocrinology 61:207–212

Watkins LR, Maier SF, Goehler LE (1995) Cytokine-to-brain communication: a review and analysis of alternative mechanisms. Life Sci 57:1011–1026

Waynert DE, Stookey JM, Schwartzkopf-Gerwein JM, Watts CS, Waltz CS (1999) Response of beef cattle to noise during handling. Appl Anim Behav Sci 62:27–42

Weesner GD, Harms PG, McArthur NH, Wilson JM, Forrest DW, Wu TJ, Pfaff DW (1993) Luteinizing hormone-releasing hormone gene expression in the bovine brain: anatomical localization and regulation by ovarian state. Biol Reprod 49:431–436

Whay HR, Waterman AE, Webster AJ (1997) Associations between locomotion, claw lesions and nociceptive threshold in dairy heifers during the peri-partum period. Vet J 154:155–161

Widmaier EP, Rosen K, Abbott B (1992) Free fatty acids activate the hypothalamic-pituitary-adrenocortical axis in rats. Endocrinology 131:2313–2318

Wingfield JC, Ramenofsky M (1999) Hormones and the behavioral ecology of stress. In: Balm PHM (ed) Stress physiology in animals, CRC Press, Boca Raton, pp 1–51

Wolfenson D, Frei YF, Snapir N, Berman A (1981) Heat stress effects on capillary blood flow and its redistribution in the laying hen. Pflugers Arch 390:86–93

Wolfenson D, Lew BJ, Thatcher WW, Graber Y, Meidan R (1997) Seasonal and acute heat stress effects on steroid production by dominant follicles in cows. Anim Reprod Sci 47:9–19

Yayou K, Ito S, Yamamoto N, Kitagawa S, Okamura H (2010) Relationships of stress responses with plasma oxytocin and prolactin in heifer calves. Physiol Behav 99:362–369

Youngren OM, el Halawani ME, Silsby JL, Phillips RE (1991) Intracranial prolactin perfusion induces incubation behavior in turkey hens. Biol Reprod 44:425–431

Zhang Y, Proenca R, Maffei M, Barone M, Leopold L, Friedman JM (1994) Positional cloning of the mouse obese gene and its human homologue. Nature 372:425–432

Chapter 12
Molecular Mechanisms of Livestock Adaptation

Vijay Kumar Saxena and Narayanan Krishnaswamy

Abstract Food producing animals in the developing world are exposed to a myriad of inevitable abiotic stressors that are fundamentally different from those animals in intensive production systems of the developed countries. The gamut of stressors that impairs the productivity includes, but not limited to, vagaries of the monsoon, scarce grazing resources, exposure to industrial contaminants. Understanding the cellular and molecular mechanisms behind the short and long-term adaptation required by tropical food animals is necessary to evaluate mitigatory measures aimed at improving productivity. Cellular proteins are affected by stressors resulting in increased population of proteins with non-native conformations due to improper folding. Heat shock proteins (HSPs) are a family of approximately a dozen proteins that are evolutionarily conserved. Many HSPs function as molecular chaperones with critical roles as regulator of protein folding and structural function. Studies done on the unicellular yeast depict the temporal variation in the gene expression profile when various stressors are used as treatment and thereby many common environment-specific response genes (CER) were identified constituting the 18–38% of the genome. CER genes constitute induced expression of classical heat shock genes, osmotic stress protectants such as polyols and trehalose, protein degradation enzyme, genes involved in increased membrane permeability and ion transport, as well as compensatory expression of isozymes or allozymes, and free radical scavengers such as superoxide dismutase, glutathione system, and cytochrome P450. Many of these genes are hypothesized to have common stress response elements (STRE) consensus sequences in their promoter

V. K. Saxena (✉)
Division of Physiology and Biochemistry, Central Sheep and Wool Research Institute, Avikanagar, Jaipur, Rajasthan 304501, India
e-mail: drvijaysaxena@gmail.com

N. Krishnaswamy
Division of Veterinary Obstetrics and Gynaecology, Indian Veterinary Research Institute, Mukteswar, Uttarakhand 263138, India

V. Sejian et al. (eds.), *Environmental Stress and Amelioration in Livestock Production*, DOI: 10.1007/978-3-642-29205-7_12,

region and are activated by common transcription factors such as Msn2 and Msn4 bringing coordinate regulation and global expression of genes under stress conditions. CER also constitute genes which are repressor genes associated with translation and protein synthesis to shunt energy in favor of large-scale ATP requirement for the chaperone function. Cells in response to stress also bring about changes in ratio of saturated lipids to unsaturated lipids in their membrane to alter flexibility as well as transport across membrane, which correspond to homoviscous adaptation. Knowledge on the molecular mechanism of environmental stress is still in its infancy and may eventually explain the biodiversity of the animal genetic resources.

Keywords HSP · Chaperones · CER · DBRP · Promoter · GAGA factors · Proteasomes

Contents

12.1 Introduction

The ability of an organism to respond to the changes in environmental temperature, pressure, osmotic variation, and the availability of nutrients determines its competitive fitness as well as survival. In the current perspective of global climate change, it is essential to understand the effect of environmental changes on the organism as well as the adaptive mechanisms in their arsenal to combat them. To have an understanding of the environmental changes at the organism level, it is essential to study the response of the cell to environmental changes because it provide clues to the molecular apparatuses enabling cells to adapt to new environments and the molecular mechanisms which have evolved to regulate the

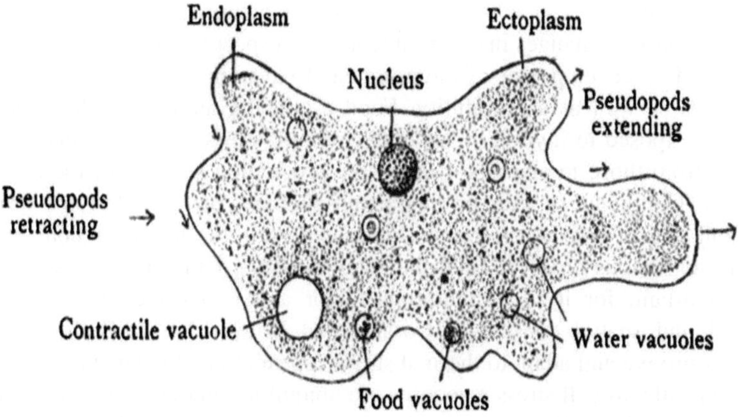

Fig. 12.1 Structural features of an amoeba

remodeling of gene expression that occurs in the new environments. Let us take, for instance, a simple unicellular organism, amoeba, a common inhabitant of ponds, ditches, and other water bodies with its plastic and pliable pseudopodia or false feet. These pseudopodia are of variable size and are capable of being protruded or retracted often with considerable speed. When amoeba is exposed to high temperature above 35°C, it ceases its activities. It responds negatively to both high and low temperature stimuli. Optimum temperature for its survival lies between 25 and 35°C. Therefore, amoeba demonstrates an adaptive capability to thermal stress from temperature range 25–35°C. If we revisit the same organism in response to the osmotic stress, the protoplasm of an amoeba is of higher concentration than the freshwater that prevails in its environment. This leads to entrance of water into the body of amoeba by endosmosis through semipermeable membrane. Not only does water enter through the plasmalemma, but also some water is formed in the cytoplasm as a result of metabolic activity and additional water gets in along with ingested food organisms. Now, the question arises how an amoeba tackles this menace of constantly in-coming osmotic water inside the cytoplasm as it is essential for the organism to get rid of this water to prevent swelling and rupture of amoeba's body. Amoeba has developed contractile vacuole system to pump out this excess water outside. Tiny membrane-bound vesicles fill water from cytoplasm to feeder vacuoles and they discharge their content in contractile vacuoles. Figure 12.1 describes the features of an amoeba.

Another example of an amoebaean species, leading a parasitic life within the intestine of humans, *Entamoeba histolytica, is* a protozoan responsible for painful intestinal ulcers leading to amoebic dysentery. This amoeba an endoparasite of the upper large intestine lacks functional contractile vacuoles as it inhabits an isotonic environment. The osmotic concentration of its cytoplasm is equal to the intestinal fluid of the host, so there is no need for osmoregulation and therefore, contractile vacuoles are absent. The examples of the two previously described amoebaean species demonstrate a need for the constant adaptation to changes in the environment. Constant changes in environment with respect to biotic as well as abiotic factors lead to development of environmental stress in animal affect production capabilities as well as survivability. Food producing animals in the developing world are exposed to myriad of inevitable abiotic stresses that are fundamentally different from those in animals in intensive production systems of the developed countries. The gamut of stressors impair productivity includes, but not limited to monsoon, scarce grazing resources, and exposure to industrial contaminants. Among the various physical environmental stressors, temperature is ecologically most important, for it is a ubiquitous factor and in most environments lacks temporal and spatial constancy (Cossins and Bowler 1987). Biologically, the ability to survive and adapt to thermal stress appears to be fundamentally essential for cellular life, as cell stress responses are ubiquitous among both eukaryotes and prokaryotes which utilize conserved heat shock proteins (HSPs) as chaperones (Lindquist et al. 1991; Parsell and Lindquist 1993). This kind of acute heat shock response is very essential for poikilothermic animals, wherein change in environmental temperature brings about changes in their body temperature. But even in euthermic organisms, in which core temperature of the body is tightly regulated, considerable variation in thermoregulatory set-point may occur during very severe environmental stress, exercise, and fever. Therefore, heat shock response are a rapid molecular mechanism; transient and short acting through production of HSPs in response to short exposure to sublethal heat stress (Feder and Hoffmann 1999). There are other methods for heat acclimation in response to persistent chronic heat (Senay et al. 1976; Sawka et al. 1996; Horowitz 1998). These mechanisms require time to develop, but are persistent. It is possible that reprogramming of gene expression could result in more persistent response. Presently, some 100 genes including those encoding HSPs are known to be upregulated by heat stress and a further 20 more are affected by cold stress. The differences occur in specific expression profile as well as cascade of interacting pathways interact via expression of common factors important for the transcription of the heat stress genes.

Cells do demonstrate changes at the level of protein as well that make them more stable to environmental cues. It is important to understand the factors important for the stability of protein. Stability of a protein depends upon the stability of its native state, its proper folding, and attainment of important weak interactions among the vital amino acids which are essential for its interaction with the substrate/interacting molecules. Temperature has impacts on both of these attributes, i.e., structure and function, and so proteins adapted to work at one temperature are inherently unable to maintain functions at temperatures far away from the optima. By the study of thermophilic and psychrophilic organisms, it has become apparent that some do not employ any special tricks, mechanisms, or stabilizing factors to prevent inactivation. However, substitution of amino acids leading to improved overall stability and functionality at temperature extremes has been documented.

12.2 Yeast Models of the Effects of the Environment on Genome Expression

As we have discussed, cellular coordination of genome expression is responsible for accommodation to changes in their environment. Therefore, in cells exposed to a stressor, gene expression is altered to reduce those genes solely involved with production to those enhancing protection of the organism from environmental extremes. Under environmental stress, a number of HSPs are induced in response to a variety stressors allowing cellular expression of properly folded protein molecules, through their function as molecular chaperones. Molecular chaperones are responsible for assisting in proper folding of proteins. Several studies conducted on yeast have identified genes whose expressions have been induced/ repressed in response to environmental changes. These studies demonstrate a number of transcriptional activators and repressors likely responsible for contributing to coordinate remodeling of gene expression. Transcriptional activators are proteins that bind DNA, promoting transcription of nearby genes. They may bind to cis-sequences near the promoter, or other locations and bring about enhanced transcription of specific genes. Most activators enhance binding of the RNA polymerase (formation of the closed complex) or transition to an open complex required for initiation of transcription. The yeast heat shock transcriptional activator Hsf 1 and the canonical sequence it binds has been elucidated (Parker and Topo 1984; Wu 1985; Kingston et al. 1987; Sorger and Pelham 1987). In the absence of heat shock, HSP90 sequesters Hsf-1 within the cytoplasm in an inactive state (Ali et al. 1998; Duina et al. 1998; Zou et al. 1998; Bhardwaz et al. 1999). Causton et al. (2001) demonstrated through yeast genomic expression that 499 genes undergo induction (216), or repression (283) in response to changes in environmental stimuli. These genes are referred to as common environmental response genes (CER). The CER involve approximately 10% of the yeast genome, suggesting large proportion of the yeast genome is remodeled during a range of environmental changes. Genes comprising a portion of the CER include those with roles in carbohydrate metabolism, cell stress, and energy production. Many CER genes are involved with glycolysis, which upon induction lead to enhanced ATP production. This is required for function of ATP-dependent molecular chaperones and other cellular machinery involved responding to cellular stressors. Genes encoding the subunits of the trehalose synthetase were found to be overexpressed (TPS1, TPS2, TPS3, and TSL1) (Jelinsky et al. 1999; Rep et al. 2000). The disaccharide trehalose, which accumulates intracellular during heat shock and stationary phases in many organisms, enhances thermotolerance and reduces the aggregation of denatured proteins. Exposure of *Saccharomyces cerevisiae* to mild heat shock (38°C) or to a proteasome inhibitor such as MG132 induces trehalose accumulation and markedly increases the viability of the cells upon exposure to a free radical generating system (H_2O_2/iron) (Benaroudj et al. 2001). Upon returning to normal growth temperature (28°C) or when MG132 is removed from the cells medium, the trehalose content and resistance to oxygen radicals decreases rapidly

(Benaroudj et al. 2001). Using mutants unable to synthesize trehalose marked increases in sensitivity to oxygen radicals compared to the wild-type cells were observed demonstrating the importance of this pathway. Trehalose is also said to prevent transition events in the lipid bilayer providing yeast the capacity to withstand desiccation to a certain limit.

A number of classical heat shock genes (HSP12, HSP26, HSP42, HSP78, HSP104, SSA4, and SSE2) are also induced in response to environmental changes. Many of these heat shock genes encode molecular chaperones that facilitate proper protein folding or maintenance of particular product conformation. Gene products that are involved in protein degradation (PHB2, RPN5, UBC5, UBC8, and YPS6) are also induced in response to specific stressors. When a cell is exposed to a variety of stresses, protein denaturation and damage occurs and these abnormal proteins accumulate within the cell. The genes encoding proteins of the protein degradation complex (proteasome) are produced to remove denatured protein aggregates. Protein sequences of denatured proteins are recognized by the destruction box recognizing protein (DBRP) because of attached moieties of another protein, ubiquitin. The ubiquitin moieties are coupled to denatured proteins by enzymes of the proteasome complex and are targeted for degradation.

Other genes induced in the CER include those involved in reducing the impact of oxidative stress. These proteins function in reduction of reactive species such as hydrogen peroxide capable of damaging proteins and nucleic acids. Genes involved in maintaining the reduced environment within cells are also induced. These genes are ion homeostatic genes involved in sequestration of ion or metal transport (like BSD2) or in thioredoxin or glutathione regulation (TTR1, YDR435C, and YCL035C). A variety of CER genes are induced commonly against a range of stressors and are referred to as general stress response genes (Tregor et al. 1998). In *S. cerevisiae*, the $(C_2H_2)_2$ zinc finger transcription factors, Msn2, and Msn4 play central role in responses to a range of stresses by activating gene transcription through a stress response element (STRE). These stress response-related genes possess STRE consensus sequences in their promoter regions: the Msn2/Msn4 transcription factors bind to these regions and activate genes relevant to the stress condition. However, it is not yet known whether the Msn2/Msn4 transcription factors bring about coordinate regulation of general stress response genes.

12.3 Common Environmental Response Genes Repressed

Genes repressed in response to the environmental stress are mostly concerned with translation of genes for cytoplasmic ribosomal protein, DNA polymerase I, II, and III, transcription, t-RNA synthetases, proteins required for processing ribosomal RNA, and a subset of translation initiation factors.

The identification of the variety of CER genes involved in stress responses suggests that these responses are aimed at production of additional energy (ATP), maintenance of environment as well as the repression of protein synthesis to ensure energy conservation, and minimize unnecessary burden on the part of the cell.

Another strategy employed by yeast cells to counteract an environmental stress is that it brings change in the ratio of saturated to unsaturated lipids in the membranes which alters fluidity, and membrane transport; these correspond to homoviscous adaptation of the organism. Unsaturated fatty acids play essential roles in the biophysical characteristics of cell membranes and determine the proper function of membrane-bound proteins. Thus, the cells ability to alter the degree of unsaturation of membrane lipids is important to cellular acclimatization and to altered environmental condition.

12.4 General response of cells to heat stress

Exposure of heat stress to the cell leads to the following important consequences:

1. Inhibition of DNA synthesis, transcription, RNA processing, and translation
2. Inhibition of progression through the cell cycle
3. Denaturation and aggregation of protein
4. Increased degradation of protein through proteasomal and lysosomal pathways
5. Disruption of cytoskeletal elements (microtubules, microfilaments, and intermediate filaments)
6. Changes in membrane permeability leading to accumulation of intracellular Na, H, and calcium ions.

In mammalian cells, non-lethal heat shock produces increased thermotolerance through enhanced expression of heat shock genes. Cellular stress responses can be induced by lipopolysaccaride (LPS) and arsenite exposure. The responses initiated by one stressor may lead to cross-protection to other stressors due to a relatively broad specificity of HSP or interaction between their activation pathways.

12.5 Heat Shock Factors

Heat shock factors are transcription factors regulating the expression of HSPs through interaction with specific DNA sequences in the promoter of HSP genes, Heat shock elements (HSE). HSE are a stretch of DNA located in the promoter region of genes containing sequential multinucleotide, copies of 5′-nGAAn-3′ and are found in both HSP genes and a variety of other genes responsible for stress tolerance. Heat shock factor-1 (HSF-1), HSF-2, HSF-3, and HSF-4 have been identified to date. (HSF-1) plays a major role in heat shock response while other members (HSF–2, HSF-4) are activated after prolonged stress or participate in normal cellular processes, embryonic development, and cellular differentiation.

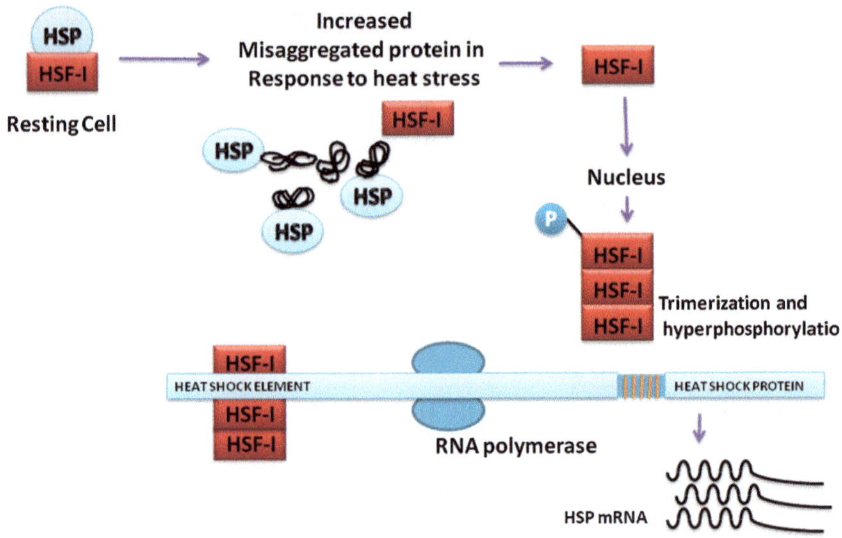

Fig. 12.2 Role of heat shock factors in transcribing HSP mRNA. In resting cell, HSF-1 including other transcription factors forms complexes with several heat shock proteins including HSP 70 or HSP 90. The initial stimulus for the release of HSF-1 appears to be recognition of hydrophobic domains of misfolded proteins by heat shock proteins. Before heat-induced activation, HSF-1 exists as monomers localized within the cytoplasm. Activation of HSF-1 results through homotrimerization, nuclear translocation, and subsequent phosphorylation of HSF-1. HSF-1 after activation binds to heat shock elements of the promoter regions leading to enhanced transcription of HSP mRNA

In resting cells, HSF-1 forms complexes with several HSPs including HSP70 or HSP90. The initial stimulus for HSF-1 appears to be recognition of the hydrophobic domains of denatured proteins, similar to the processes occurring during heat stress. When cells are exposed to heat stress, increased numbers of misfolded proteins are accumulated and it has been postulated that competitive release of transcription factor occurs from HSPs in the cytoplasm, as HSPs are bound increasingly to misfolded proteins. Before heat-induced activation, HSF-1 exists as monomers localized within the cytoplasm. Activation of HSF-1 results through homotrimerization, nuclear translocation, and subsequent phosphorylation of HSF-1. Once these prerequisite events occur, binding of heat shock elements to the promoter regions can occur (Fig. 12.2). The nucleosomal structure of DNA segment containing the heat shock element is reorganized by special protein called GAGA factor (Tsukiyama et al. 1994).

Binding of HSF1 to an HSE consensus site does not invariably induce transcription. For instance, in LPS stimulated human monocytes, HSF-1 represses the transcription of interlukin 1-β. The net effect of binding to HSE depends upon the state of HSF-1. In some systems, hyperphosphorylation of HSF-1 leads to decreases in transcriptional activity at 37°C. HSF-1 is hyperphosphorylated at serine residues by ERK-1 family, mitogen activated protein kinase (MAPK), by protein kinase C-α and C-ζ, and by glycogen synthase kinase 3-α. All of these phosphorylases inhibit transcriptional activity. HSF-1 is also hyperphosphorylated by c-Jun NH_2-terminal kinase (JNK), leading to activation under certain conditions while causing inhibition under other conditions. These findings suggest that the key intracellular signal transduction pathways may regulate or coordinate HSF-1 mediated transcription.

Many HSPs genes recruit active DNA-dependent RNA polymerase II even in the absence of HSFs. This pausing polymerase transcribes a small segment of DNA until it reaches TATA box, where it get arrested by binding to the initial complex of TATA box binding general transcription factors. Binding of HSF timers to the HSE sets the polymerase active which can proceed to complete the transcription of heat shock RNAs.

From the preceding discussion, it is logical to ask if the subsequent steps of protein synthesis (RNA splicing, nuclear transport, and translation) are blocked, then, how are the synthesis of heat shock RNA and its translation into HSP is completed? Evolutionarily, the cell has developed a variety of strategies to allow induction of heat shock RNAs, at the expense of other cellular proteins. Various strategies have been developed to circumvent these problems. Primary transcripts of these heat shock RNAs usually do not contain introns or the open reading frame encoding the protein itself, begins after the intron and initialization may proceed from the intron as well. Exons and introns refer to specific nucleotide base sequences in the genetic code that are involved in producing proteins. Exons are the DNA bases that are transcribed into mRNA and eventually code for amino acids in the proteins. Introns are DNA bases, which are found between exons, but are not transcribed. Genes which contain introns are known as interrupted genes. An intron within a gene is removed by RNA splicing to generate the final mature RNA product of a gene. After RNA splicing, in translation, messenger RNA (mRNA) is decoded to produce a specific polypeptide according to the rules specified by the trinucleotide genetic code. The region of the nucleotide sequences from the start codon (ATG) to the stop codon is called the open reading frame. It is a sequence of DNA that starts with start codon "ATG" (not always) and ends with any of the three termination codons (TAA, TAG, TGA). Depending on the starting point, there are six possible ways (three on forward strand and three on complementary strand) of translating any nucleotide sequence into amino acid sequence according to the genetic code.These are called reading frames. Since introns are mostly absent in heat shock RNAs, they can quickly bypass the steps required for splicing and mRNA processing. Heat shock RNAs also utilize special route, avoiding those transcriptional factors, which become inactivated during stress.

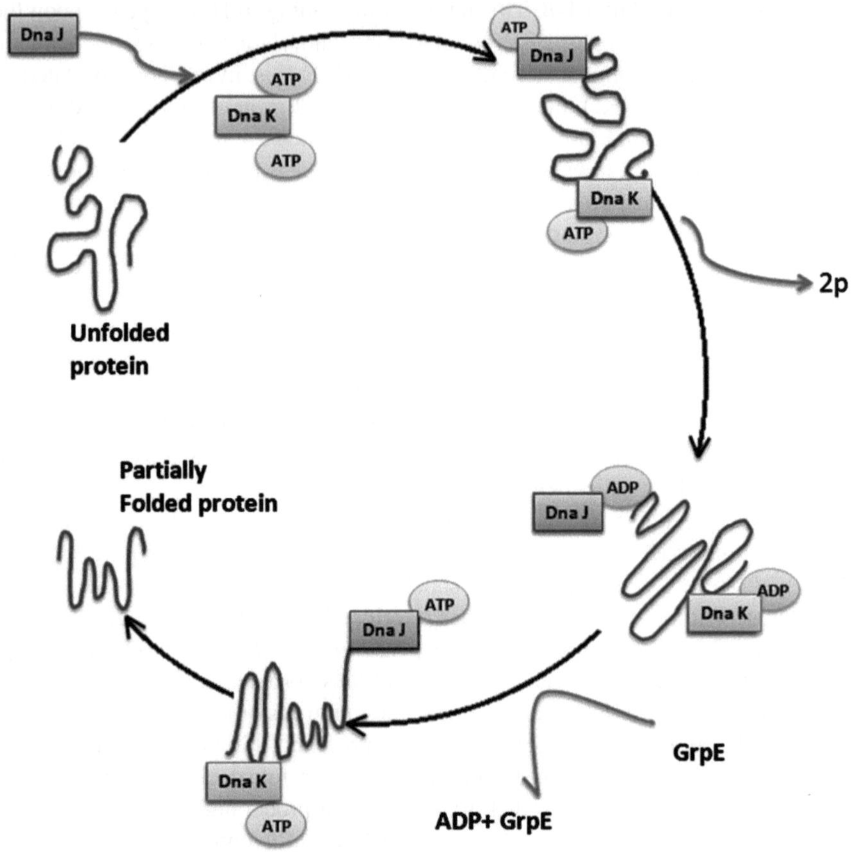

Fig. 12.3 DnaK and DnaJ Chaperone system of *E. coli*. A prokaryotic protein, DnaK, cooperatively works with DnaJ and GrpE, and these three proteins constitute the DnaK chaperone system. DnaJ and DnaK are molecular chaperones. DnaJ interacts with polypeptide chains prior to DnaK. DnaJ facilitates the ATPase activity of DnaK, by which high affinity property of DnaK for substrates is brought about. GrpE promotes ADP-ATP nucleotide exchange of DnaK and substrate release from DnaK

12.6 Heat Shock Proteins

HSPs were originally identified as proteins whose expression was markedly increased by heat shock. Several HSPs play important functions in normal cellular physiology. Induction of HSPs in mammalian cells starts within minutes after initiation of thermal stress, with peak expression upto several hours later.

HSPs possess three cardinal biochemical features which are given below.

12.6.1 Chaperonin Activity

HSPs with this function prevent misaggregation of denatured proteins and assist in proper refolding of denatured protein to native conformation. Even in unstressed cells, some of the HSPs function as chaperones. Some of the prototypical chaperonin HSP are HSP40, HSP60, HSP70, and HSP90. HSP70, which generally have a molecular weight near 70,000 daltons are more abundant in cells stressed by elevated temperatures. HSP70 proteins bind to region of unfolded polypeptide that are rich in hydrophobic residues, preventing inappropriate aggregation. These chaperonin thus "protects" both proteins subject to denaturation by heat and the new peptide molecule being synthesized. HSP70 proteins also block the folding of certain protein that must remain unfolded until they have translocated across a membrane. HSPs in the process of assisting proper folding of misaggregated polypeptide bind to and release polypeptides in a cycle that uses ATP hydrolysis for energy and co-chaperonin like HSP40. HSP 70 and HSP 40 chaperonres are the homologs of the DnaK and DnaJ chaperones of *E. coli.* DnaJ stimulates ATP hydrolysis by DnaK. DnaK-Dna J complex bind tightly to the unfolded peptide and assist in its partial folding process (Fig. 12.3).

12.6.2 Regulation of Cellular Redox State

HSPs are also involved in regulation of cellular redox states, of which the best example is HSP32, better known as Heme oxygenase enzyme (HO-1). Heme is found in hemoglobin, a principal component of red blood cells. This enzyme catalyzes the breakdown of heme to billiverdin, carbon monoxide, and free iron, which is rapidly incorporated into ferritin. Biliverdin is subsequently converted to billirubin, which is a potent antioxidant, with cytoprotective effects. The release of free iron by Heme oxygenase-1 also leads to increased expression of ferritin, which is thought to have cytoprotectant activity as sequestering peroxidant-free iron (Otterbein et al. 2000).

12.6.2.1 Regulation of Protein Turn-Over

Proteins of the proteasome complex such as ubiquitin are upregulated by heat shock, and they help in the degradation of misfolded and degraded proteins. Ubiquitin functions as a molecular tag to mark the proteins for degradation by proteasomes.

12.7 Adaptive Responses of Organism to Heat Stress

Thermal stress in the animals may provoke it to apply two main adaptative mechanisms depending upon the duration of exposure to heat stress.

12.7.1 The Heat Shock Response

A rapid molecular mechanism, transient and short acting which emerges via production of HSPs, subsequent to exposure of the cells to sublethal stress.

12.7.2 Heat Acclimation

This mechanism is switched on in response to persistent exposure to chronic heat. Acclimation requires longer exposure to sublethal thermal stress and takes time to develop. It is a "within lifetime" mechanism, reversible, and may involve a genetic basis, although not necessary (Bennet 1997). It may lead to development of new phenotypes due to remodeling of gene expression. It was classically considered to be an autonomically controlled array of physiological mechanisms that work in concert to enhance heat endurance. The criteria for its expression are reduced metabolic and heart rates, as well as body temperature, lower temperature thresholds for activation of heat dissipation effectors, and increased cardiovascular reserves and capacity of evaporative cooling systems (Senay et al. 1976; Sawka et al. 1996; Horowitz 1998). The physiological mechanisms of the heat acclimation and the animal's responses and adaptation to chronic heat exposure have been discussed in earlier chapters.

12.8 Effect of Thermal Extremes on the Function of Protein

Proteins, the building blocks of life, are the most abundant biomolecules of the cell that encompass hormones, enzymes, antibodies, transporters, muscle fibers, cytoskeletal proteins, lens of the eye, oxygen carrying protein, and myriad of many other proteins having distinct biological activities. Among these protein products, enzymes are the most varied and specialized. The function of an enzyme depends upon the careful balance between structural stability and flexibility.

The spatial arrangement of all atoms in a protein is called its conformation. The possible conformation of a protein includes any of the structural orientation possible by rotation about single bond without breaking the covalent bond. Indeed, a protein has many conformational states with each conformation varying in level of local unfolding. The need for multiple conformational states reflects the changes that should take place in a protein, as it binds to another molecule or substrate during enzyme catalysis. A protein is set to have achieved its native state under a given set of conditions when it attains a conformation of highest stability that corresponds to minimum Gibb's free energy.

An enzyme's activity in solution is most accurately represented as a sequential distribution of the microstates with conformation varying in number of weak interactions stabilizing them, local unfolding, and molecular breathing of the amino acids at critical positions. Enzymes reorient themselves by undergoing conformational shifts as they interact with the substrate or cofactors to bring chemically active species together for binding. Thus, molecular flexibility is also very important for the function of enzymes.

When an animal is exposed to thermal stress due to increase in environmental temperature, atoms constituting the proteins undergo thermal fluctuations affecting the flexibility and stable conformational state of the protein or enzyme. Enzymes do need certain amount of thermal energy from their environment to bring about the change in conformational states. That is why when the environment is too cold, the activity of enzyme decreases. If the temperature goes above certain limit, it is also deleterious for the function of protein as the weak interaction is very critical for the maintenance of structural conformation and structural flexibility gets destroyed, leading to loosening of active or its disfigurement, such that the substrate or cofactor may not be able to bind to the active site. At extremes of temperature, protein will get denatured which is detrimental to the growth and survival of cell.

A number of studies have been conducted on the effect of temperature on orthologs, which are nothing but, the protein in different species that evolved from a common ancestral gene by speciation. Normally, orthologs retain the same function in the course of evolution in disparate taxa living across broad range of temperatures, have shown that the three-dimensional structure is remarkably similar. Proteins adapted to function to different regimens of temperature do not adapt by bringing very remarkable change in their three-dimensional structure. It also suggests that adaptation to modified thermal regimes does not involve new structural components or rearrangements of secondary structural components. The next focus got shifted to study the changes in the primary structure of protein during adaptation to temperature's extremes. Haney et al. (1999) compared the genome of one hyperthermophilic archaebacterium, *Methanococcus janaschii* (grows at temperature of 85°C) with sequences from its mesophilic congeners such as *M. voltae*, *M. vanniellie*, and *M. plaudes*, by doing multiple sequences alignment of 115 proteins, which revealed the following facts:

1. A decrease in uncharged polar residues mainly in favor of non-polar amino acids which participate in hydrophobic interaction.
2. An increase in charged residues, which may be involved in large number of stabilizing salt bridges within molecular structure.

3. An increase in the residue hydrophobicity, which is expected to stabilize warm adapted protein as the hydrophobic effect becomes stronger with increase in temperature.
4. Increased residue volume, which may further increase stability due to tight packing of the residues and reduced rotational entropy of the unfolded form.

It is therefore concluded that thermophiles undergo substitution to replace uncharged polar residues by non-polar residues to have increased hydrophobic interaction force for restoring stability of the molecule at higher temperatures.

At low temperature, the same general changes in amino acids composition have been found to apply but by reversing the changes described for thermophiles in general.

1. Fewer ionic interaction in psychrophilic proteins leads to less structural stability (Feller and Gerday 1997).
2. Intramolecular hydrogen bonds are less numerous in psychrophilic proteins.
3. Reduction in hydrophobicity of proteins.
4. Increased number of polar or charged groups on protein surfaces in psychrophiles.
5. Psychrophilic proteins are more likely to have more of unfavorable α-charged dipole interaction, thus destabilizing the secondary structure. The amino acids in α helix are arranged in a right-handed helical structure where each amino acid residue corresponds to a 100° turn in the helix (i.e., the helix has 3.6 residues per turn), and a translation of 1.5 Å (0.15 nm) along the helical axis. A helix has an overall dipole moment caused by the aggregate effect of all the individual dipoles from the carbonyl groups of the peptide bond pointing along the helix axis. This can lead to destabilization of the helix through entropic effects. As a result, α helices are often capped at the N-terminal end by a negatively charged amino acid, such as glutamic acid, in order to neutralize this helix dipole. Now, if positively charged amino acids such as arginine, lysine, and histidine are abundant at the N-terminal end, it will lead to helix destabilization, in turn will reduce the stability of protein.

In psychrophilic protein to counter compromised flexibility due to nonavailability of essential thermal energy at low temperature, proteins adapt to have fewer interactions stabilizing their structure. The enhanced flexibility necessary for the protein to function at ultra-low temperatures necessitates a reduction in protein stability and this is achieved by fewer ionic interactions, less numerous intramolecular hydrogen bonds, and decrease in hydrophobicity which would reduce stabilizing effect of hydrophobic interaction.

It has also been noticed that cells employ certain extrinsic stabilizers to alter stability as well as phenotypic plasticity. These extrinsic stabilizers are low molecular weight stabilizing solutes known as "compatible solutes". These extrinsic stabilizers are better than random amino acid substitution as been discussed by the following two ways:

1. They are able to occupy broad thermal niches by modifying concentrations of thermostabilizing solutes in response to transient change in temperatures.
2. They can inhabit a newly available niche without any need to modify thermally sensitive proteins in their repertoire, a process which may take thousands of years.

The same level of stabilizing solutes would have comparable effect on all the proteins as these compounds interact with the protein in the way generally independent of specific physiochemical attributes of the protein being stabilized.

Hensel and Koing (1988) demonstrated that the 2, 3 bi-phoshoglycerate acts as an enzyme stabilizer in vitro in methanogenic bacteria and is accumulated when growth temperature of the bacterium is increased. Other stabilizing solutes that have been shown to increase in concentration in certain species at higher growth temperatures include α-mannosyl glycerate, diglycerol phosphate, and inorganic ions.

Another category of stabilizing effect of intrinsic proteins in both high temperature and cold temperature extremes are HSPs acting as chaperones. Their properties, mechanism of action have already been dealt in earlier part of this chapter. Their response is ubiquitous, indicating the importance of HSP as cellular protector of proteins against denaturation at suboptimal temperature.

12.9 Conclusions

Short-term adaptability of the cells to the thermal stress, which is the focus of the chapter reveals that the activation of HSF by hyperphosphorylation and trimerization is the primary trigger to the onslaught of thermal attack, which in turn, upregulates the HSPs, the mainstay proteins to maintain the cellular homeostasis. Upregulation of HSF and HSPs under elevated temperature corresponds with a concomitant downregulation of the transcription and translation of the genes involved in the growth, metabolism, and differentiation. HSPs ensure the constitutive cell functions by virtue of their chaperonin activity, maintaining the normal redox state and scavenging the degenerated proteins through ubiquitin. Conformational stability of the structural and functional proteins at thermal extremes is achieved by substitution of non–polar aminoacids in lieu of uncharged polar residues that strengthens the hydrophobic interactions. Stabilization of the salt bridges by an increase in the charged residues as well as increased residue volume also contributes for maintaining the conformation. Cells also secrete soluble low molecular weight extrinsic stablilizers which increase the stability of the cellular proteins. These thermostabilizing 'compatible solutes' work on a broad range of termperature fluctuations. The summation of the acute response of the cell to the elevated temperature is termed as heat shock response. Heat acclimation, which is the long-term response of the cell is induced by chronic and sustained exposure to the thermal stress and is characterized by reduced metabolic and heart rates, as well as body temperature, lower temperature thresholds for activation of heat dissipation effectors and increased cardiovascular reserves, and capacity of evaporative cooling system. The cellular and molecular mechanisms of the heat acclimation as well as the mechanistic link between the short- and long-term adaptability to heat stress are clearly the area of future research, which may pave way for designing mitigatory strategies.

References

Ali A, Bharadwaj S, O'Carroll R, Ovsenek N (1998) Hsp90 interacts with and regulates the activity of heat shock factor 1 in Xenopus oocytes. Mol Cell Biol 18:4949–4960

Benaroudj N, Lee DH, Goldberg LA (2001) Trehalose accumulation during cellular stress protects cells and cellular proteins from damage by oxygen radicals. J Biol Chem 276:24261–24267

Bennett AF (1997) Adaptation and evolution of physiological characters. In: Dantzler WH (ed) Handbook of physiology, comparative physiology. Oxford University Press, Oxford, pp 3–16 (American Physiological Society)

Bhardwaj S, Ali A, Ovsenek N (1999) Multiple components of the HSP90 chaperone complex function in regulation of the heat shock factor 1 in vivo. Mol Cell Biol 19:8033–8041

Causton HC, Ren B, Koh SS, Christopher T, Harbison KE, Jennings EJ, Lee TI, True HI, Lander ES, Young RA (2001) Remodeling of yeast genome expression in response to environmental changes. Mol Biol Cell 12:323–337

Cossins AR, Bowler K (1987) Temperature biology of animals. Chapman and Hall, London

Duina AA, Kalton HM, Gaber RF (1998) Requirement for Hsp90 and a CyP-40-type cyclophilin in negative regulation of the heat shock response. J Biol Chem 273:18974–18978

Feder ME, Hoffman GE (1999) Heat-shock proteins, molecular chaperones, and the stress response: evolutionary and ecological physiology. Annu Rev Physiol 61:243–282

Feller G, Gerday C (1997) Psychrophilic enzyme: molecular basis of cold adaptation. Cell Mol Life Sci 10:830–841

Haney PJ, Badger JH, Buldak GL, Reich CI, Woese CR, Olsen GJ (1999) Thermal adaptation analyzed by comparison of protein sequences from mesophilic and extremely thermophilic Methanococcus species. Proc Natl Acad Sci U S A 96:3578–3583

Hensel R, Koing H (1988) Thermoadaptation of methanogenic bacteria by intracellular ion concentration. FEMS Microbiol Lett 49:75–79

Horowitz M (1998) Do cellular heat acclamatory responses modulate central thermoregulatory activity? NIPS 13:215–218

Jelinsky SA, Samson LD (1999) Global response of *Saccharomyces cerevisiae* to an alkylating agent. Proc Natl Acad Sci U S A 96:1486–1491

Kingston RE, Schuetz TJ, Larin Z (1987) Heat-inducible human factor that binds to a human hsp70 promoter. Mol Cell Biol 7:1530–1534

Lindquist S (1991) The heat shock response. Annu Rev Biochem 55:1151–1191

Otterbein LE, Choi AM (2000) Heme oxygenase: colors of defense against cellular stress. Am J Physiol Lung Cell Mol Physiol 279:1029–1037

Parker CS, Topol J (1984) A drosophila RNA polymerase II transcription factor specific for the heat-shock gene binds to the regulatory site of an hsp70 gene. Cell 37:273–283

Parsell DA, Lindquist S (1993) The function of heat shock protein in stress tolerance: degradation and reactivation of damaged protein. Annu Rev Genet 27:437–496

Rep M, Krantz M, Thevelein JM, Hohmann S (2000) The transcriptional response of *Saccharomyces cerevisiae* to osmotic shock. J Biol Chem 275:8290–8300

Sawka MN, Wenger CB, Pandolf KB (1996) Thermoreg-ulatory responses to acute exercise-heat stress and heat acclimation. In: Fregly MJ, Blatteis CM (eds) Hand-book of physiology, Sect. 4: Environmental physiology, Oxford University Press, Oxford, pp 157–187 (American Physiological Society)

Senay LC Jr, Mitchell D, Wyndham CH (1976) Acclimatization in a hot, humid environment: body fluid adjustments. J Appl Physiol 40:786–796

Sorger PK, Pelham HRB (1987) Purification and characterzation of a heat shock element binding protein from yeast. EMBO J 6:3035–3041

Treger JM, Schmitt AP, Simon JR, McEntee K (1998) Transcriptional factor mutations reveal regulatory complexities of heat shock and newly identified stress genes in Saccharomyces cerevisiae. J Biol Chem 273:26875–26879

Tsukiyama T, Becker P, Wu C (1994) ATP-dependent nucleosome disruption at a heat-shock promoter mediated by binding of GAGA transcription factor. Nature 367:525–532

Wu C (1985) An exonuclease protection assay reveals heat-shock element and TATA-box binding proteins in crude nuclear extracts. Nature 317:84–87

Zou J, Guo Y, Guettouche T, Smith DF, Voellmy R (1998) Repression of heat shock transcription factor HSF1 activation by HSP90 (HSP90 complex) that forms a stress-sensitive complex with HSF1. Cell 94:471–480

Chapter 13
Genetic Adaptability of Livestock to Environmental Stresses

Soumen Naskar, Gopal R. Gowane, Ashish Chopra, Chandan Paswan and Leslie Leo L. Prince

Abstract The concept of adaptability revolves around fitness describing relative ability of an individual to survive and reproduce next generation to ensure continued survival of the population and is the result of natural selection over many generations. Current trend in genetic selection has severely eroded the genetic base ignoring the diversity of the production milieu, importance of adaptation, production of multiple products and social value of the livestock. The problem has been compounded with non-capturing of environmental costs though animal genetic resources (AnGR) on extensive and intensive scale are affected by direct impacts of climate change. Unplanned genetic introgression and crossbreeding has contributed to the greatest extent toward the loss of indigenous breeds. The genetic mechanism influencing fitness and adaptation is not well explored and adaptation traits are usually characterized by low heritability. Further, it may be difficult to combine the adaptation traits with high production potential as there seem to be different physiological and metabolic processes involved. Though decision regarding matching genotypes with environment or vice versa will be situation specific, the low and intermediate level of animal production in many parts of the world suggests that increased yields and efficiency will be more environmentally sustainable than extensive goals ensuring genetic diversity, environmental soundness, animal health and welfare, and social viability. Breeding for climate change adaptation or mitigation will not be necessarily different from existing

S. Naskar (✉)
National Research Centre on Pig, Rani, Guwahati, Assam 781131, India
e-mail: snrana@gmail.com

G. R. Gowane · A. Chopra · C. Paswan · L. L. L. Prince
Division of Animal Genetics and Breeding, Central Sheep and Wool Research Institute, Avikanagar, Via-Jaipur, Rajasthan 304501, India

V. Sejian et al. (eds.), *Environmental Stress and Amelioration in Livestock Production*, DOI: 10.1007/978-3-642-29205-7_13,

programs. However, the problems associated with measuring the phenotypes relevant to adaptation have to be overcome. Breeding indices should be balanced to include traits associated with heat resilience, fertility, feed conversion efficiency, disease tolerance and longevity in addition to higher productivity, and give more consideration to genotype by environment interactions (GxE) to identify animals most adapted to specific conditions and natural stratification of breeds and species by climatic zones. Favorable correlation suggests that if major importance is placed on performance traits in stressful environments, adaptability traits would not be compromised and thus the most productive and adapted animals for each environment need to be identified for breeding purposes. Recent successes like slick hair gene in cattle, halothane gene in pig asks for extensive efforts for finding significant quantitative trait loci (QTL) for stress and exploitation of heat shock proteins (HSP). Implementation of marker-assisted breeding value estimation (MA-BVE) using dense genome map for highest possible accuracy will be a welcome step. There is a need of extensive study of interaction among the drivers of changes of climate and livestock, studying it in a composite manner. Appropriate organizational structures and adequate funding to support a climate resilient animal agriculture will be vital.

Keywords Adaptation · AnGR · Farming system · Genetic improvement · Resistance/tolerance · Sustainability

Contents

13.1 Understanding Adaptability

13.1.1 Why Variation in Adaptability Exists

Population size increases exponentially in most living species whereas resources that support their existence grow in arithmetic fashion. Resources such as water, food, and energy set the stage for competition among various populations. Competition for resources occurs between and within species resulting in natural selection (Jenkins 2004). The concept of adaptability revolves around 'fitness' describing the relative ability of an individual to survive and reproduce the next generation to ensure continued survival of the population. While some individuals of the population are allowed to reproduce at typical rate others are not due to different limiting factors such as nutrition and environment. Changes that occur against environmental challenges to make an animal 'fit' may be reflected at different levels like phenotypic changes devoid of genetic origin, phenotypic adjustments attributable to genetic variation in the population (phenotypic plasticity; de Jong and Bijma 2002), mutations that provide an advantage to future genetic variation pool (adaptive mutation; Rosenberg 2001) or through behavioral adjustments (Barnes and Partridge 2003).

Evolution, challenging environment, competitive interaction for feed, grazing tract, disease challenges, and civilization needs to facilitate relatively small group of animals to become separated within defined geographical and climatic niche. These small isolated groups in their new local environment continue to produce and reproduce for generations resulting in loss of genetic variation due to chance, otherwise called genetic drift (Hammond 1947). Consequently, generation of local population with physiological changes compatible with survival and reproductive success is ensued. Loss of variability in small population is not necessarily due to chance only but also by 'founder effect' whereby the entire population might have started by a single female (Hedrick and Kalinowski 2000). Adaptation to local environment through fixation of favorable gene and gene combinations by selection may create more genetic variation across population and may increase inbreeding as well. The key factor that determines adaptability as measured through survival and reproduction is inherent genetic variation which interacts with environmental constraints and creating phenotypic variation.

To illustrate the case of variation in adaptability, there are numerous examples like zebu cattle that is uniquely suited to hot and humid climates because of its smooth coat, primary hair follicles, improved sweat, and sebaceous glands, and better ability to lose moisture by evaporation than *Bos taurus* cattle (Turner 1980). The ability of zebu cattle to maintain thermal equilibrium necessary for normal function and performance is also attributed to coat color, pigmentation, conformation, genetic adaptation to the source of nutrition (forages), and resistance/tolerance to pests and diseases (Bonsma 1973). Variation in adaptability between *Bos indicus* and *B. taurus* cattle may be due to their origin in distinct climates in which *B. indicus* might have acquired thermotolerant genes (Hansen 2004). Among temperate breeds, Jersey dairy cows are more resistant to heat stress than Holstein cows (Sharma et al. 1983). Typically, goats are regarded to be the best adapted species to harsh environments (Silanikove 2000b). In addition, sheep and goat are more tolerant to heat than cattle (Silanikove 2000a, b; Khalifa et al. 2005) which has high metabolic rate and poorly developed water retention mechanism (Bernabucci et al. 2010).

The harsh effect of climate change is expected to have maximum impact on vulnerable pastoral communities whose livestock production systems are located in drylands (Oseni and Bebe 2010). A study conducted in Malabot pastoral communities, North Horr; Ngurman pastoral communities, Kajiado; and Rift valley province of Kenya ranked drought, epidemic diseases (associated with climatic variability), heat stress and seasonal floods in decreasing importance with regards to deleterious impacts on livestock production system.

Response to thermal challenges by animals has been in the form of acclimation, acclimatization, adaptation (both genetic and phenotypic), and these have been defined by the International Commission for Thermal Physiology (ICTP 2001). A physiological and behavioral change that reduces environmental stress and enhances endurance during the lifetime of an organism is called acclimation whereas acclimatization is a response to changes in its natural climate (e.g. seasonal). Adaptation can be changes that reduce the physiological strain imposed by stressful components of the total environment within an organism's lifetime (phenotypic) or result of genetic selection in a species or subspecies (genotypic). Animals are considered acclimated to a given ambient when body temperature returns to pre-stress levels (Nienaber et al. 1999). Cellular and systemic responses associated with acclimation are coordinated, requiring days or weeks and not homeostatic (Bligh 1976). Among different environmental stressors, the most important is ambient temperature (Horowitz 2002) and there are no distinct terms that refer genetic adaptation to climate. Consequently, acclimation, acclimatization, or phenotypic adaptation is often used in the same context in livestock production.

13.1.2 Heat Tolerance and Critical Temperatures

Environmental factors such as temperature, humidity, wind velocity, and radiation determine comfort and stress levels of animals. Different biometeorological indices have been developed to ideally predict the set and critical points when animals start experiencing heat stress (Bohmanova et al. 2007; Vitali et al. 2009). Most temperature and humidity indices (THIs) use easily measurable and accessible dry bulb temperature and humidity. This vary for different geographical locations based on values allocated to relative humidity. Although solar radiation is a very strong contributor to stress especially in the tropics, it is difficult to measure and its effect is widely variable because of coat characteristics of animals. Therefore, it has been suggested that inclusion of wet bulb variable for temperature, humidity, wind velocity, and solar radiation will be an improved choice for pasture and feedlot-based systems (Bernabucci et al. 2010).

Temperature-humidity index (THI) values have so far been categorized into mild, moderate, and severe on dairy cows arbitrarily (Armstrong 1994). It has been speculated that as THI reaches 72, milk synthesis starts decreasing (Armstrong 1994). But recent studies (Zimbelman et al. 2009) suggest that this value may be lower in high yielding dairy cows. High milk producing breed are more susceptible to heat stress which may be attributed to vulnerability in reducing endogenous and metabolic heat production and increase heat dissipation (Kadzere et al. 2002; Collier et al. 2005). Current trend in which genetic selection in dairy cow is primarily influenced by milk yield is likely going to give rise to increase in 'elite' germplasm with increased susceptibility to heat stress. Genetic adaptation to adverse environmental conditions including heat stress is a slow process and is the result of natural selection over many generations (Ames and Ray 1983).

13.1.3 Characterization of Farm Animal Performance

Domestic AnGR supply nearly one-third of agricultural production and human food needs (FAO 1999a, b). Developing countries are more dependent on domestic animals for survival and economic development. In addition to providing protein especially to women and children (Delgado et al. 1999; Alderman 1994), AnGR provides non-food items such as hides, skin, wool, and manure for farm energy and draft power. AnGR contributes considerably to environmental health especially in mixed farming system (De Hahn et al. 1997). There is a considerable contribution of AnGR to socio-cultural needs of certain regions, e.g. use of Muturu cattle in Nigeria (Rege and Gibson 2003) or Mithun (*Bos frontalis*) in Nagaland state of India for social functions. Unfortunately, characterization of farm animal performance has been performed with a bias. For example, there is often attempt to compare the performance of temperate and tropical breeds of livestock without giving due consideration of their diverse origins. Low milk or meat producing native livestock may be for adaptive strategy and not necessarily inferiority while high producing breeds from developed countries may have been the consequence of years of artificial selection for product-oriented targets for human needs (da Silva 2007).

Tropical livestock species should have efficient mechanism to dissipate excess heat through skin and respiration, and avoid incoming thermal energy from the environment. Similarly, animals living in extremely dry environments and deserts must have protective mechanisms against evaporative water loss and intense solar radiation. Conversely, animals living in extremely cold regions need to protect and preserve their body heat. Morphological characteristics of the skin viz. color, thickness, concentration of sweat glands and hair coat viz. thickness, number of hairs per unit area, length and diameter of the hair, angle of the hair to skin surface, may change according to the needs required to survive in a particular environment. If cattle of temperate breeds are transferred to tropical regions, their coats tend to reduce thickness significantly to improve transfer of metabolic heat (Da Silva et al. 1988; Da Silva 2000). Respiratory heat loss (sensible and latent) is high in sheep domesticated in tropical regions (Da silva 2002). Skin pigmentation is another important trait animals use to protect deep tissues against solar short wave radiation in tropical regions (Hansen 1990; Gebremedhin et al. 1997; Hillman et al. 2001). Typically, dark coated animals receive more heat loads than light colored ones (Hansen and Lander 1988). Thus, light colored coat is more desirable in tropical regions (Goodwin et al. 1997) although contradictory results also exist. However, while European cattle breeds have almost the same level of pigmentation in their hair and skin beneath, tropical cattle have lighter coat above highly pigmented skins. A credible exception is the Jersey breed whose pigmentation is independent of the hair color which may partly explain its wide adaptability in tropical countries (Da Silva 2007). Recent tools like infra-red thermometry, infrared digital camera, radio-telemetry, and data-loggers have precisely captured body temperature variations among different breeds of animals (Nienaber et al. 1999; Collier et al. 2003; Da Silva 2007).

The ability of tropical ruminants to walk long distances without feed and water in search of scarce feed resources and cope with underfeeding remains largely unexplored. In the same vein, resistance/tolerance to numerous infectious diseases and parasitic infestations notably trypanosomiasis, dermatophilosis, or foot rot, tick-borne diseases, and internal parasites afford them distinct advantage which will be of increased importance in changing climatic conditions. Efficient conversion of low grade feed resources to high quality animal proteins and enhanced immunity to diseases, in addition, is evidence of adaptation by these animals living in the tropics. These attributes are discussed in detail elsewhere in this chapter.

13.1.4 Genetic Manipulation in Livestock Production

Exploiting genetic variation to breed superior animals is often misunderstood and sometimes thought to be panacea to solve all problems associated with unfavorable environments. Genetic adaptation is relatively a slow process which is subject to ever-changing geological, biological, and climatic conditions (Dobzhansky 1970). A dichotomy exists in the sense that genetic adaptation and fitness are synonymous in terms of evolution, implying that only those who survive and reproduce for next generations in a given set of environmental challenges are fit and thus better adapted. A given breed of livestock species may have evolved over many generations for particular agro-ecological system and this knowledge is often ignored due to excessive focus on short-term strategies and benefits. Indiscriminate use of exotic breeds to improve indigenous stocks that ignored natural stratification by climatic zones has led to inappropriate production system, severe erosion, reduced adaptation of indigenous germplasm, and loss of indigenous genetic resources in some regions (Rege and Gibson 2003). Emphasis on improvement of a single trait like milk production, widespread use of artificial insemination, and embryo transfer and intense marketing strategy has narrowed genetic diversity of Holstein–Friesian (Wickham and Banos 1998). It is estimated that about 16% of uniquely adapted breeds have gone extinct (Hall and Ruane 1993) and a further 32% are at risk of becoming extinct (FAO 2000). This alarming trend alludes to the fact that AnGR has not been properly integrated into climate change adaptation and mitigation strategies in a way that it truly enhance agricultural biodiversity even with the knowledge that livestock are relatively more affected than crops. Man's concept of productivity in terms of milk, meat, and wool production may not synchronize with nature's need and sometimes may be in direct conflict. For example, a cow should be able to produce enough milk to feed its calf especially in time of distress and not necessarily to satisfy human needs. In fact, selection for extreme productive functions that ignores environmental conditions in most cases lead to sharp decline in natural fitness. Thus, natural adaptive ability developed in native stock over many generations is an asset that deserves reassessment (Ames and Ray 1983). Widespread use of Brahman cattle in crossbreeding programs in USA (Ames and Ray 1983), importation, development, and re-exportation of African Boran and Tuli cattle breeds (Rege and Gibson 2003), resilience of Sahiwal breeds in India, Kenana, Butana, and N'dama cattle breeds are some examples that illustrate the importance of adaptability.

The need to integrate adaptive traits e.g. heat tolerance in the production environment was first proposed by Rhoad (1940). Da Silva (1973) estimated the genetic variation of adaptive traits in Brazilian beef cattle and observed that increase in body temperature after exposure to sun during hottest period of the day had a moderate heritability coefficient (0.443) and high negative genetic correlation with average daily weight gain (−0.895). Heritability of sweating rate (0.222), skin pigmentation (0.112), hair coat pigmentation (0.303) and thickness (0.233), and hair length (0.081) of Jersey cattle in tropical region was found to be low to medium (Da Silva et al. 1988). Moderate heritability of many adaptive traits suggest that there is no reason why genetic manipulation to improve adaptability and productivity is not possible. Perhaps, a trade-off between 'economic' and adaptive traits may be necessary in certain regions of the world especially in tropical countries.

13.1.5 Farming System Evolution and Industry Trends

Evolution of farming system has played an important role in creating variation in adaptability. Farmers share gene pools within extremely limited geographical areas which results in few breeds becoming predominant and prominent as compared to diverse AnGR with varied production systems. Factors like narrow choice of genetic base for better 'command and control' with inclination to meet globally integrated markets, commodity-specific large-scale organized production trend (Hoffmann 2010) have made a new suborder. Organized livestock sector has made animal breeding a 'high-tech' exercise. However, pastoralists appear to be the most important catalytic agent for breeding animals adapted to extreme temperatures and humidity although their involvement in animal breeding is limited as has been witnessed in Africa or Central Asia (Blench 2005).

Moreover, a great difference exists between large- and small-scale farming in the sense that environment has been made to match the genotypes for large-scale farming assured of reliable infrastructure and predictable price of inputs, small scale mainly in the form of subsistence farming has evolved by imbibing genotype for the environment. The problem has been compounded by two factors. One is the failure to define fair value for AnGR beyond product/functions (e.g. milk, meat, etc.) to disease tolerance and adaptive attributes. The other is using crossbreeding/breed replacement as a tool predominantly in developing countries with the desire to improve output of specific product as a response to market forces ignoring other functions relevant to local environment or local community (Rege and Gibson 2003). Thus it is expected that while altered livestock production regime to changing environmental needs will be on anvil, more system research in the form of breed characterization, incorporation of adaptation traits to production environment will be required.

13.2 Genetics and Adaptive Traits for Harsh Environment

Evolution is a continuous process which molds and remolds combinations of genes in living things to adapt to changing environment. A trait is an aspect of the developmental pattern of the organism. An adaptive trait, therefore, is an aspect of the developmental pattern which facilitates the survival and/or reproduction in animals in a certain succession of environments (Dobzhansky 1956). Fitness is an important trait for any species which centers on the adaptation of individual species in a given environment to propagate their genes in subsequent generations. Livestock keepers throughout the world have been practicing animal husbandry and breeding in unfavorable environment for centuries. As a result, many breeds of the unsympathetic environment have developed many adaptive traits that enhance survivability. Today with advancement in the scientific acumen, efforts are needed for enhancing the competence of breeding for harsh environments and also to make the future planning and execution for developing strains which can withstand the unforeseen climatic changes.

13.2.1 Climatic Stress Resistance

Climatic stress of hot and cold extremes has helped us to develop the hardy breeds which are now resistant to the stressful environment. Raised under extensive systems, Awassi ewes showed better adaptation traits (lower thyroxine, rectal temperature (RT), and pulse rate levels) than those raised under intensive management systems (Hamadeh et al. 1997). Many parts of the Saharan and sub-Saharan areas of the world such as Gobi desert of China and the great Thar Desert of India continually face these kinds of challenges. However, we find livestock species such as camel and sheep living in extreme environmental conditions and thriving well. Nomadic tribes of the western Rajasthan and Bedouin herders of Saudi Arabia are few of well-known herders who maintain animals tolerate extreme climatic stress. The trend of globalization has distributed the germplasm throughout the world, which may not be suitable for future climatic changes. Developing countries which still have a high proportion of the population in rural areas are better placed for a future when localization will replace globalization as the basis of sustainable lifestyles. However, developing countries still remain more vulnerable to the effects of climate change due to their high reliance on natural resources, very limited capacity to adapt institutionally and financially, and high poverty levels (Thornton et al. 2006).

13.2.2 Selection of Heat Tolerant/Genetic Traits for Adaptation

Selection is the only tool available with the breeders which effectively enhances the performance of the flock for the traits being selected. However, the rate of genetic improvement is slow with traditional breeding tools; there is a need to look into the genetic potential of existing species for heat tolerance. Climate change is a slow process which allows animals adapt to the prevailing environmental conditions. Breeding for climate change adaptation or mitigation will not be necessarily different from existing breeding programs; however, the problems related to measuring the phenotypes relevant to adaptation have to be overcome. Breeding indices should include traits associated with thermal tolerance, low quality feed and disease resistance, and give more consideration to genotype by environment interactions (GxE) to identify animals most adapted to specific conditions (Hoffmann 2010). Good environment favors high production whereas bad environment hampers it. Production of any kind, milk or meat characteristics, or wool traits require congenial environment to have better genotype and environmental interactions. Selecting animals for heat tolerance needs a new understanding by livestock holders and development agencies. A fresh project at mass scale with government's efforts will add much more meaning to it.

In a report by the FAO's Committee on Genetic Resources for Food and Agriculture (CGRFA 2009), it is noted that the management of animals under natural selection by pastoralists in marginal areas plays an essential role in their adaptation and fitness in such environments. Genetic variability for heat tolerance exists between breeds which are more in locally adapted breeds than exotic breeds, characterized by metabolic heat production and ability to lose heat and this variability within and between breeds needs to be exploited (Renaudeau et al. 2010). Sansthan and Rollefson (2005) reported that indigenous livestock management practices through social and rational breeding mechanisms among pastoral communities in India have contributed to breed adaptation to unfavorable environments. There are many Indian breeds of cattle (*B. Indicus*) which perform equally well in the hot climate, as they have been unknowingly selected for their ability to survive in unfavorable environments. Patagonia region (Argentina) is known for its harsh environment, poor rangeland and extreme climatic conditions but goat flocks are distributed on the north and west of this region. In north and central Neuquén, small holders breed local Criollo goat mainly for meat production. Lanari et al. (2005) reported that directional selection practices by herdsmen were the main factors in adaptation and micro-evolution of Criollo goat populations in their native environments.

Genetic mechanisms influence fitness and adaptation. Adaptation traits are usually characterized by low heritability. Selection on the basis of observation of heat stress seems to be difficult and costly affair. Molecular markers for this trait, however, can be exploited. Heat shock proteins 70 (Hsp70) are ubiquitously expressed proteins which protect animals against heat shock (HS) or stress, whether extreme hot or cold. These proteins virtually exist in all living organisms including microorganisms. They are an important part of the cell's machinery for protein folding, and help to protect cells from stress. Finding the polymorphism at genetic level for Hsp70 and their correlation with the observed heat stress resistance are useful tools for discovering sturdy animals. Molecular techniques like RFLP or sequence-based typing (SBT) for these genes can be used in combination with the traditional breeding to discover or evaluate traits of interest. This approach will definitely lead to production of new generation of animals which will be more heat tolerant without much compromise for production traits. Selection for heat tolerance based on RT measurements, skin characteristics, and inclusion of a THI in the genetic evaluation models is promising. Genetic antagonisms among heat tolerance, milk yield, and reproductive performance indicate that it may be difficult to combine the traits desirable for adaptation to high temperature environments with high production potential in the dairy sector (Ravagnolo and Misztal 2002). Such genetic antagonisms between adaptation to high temperature environments and high production potential seem to be less prominent in beef cattle where improved characterization of adaptive traits, use of reproductive technologies and molecular markers, and strategic crossbreeding are being used in breeding programs.

13.2.3 Impact of Stress on Production Traits

AnGR on extensive and intensive scale are affected by direct impacts of climate change such as catastrophic events, disease epidemics, physiological stress conditions, and water availability. Indirect impacts of climate change on AnGR include fodder quality, fodder quantity, and host-pathogen interaction. Climate change could indirectly affect livestock and dairy industries by affecting the availability and price of crops used for animal feed (Wolfe et al. 2008). Stress elicits multiple physiological responses. In response to stressful events, metabolism gets stimulated and products of energy and protein metabolism may be altered (Loerch and Fluharty 1999). Stress also suppress appetite, reduce growth rate, alter digestive and rumen function, and compromise immune function (Loerch and Fluharty 1999), suggesting its role in driving the energy balance of the body potentially in different direction. Stress has been shown to negatively impact live animal performance. Steers that have been heat stressed were reported to have lower ADG, lower dry matter intake, and decreased feed efficiency than those that received shading to mitigate heat stress (Mitlohner et al. 2001, 2002). In an experiment on flighty steers which had a greater response to handling (the stressor), Brown et al. (2004) found higher feed to gain ratios, and exit velocity was negatively correlated with ribeye area. An experiment for fine wool production by crossbreeding was adopted in the semi-arid region of Rajasthan, India in which Rambouillet and Russian Merino were crossed with the local breeds. However, it was seen that the environment in the semi-arid region posed constraints for this genotype to fully exhibit its potential. These animals were then shifted to a sub-temperate region (Mannavanur) where already a flock of Bharat Merino sheep was maintained (Gowane et al. 2010). It reflects the importance of climate on the productivity of animals and the extent to which it can affect breeding programs.

Heat stress in dairy cattle can have a long-term effect (weeks to months) on both milk production and birthing rates (Klinedinst et al. 1993). In dairy cows temperature optimum for maximum milk production in US is between 4.5 and 23.8°C. Humidity also is an important factor as heat stress can occur at temperatures as low as 23.8°C with a relative humidity more than 65%, or at temperatures of 26.7°C at relative humidity >30% (Wolfe et al. 2008). Heat stress can impact animal production and profitability in dairy cattle by lowering feed intake, milk production, and reproduction. Climate change is also affecting survivability tactics and patterns of the birds. Clutch size is an important life-history trait because it sets a hard upper limit on offspring production. It is also the easiest life-history trait to measure, and data on clutch size variation often are available when data on other life-history traits are not. According to the report of Winkler et al. (2002), tree swallows have advanced their mean date of clutch initiation (lay date) by ~9 days over the past 30 years across North America, apparently in response to climate change. To understand the relationship between climate change and lay date, von Haartman (1982) suggested a dichotomy in the way that birds adjust their clutch sizes to the timing of breeding. Taking this as a precedent we can visualize the similar effects on all backyard poultry or other birds. However, intensification of the poultry breeding and its industrialization has provided the microenvironments for the birds, which hardly get any chance to interact with macroenvironment. This by large alludes to the fear of major impact of climate change on poultry industry.

Milk productivity in dairy farming is a reflection of interactions between the production potential of the livestock, nutritional knowledge, veterinary care, availability of technologies, investment, active extension service, research as well as readiness to agree and accept innovations. Productivity enhancement depends upon definition of limiting factors at each stage and execution of appropriate corrective steps. The specific response of an animal to stress is variable given that some animals adapt or cope with stress more effectively than others. Identification or measurement of an animal's response to stress may provide a tool for identification of cattle that are not negatively impacted by stress (Falkenberg 2006). Bonsmara cattle have been selected for production and adaptation to subtropical environments and have been shown to produce tender beef (Strydom 1994). It has been hypothesized that these cattle may adapt to stress more effectively than other breeds of cattle due to their unique selection history (Falkenberg 2006). Similarly there are genetic resources which are more resistant to stress and perform equally better in varying environments.

How do we minimize heat stress which is a possible immediate outcome of climate change in livestock? Wolfe et al. (2008) discusses several strategies for minimizing the impact of heat stress. Lower cost measures such as reduction in overcrowding, lesser time in hot holding areas, and provision of shades in sunny areas were suggested. Moderate cost measures include improved ventilation and fan systems, improved insulation, and installation of misters or sprinklers for cooling. Higher cost measures include new building design and construction and installation of thermostat controlled air conditioning systems (Wolfe et al. 2008). Some of the increased costs for cooling in summer could be compensated for by reduced heating requirements in winter. Chase (2011) suggested nutritional modifications can lessen the effects of climate change on production performance. The modification of feeding schedules and feed rations like feeding during cooler parts of the day and increasing the proportion of easily digestible feeds is recommended. Water consumption of dairy cattle can increase by 20–50% on hot, humid days, therefore unlimited supply of water to the cattle is essential (Wolfe et al. 2008).

13.2.4 Grazing and Pastoral Systems

Nutritional stress is another important aspect of the climatic change. Transboundry animals pass and trade are managed by the nomadic tribes making them prominent stakeholders of the indigenous livestock. Given current trend of climate change, there is possibility that the grazing resources will shrink and pastoral system of animal rearing may go extinct. The stress of traveling to new locations in search of nutritional resources during hot weather and competing with other animals has some negative impact on animals and will become less attractive as grazing resources shrink. Almost all the ruminants are involved in transboundary migration for searching of food resources in developing tropical countries and climate change will surely have impact on this production system.

Alternative strategies for grazing such as evening to night grazing and early morning grazing is important to animals reared in tropical, to avoid burning temperatures of hot sun during afternoons. Ability of animals to reach resources and rehydrate is very important and this leads to indirect selection of breeds with taller legs that cover more distance in short time. Scarcity of resources in the field compelled farmers to take their sheep on migration as an alternative strategy. Migration became common during drought season (Rathore 2004). Malpura sheep was crossed with the Marwari sheep from arid region, which resulted in the production of crossbreds which are sturdier, have long legs and are best for migration purpose. These sheep are now preferred whenever farmers need to go on migration (Gowane and Arora 2010). In the future when scarcity of grazing resources may be common and migration for food will be a necessity, these breeds may probably be common.

13.2.5 Reproduction Seasonality

World's climate is changing rapidly and there is a concern that many species may face extinction if new seasonal strategies are not evolved (Bradshaw and Holzapfel 2006). Seasonal breeding evolved so that animals can conceive at a particular period in the year to facilitate birth in the spring. In most seasonal species, the timing of reproduction is controlled by day-length photoperiod. Moving sheep or horses across the equator will result in a reversal of breeding patterns after a period of adjustment. In areas closer to equator, photoperiodic animals may cycle through most of the year. The hormonal signal by which day-length acts to regulate reproduction is melatonin, produced by the pineal gland. Melatonin is produced only at night because light striking the retina sends a nerve impulse to the pineal and inhibits melatonin synthesis (Rosa and Bryant 2003). Therefore, melatonin is secreted for a long time when day-length is short and for a shorter period of time when days are long. Where and how melatonin acts to signal day-length to the brain is unclear but some change occurs in GnRH and LH secretion. The frequency of GnRH and LH pulses decreases under the non-breeding season photoperiod. In mares this occurs whether the ovaries are present or not (reviewed by Rosa and Bryant 2003). This is called steroid independent effect of photoperiod. In ewes this only occurs if ovaries are present and producing estrogen. This is called steroid dependent effect of photoperiod. Estrogen becomes potent inhibitor to GnRH and LH during the anestrous season in ewes. This is related to the changing patterns of melatonin secretion but the exact relationship is still unknown. Considering the fact that there are breed differences in the reproductive response of sheep to changes in photoperiod and that seasonal reproduction is driven by the pattern of melatonin secretion, it could be expected that breed differences would be due to different genetic abilities to secrete melatonin (Rosa and Bryant 2003). However, this does not seem to be the case since Lincoln et al. (1990) report examples in which the same pattern of melatonin secretion (i.e. the duration of the period of elevated levels reflects the period of darkness) is found among breeds which differ in the degree of seasonality of reproduction. Therefore, it may not be the melatonin signal which differs between breeds but the way the signal is translated in the brain (Rosa and Bryant 2003).

Heat stress can decrease sperm production in males and cause increased embryo mortality if it occurs very early in development. Shading and or cooling systems must be provided to avoid heat stress. Overcoming the seasonality of reproduction in sheep allows the lactation (suckling lambs) during pasturing. Development of sheep with a highly reduced seasonality of breeding has been performed using spring fertility measurements (Notter and Cockett 2005). However, response to selection using spring fertility remains limited due to very low levels of heritability (<0.1) and ewe age effects on fertility and response to selection. On Mérinos d'Arles ewes, Hanocq et al. (1999) reported 28% spontaneous ovulatory activity in April and consistent heritability and repeatability estimates (0.37 and 0.20, respectively). Pelletier et al. (2000) found that ovulatory activity was associated with the polymorphism of the gene that encodes melatonin receptor. This polymorphism could not be used as genetic marker for selection of out-of-season breeding ability in sheep. All latitudes find some mammals reproducing seasonally, even in the deep tropics. The energetic costs of reproduction are huge, and when the cost to gain ratio of foraging varies seasonally, births must occur when the muscle and thermoregulatory costs of obtaining food are most favorable (Bronson 2009). The mammals that will probably be most vulnerable to climate change will be the longer lived species at the mid to higher latitudes whose reproduction is rigidly controlled by photoperiod. The future will probably see shifts in species' ranges to higher latitudes as the climate warms. Probably the environmental conditions of the tropics today are the conditions of the higher latitudes of the future (Bronson 2009). Proper management of the livestock for regulation of stress related to reproduction may help us to maintain them in the current reproductive status. Proper housing, nutritional management, and seasonal variation buffering will probably add an extra expenditure on part of livestock holder in future.

13.2.6 Disease Challenges and Associated Tolerance/Resistance

Livestock diseases adversely affect animal production throughout the world. Changes in climatic patterns and seasonal conditions may affect disease behavior in terms of spread pattern, diffusion range, amplification, and persistence in novel habitats. Pathogen invasion may result in the emergence of novel disease complexes, presenting major challenges for the sustainability of future animal agriculture at the global level (de La Rocque et al. 2008). The geographical expansion of vector-borne infectious diseases to higher elevations and higher latitudes will be expected. In Himalayan ranges in India (Kumaon hills), it was impossible to find mosquitoes few years back but with current trend in climate change, mosquitoes are commonly found in many parts of this region. This will definitely have effect on transmission and course of diseases such as Malaria. The change in the temperature profile of any region can harbor the population of new vectors for new diseases, and it is probable to see low altitude marked diseases in high altitude areas.

Climate strongly affects agriculture and livestock production and influences animal diseases, vectors and pathogens, and their habitat. Temporal and geographical distribution of vector-borne infectious diseases such as bluetongue, West Nile fever, vesicular stomatitis, and New World screw worm in South America are likely to change (Pinto et al. 2008). Better prevention and intervention measures will be required to check the impact of climate change assisted animal husbandry calamities in South America. In Africa, the impact of climate change might be devastating if not managed properly. While direct impacts of possible climate change in Africa include increased ambient temperature, floods, and droughts, indirect impacts may be the consequence of reduced availability of water, forage, and changes in the environment that promote the spread of contagious diseases through increased contact among animals, or increased survival or availability of the agent or its intermediate host (Van den Bossche and Coetzer 2008). The potential vulnerability of the livestock industry will depend on its ability to adapt to climate changes. The Asian continent, because of its size and diversity, may be affected significantly by the consequences of climate change, and its new status as a 'hub' of livestock production gives it an important role in mitigating possible impacts of climate variability on animal health (Forman et al. 2008). In Asia, animal health may be affected by climate change in four ways: heat-related diseases and stress, extreme weather events, adaptation of animal production systems to new environments, and emergence or re-emergence of infectious diseases, especially vector-borne diseases critically dependent on environmental and climatic conditions (Forman et al. 2008). Strategies to mitigate these challenges include veterinary care, controlling zoonoses, and major emphasis on prediction of possible threats.

Veterinary care is the costly and recurrent affair for tackling the probable disease challenges due to climate change. Disease resistant breeds will reduce the cost of veterinary care and production losses in the future. Breeds that are reported to be resistant or tolerant to trypanosomiasis, tick burden, tick-borne diseases, internal parasites, dermatophilosis, or foot rot will be most preferred one. Disease resistance depends much on the type of pathogen and the hosts' resistance or tolerance mechanisms. Breeding for disease resistance is influenced by the availability and costs of alternative treatment (e.g. vaccines, drugs) and antimicrobial resistance of pathogens. Potential exists for genetic improvement of disease resistance with the use of molecular markers and marker assisted selection.

An important approach therefore to alleviate the effects of diseases emanated from climate change is to develop disease-resistant breeds. There are several advantages for incorporating genetic element in the disease management strategies as recognized by FAO (1999a, b):

1. The permanence of genetic change once it is established;
2. The absence of the need for purchased inputs once the effect is established;
3. The effectiveness of other methods is prolonged as there is less pressure for the emergence of resistance;
4. The possibility of broad spectrum effects (increasing resistance to more than one disease);
5. The possibility of having less impact on the evolution of macroparasites such as helminths, compared to other strategies such as chemotherapy or vaccination; and
6. Adding to the diversity of disease management strategies.

The maintenance of diversity of genes underlying resistance provides an important resource for combating effects of possible future pathogen evolution (FAO 2007a, b). Crossbreeding for introgression of the disease-resistant genes in the new population is a strategy or alternatively select within a population with highly resistant individuals. Compilation by FAO (2007a) reports resistance or tolerance to specific diseases and parasites in several breeds across livestock species such as buffaloes, cattle, goat, sheep, pig, horse, and deer. Document revealed many breeds resistant to one or more type of diseases such as N'Dama cattle for infection against ticks of various species, *Anaplasma marginale and Haemonchus contortus*. Sahiwal cattle against *Theileria annulata*, etc. Several workers have tried to incorporate resistant genes to these diseases in their experimental flocks. Studies have shown that Trypano-tolerant breeds are more productive than susceptible animals under moderate to high tsetse challenge (Agyemang et al. 1997). Similarly many breeds which are resistant to diseases are also better in production profile as compared to other breeds in the similar environment (Baker 1998). This property of disease-resistant breeds make them superior to their counterparts and help them become the breed of choice for introgression of the favorable genes in the other populations through genetic intervention as a measure of protection against the future environmental threat.

13.2.7 Variation in Beef Cattle Adaptability

Adaptation to the environment in which any animal is being reared for production is very important. Genetic Slippage (Dickerson 1955), a phenomenon where performance of the individuals reared in a specific environment degrades when forced to perform in different environment, is of utmost importance in case of beef cattle production system. In many of the animal breeding experiments this feature as an outcome was visible. With the knowledge of changing climate and its possible effects on the animal production system, the beef cattle adaptability became an issue of concern where balanced genetic potential for adaptation, production, and product quality within specific environments is anticipated. Cattle were domesticated in western Asia before 10,000 years (Hohenboken et al. 2005), and then they were migrated to several regions of the world, for utilization. Now they are all over the world in Europe, Asia, Africa, America, and many other parts. Domestication of the cattle was general and it hardly had any stress quotient involved when animal production system was concerned. With advent of the scientific knowledge and more demand for quantum and quality of the product, the stress level has increased with more pressure on animal to adapt to the changing environment. Beef cattle production demands more input and output on part of rearer and reared, respectively. Change in the climate will have changing conditions and therefore it will be a tough testing time for already less adapted cattle. Genotype and environment interaction has a lot of role to play in beef cattle production system. Burns et al. (1979) have asserted that genetic adaptation to local environments is important in commercial beef cattle production. Evidence supporting these recommendations was provided by their classical experiment to investigate GxE (Hohenboken et al. 2005) which makes it clear that why adaptation is a critical problem in beef cattle production system.

Beef cattle production system demands intense selection, which is done at the cost of many other things. Selection to increase sustained annual production selects automatically for traits important to adaptation. In recent decades, however, the application of refined knowledge of inheritance, improved information technology and advanced reproductive techniques has allowed dramatic increases in selection intensity and selection response. Rapid response to intense selection for increased product can seize resources formerly utilized to support reproduction and survival. In the coming future, this condition will be more aggravated as adaptation to the changing climate will be next to impossible task at least for beef cattle. Improved adaptation enhances financial well-being of beef cattle producers. Potential benefits from improved beef cattle adaptation include enhanced animal well-being, increased profitability for beef cattle producers, more desirable products for beef consumers, enhanced resource conservation, and more effective forage resource utilization (Hohenboken et al. 2005). Beef cattle production can not be profitable unless cattle are productive, efficient, and produce a desirable end product. Selection to improve traits contributing to these ends is desirable if not required. In addition, cattle that are genetically adapted to their environment incur lower costs than unadapted but otherwise comparable cattle. Profitability of beef cattle production would be enhanced by including locally rational measures of adaptability in industry selection schemes and breeding objectives. This may include identifying indicator traits, DNA markers, and genes important to adaptation (Hohenboken et al. 2005). The money saved on producing the livestock product incurs more returns to the livestock holder.

13.2.8 Environmental Manipulation

The changing scenario of animal production system and climate change needs an overall approach for tackling the situation. Animal environment is defined as any external factor that influences the productive response of farm animals. This would include, but is not limited to, effective ambient temperature, photoperiod, sound, altitude, environmental contaminants, psychological restraint, and management system (Ames and Ray 1983). Proper management practices such as adoption of husbandry practices like thermal control for microenvironment, veterinary aid at hand with all new technologies, fodder production, and availability at all time, etc. Access to such technologies will determine the ability of producers to adapt their herds to the physiological stress of climate change. Historically, animals have been provided protection from climatic variables by sheds, barns, and other buildings as a matter of humane treatment but, more recently, there has been increased use of environmental control to enhance rate and efficiency of production (Ames and Ray 1983). Intensive livestock production systems have more potential for adaptation through the adoption of technological changes. This may keep the high-output breeds in such production systems. However, this type of system will be more input demanding and therefore will be in the limited hands.

13.2.9 Impact of Climate Change on AnGR

Livestock keepers and breeders utilize AnGR for food and agriculture. Livestock keepers remain the main custodians of AnGR diversity. Livestock production both contributes to and is affected by climate change (FAO 2010). Geometric human population growth with limited food resources has new threats in waiting for us. Demand for meat, milk, and other animal product is dramatically increasing. The sustainable use, development, and conservation of AnGR for food and agriculture will make a vital contribution to hunger quotient of human race. A diverse resource base is critical for human survival and well-being, and a contribution to the eradication of hunger. AnGR are crucial in adapting to changing socio-economic and environmental conditions, including climate change. They are the animal breeder's raw material and among the farmer's most essential inputs. They are essential for sustainable agricultural production (FAO 2007a, b).

The climate change poses a great threat to the AnGR. Threats are numerous starting from effect on individual animals to the population as a whole. The change in the pattern of disease occurrence and distribution, geographical disease barrier erosion, heat stress as a common factor demanding more heat tolerant individuals, scarcity of the food resources and water as a general factor, overall decline in the production level of the animals as a compensatory mechanism for stress tolerance, reproductive disturbances, and loss of animal genetic diversity at population level will be few among many effects of climate change on AnGR. The overall increase in the temperature will have its impact on the physiological stress and thermoregulatory control, nutrition, and disease status of the animals. Species and breed differences in reaction to these stressors have been reviewed by Hoffmann (2010). Substantial differences in thermal tolerance lie between species; however, there are also differences between breeds of a species. *B. indicus* is generally more heat resistant than *B. Taurus*. In general, the high-output breeds originating from temperate regions that provide the bulk of market production today are not well adapted to heat stress, partly because of increased metabolic heat production. There are also breed differences in disease resistance that may become relevant to buffer against the projected rapid spread of pathogens or even small spatial or seasonal changes in disease distribution. The expected increase and often novel disease pressure will favor genotypes that are resistant or tolerant to the diseases in question. Many breeds are characterized as the resistant or tolerant to a specific or a group of diseases. Some are experimentally proven; however, many are based on the anecdotes and need verification. Information on these characteristics along with traits like heat tolerance, greater feed conversion efficiency, etc. will help the breeders to formulate the future breeding strategies. Adaptation to the climate change is another phenomenon. Either breeders can adapt their animals' genetics to the changed environment, or adapt the production environment while maintaining their breeds. Both the ways strategy is not so simple. Changing genetics of the animals is time consuming and more input demanding job, however, once done, it is supposed to be the permanent solution for the problem. Changing production environment is also a costly affair and needs constant pumping of resources. Although costly but it will be the choice of management in livestock sector where intensive livestock practices such as commercial poultry production, beef cattle production, etc. are followed.

13.2.10 Factors Accelerating Erosion of Livestock Biodiversity

Genetic erosion is a process whereby an already limited gene pool of an endangered species of plant or animal diminishes even more when individuals from the surviving population die off without getting a chance to meet and breed with others in their endangered low population. Genetic erosion occurs because each individual organism has many unique genes which get lost when it dies without getting a chance to breed. Low genetic diversity in a population leads to a further diminishing gene pool, inbreeding and a weakening immune system resulting in eventual extinction. The 1992 United Nations Conference on Environment and Development (Earth Summit) held in Rio de Janeiro was an important landmark. The Convention on Biological Diversity (CBD), signed in Rio by 150 governments, committed the nations of the world to conserve their biodiversity, to ensure its sustainable use, and to provide for equitable sharing of the benefits arising from its use. By 2005, 188 countries had become parties to the CBD (FAO 2007a, b). Agricultural biodiversity is the product of thousands of years of activity during which humans have sought to meet their needs in a wide range of climatic and ecological conditions. Well-adapted livestock have been an essential element of agricultural production systems, particularly important in harsh environments where crop farming is difficult or impossible (FAO 2007a, b).

Several factors are in play for the erosion of the livestock biodiversity. Focus on very few economically important breeds is one reason. Few breeds such as

Holstein–Friesian, Jersey, etc. are given too much importance as they contribute directly to the exchequer. World's best genetic improvement programs are employed for these few chosen breeds and many known and thousands of unknown breeds in least developed and developing countries go unsung. As a result of this, many breeds lose their potential to compete with other breeds and ultimately fall in the category of endangered population. The erosion of AnGR is a long-term threat to ensuring food security and rural development. The population status of many breeds is still unknown, and the problem may thus be underestimated. Most developing countries and some developed countries do not currently have AnGR conservation strategies or policies in place. Without strategically planned interventions, using both in situ and ex situ conservation, erosion will continue and may accelerate.

Factors accelerating erosion of livestock biodiversity also belong to all sorts of human activity areas (Blench 2005). Development interventions constitute the major factor in developing countries. Here the preference is given to the high-input, high-output breeds developed for benign environments of the developed countries. Commercial interests in those countries promote use of relatively temperate-adapted breeds and create unrealistic expectations in developing countries. This works as a double-edged sword, first it criticizes the local indigenous livestock as inferior and next it dilutes the native livestock with foreign genes, least adapted for such environment. In many developing countries this activity has created the never ending series of livestock erosion problems; India is the best example for this phenomenon. Blench (2005) emphasizes on specialization as the factor for erosion of livestock biodiversity. Emphasis on a single productive trait, e.g. dairying, leading to exclusion of multipurpose animals. Many Indian cattle breeds are developed since ages as the multipurpose animals. Bullocks are used as the work animals for draft purpose and cows for the milch purpose; however, with advent of mechanical farming system the breeds of such importance are loosing their place in the competition for existence. These days' cows with more milk production efficiency are given importance and therefore the dual purpose or multipurpose breeds are becoming endangered. Worldwide there is a consensus on one issue among several experts that genetic introgression is the major factor for erosion of livestock biodiversity. Crossbreeding and accidental introgression have contributed to the greatest extent toward the loss of indigenous breeds. Economic change in the domestic and international market has posed a lot of challenges for the livestock products. The synthetic products are out-competing the natural products, which in turn are devaluating the livestock resources.

In purview of the climate change, the livestock biodiversity has lot of challenges to face in the near future. Scarcity of food and water resources, change in the temperature and humidity, walking stress, production, and reproduction stress will add a lot to the problems of the livestock management in the coming era. Therefore, the climate or vegetation change may make a breed unviable in a particular habitat, leading to loss of livestock biodiversity. Natural disasters such as floods, drought, and epizootics are also affecting the livestock population

(Blench 2005). These disasters will preferentially affect remote or isolated human and livestock populations.

13.3 Using Adaptability to Improve Profitability

Domestication of the animals and the agriculture laid the foundation of the human civilization. Food and livelihood security were provided by the biodiversity in animals and agriculture. Biodiversity, in general, is the form of life available in an ecosystem or in the world as whole. It is the variation in the genetic composition of the living organism. Livestock biodiversity is the variation in the livestock species. Livestock biodiversity is the resultant of evolutionary forces (mutation, migration selection, genetic drift, and founder effect), geographical isolation, and adaptation. About 200 years ago, the concept of breeds emerged. But, today's breed with their unique combination of genes would not have emerged without continuous active managements and selection by farmers and pastoralists over 12,000 years since the first livestock species were domesticated (CGRFA 2009). The most decisive factor in selection of animals in early days was adaptation to a given environment (Gillespie 1997). Adaptation is assessed based on the ability of animal to survive, reproduce, and produce successfully in a particular environment (Gillespie 1997). Over the past few decades the selection pressure has increased to increase the production via the application of modern quantitative genetics methods with little regard to the preservation of genetic diversity. During the last decades, development of and increased focus on more efficient selection programs have accentuated genetic improvement in the number of breeds. Artificial insemination, embryo transfer technology, better veterinary health facilities, and improved transportation facilities have facilitated the dissemination of the genetic material. As a result, high productive breeds of livestock have replaced local ones across the world. This trend of breed replacement and breed dilution through crossbreeding has resulted into the breed erosion (FAO 2007a, b). According to the FAO, 20% of the roughly 7,600 breeds reported worldwide belonging to the 18 mammalian species and 16 avian species are at risk and 62 breeds became extinct within the first 6 years of the century (FAO 2007a, b).

The livestock sector is changing rapidly in response to a variety of the drivers. Globally, human population is expected to increase from 6.2 billion today to 9.2 billion by 2020. Global demand of livestock products will increase significantly in the coming decades (Delgado et al. 1999). Significant changes in physical and biological systems have already occurred on all continents and in most oceans, and most of these changes are in the direction expected with warming temperature (Rosenzweig et al. 2008). There does not seem to have been a great deal of work done on the direct impacts of climate change on heat stress in animals, particularly in the tropics and subtropics. As has been observed by Easterling and Apps (2005), that a lack of appropriate physiological models that relate climate to animal physiology rather limited the confidence that could be placed in predictions of impacts, referring it to “a major methodological void”. It is clear, however, that warming will alter heat exchange between animal and environment, feed intake, mortality, growth, reproduction, maintenance, and production potentially (Standing Committee on Agriculture 1990). The potential for widespread genetic devastation in the future as a result of inexorably rising temperatures is great. The impacts of such losses are difficult to imagine, and the problems that will be caused by the loss of genes for disease and pest resistance, for environmental adaptation, and for other desirable traits in both plants and animals, cannot be over-stressed. Animal and plant genetic resources are the ultimate non-renewable resource; once gone, they are gone for good. Their importance is critical, but the complexity of ecosystems means that it is extremely difficult to assess the impacts of climate change on biodiversity. Animal and plant genetic resources have extremely high value, and the costs associated with their loss may be simply enormous, but to all intents and purposes neither their value nor the costs of their loss can be realistically quantified. Given that this situation is not likely to change very rapidly in the future, it makes much sense for any consideration of climate change and biodiversity to emphasize conservation as well as attempting to fill critical knowledge gaps to enable realistic assessments to be carried out on the likely impacts of adaptation and mitigation activities on biodiversity and other aspects of sustainable development (IPCC 2002). As CGRFA (2007) notes, pastoralists and smallholders are the guardians of much of the world’s livestock genetic resources. This poses particularly challenging problems for conservation, but there is a great deal that can and must be done. Therefore, there is dire need of characterization, conservation, and scientific breeding of the biodiversity of livestock as a credit for unseen and unpredictable future to have sustainable development, livelihood, and food security of the world.

It has been realized time and again that high producing, genetically altered or genetically engineered animals are capable of performing well only when there is assured input, sound management system, well defined long-term policy planning, and short-term development plan that are predominant in many developing, specially tropical, countries.

A fact that has consistently been underrated is that most of the tropical breeds of livestock are multipurpose with products ranging from milk, meat, wool, draft, organic manure, farm energy to hide, skin, etc. Many a fitness characters possessed by these breeds which are either absent or possessed to a lesser degree by the temperate breeds include lower metabolic rate that generates less heat, reduced panting but more ready sweating that conserves energy, higher intake of poor quality feed, higher digestibility and efficiency of feed conversion, feed intake less affected by high environmental temperature, greater ability to retain feed and water in the large intestine, low water requirement and better resistance to ticks, insects, and some diseases (Wilson 2009).

To cite some of the commercially successful and thriving indigenous breeds of livestock well adapted to often harsh production environment include Sahiwal cattle of India (Ogilvie 1947), Kenana cows of Sudan (Alim 1960), Nandi cattle in Kenya (Wilson 2009). Some prized tropical beef breeds are Caracu derived from

Criollo, Indubrasil or Nellore (originally derived from Ongole cattle of India) in Brazil (Cardellino 2000), Brahman cattle (originally from India) in USA, Boran in Ethiopia and Kenya (Haile Mariam et al. 1998; Homann et al. 2005), Mashona and Tuli cattle in Botswana and Zimbabwe (Moyo 1990; Homann et al. 2005). Similarly, Haryana and Kankrej cattle in India and Afrikander cattle in South Africa (Wilson 2009) are renowned for draft or pack purpose. Among the sheep breeds, Chios and Awassi breeds of Near East for milk purpose, Sudan Desert for dual purpose (milk and meat) (Wilson 2009), Awassi breed of Near East as multipurpose (Amin and Peters 2006; Wilson 2009) and Chios breed of Near East and eastern Mediterranean region (Hatzimimaogiou et al. 1990) are front runner in productivity. Boer goat of South Africa (SASBA 2004), Jamunapari, Beetal, and Black Bengal goat of India, Damascus goat in Cyprus (a native of Syria and Lebanon) (Mavrogenis et al. 2006; Wilson 2009) are widely used for profitable goat farming.

It is interesting to see that since 1960s, modern single purpose breeds and modern methods of husbandry have penetrated remotest parts of the developing counties of tropics (Wilson 2009). Against this backdrop, a weak undercurrent nudged the conservation or preservation of native breeds since 1970s which was initially confined to richer and developed countries e.g. Rare Breeds Survival Trust in UK (Alderson 1990) then to developing countries though mainly in the form of academic interest (Wilson 2009).

Crossbreeding of indigenous livestock with exotic one has been widely practiced for improving productivity pending detailed documentation of their production potential. There have been many successful cases where well-defined breeding policy, nutrition, management, and health coverage were ensured like as has been cases in highlands of eastern Africa specially Kenya with Ayrshire, Friesian, Jersey, and Guernsey cattle crossed with zebu animals (Bebe et al. 2003). For example, Australian Milking Zebu (AMZ), a cross of Sahiwal and Red Sindhi bulls with Jersey cows, was developed during 1960s, with milk production over 2,100 kg per lactation and 4.5% fat (Stephens 2006). But more remarkable is it because of heat tolerance and tick resistance. Similarly, Australian Friesian Sahiwal (AFS) whose development is of recent origin has better performance and fertility than AMZ (Stephens 2006). Development of Senepol cattle (lesser known as Nelthrop) in Caribbean Islands combined N'dama traits of heat tolerance with higher milk production, good meat quality, and docility of Red Poll's (Wilson 2009). Santa Gertrudis, a composite breed of 3/8 zebu and 5/8 Shorthorn developed in Texas, USA, is an excellent beef breed that has found wide acceptability in tropical and subtropical countries also. The Droughtmaster, a composite of mainly Brahman and Shorthorn developed in Queensland, Australia, produces quality beef with good reproductive performances. It is also credited for docility, heat and drought tolerance, and tick resistance. Jamaica Hope, developed through crossing of native cows with Jersey bulls and subsequently Sahiwal bulls of India, have had mixed success. The Droper sheep, derived from British Dorset Horn and Persian Blackhead in South Africa, is one of the few examples of sustained success in small ruminant development with high prolificacy, rapid growth rate but without wool that has found wide acceptability (Wilson 2009).

Going by the present trend, it is believed that food production systems in developing countries are moving into marginal areas for which sustainable farming systems and the well-adapted species and breeds of livestock species are yet to emerge (Wilson 2009). There are some exceptions like Barca people of Eritrea who bred their cattle for centuries for milk production and docility (Wilson 2009) or the WoDaaBe herders in Niger where cattle breeding system has consistently worked toward securing reliable exploitation of structural unpredictability through three pronged approach, viz. by organizing their cattle into matrilineal lineages, exploiting the capacity for both genetic and extra-genetic inheritance in their cattle breeding population and relying on lineage duration rather than peak productivity as the primary criterion for selection (Krätli 2008).

13.3.1 Characterization of Livestock Breeds

Livestock currently provides 43% of global agricultural output in value terms, with a projected increase (FAO 2010). Traditional livestock systems based on local breeds contribute to the livelihoods of 70% of the world's rural poor (Hoffmann 2010). Despite the importance of livestock to poor people and the magnitude of the changes that are likely to befall livestock systems, the intersection of climate change and livestock in developing countries is a relatively neglected research area. Little is known about the interactions of climate and increasing climate variability with other drivers of change in livestock systems and in broader development trends (Thornton et al. 2009). Effective management of farm animal genetic resources (FAnGR) requires comprehensive knowledge of the breed characteristics, including data on population size and structure, geographical distribution, the production environment, and within- and between-breed genetic diversity. Integration of these different types of data will result in the most complete representation possible of biological diversity within and among breeds, and will thus facilitate effective management of AnGR (Groeneveld et al. 2010).

'Breeds' are cultural concepts rather than physical entities, and these concepts differ from country to country, making characterization only at the genetic level rather difficult (Boettcher et al. 2010). But FAO uses the following broad definition of the breed concept, which accounts for social, cultural, and economic differences between animal populations, and which can therefore be applied globally in the measurement of livestock diversity:

> Either a sub-specific group of domestic livestock with definable and identifiable external characteristics that enable it to be separated by visual appraisal from other similarly defined groups within the same species or a group for which geographical and/or cultural separation from phenotypically similar groups has led to acceptance of its separate identity (FAO 1999a, b).

13.3.1.1 Phenotypic Characterization

"Phenotypic characterization of AnGR" is used to refer to the process of identifying distinct breed populations and describing their characteristics and those of their production environments. In this context, the term "production environment" is taken to include not only the "natural" environment but also management practices and the common uses to which the animals are put, as well as social and economic factors such as market orientation, niche marketing opportunities, and gender issues. Recording the geographical distribution of breed populations is here considered to be an integral part of phenotypic characterization (FAO 2010).

As per the guidelines issued by CGRFA (2011), the phenotypic characterization is further divided into primary and advanced characterization. Primary characterization is used to refer to activities that can be carried out in a single visit to the field (e.g. measurement of animals' morphological features, interviews with livestock keepers, observation and measurement of some aspects of the production environment, mapping of geographical distribution). The term "advanced characterization" is used to describe activities that require repeated visits. These include the measurement of productive (e.g. growth rate, milk production) and adaptive (e.g. resistance or tolerance to specific diseases) capacities of breeds in specified production environments.

Based on the types of background data available there are two approaches to characterization. Exploratory approach is followed if no background data related to existence of distinct breed is available. Confirmatory approach is followed when some base line data is available. This is used to validate breed identity. Different types of data collected in phenotypic characterization study are:

1. Geographical distribution, population sizes, and structures
2. Phenotypic characteristics (physical features and appearance)
3. Economic traits (e.g. growth, reproduction, and product yield/quality)
4. Productive and adaptive attributes of the breeds
5. Information on the breeds' origin and development
6. Images of typical adult males and females, as well as herds or flocks in their typical production environments
7. Any known functional and genetic relationships with other breeds within or outside the country
8. Biophysical and management environment (s) in which the breeds are maintained
9. Responses of the breeds to environmental stressors, such as disease and parasite challenge, extremes of climate and poor feed quality, and any other special characteristics of the population in terms of adaptation
10. Relevant indigenous knowledge of traditional management strategies used by communities to utilize the genetic diversity in their livestock.

Geographical distribution of the animals helps in categorizing breed as local or transboundary breed. In a recently developed classification (FAO 2007a, b), breeds present in only one country are termed local breeds and those present in more than one country are termed transboundary breeds, the latter being further differentiated into regional and international transboundary breeds depending on the extent of their distribution. Demographic data are fundamental to the assessment of the risk status of livestock breeds—a key step in the strategic planning of AnGR management. Risk status depends on several factors. First, it is linked to the size and structure of the population. Effective population size (Ne) is the preferred measure for the assessment of risk status (FAO 1992; Gandini et al. 2004). In changing scenario of climatic condition, it is advisable to collect additional information on non-biotic (climate, water scarcity, seasonal feed scarcity, etc.) stressors. These may be decisive factors in tolerating the environmental shift.

Production recording helps in evaluating the production potential of the animals in different environments. It gives fare idea of genotype environment interactions. It is required to compare breeds performance and to predict future performance.

13.3.1.2 Molecular Characterization of Breeds

It is the method to explore polymorphism at molecular and protein level to study the genetic variation within and among the breeds. Because of the instability and poor polymorphism of protein, DNA markers are the markers of choice. DNA markers can be direct/functional or indirect/neutral markers. Direct markers are the segment of DNA present within the gene of interest or gene itself is used as marker. Indirect markers are linked marker not within gene of interest. Most extensively used DNA marker for study of breed diversity is microsatellite marker. It is repeated DNA segments of 6–7 base pairs and neutral marker (Groeneveld et al. 2010). Ideally 30 markers assigned by the FAO should be used for characterization of populations. It gives more accurate result and also opportunity to compare the results with other within and between breeds if has been carried by using similar set of markers. Single nucleotide polymorphisms (SNPs) are now frequently used markers as these are based on the difference of one base pair and give more accurate result. Mitochondrial DNA (mtDNA) markers are maternal markers used in identification of wild ancestors, localization of domestication centers, and reconstruction of colonization and trading routes (Bruford et al. 2003; Ajmone-Marsan et al. 2010; Groeneveld et al. 2010). Y-chromosomal markers are Y-chromosomal variation; it is used as a powerful tool to trace gene flow by male introgression (Petit et al. 2002). We are witnessing the revolution in genetics with the advent of high throughput affordable genome sequencing Ajmone-Marsan and GLOBALDIV 2010. Molecular characterization is carried out to study genetic variability, phylogenetic relationship, gene mapping, and marker-assisted selection and to develop DNA repository for future use.

Breed is the functional unit of conservation and characterization is used to assess the values of the animals for its present and future use. Characterization helps in formulating breeding and conservation plan.

13.3.2 Conservation of the AnGR for Changing Climate

The International Union for the Conservation of Nature and Natural Resource's (IUCN) World Conservation Strategy has defined the need for conservation as

> the management for human use of the biosphere so that it may yield the greatest sustainable benefit to present generations while maintaining its potential to meet the needs and aspirations of future generations. Thus conservation is positive, embracing preservation, maintenance, sustainable utilization, restoration and enhancement of the natural environment (IUCN 1980).

There are two methods of conservation in situ and ex situ. In situ conservation is rearing of the animals in their natural habitat. Ex situ conservation is rearing or preserving animals away from adopted habitat. Livestock breed conservation should be formulated based on the prediction of future drivers of change. It is predicted that climate change may be accompanied by natural catastrophes like flood, earthquake, emergence of new disease, etc. There is risk of local rare breed being lost in localized disasters (Carson et al. 2009). To secure against such disasters, it is necessary to conserve these breeds using ex situ method of conservation like cryo-conservation of genetic material in gene bank. Most of the conservation works are limited to developed country but the future security of biodiversity lie in developing world. In developing countries, few breeds of major species are covered by conservation programs, and the focus is typically on in vivo conservation (FAO 2007a, b). Ex situ conservation measures should be encouraged to prepare for uncertain consequences of climate change so that valuable genotypes can be restored if needed.

There are two arguments for conservation of animals viz. cultural and insurance (Simianer 2005a, b). Animal breeds development has parallel history with human cultural development (Hanotte et al. 2002). It has political and religious importance; an example is the king's privilege to keep a herd of white N'guni cattle in South African Zulu tribes (Felius1995). Little is known about the interactions of climate variability and other drivers of livestock systems (Thornton et al. 2009). Therefore, genetic diversity of the animals can be considered as the insurance against future change (Smith 1984).

13.3.3 Livestock System and Breed Conservation

Global livestock sector has grown into dichotomy. Intensive systems are where animals are kept under intensive commercial production system. These are well characterized breed of animals reared for specific purpose. Custodian of two-third of the livestock population is smallholders and pastoralist in developing tropical countries of Asia and Africa. Their environment is characterized by hot and humid, hot and dry, high and low altitude, scarce water and feed resources, high disease incidence (Solkner et al. 1998). These animals are not well characterized but have ability to deliver different kind of the products. They have immense range of biodiversity. Conservation and breeding for high production have conflicting interest, oriented toward market demand least caring for conservation of biodiversity. Most livestock keepers cannot afford and are not willing to safeguard animal biodiversity without appropriate incentives (FAO 2009; LPP et al. 2010). Many European countries have established the mechanism for supporting conservation activities (Hoffmann and Scherf 2010). Most countries pay breeders of breeds at risk incentive, if breeders are a member of the relevant breed society and follow approved breeding programs. The United Kingdom provides support to breeders for grazing animals (Small and Hosking 2010). Other than monetary benefit, emphasize should be given for the recognition of the breeders' contribution to breed conservation by society (Hiemstra et al. 2010). Promoting marketing and extra price to products derived from local breeds offer important opportunities to promote conservation (LPP et al. 2010).

Fundamentals of breeding system to mitigate changing environment will remain same but we need to shift our priorities. Being genetically diversified, adapted to environment, resilient to disease and temperature gradient local breeds may be favored for breeding. Along with the high production of animals produce we should make sure that it is sustainable for food security and livelihood. Except some high producing breeds of animal others are either poorly characterized or not characterized at all. Characterization of animal resources is the need of the hour. Before going for major change in policy to mitigate climatic change there is need of extensive study of interaction among drivers of changes of climate and livestock. Model of study to asses the climate change will be more effective once it is done in composite manner, considering not only livestock but also all related factors which affect it directly or indirectly. In coming future there will be need to consider several new traits like FCR, disease resistance, resilient to heat; then only selection of animal for sustainable production will be fruitful.

13.3.4 GxE Interaction

The property of the organisms to develop systematically different phenotypes in different environments is called phenotypic plasticity (de Jong and Bijma 2002) which is obviously a genetic phenomenon revealed phenotypically through local and global expression of genotypes. A variation in plasticity between different genotypes exists when the difference in phenotypic value between animals among different environments is not constant. This variation in phenotypic plasticity between genotypes results in the GxE (Strandberg 2006). Statistically GxE interaction can be put like the interaction of the genotype and environment is greater than the sum of the individual parts, which means they are not additive. Biologically we know that environment controls the expression of the genes in the living system, which may be up-regulated or down-regulated depending on the need of the expression of the genotype. In GxE interactions, the trait is partly influenced by different genes in different environments indicating the biased expression patterns of the genes to varying environment. Weak GxE interactions result in non-identical differences between animals' breeding values whereas, strong GxE interactions result in significant re-ranking of animals' breeding values in the differing environments (Wallenbeck et al. 2009). Strong GxE interactions result in large over-prediction of economic outputs if these interactions are not accounted for in the breeding program (Dominik and Kinghorn 2008). Due to the varying magnitude of phenotypic change between environments there is a re-ranking of individuals for trait expression and, consequently, selection for superior performance in one environment will not necessarily result in enhanced performance under other sets of environmental conditions (Falconer and Mackay 1996). In a situation where GxE interactions result in re-ranking of animals in the genetic evaluation system, then consequences can be grave if animals are used in totally different environment. We can also plan the breeding strategy based on the estimate of genetic correlations between two environments for a trait. Robertson (1959) suggested that genetic correlations of less than 0.8 were indicative of a significant GxE. More recently, Smith and Banos (1991) and Mulder and Bijma (2006) investigated GxE interactions in dairy cattle populations and determined that if genetic correlations between environments are greater than 0.8–0.9, genetic gain can be increased by selection across environments.

Here we review few experiments where GxE interaction has been studied for different livestock species to have an insight into what exactly it means. Existence of the GxE interaction during the fattening period for tropical Creole beef cattle was confirmed (Vallee 2007). The fattening study in two different environments, intensive and pasture feeding regime have shown that the performance of individuals change with the environmental conditions which also resulted in re-ranking of individuals with varying estimates of heritability. Other studies (Menendez Buxadera and Mandonnet, 2006; Ceron-Munoz et al. 2004) also proved the importance of the GxE interaction in the tropics. In another experiment, Wallenbeck et al. (2009) evaluated GxE for growth rate and carcass leanness in organic and conventional pig production environments. They reported weak GxE for both growth rate and carcass leanness in organic and conventional pig production environments, and some re-ranking of boars' breeding values between environments. In situations where there are weak GxE interactions in organic and conventional production systems, Boelling et al. (2003) argued that it is possible to combine data from both systems and operate a shared breeding program. If there are strong GxE interactions, information from one system would be of little value in connection with the other and separate breeding programs should be preferred (Wallenbeck 2009). Intensive production system has always favored the layer and broiler industry. Genotype x environment interactions have been detected for egg production in layer hens and body weight traits in broiler chickens (Mathur and Horst 1994; Settar et al. 1999). In an experiment in turkeys (*Meleagris gallopavo*) by Case et al. (2010), GxE interaction was studied for egg production. Study indicated that seasonal environmental influences affect egg production, fertility, and hatchability phenotypes to a different extent suggesting a GxE, which was supported by the genetic correlation below 1.0 for each trait expressed in different seasons. Case et al. (2010), therefore, concluded that the egg production, fertility, and hatchability in turkeys could be considered as 2 distinct traits based on season of lay.

The GxE interaction becomes all the more important because of the globalization of breeding program with increased trade in frozen semen and embryos. The INTERBULL system evaluates bulls of six major dairy cattle breeds, viz. Ayrshire, Brown Swiss, Guernsey, Holstein, Jersey, and Simmental, on an international basis wherein 30 countries of 4 different continents are included in the program representing different production system, for example model specialized grazing conditions are represented by New Zealand whereas Australia and South Africa represent the tropical parts of the world. Since the daughters of the same bulls are spread over many a countries and environments, it becomes possible to estimate the genetic correlations of the performances. This takes care of the GxE interactions that exist between varying production system of different regions while estimating the breeding values of the bulls. The program has subsequently been expanded to include calving traits, fertility, and longevity and mastitis resistance as components to production and conformation traits. Applying GxE correlations, this program has so far been successful in selecting bulls according to breeding objectives of each region and country which in turn has assured the maintenance of larger genetic diversity.

The coming era for the animal husbandry will see lot of challenges. Changing climate and therefore changing resources will bring altogether new set of problems for the livestock and livestock keepers to tackle. The genes which are favorable to individuals today may not be of much importance to the animals for tomorrow, and therefore importance of plasticity of the genome will be important. Genotype and environment interaction have a lot to display in the near future for animal husbandry practices.

13.4 Animal Breeding Goals for Sustainable Systems

> Our oldest domesticated animals are still capable of rapid improvement or modification (Darwin, Origin of the species 1859)

An increase in the world population will emphasize an increase in production and productivity per hectare and higher efficiency per unit of product. Regional increases in human welfare will emphasize the need for producing products of high quality, according to local cultural/social requirements (e.g. meat quality). Torp Donner and Juga (1997) reviewed studies suggesting that under current low and intermediate production levels, increased yields and efficiency will be more environmentally sustainable than extensive goals. However, very intense production systems relying on fossil fuels involved in production and transportation carry with them some ecological and economical risks (Heitschmidt et al. 1996).

Table 13.1 Probable characteristics of future agricultural systems and potential animal breeding strategies

Characteristic of development	Animal breeding strategy
Technical and ecological aspects	
Increased human food requirement (larger population and more welfare)	Increase production and productivity; higher efficiency per unit product; increased intake and utilization of non-human-food components; improve product quality
Higher energy and nutrient costs; more use of marginal land	Improve utilization of local feeds; reduce costs by improved health, fertility, and other functional traits; increase intake of (bulky) roughage, and adaptation to low-energy-input systems
Diversification in systems adapted to specific locations and conditions	Reduce environmental sensitivity of animals (increased robustness and capacity of adaptation); diversification of breeding goals
Regulations on compounds such as nitrate and phosphate	Increase biological efficiency in broader terms (not only energy, but also protein and minerals/ elements)
Reduced use of chemical medications	Improve genetic disease resistance in general and tolerance to particular infections and parasites
Genetic and biotechnology methods	Introduce more risk-averse strategies after high-level ethical considerations; aim at low inbreeding and maintaining genetic diversity
Cultural/social and personal aspects	
Concerns about animal welfare	Improve tolerance to metabolic stress; improve health, fertility, and longevity; improve/maintain adaptation to improved management systems (e.g. floor systems for hens).
Use of intellectual property rights	Alliances and cooperation; competitive associations
Increased concern about animal-mediated human diseases	Improve genetic disease resistance in general and tolerance to particular infections and parasites
Privatization of breeding companies, international trade, increased competition	Alliances and co-operation; competitive associations with local or market-oriented and diverse breeding goals, including cultural/social aspects and recognition of personal preferences
Concerns about loss of historical, cultural breeds, and genetic diversity	Develop conservation programs for breeds not under selection (in situ and ex situ); maintain or increase effective population sizes of active breeding populations, and aim for broad breeding goals

Source Olesen et al. (2000)

This illustrates the need for whole-system analysis when considering sustainability in animal breeding. An overall definition of sustainable animal breeding is

> the extent to which animal breeding and reproduction, as managed by professional organisations, contribute to the maintenance and good care of AnGR for future generations (Gamborg and Sandøe 2005).

Francis and Callaway (1993) summarized the elements of sustainability:

Table 13.2 Important traits to include in a breeding goal according to the constraints on feed resources and environmental stress of the production system

Environmental stress	Constrained feed resources	Unconstrained feed resources
High	Adaptability	Adaptability
	Feed efficiency	Productivity
Low	Feed efficiency	Productivity
	Product quality	Product quality

Source Amer et al. (1998)

1) Resource efficiency: most efficient possible use of non-renewable resources and, whenever possible, substituting local renewable resources for those imported from outside the farm;
2) Profitability: economically profitable in both the short and the long term;
3) Productivity: maintaining and enhancing the productivity of all basic resources rather than destroying or degrading them;
4) Environmental soundness: minimal negative impact both on the farm and beyond the farm borders;
5) Social viability: equitable systems favoring owner/operator farms, contributing to viable rural economy, infrastructure, and community, supporting and integrating with overall society.

Key words characterizing sustainable animal breeding are product quality, genetic diversity, efficiency, environment, and animal health and welfare. We need animals that can contribute to optimizing the system. Where land is restricted, we should rather focus on animals that can contribute to increase the farms' production per hectare rather than focusing on increased production per animal. Breeding objectives should be concerned with the individual producer's interest because the producer's primary reason for buying certain breeding stock at a certain price will be based on an assessment of how the animals will contribute to the efficiency of a farm (Harris 1970). For example, in temperate zones, breeding objectives for intensive milk production have, therefore, been developed for producers rather than for national programs; hence, emphasis is put on profit maximization (Pearson 1986). The socio-economic (market) attitude of the decision maker influences the perspective (ontology and epistemology) to be considered and is, therefore, crucial in defining the breeding goal. A common interest for the society as a whole may not be a sufficient incentive for an individual farmer making breeding decisions (Brascamp et al. 1998).

Olesen et al. (2000) listed probable characteristics of future agricultural systems relevant to animal breeding. These are given together with potential animal breeding strategies in Table 13.1. In order to promote breeding goals based on a holistic, long-term perspective, including many of the characteristics of Table 13.1, additional governmental policies are required. Legislation (as the environmental legislation in Western Europe), taxes, and user charges may contribute to such incentives for the individual farmer, especially when, for example, the environmental costs are not captured in the marketplace (Goodland and Daly

1996). Naylor et al. (1998) also stressed the need for incentives through regulation, pollution taxes, or reduction of financial subsidies to improve the efficiency and reduce the environmental impacts of shrimp and salmon farming.

A clear understanding of farmers' perceptions of traits related to animal adaptation to extreme environments is crucial for the design of breeding strategies for climate change adaptation for pastoral communities. Many of the potential animal breeding strategies mentioned (Table 13.1) refer to a broader definition of breeding goals, not aiming at higher production levels per animal only, but balancing higher productivity with improved functional traits such as health, fertility, and feed intake capacity.

13.4.1 Breeding Goal According to the Constraints on Feed Resources and Environmental Stress of the Production System

Genetic improvement is a biological and technological development. The essence of these developments is to improve the efficiency of a production system, saving inputs of production factors per unit product and a change toward the use of cheaper production factors. Saved production factors can either be used in the system where they are saved (and thus expand product output of this system) or can be transferred to another system (e.g. via the market) (Willer 1967). Different constraints on the production system give rise to alternative uses and, therefore, alternative values of saved production factors. Therefore, breeding for high-input systems in only the western and industrialized world is not sufficient, nor is it culturally and socially acceptable simply to transfer such high-input systems to developing countries together with the breeding stock. Table 13.2 summarizes important traits that should be emphasized in a breeding goal according to the environmental stress and feed constraints of the production system. When the feed resources are constrained, feed efficiency traits become more important, whereas adaptation and functional traits are more critical for systems with high environmental stress. Again, adaptation to local natural and social conditions is important. With fewer constrained feed resources and low environmental stress, product quality becomes more important.

13.4.2 Community-Based Breeding: A New Approach to Genetic Improvement

One such approach that has recently stimulated global interest is a community-based breeding strategy. Indigenous livestock management practices through social and rational breeding mechanisms contributed to breed adaptation to harsh environments. Pastoral communities have clear-cut breeding objectives which are multi-faceted, and include the animal's ability to survive in harsh environments. In such situations farmer's participation become beneficial while preparing any breeding project. These programs take into account, from inception, the farmers' decisions and participation which are determinant for their success. These include proper consideration of farmers' breeding objectives, infrastructure, participation, and ownership (Sölkner et al. 1998). Designing a community-based breeding program is much more than genetic theories and increased productivity. It is a matter of infrastructure, community development and an opportunity for improved livelihood of livestock owners through better animals and markets for their products.

13.4.3 Global Approach of Adaptive Traits for Harsh Environment

To address the undesirable effect of environmental stressors, the most productive and adapted animals for each environment need to be identified for breeding purposes. This in turn will ensure increased food production without further expansion of animal numbers with subsequent effect of land degradation (Philipsson et al. 2006). However, increasing productivity including fitness and adaptive traits with minimized environmental impact and maintaining diversity at the same time is not an easy task. Bridging socio-economic demands, often with limiting resources in low-input production system additionally need to consider sustainability issue and maintaining genetic diversity. It has been estimated that 50% of the total genetic variation is accounted for by variation among breeds within species (Hammond 1994). With each livestock species generally performing particular functions, focus of conservation of AnGR has to be within species (Rege and Gibson 2003). It is to be remembered that conservation of AnGR is not justified for the sake of biological resources only, but to contribute to human livelihoods. As it has been normally observed that the breeding programs in the hands of farmers' cooperative, often with governmental support, have been successful for most of the livestock species throughout the world (Ahuya et al. 2004). Continuity of this success trend as well as their participation can be ensured provided the breeding program is flexible, integrated, simple, and reliable. The principle of 'KISS' (keep it simple to be sustainable) has been emphasized by Philipsson et al. (2006). Since there is a natural stratification of livestock breeds by climatic zones, strategy shall be to identify a suitable breed from the available indigenous breeds for which population structure, detailed phenotypic descriptions, and relationship between the breeds must be known. It has also been observed that improving indigenous breeds and progressively substituting the breed with another indigenous breed as the strategies for genetic improvement of cattle in the tropics have largely been ignored (Payne and Hodges 1997; Philipsson et al. 2006). In addition, the additive genetic variation which seems to be large within the indigenous breeds has not been exploited optimally.

As has been discussed earlier, whether to select animals for important traits e.g. production, reproduction along with adaptive traits in harsh environment, is debatable. It is evident that physiological adaptability is expressed in performance, hence the question arises whether selection of animal on the basis of performance in given environment (stressors) alone will be sufficient or not. Favorable correlation suggests that if we place major importance on performance traits (e.g. reproduction, growth etc.) in stressful environments, adaptability traits would not be compromised (Burrow et al. 1991) and will lead selection of most favorable animals (McDowell 1972; Turner 1984). Addition of adaptive traits as such might not be that rewarding because firstly, estimates of heritability of adaptability traits in most of the population is not available; secondly, since genetic correlation between adaptive and productive traits is normally not very high, genetic progress for individual traits may slow down with inclusion of additional traits. Analysis of the history of breeding schemes of developing countries suggests that the breeding schemes shall be kept as simple as possible avoiding complicated selection criteria (Kosgey et al. 2006).

13.4.3.1 Breeding for Climate Change Adaptation and Mitigation

The IPCC has defined “Adaptation” as “initiatives and measures to reduce the vulnerability of natural and human systems against actual or expected *climate change* effects”. There are many ways in which producers can adapt to climate change.

Breeders can follow two alternative strategies: adapt the environment to the need of the animals as is the case in industrial animal production systems or keep animals that are adapted to the respective environment as is the case in low-input smallholder and pastoral systems.

Although the direct effects of climate change on the animals are likely to be small as long as temperature increases do not exceed 3°C (Easterling et al. 2007), projections suggest that further selection for breeds with effective thermoregulatory control will be needed. This calls for the inclusion of traits associated with thermal tolerance in breeding indices, and more consideration of GxE to identify animals most adapted to specific conditions. Breeding for climate change adaptation will not be fundamentally different from existing breeding programs; however, the problems related to measuring the phenotypes relevant to adaptation have to be overcome. In the past decade, breeding goals in many commercial breeding programs have broadened through changes in selection indices and aim to improve production, longevity, and functional traits simultaneously—for example in dairy cattle, pigs, and layer chickens (Wall et al. 2008). As a result, correlations between breeding values with broader indexes that include functional traits are lower than those with production traits only (Mulder et al. 2006).

Correlations between the performances of genotypes in different environments are less than unity because of GxE, differences in trait definitions, and differences in data collection and analysis procedures (Zwald et al. 2003). Such correlations between breeding values are lower in high temperature countries suggesting that heat stress plays an important role in GxE (Zwald et al. 2003). Biologically important GxE is assumed if the correlations between the performance of the same genotype in different environments are below 0.8 (Robertson 1959). A single breeding program with progeny testing of sires in different environments and applying index selection to simultaneously improve performance in those environments is recommended for genetic correlations between environments higher than 0.6. At lower genetic correlations between environments, environment-specific breeding programs are necessary to breed for special adaptability (Mulder et al. 2006).

Heat Tolerance

While substantial differences in thermal tolerance lie between species, there are also differences between breeds of a species. Ruminants generally have a higher degree of thermal tolerance than monogastric species.

Finocchiaro et al. (2005) proposed the use of heat-resistant individuals in a sheep breeding program as a main strategy to improve animal welfare and productivity in hot climates. Various physiological and blood parameters differ between local and exotic cattle breeds in Brazil (McManus et al. 2008). Collier et al. (2008) suggested that there is some opportunity to improve heat tolerance through manipulation of genetic mechanisms at cellular level. Selection for heat tolerance in high-output breeds based on rectal temperature measurements and inclusion of a THI in genetic evaluation models are promising. Different parameters, such as THI or dry bulb temperature measurements, are used as indicators for heat stress (Finocchiaro et al. 2005; Bohmanova et al. 2007; Dikmen and Hansen 2009). Different THI definitions were found prefable in the US, depending on the extent of natural and artificial evaporative cooling (Freitas et al. 2006; Bohmanova et al. 2007). The genetic variance due to heat stress was substantial at high THI (Ravagnolo and Misztal 2002).

However, in the dairy sector it may be difficult to combine the traits desirable for adaptation to high temperature environments with high production potential because there seem to be different physiological and metabolic processes controlling heat tolerance and milk yield on one hand, and heat tolerance and reproductive performance on the other (Ravagnolo and Misztal 2002). In beef cattle, the genetic antagonisms between adaptation to high temperature environments with high production potential seem to be more limited than in dairy, and improved characterization of adaptive traits, use of reproductive technologies and molecular markers, and strategic crossbreeding are being incorporated into programs, for example in the Australian Beef Cooperative Research Centre (Prayaga et al. 2006).

Productivity and Feed Efficiency

Increasing productivity is a condition *sine qua non* for all production systems because of the need to make efficient use of the available inputs and to reduce the livestock sector's environmental footprint. In addition to selection for increased production per se, any selection that reduces mortality and increases early maturity, fertility, and longevity tends to contribute to reducing greenhouse gas (GHG) emissions per unit of output (Hoffmann 2010). Breeding for high performance and improved FCR, and reduced mortality due to better hygienic management, have significantly reduced the amount of feed (and land needed to produce this feed) per unit of product—more in monogastrics and in dairy cattle than in beef cattle, or sheep (Flock and Preisinger 2002; Capper et al. 2009). Jones et al. (2008) investigated the role of genetic improvement in a Life-Cycle-Analysis model in the UK and found that the annual reduction in GHG emission ranged from 0.8% in pigs and dairy cattle to 1.2 and 1.3% in broilers and layers, respectively. The largest contributions in broilers came from improved FCR, and in pigs from improvements to growth rate and fertility. Genetic gain in milk performance has considerably reduced the environmental impact of diary production in the USA (Capper et al. 2009). Future options for selection in ruminants lie in the host components of rumen function, in post-absorption nutrient utilization and in disease resistance. In pigs and poultry, the genetic variation in digestion parameters can be exploited (Warkup 2007). In addition to potential for fertility improvements in ruminants, for example decreasing the age of first-calving in zebuine cattle, there is sufficient genetic variability in feed intake, independent of liveweight and daily gain (Flachowsky and Brade 2007) to permit selection for this trait. Assuming that future dairy systems may become more reliant on pasture than grain feeding, Hayes et al. (2009) proposed to select sires whose daughters will cope better with low feeding levels and higher heat stress. They identified markers associated with sensitivity of milk production to feeding level and THI in Jersey and Holstein. Because feed-efficient animals are also more cost-effective and productive, the Australian beef industry now includes net feed efficiency as an integral part of its breeding program (Beef CRC 2008). Alford et al. (2006) calculated that CH_4 could be reduced by up to 16% in 25 years if residual feed intake (RFI) were included in beef selection programs. Initial costs to identify individuals with low RFI are however, high, particularly in grazing animals (Arthur et al. 2004). Because of the above-mentioned differences in feed quality, productivity improvements in pasture-fed ruminants in the tropics will result in higher relative CH_4 reductions than in ruminants grazing more digestible temperate pastures (McCrabb and Hunter 1999). Possible synergies between plant and animal breeding need to be better developed (FAO 2008).

Disease Resistance

Experiments in domestic species have shown that there are often genetic differences in response to disease challenge (Bishop et al. 2002). Some of this variation is caused by single genes and some by multiple genes each with small effect. There is potential for genetic improvement of disease resistance, and various commercial breeding programs already include resistance against helminthosis, ticks, mastitis, *E. coli*, or scrapie. FAO (2007a, b) lists breeds, mainly from developing countries, that are reported to be resistant or tolerant to trypanosomiasis, tick burden, tick-borne diseases, internal parasites, dermatophilosis, or foot rot (59 cattle breeds, 33 sheep breeds, 6 goat breeds, 5 horse breeds, and 4 buffalo breeds). Again, many of these reports are based on anecdotal evidence rather than scientific studies, and the underlying physiological and genetic mechanisms are not well understood. Various studies have been undertaken to map genes (e.g. Regitano et al. 2008) and study gene expression (e.g. Berthier et al. 2008) in relation to these diseases, but no reports verifying causal mutations have been produced.

Extensive research on the genetics and breeding for worm resistance has been carried out in Australia, New Zealand, South Africa (de Greef 2009), and also in India (Singh et al. 2009). Tick counts, fecal worm egg counts (FEC), rectal temperatures, and coat scores have been used as indicator traits of adaptability of beef cattle to assess the suitability of particular genotypes to tropical environment. Singh et al. (2009) evaluated Avikalin (Rambouillet x Malpura) breed for fecal egg count at naive and exposed stage of natural infection (predominantly *Haemonchus contortus*) and two divergent lines were created by selecting progenies from sires with low (R line) and high (S line) mean fecal egg count (FEC). The performance of selected animals from both the lines was compared over the years and no untoward effects on different performance traits were observed. Breeding for disease resistance depends very much on the type of disease and the hosts' resistance or tolerance mechanisms, the availability and costs of alternative treatment (e.g. vaccines, drugs) and antimicrobial resistance of pathogens. In any case, the importance of molecular markers and marker-assisted selection in such breeding programs will increase (Bishop et al. 2002; Prayaga et al. 2006).

Studies in sheep to detect quantitative trait loci (QTL) for nematode resistance or detect associations with candidate genes are now well advanced in NewZealand, Australia, Kenya, US, and Europe (UK, France, Italy, and Spain) although results are not readily available in the public domain (Bishop and Morris 2007) and much success has not been achieved in this area (e.g. Marshall et al. 2009). Several studies have also looked at associations between specific genes or markers and FEC. Coltman et al. (2001) found significant associations with a microsatellite within the interferon gamma gene in feral sheep, and various associations with microsatellites in or near the major histocompatibility complex (MHC) have been observed (Schwaiger et al. 1995; Janssen et al. 2002).

In many a studies, somatic cell count (SCC) has been used as a marker to select for increased resistance to mastitis though recent estimates of heritability for SCC are found to be generally low ranging from 0.1 to 0.2 (Legarra and Ugarte 2005;

Bishop and Morris 2007). The review by Mrode and Swanson (1996) summarized many genetic estimates, concluding that the heritability of mastitis incidence in dairy cattle is low ($\sim$0.04), as also is the heritability of SCC (0.11 $\pm$ 0.04), but the genetic correlation between the two is high at $\sim$0.70. Attention is now turning to the mapping of QTL for SCC in dairy ewes, in both experimental crosses and commercial breeding programs, as described by Barillet et al. (2005).

Trypano-tolerant N'Dama cattle respond more rapidly and with a greater magnitude to infection compared to the trypano-susceptible Boran cattle. Gorman et al. (2009) showed major gene expression differences exist between cattle from trypano-tolerant and trypano-susceptible breeds. Using bovine long oligonucleotide microarrays and real-time quantitative reverse transcription PCR (qRT-PCR) validation, they analyzed peripheral blood mononuclear cells (PBMC) gene expression in naive trypanotolerant and trypanosusceptible cattle experimentally challenged with *Trypanosoma congolense* across a 34 day infection time course. Trypanotolerant N'Dama cattle displayed a rapid and distinct transcriptional response to infection, with a 10-fold higher number of genes differentially expressed at day 14 post-infection compared to trypano-susceptible Boran cattle.

Resource Allocation Theory

The resource allocation theory argues that under selection within a particular environment, the resources used by the animal are optimally distributed between the important traits for breeding and production within that environment (Beilharz et al. 1993). This implies that any additional selection mediated increase in performance of a production-related trait, without a concurrent increase in resources, must lead to decline in other traits due to a re-allocation of resources. The decrease in these traits is proportional to the heritability of the "allocation factor", defined as the proportion of resources devoted to production vs. fitness (Van der Waaij et al. 2004).

13.4.4 Genes Involved in Stress

13.4.4.1 Heat Shock Factors

A transcription factor family known as the heat shock factors (HSF) has been implicated as important first responders during the onset of elevated cell temperature (Page et al. 2006). They regulate HSP expression through interaction with a specific DNA sequence in the promoter—the heat shock element (HSE). The HSE is a stretch of DNA located in the promoter region of susceptible genes containing multiple sequential copies (adjacent and inverse) of the consensus pentanucleotide sequence 5′-nGAAn-3′ (Morimoto 1998) and has been found in both HSPs and in a number of other genes. These transcription factors coordinate the cellular response to thermal stress and affect expression of a wide variety of genes including HSPs (Akerfelt et al. 2007). Three HSFs have been identified in mammalian systems: HSF-1, HSF-2, and HSF-4 (Morimoto 1998). A fourth HSF, HSF-3, is present in avian species but not in humans. The HSF1 gene has been mapped to chromosome 14 in cattle (Winter et al. 2007). HSF-1 is involved in the acute response to heat shock; after activation by thermal stress, HSF-1 is found primarily in the nucleus in trimeric form, concentrated (in human cell lines) in granules (Sarge et al. 1993). It is this activated trimeric form of HSF-1 that binds to the HSE and is involved in increased HSP gene transcription during heat stress (Sarge et al. 1993), hence primarily responsible for inducing HSP gene expression during hyperthermia (Pirkkala et al. 2001).

13.4.4.2 Heat Shock Proteins

HSPs were originally identified as proteins whose expression was markedly increased by heat shock (Lindquist 1986). Several HSPs are expressed even in unstressed cells and play important functions in normal cell physiology. Although the intensity and duration of the heat stimulus needed to induce HSP expression vary considerably from tissue to tissue, a typical in vitro exposure involves heating mammalian cells to 42–45°C for 20–60 min and then reverting them to normothermic temperatures (37°C). The HSPs are traditionally classified by molecular weight. As proteins, HSPs possess three principal biochemical activities. The first activity is chaperonin activity (Georgopoulos and Welch 1993). HSPs with this function help prevent misaggregation of denatured proteins and assist the refolding of denatured proteins back into native conformations. The prototypical chaperonin HSPs are the members of the HSP40, HSP60, HSP70, and HSP90 families of proteins. The second activity is regulation of cellular redox state, of which the best example is HSP32, better known as heme oxygenase-1 (HO-1) (Otterbein and Choi 2000). The third principal biochemical activity of HSPs is regulation of protein turnover (Parsell and Lindquist 1993). An example is ubiquitin, which is expressed in unstressed cells, up-regulated by heat shock, and that serves as a molecular tag to mark proteins for degradation by proteasomes.

A study conducted by Basiricò et al. (2011) was aimed to investigate the association between inducible Hsp70.1 SNPs and heat shock response of PBMC in dairy cows. Results indicated that the presence of SNPs (C/- and G/T) in the 5'-UTR region of inducible Hsp70.1 ameliorates heat shock response and tolerance to heat of bovine PBMC. These mutation sites may be useful as molecular genetic markers to assist selection for heat tolerance.

13.4.4.3 Halothane Gene in Pigs

The porcine stress syndrome (PSS) gene is found to cause malignant hyperthermia (heat shock) when pigs are exposed to environmental stressors. This gene is commonly referred to as the halothane gene because the PSS condition can be triggered by exposing pigs to the anesthetic, halothane gas. The PSS condition is an inherited disorder in swine. Inheritance of the PSS gene is at a single locus and there are two alleles, one dominant (N) and one recessive (n). Swine inheriting the n allele are more sensitive to stress. The nn pigs are known as stress susceptible and typically produce PSE meat, whereas NN ones produce lesser (Monin et al. 1999). PSE meat is characterized by its pale color, lack of firmness, fluid (exudates) dripping from its cut surfaces and denaturing of muscular proteins (Franck et al. 2003). The nn pigs also develop malignant hyperthermia which is a muscle disease characterized by an abnormal response to stress, exercise, and anesthetics (Klip et al. 1986; Monin et al. 1999). In most purebred pedigrees this trait is selected against and not allowed in the breed.

Khazzaka et al. (2006) highlighted the relationship between HAL genotypes, Hsp70, and Ca^{2+} responses to a heat stress. HAL genotypes significantly affected the Hsp70 induction after HS. The concentration of Hsp70 increased significantly in NN pigs' blood, 2 h after the heat stress start whereas this significant increase occurs from 1 h after the stress induction starts in Nn ones. The Hsp70 expression in nn animals did not differ significantly from the basal concentration after heat stress. Normal (NN) and heterozygote (Nn) had increased levels of Hsp70 in blood after heat stress indicated that the presence of one allele N probably plays a role in promoting cell survival in response to stressful conditions.

13.4.4.4 Slick Hair Gene in Cattle

Evidence was found that supports the existence of a major gene (designated as the *slick hair* gene), dominant in mode of inheritance, that is responsible for producing a very short, sleek hair coat (Olson et al. 2003). Cattle with slick hair were observed to maintain lower RTs. The gene is found in Senepol cattle and Criollo (Spanish origin) breeds in Central and South America. This gene is also found in a Venezuelan composite breed, the Carora, formed from the Brown Swiss and a Venezuelan Criollo breed. Data from Carora × Holstein crossbred cows in Venezuela also support the concept of a major gene that is responsible for the slick hair coat of the Carora breed. The effect of the *slick hair* gene on RT depended on the degree of heat stress and appeared to be affected by age and/or lactation status. The decreased RT observed for slick-haired crossbred calves compared to normal-haired contemporaries ranged from 0.18 to 0.4°C.

13.4.4.5 Nramp1 Gene

Natural resistance associated macrophage protein 1 (*Nramp*1, previously known as *Lsh/Ity/Bcg* gene) has been identified as a major gene in many species (Vidal et al.

1993). *Nramp*1 gene is expressed in late endosomes of macrophages. Functional studies have revealed that *Nramp*1 regulates antimicrobial activity of macrophages (Barton et al. 1995). In mouse, a point mutation in the coding region of the gene *Nramp*1 causes a single amino acid change from Gly to Asp at position 169 resulting in a susceptibility phenotype of mice in the early phases of infection with *S. enterica* serovar *Typhimurium*, *Leishmania donovani*, or various species of *Mycobacterium* (Vidal et al. 1995). Recently $(GT)_n$ microsatellite polymorphism at 3′-UTR in *Nramp1* has been found to be associated with resistance/susceptibility to *Brucella abortus* (Adams and Templeton 1998; Barthel et al. 2001), Salmonella, Paratuberculosis (Pinedo et al. 2009) infection in cattle and buffalo (Ganguly et al. 2008).

13.4.4.6 Other Genes

The number of reported genes involved in the cell stress response to heat is rapidly increasing, as more is learned about the cellular processes affected by HS and those that interact with HSPs. Additionally, gene chip arrays, which allow researchers the ability to investigate the expression of thousands of sequences simultaneously, are likely to substantially increase the number of genes known to be involved in the cellular response to HS. Several gene chip array experiments have been performed that specifically examined the role of HS on gene expression. An experiment reported in 1996 (Schena et al. 1996) used an array containing 1,000 genes to examine changes in gene expression in human T cells and demonstrated the feasibility of using this technology to identify new candidate heat responsive genes. Approximately 50 genes not traditionally considered to be HSPs have been found by Sonna et al. (2002) to undergo changes in expression in response to heat stress. Several genes with well established physiological functions have been reported to be induced by cold stress. Two genes likely to play important roles in cell physiology during cold exposure are cell cycle proteins p53 and p21, which are induced by exposure to temperatures that produce cell cycle arrest (Matijasevic et al. 1998; Ohnishi et al. 1998). Importantly, p53 appears to play an integral role in the cold-induced cell cycle arrest, as cells lacking a wild-type p53 gene escape this arrest both in human fibroblasts (Matijasevic et al. 1998) and in glioblastoma cell lines (Ohnishi et al. 1998).

Liu et al. (2010) examined the genetic polymorphism of the *ATP1A1* gene in 160 Holstein cows using strength single-strand conformation polymorphism and DNA sequencing methods. *G* to *A* at position 14,103 in exon 14 and *C* to *T* at position 14,242 in intron 14 of the bovine *ATP1A1* gene were identified having a significant correlation between *ATP1A1* gene polymorphism and the coefficient of heat tolerance ($P < 0.01$) and with respiratory rate ($P < 0.01$). Genotype *AC* was the most favorable genotype for heat tolerance.

In poultry there are several genes that affect heat tolerance. Some genes, such as the dominant gene for naked neck (Na), affect the trait directly by reducing feather cover, while others, such as the sex-linked recessive gene for dwarfism (dw), reduce body size and thereby reduce metabolic heat output. The frizzle (F) gene may be useful in stocks that have to perform under hot humid conditions. This gene reduces the insulating properties of the feather cover (reduced feather weight) and make it easier for the bird to radiate heat from the body.

13.4.4.7 QTL for Stress

A QTL is a genomic region that is associated with or modulates the variation in a measurable phenotype (Williams et al. 1998). A QTL analysis is an invaluable tool for examining the genetic bases for quantifiable and complex phenotypic measures. In understanding genetic variation associated with complex phenotypes, it is believed that a multitude of genetic loci where each share a relatively small modulation of such traits (Lande 1981) are involved with some of these loci expected to have a more substantial effect on observed phenotypes (Lai et al. 1994). In particular, certain QTLs may account for a comparatively large portion of the variance in different complex traits, which could be analyzed by studying the pattern derived when separating alleles from opposing spectrums at a major QTL (Williams et al. 1998).

13.4.4.8 Stress and Anxiety QTL

Numerous studies have performed QTL analyses to investigate genomic regions that may be associated with stress and anxiety phenotypes. As stress and anxiety are complex phenomena, researchers have examined various behavioral and physiological measures of stress and anxiety using mouse or rodent models. When examining the genetic basis of stress and anxiety using QTL analyses, numerous researchers consistently found significant QTLs on Chromosme 1, 4, and 15. Significant loci on Chromosome 15 have been shown to influence anxiety-related behaviors (Flint et al. 1995; Turri et al. 2001), emotionality (Flint 2002), and restraint stress (Tarricone et al. 1995), as well as avoidance behavior (Turri et al. 2001), suppression of activity in anxiogenic or threatening environments (Flint 2001; Henderson et al. 2004), stress-induced hyperthermia (Thifault et al. 2008), and acoustic startle response (Joober et al. 2002). With regard to significant QTLs on Chromosome 4, significant association have been found to influence stress responsiveness, including corticosterone levels 7 h after EtOH administration (Roberts et al. 1995), stress-induced hypothermia (Thifault et al. 2008), and tail-suspension test induced hyperthermia (Liu et al. 2007).

Many studies have found significant QTLs on Chromosome 5 associated with stress and anxiety-related phenotypes, including restraint stress open field rearings (Tarricone et al. 1995), corticosterone levels 1 and 7 h following EtOH administration (Roberts et al. 1995). In addition, a study by Dai et al. (2009) revealed that the GABAA receptor $\alpha 2$ (*Gabra2*) gene on Chromosome 5 plays a critical role in the stress response.

Collier et al. (2006) evaluated the effects of thermal stress on mammary development and gene expression by microarray technique using a bovine mammary epithelial cell collagen gel culture system. Acute thermal stress was characterized by inhibition and regression of the ductal branches. Gene expression profiling revealed an overall upregulation of genes associated with the stress response and protein repair. In contrast, genes associated with cellular and mammary epithelial cell-specific biosynthesis, metabolism, and morphogenesis were generally downregulated by thermal stress.

Lilja et al. (2000) identified four QTL with significant effect, in F2 generation of wild boar—Yorkshire intercross, one influencing stress-induced alterations in number of neutrophils (chromosome 8), one influencing spontaneous proliferation after stress (chromosome 2), one influencing mitogen-induced IL-2 activity after stress (chromosome 6) and one influencing stress-induced alterations in mitogen-induced IL-2 activity (chromosome 12). The stress protocol induced a decrease in number of circulating neutrophils and in spontaneous proliferation in vitro, whereas phagocytic capacity, mitogen-induced proliferation and spontaneous IL-2 activity increased.

In poultry as an alternative to vaccination control, increased genetic resistance to Marek's disease (MD) represents an attractive solution for lowering disease outbreaks. Genetic mapping of QTL affecting susceptibility to MD virus-induced tumors was conducted by Vallejo et al. (1998) who for the first time reported the mapping of non-MHC QTL involved in MD susceptibility in chickens. Two significant and two suggestive MD QTLs were detected in four chromosomal regions, explaining 11–23% of the phenotypic and 32–68% of the genetic MD variation. A genome-wide scan using 119 microsatellite loci allowed Zhu et al. (2003) to map QTLs associated with resistance to avian coccidiosis to GGA1. QTL associated with immune response to SRBC, Newcastle disease virus and *E. coli*, and with survival were investigated by Yonash et al. (2001). Three markers were shown to be significantly related to these traits.

The genetic variation in many aspects of host resistance to nematodes is well documented. However, the use of molecular information (mainly QTL) that has long been advocated as a promising tool to improve difficult traits, contributed little to genetic improvement in livestock breeding schemes up to now (Bijma 2009). Reasons are that QTL have often been detected in experimental crosses, the number of QTL soon becomes impractically large, linkage phase between markers and QTL may change over time, and QTL effects may change over time. A major step forward will come from the implementation of marker-assisted breeding value estimation (MA-BVE) using dense maps covering the entire genome (Meuwissen et al. 2001). Furthermore, microarray studies do have the ability to detect genes differentially expressed between 'resistant' and 'susceptible' animal, with pathways implicated in these differences including the development of acquired resistance and the structure of the intestinal smooth muscle (Diez-Tascon et al. 2005). The interest in MA-BVE approach is solely in the breeding value of the candidates with the objective to estimate the breeding value with the highest possible accuracy using all phenotypic and genomic information. There is no interest in the location or effect of individual QTL. Using a mixed model, the method gives an estimate for each marker haplotype and the breeding value of an individual is the sum of the effects of its markers or haplotypes (Bijma 2009).

13.4.5 Developing Breeding Schemes to Assist Mitigation

In today's world, livestock production systems have a dual role, not only in food production, but also in the provision of public goods objectives including biodiversity and landscape value. However, agriculture also generates external costs or negative public goods; for example, diffuse pollution to air and water. Mitigating GHG emissions from livestock is increasingly recognized as a necessary part of meeting worldwide climate change obligations.

Essentially there are five main routes, identified by Wall et al. (2010), through which genetic improvement can help to reduce GHG emissions:

13.4.5.1 Improved Productivity and Efficiency

Typically, selective breeding can achieve annual rates of response of between 1 and 3% of the mean in the trait (or index) under selection (see Simm et al. 2004 for a review). Selection for productivity and efficiency helps mitigate GHG production in two ways. Firstly, higher productivity generally leads to higher gross efficiency as a result of diluting the maintenance cost of the productive (and non-productive) animals, for example the select line of Holstein dairy cows in long running Langhill experiment have 17% higher yield per lactation and a 14% higher gross efficiency (Veerkamp et al. 1995). Secondly, a given level of production (e.g. a national milk quota) can be achieved with fewer high yielding animals and followers.

13.4.5.2 Reduced Wastage

Many fitness traits have been shown to have a genetic component and so there is scope to improve them via genetic selection. Selection for fitness traits (e.g. lifespan, health, fertility) will help to reduce emissions by reducing wastage of animals. Improving lifespan in dairy cows and maternal line animals (i.e. ewes and beef cows) will reduce wastage by reducing the number of followers. Improving health and fertility will reduce involuntary culling rates. This reduces emissions from dairy, beef, and sheep systems (increased maternal survival) by reducing the numbers of followers required.

13.4.5.3 Direct Selection to Reduce Emissions

Direct selection for methane emissions and their reduction would ideally be based on direct measurement of methane emissions. It is important to note that direct measurement of all sources of methane emissions from individual animals (exhaled by the animal due to enteric fermentation, manure management, and flatulence) may prove difficult. In pig, there is the example of the 'Enviro Pigs'

(Trade marked) developed at Guelph in Canada which have been genetically modified to excrete 60% less phosphorus.

13.4.5.4 Developing New Indices for Selection on Emissions

Broader breeding goals have become the norm in many livestock species, i.e. selection is usually on a combination of production and "fitness" (health, fertility, longevity) traits. Breeding goals can be built in a number of ways including the popular method of creating an index by weighting traits by their relative economic value (REV). But, the 'valuation' of traits may be complex as there are several different scientific approaches. These include the use of restricted or desired gains, the use of REVs, the adaptation of economic values based on 'conjoint' analyses and total farm modeling, where trait changes are related to environmental impact.

13.4.5.5 Exploiting Genetic Resources

There is a growing awareness about the potential long-term importance of domestic AnGR. One of the key reasons for maintaining these resources is that a 'pool' of genes and gene combinations will be available for the future, acting as an insurance policy if 'modern' genotypes fail in a particular environment or have low gene frequencies in a desired trait.

13.5 Future Considerations

13.5.1 Resource Conservation

Future of the animal production system has lot of challenges to face. Changing climate, temperature variation, genetic erosion, mechanization of the agriculture and heavy industrialization will have a lot of impact on deciding and framing the future policies regarding animal husbandry practices worldwide. Genetic adaptability of livestock to harsh climate will probably see the narrowing of the genetic resources, as fewer breeds with more tolerant genes will have upper hand in the era of changing climate. Apart from this, in the environmentally protected husbandry practices, where intensification of the animal production system is practised, there is a chance that only few breeds with higher genetic potential for production traits will be retained. Multinational companies with specialized breeding objectives as in case of poultry breeding will certainly have a very limited and narrowed down genetic diversity. This is so because a breed with high genetic merit for a particular trait will be more beneficial to maintain for an entrepreneur than to maintain a stock of genetically diverse group of animals. This is clearly seen in poultry industry where White leg horn breed of poultry is preferred by many companies for their high egg laying capacity. Within poultry stocks, very large population sizes are coupled with intense selection. Effective population sizes remain substantial (Warren 1996), but rates of inbreeding are generally not reported (Notter 1999). Similar threats are faced in pig industry. Reduced breed diversity in U.S. swine seems inevitable. The industry which has moved toward a single confinement production system and a single lean pork market (Notter 1999) will pose similar threats. Certainly it augurs there is a threat well speculated that this kind of monotonous genome rule will always be susceptible to the sudden wipe out of the resources due to an unpredictable epidemic. We have already experienced this kind of havocs in recent past as in poultry industry where whole population of birds was at stake due to bird flu. One major reason behind these kinds of threats is loss of genetic diversity or plasticity of the population.

Demand for the variability of the genetic resources is in waiting and its importance is really much sought after. Diversity of the production milieu, the importance of adaptation to the variety grazing conditions, the production of multiple products such as meat, milk, and fiber, medicinal products and special ethical values from particular animals, social value of the livestock to particular community requires and asks for the use and maintenance of various genetic resources. Although intensification of the animal production system seems to take us toward reduced genetic diversity, but efforts to maintain it are urgently required. Notter (1999) suggested some of the recommendations for maintenance of the genetic diversity in US, which can be modified and made suitable for whole animal genetic diversity in the world. Development of the appropriate organizational structures and adequate funding levels to support a worldwide animal germplasm system can be a first step. Efforts by FAO and many other government and non-government agencies are already in pipeline to raise the funds and maintain the livestock genetic biodiversity. However, still major thrust in some Asian and African countries is required, which of course needs a worldwide support financially as well as intellectually. Another major step in this regard as Notter (1999) puts is the close monitoring for the rates of genetic change and inbreeding in intensively managed, highly selected populations and to develop efficient, strategic mechanisms for conservation of genetic diversity. This strategy is very important and in many farms where intensive selection is being practised, the monitoring of the inbreeding rates is well accounted for. Apart from this global awareness regarding genetic diversity of animals, and understanding the needs of real protector of the livestock diversity that is farmer himself in developing countries is very essential. Programs which mutually benefit the farmers and the objectives of maintenance of animal genetic diversity must go hand in hand.

13.5.2 Animal Welfare, Intensification, and Logical Decision-Making

During last five decades or so especially in the western part of the world, intensification of the animal production system has taken place, which has definitely perpetuated to rest of the world, but not with the same zeal. Poultry, rabbit, pig were the species which were benefited with the concept of intensification, whereas large animals took a pretty long time to enjoy these benefits. In the south and eastern part of the world, bovine farming is still an extensive system task, and in very few places it has been intensified. Similarly for sheep and goat the scenario of the production system has not changed till now, and by large they are reared on extensive system of management. In the event of adaptability of livestock to harsh environment, the intensification of the animal production system has a greater role to play along with several other issues of animal welfare and decision making in the animal production system.

Adaptability of the livestock to the environmental stresses is a complex phenomenon. Various mechanisms suggested for this task such as environmental manipulation, animal system modification, and genetic manipulation leads to several debates regarding animal ethics and welfare. The animal production system for adaptation in the harsh environment has lot of facets. As far as environmental manipulation at micro level is concerned, confinement of the animals is very important aspect. Changing temperature and increasing stress factors add to the hurdles in the production system. Therefore, the confinement of the animals to a well-suited place which is controlled for environmental factors can be a better alternative. Also the factors responsible for care and management of the animals such as detection and treatment of the diseases and prevention of the major diseases along with proper feed and fodder supply will serve to ameliorate the situation. The world is divided in broadly two parts, one where resources are not the problem in animal production system and most of the western part of the world falls in this category. The poor countries and developing part of the world are still held up with the problems of too many mouths and less food. So in this particular part of the world animal production system cannot be improved from the extensive style to intensive style, excluding the poultry or similar profit fetching sectors.

In the other strategy where genetic manipulation of the animals for making them more adaptive for environmental stresses is concerned, the task is again uphill. As discussed in the manuscript, the change brought up by this method will be permanent and hopefully fruitful. Many projects for developing the more resistant and tough-hardy breeds with no loss of productivity will be blessings in the coming future.

Decision making for animal production system is very important characteristic where a producer craves for the profit and he also needs to keep in his mind the animal ethics and welfare issues. In such a scenario, the future of the animal production system especially for the intensification for specially derived traits will depend much on how logically we make decisions. Inclusion of the animal welfare issues in the intensive animal production system were suggested by Fraser (2005). He suggested that strong animal care values can be encouraged and sustained in the intensive animal farming. Producers may also be protected from the market pressures which force the cost cutting strategies leading to animal welfare violations. Organic concept of farming with open non-confinement animal production system can also be an alternative where farmers of traditional agrarian mentality can take the lead. Logical decision-making provides a structured and sequenced approach to decision making. Such an approach can help to ensure discipline which subsequently builds consistency into the decision-making process. Well structured plan for implementation of the animal welfare program in the intensive production systems will lead to better future of inclusive interest.

13.5.3 Research Needs

With regards to future considerations in animal production systems, there are several research issues which must be given due importance. Coming future will have a burning basic issue to solve in front of all the research and development agencies, and that will be food resources. Especially in developing and least developed part of the world, the increasing human population and decreasing food reserves, along with challenges posed by climate change will have a war-like situation, for which we need to search solutions right now. Change in the structure of the population, variability crisis in AnGR, food resource scarcity for livestock as a result of vanishing of common property resources (CPR) will add much to the problems associated with food animal husbandry practices in developing countries. However, for developed part of the world, the situation may not be so severe. Building and maintenance of the CPRs with local bodies and their sustainable use seems to be a very important area of research. Many economists and agro-foresters do work on this issue, but government forced agencies with will and capital needs to work on this issue to have better future of animal husbandry in developing countries. Apart from this, researchers also see a major problem associated with the animal husbandry sector, and that is production of the GHG. The major research force must help to strike a balance between emissions of the GHG from the livestock in coming future. There is one more challenge due to climate change in coming future; it may bring changes in disease epidemiology. They are likely to increase the importance of genetic resistance and tolerance in disease control strategies. So epidemics of the disease, its re-emergence in new parts of the world as suggested already in this chapter, its prevention and cure pose a new area of research in coming future. Intensification of the animal production system will demand a lot of technological know-how, and therefore it will also be an area to be explored by researchers keeping pace with climate change. One of the most important researchable issues will be maintenance of the global animal genetic diversity.

13.5.4 Challenges and Opportunities

Future has a lot of surprises in waiting for us. As discussed throughout this chapter, future of the animal husbandry keeping in view the changing climate and adaptation for the environmental stress will be challenging both for the animals as well as livestock holder and common man. Livestock breeding and production systems are complex and knowledge intensive. The question posed by Hoffmann (2010) can be put forth here again as the biggest challenge in coming future. *How can AnGR cope with and adapt to climate change while continuing to contribute to food security and rural livelihoods*? Now this remains the major challenge ahead of us, around which all other issues revolve. With change in the climate, the temperate countries think of introducing the genetic component from tropical animals into their breeds by strategic crossbreeding. Now this seems to be a big challenge as this is not a small task to be done. However, it has a big opportunity of taking the whole of animal husbandry to a new scale, where animals will have better adaptive capacity in the changing climate. Such changes in the species or breed mix may potentially lead to a reverse in the current flow of genetics (Hoffmann 2010). Another major challenge with regards to climate change and adaptation to environmental stress is disease management, which will be the real punishment for farmer as well as exchequer. Rural livelihood is totally dependent on the agriculture and animal husbandry practices, and in most of the places livestock is integral to the socio-economic and cultural way of life of the rural folk. Sustainability of the livestock therefore is very much important for maintaining the lifeline of rural society. Nevertheless future has lot of challenges for us, but it has also hidden opportunities for experiencing altogether new facet of the animal husbandry practices. We have a chance to mitigate the effects of the climate change, and to sustain the AnGR. Need to search for alternate feed resources for animals, heat tolerance, disease resistance, disease control, strategic breeding, CPRs, and many such issues will bring lot of knowledge led revolution throughout the world which can change the shape of future animal husbandry practices.

References

Adams LG, Templeton JW (1998) Genetic resistance to bacterial diseases of animals. Rev Sci Tech Off Int Epizoot 17:200–219

Agyemang K, Dwinger RH, Little DA, Rowlands GJ (1997) Village N'Dama cattle production in West Africa: six years of research in the Gambia, International Livestock Research Institute, Nairobi, Kenya, and International Trypanotolerance Centre. Banjul, The Gambia, p 131

Ahuya CO, Okeyo AM, Mosi RO, Murithi FM (2004) Growth, survival and milk production performance of Toggenburgs and their crosses to East African and Galla goat breeds in the eastern slopes of Mount Kenya. In: Smith T, Godfrey SH, Buttery PJ, Owen E (eds) The contribution of small ruminants in alleviating poverty: communicating messages from research. Proceedings of the third DFID Livestock Production Programme Link Project

(R7798) workshop for small livestock keepers, Izaak Walton Inn, Embu, Kenya, 4–7 Feb 2003. Natural Resources International Ltd, Aylesford, Kent, UK, pp 40–47

Ajmone-Marsan P, Garcia JF, Lenstra JA (2010) On the origin of cattle: how aurochs became cattle and colonized the world. Evol Anthropol: Issues News Rev 19(4):148–157

Ajmone-Marsan P, GLOBALDIV Consortium (2010) A global view of livestock biodiversity and conservation—GLOBALDIV. Anim Genet 41:1–5

Akerfelt M, Trouillet D, Mezger V, Sistonen L (2007) Heat shock factors at a crossroad between stress and development. Ann N Y Acad Sci 1391:1–13

Alderman H (1994) Why should it matter what commodity is the source of agricultural profits? Dairy development in India. In: Von Braun J, Kennedy E (eds) Agricultural commercialization, economic development and nutrition. The Johns Hopkins University Press, Baltimore

Alderson L (1990) The work of the Rare Breeds Survival Trust. In: Alderson L (ed) Genetic conservation of domestic livestock. CAB International, Wallingford, pp 32–34

Alford AR, Hegarty RS, Parnell PF, Cacho OJ, Herd RM, Griffith GR (2006) The impact of breeding to reduce residual feed intake on enteric methane emissions from the Australian beef industry. Austr J Exp Agri 46:813–820

Alim KA (1960) Reproductive rate and milk yield of Kenana cattle in Sudan. J Agric Sci Cambridge 55:183–188

Amer PR, Mpofu N, Bondoc O (1998) Definition of breeding objectives for sustainable production systems. In: 6th World Congress of Genetics Applied to Livestock Production, vol 28, pp 97–104

Ames DR, Ray DE (1983) Environmental manipulation to improve animal productivity. J Anim Sci 57:209–220

Amin MY, Peters KJ (2006) Awassi sheep production and the development of breeding program options in Syria. In: Prosperity and Poverty in a Globalised World – Challenges for Agricultural Research (Tropentag, 11-13 October 2006). University of Bonn, Germany

Armstrong DV (1994) Heat stress interaction with shade and cooling. J Dairy Sci 77:2044–2050

Arthur PF, Archer JA, Herd RM (2004) Feed intake and efficiency in beef cattle. Overview of research and challenges for the future. Austr J Exp Agri 44:361–169

Baker RL (1998) Genetic resistance to endoparasites in sheep and goats. A review of genetic resistance to gastrointestinal nematode parasites in sheep and goats in the tropics and evidence for resistance in some sheep and goat breeds in sub-humid coastal Kenya. FAO/UNEP Anim Genet Res Info 24:13–30

Barillet F, Arranz JJ, Carta A (2005) Mapping quantitative trait loci for milk production and genetic polymorphisms of milk proteins in dairy sheep. Genet Sel Evol 37:109–123

Barnes AI, Partridge K (2003) Costing reproduction. Anim Behav 66:199–204

Barthel R, Feng J, Piedrahita JA, McMurray DN, Templeton JW, Adams LG (2001) Stable transfection of the Bovine NRAMP1 gene into murine RAW264.7 cells: Effect on Brucella abortus survival. Infect Immun 69(5):3110–3119

Barton CH, Whitehead SH, Blackwell JM (1995) Nramp transfection transfers Ity/Lsh/Bcg-related pleiotropic effects on macrophage activation: influence on oxidative burst and nitric oxide pathways. Mol Med 1:267–279

Basiricò L, Morera P, Primi V, Lacetera N, Nardone A, Bernabucci U (2011) Cellular thermotolerance is associated with heat shock protein 70.1 genetic polymorphisms in Holstein lactating cows. Cell Stress Chaperones 16(4):441–448

Bebe BO, Udo HMJ, Rowlands GJ, Thorpe W (2003) Smallholder dairy systems in the Kenya highlands: breed preferences and breeding practices. Livest Prod Sci 82:117–127

Beef CRC Cooperative Research Centre for Beef Genetic Technologies (2008). http://www.beef.crc.org.au/

Beilharz RG, Luxford BG, Wilkinson JL (1993) Quantitative genetics and evolution: is our understanding of genetics sufficient to explain evolution? J Anim Breed Genet 110:161–170

Bernabucci U, Lacetera N, Baumgard LH, Rhoads RP, Ronchi B, Nardone A (2010) Metabolic and hormonal acclimation to heat stress in domesticated ruminants. Animal 4(7):1167–1183

Berthier D, Chantal I, Thevenon S, Sakande H, Maillard JC, Bengaly Z, Piquemal D, Marti J, Cuny G (2008) Study of bovine trypanotolerance by whole transcriptome analysis. Ann N Y Acad Sci 1149:71–76

Bijma P (2009) Maintaining fitness by within breed selection. In: van der Werf JHJ, Graser HU, Frankham R, Gondoro C (eds) Adaptation and fitness in animal populations. Evolutionary and breeding perspectives on genetic resource management, The Netherlands

Bishop S, de Jong M, Gray D (2002) Opportunities for incorporating genetic elements into the management of farm animal diseases: policy issues. CGRFA background study FAO, Rome, p 18

Bishop SC, Morris CA (2007) Genetics of disease resistance in sheep and goats. Small Rumin Res 70:48–59

Blench R (2005) Conservation of indigenous livestock: sustaining biodiversity for current and future generations. In: Presentation in CGIAR annual general meeting 6th Dec 2005, Marrakech, Morocco. http://www.cgiar.org/meetings/agm05/stakeholders_docs/agm05_stake_4a_blench.pdf

Bligh J (1976) Introduction to acclimatory adaptation—including notes on terminology. In: Bligh J, Cloudsley-Thompson JL, Macdonald AG (eds) Environmental physiology of animals. Wiley, NY, pp 219–229

Boelling D, Groen AF, Sorensen P, Madsen P, Jensen J (2003) Genetic improvement of livestock for organic farming systems. Livest Prod Sci 80:79–88

Boettcher PJ, Tixier-Boichard M, Toro MA, Simianer H, Eding H, Gandini G, Joost S, Garcia D, Colli L, Ajmone-Marsan P, Consortium GLOBALDIV (2010) Objectives, criteria and methods for using molecular genetic data in priority setting for conservation of animal genetic resources. Anim Genet 41:64–77

Bohmanova J, Misztal I, Cole JB (2007) Temperature humidity indices as indicators of milk production losses due to heat stress. J Dairy Sci 90:1947–1956

Bonsma JC (1973) Crossbreeding for adaptability. In: Koger M, Cunha TJ, Warnick AC (eds) Crossbreeding beef cattle, Series 2, University of Florida Press Gainesville, Florida

Bradshaw WE, Holzapfel CM (2006) Evolutionary response to rapid climate change. Science 312:1477–1478

Brascamp EW, Groen AF, De Boer IJM, Udo H (1998) The effect of environmental factors on breeding goals. 6th World Congress of Genetics Applied to Livestock Production 27:129–136

Bronson FH (2009) Climate change and seasonal reproduction in mammals. Philos Trans R Soc 364:3331–3340

Brown EG, Carstens GE, Fox JT, White MB, Randel RD, Holloway JW (2004) Relationships between temperament and performance traits in growing calves. Beef Cattle Research in Texas

Bruford MW, Bradley DG, Luikart G (2003) DNA markers reveal the complexity of livestock domestication. Nat Rev Genet 4:900–910

Burns WC, Koger MW, Butts T, Pahnish OF, Blackwell RL (1979) Genotype by environment interaction in Hereford cattle: II. Birth and weaning traits. J Anim Sci 49:403–409

Burrow HM, Gulbrabsen B, Johnson SK, Davis GP, Shorthose WR, Elliott RF (1991) Consequences of selection for growth, and heat resistance on growth feed conversion efficiency, commercial carcass traits and meat quality of zebu crossbred cattle. Austr J Agric Res 42:1373–1383

Capper JL, Cady RA, Bauman DE (2009) The environmental impact of dairy production: 1944 compared with 2007. J Anim Sci 87:2160–2167

Cardellino RA (2000) Animal genetic resources in southern Brazil. Arch Zootec 49:327–331

Carson A, Elliot M, Groom J, Winter A, Bowles D (2009) Geographical isolation of native sheep breeds in the UK—evidence of endemism as a risk factor to genetic resources. Livest Sci 123:288–299

Case LA, Kelly MJ, Miller SP, Wood BJ (2010) Genotype × environment interaction as it relates to egg production in turkeys (*Meleagris gallopavo*). J Anim Sci 88:1957–1966

Ceron-Munoz MF, Tonhati H, Costa CN, Rojas-Sarmiento D, Echeverri DM (2004) Factors that cause genotype by environment interaction and use of a multiple-trait herd-cluster model for milk yield of Holstein cattle from Brazil and Colombia. J Dairy Sci 87(8):2687–2692

Commission on Genetic Resources for Food and Agriculture (CGRFA) (2007) The state of the world's animal genetic resources for food and agriculture. FAO, Rome, p 523

Commission on Genetic Resources for Food and Agriculture (CGRFA) (2009) Contributions of smallholder farmers and pastoralists to the development, use and conservation of animal genetic resources. CGRFA/WG-AnGR-5/09/Inf.4

Commission on Genetic Resources for Food and Agriculture (CGRFA) (2011) Draft guidelines on phenotypic characterization

Chase LE (2011) Climate change impacts on dairy cattle. Retrieved online from www.climateandfarming.org/pdfs/FactSheets/III.3Cattle.pdf

Collier RJ, Coppola C, Wolfgram A (2003) Novel approaches for the alleviation of climatic stress in farm animals. In: Interactions between climate and animal production. EAAP Technical Series No. 7. Wageningen Academic Publ., Wageningen, pp 61–71

Collier RJ, Baumgard LH, Lock AL, Bauman DE (2005) Physiological limitations, nutrient partitioning. In: Sylvester-Bradley R, Wiseman J (eds) Yield of farmed species: constraints and opportunities in the 21st century. Nottingham University Press, Nottingham, pp 351–377

Collier RJ, Stiening CM, Pollard BC, Vandal MJ, Baumgard LH, Gentry PC, Coussens PM (2006) Use of gene expression microarrays for evaluating environmental stress tolerance at the cellular level in cattle. J Anim Sci 84:1–13

Collier RJ, Collier JL, Rhoads RP, Baumgard LH (2008) Invited review: Genes involved in the bovine heat stress response. J Dairy Sci 91(2):445–454

Coltman W, Pilkington J, Krunk LEB, Wilson K, Pemiberton JM (2001) Positive genetic correlation between parasite resistance and body size in a free-living ungulate population. Evolution 55(10):2116–2125

Da Silva RG (1973) Improving tropical beef cattle by simultaneous selection for weight and heat tolerance. Heritabilities and correlations of the traits. J Anim Sci 37(3):637–642

Da Silva RG, Arantes-Neto JG, Holtz-Filho SV (1988) Genetic aspects of the variation of sweating rate and coat characteristics of Jersey cattle. Braz J Genet 11:335–347

Da Silva RG (2000) Introduction to animal bioclimatology. Nobel-FAPESP, São Paulo, p 286

Da Silva RG (2002) Radiant heat load: reviewing its evaluation method. Engenharia Rural 13: 77–84

Da Silva RG (2007) Weather and climate and animal production. In: The guide to agricultural meteorological practices, WMO No. 134. http://www.agrometeorology.org/files-folder/repository/gamp_chap_11.pdf

Dai J, Wang X, Chen Y, Wang X, Zhu J, Lu L (2009) Expression quantitative trait loci and genetic regulatory network analysis reveals that Gabra2 is involved in stress responses in the mouse. Stress, first, pp 1–23

de Greef J (2009) Actual standing and perspectives for the sustainable use and development of parasite resistant or tolerant breeds in developed regions: Australia and NZ as an example. In: Presentation, joint FAO/INRA workshop on animal genetic resources and their resistance/tolerance to diseases, with special focus on parasitic diseases in ruminants, Paris

de Hahn C, Steinfeld H, Blackburn HW (1997) Livestock environment interactions: finding a balance. In: Report of a study coordinated by FAO, USAID and the World Bank, FAO, Rome, p 115

de Jong G, Bijma P (2002) Selection and phenotypic plasticity in evolutionary biology and animal breeding. Livest Prod Sci 78:195–214

de La Rocque S, Rioux JA, Slingenbergh J (2008) Climate change: effects on animal disease systems and implications for surveillance and control. Rev Sci Et Tech (Int Office Epizoot) 27(2):339–354

Delgado C, Rosegrant M, Steinfeld H, Ehui S, Courbois C (1999) Livestock to 2020: the next food revolution. Food, agriculture and the environment discussion paper 28, International Food Policy Research Institute/Food and Agriculture Organization of the United Nations/International Livestock Research Institute (IFPRI/FAO/ILRI), Washington, U S A

Dickerson GE (1955) Genetic slippage in response to selection for multiple objectives. Cold Spring Harb Symp Quant Biol 20:213–224

Diez-Tascon C, Keane OM, Wilson T, Zadissa A, Hyndman DL, Baird DB, McEwan JC, Crawford AM (2005) Microarray analysis of selection lines from outbred populations to identify genes involved with nematode parasite resistance in sheep. Phys Genom 21:59–69

Dikmen S, Hansen PJ (2009) Is the temperature-humidity index the best indicator of heat stress in lactating dairy cows in a subtropical environment? J Dairy Sci 92:109–116

Dobzhansky T (1956) What is an adaptive trait? Am Nat 90(855):337–347

Dobzhansky T (1970) Genetics of the evolutionary process. Columbia Universty Press, NY

Dominik S, Kinghorn BP (2008) Neglecting genotype x environment interaction results in biased predictions from selection index calculations. Livest Sci 114:233–240

Easterling WE, Aggarwal PK, Batima P, Brander KM, Erda L, Howden SM, Kirilenko A, Morton J, Soussana JF, Schmidhuber J, Tubiello FN (2007) Food, fibre and forest products. In: Parry ML, Canziani OF, Palutikof JP, van der Linden PJ, Hanson CE (eds) Climate change 2007: impacts, adaptation and vulnerability. Contribution of working group II to the fourth assessment report of the intergovernmental panel on climate change. Cambridge University Press, Cambridge, pp 273–313

Easterling W, Apps M (2005) Assessing the consequences of climate change for food and forest resources: a view from the IPCC. Clim Change 70:165–189

Falconer DS, Mackay TF (1996) Introduction to quantitative genetics, 4th edn. Longman Group, Essex

Falkenberg SM (2006) Stress response effects on growth, carcass characteristics, and tenderness in Bonsmara-influenced steers. M. Sc. thesis, Submitted to the Office of Graduate Studies of Texas A and M University

FAO (1992) The management of global animal genetic resources. In: Hodges J (ed) Proceedings of an expert consultation, animal production and health. FAO, Rome, pp 1–263

FAO (1999a) Opportunities for incorporating genetic elements into the management of farm animal diseases: policy issues. In: Bishop S, de Jong M, Gray D (eds) Background study paper number 18. Commission on Genetic Resources for Food and Agriculture. Rome

FAO (1999b) The global strategy for the management of animal genetic resources: executive brief. Initiative for Domestic Animal Diversity, Rome

Food and Agriculture Organisation of the United Nations (FAO) (2000) World watch list for domestic animal diversity, 3rd edn. In: Scherf B (ed) FAO, Rome

FAO (2007a) The state of the world's animal genetic resources for food and agriculture. In: Rischkowsky B, Pilling D FAO, Rome. http://www.fao.org/docrep/010/a1250e/a1250e00.htm

FAO (2007b) Global plan of action for animal genetic resources and the interlaken declaration. Rome. http://www.fao.org/docrep/010/a1404e/a1404e00.htm

FAO (2008) The genetic improvement of forage grasses and legumes to enhance adaptation of grasslands in climate change. Rome, FAO. ftp://ftp.fao.org/docrep/fao/010/ai779e/ai779e00.pdf

FAO (2009) Livestock keepers guardians of biodiversity. Animal production and health paper Rome, p 167

FAO (2010) The state of food and agriculture 2009: livestock in the balance. http://www.fao.org/docrep/012/i0680e/i0680e.pdf

Felius M (1995) Cattle breeds—an encyclopedia. Wageningen University Press, Wageningen, p 799

Finocchiaro R, van Kaam JBCHM, Portolano B, Misztal I (2005) Effect of heat stress on production of Mediterranean dairy sheep. J Dairy Sci 88(5):1855–1864

Flachowsky G, Brade W (2007) Potenziale zur Reduzierung der Methan-Emissionen bei Wiederkäuern. Züchtungskunde 79(6):417–465

Flint J, Corley R, DeFries JC, Fulker DW, Gray JA, Miller S, Collins AC (1995) A simple genetic basis for a complex psychological trait in laboratory mice. Science 269:1432–1435

Flint J (2001) Is this mouse anxious? The difficulties of interpreting the effects of genetic action. Commentary on Belzung "The genetic basis of the pharmacological effects of anxiolytics" and Olivier et al. "The 5-HT(1A) receptor knockout mouse and anxiety". Behav Pharmacol 12(6–7):461–465

Flint J (2002) Genetic effects on an animal model of anxiety. FEBS Lett 529:131–134

Flock DK, Preisinger R (2002) Breeding plans for poultry with emphasis on sustainability. In: 7th World Congress on Genetics Applied to Livestock Production, Aug 2002. Montpellier, France, pp 19–23

Forman S, Hungerford N, Yamakawa M, Yanase T, Tsai HJ, Joo YS, Yang DK, Nha JJ (2008) Climate change impacts and risks for animal health in Asia. Revue Scientifique et technique (Int Off Epizoot) 27(2):581–597

Francis CA, Callaway MB (1993) Crop improvement for future farming systems. In: Callaway MB, Francis CA (eds) Crop improvement for sustainable agriculture. University of Nebraska Press, Lincoln, pp 1–18

Franck M, Svensson M, Von Seth G, Jossel A, Figwer Ph, Poirel MT, Monin G (2003) Effect of stunning conditions on occurrence of PSE defects in hams of RN +/RN- pigs. Meat Sci 64(4): 351–355

Fraser D (2005) Animal welfare and the intensification of animal production an alternative interpretation. FAO, pp 8–17

Freitas M, Misztal I, Bohmanova J, West JW (2006) Utility of on- and off-farm weather records for studies in genetics of heat tolerance. Livest Sci 105:223–228

Gamborg C, Sandøe P (2005) Sustainability in farm animal breeding: a review. Livest Prod Sci 92:221–231

Gandini GC, Ollivier L, Danell B, Distl O, Georgoudis A, Groeneveld E, Martyniuk E, van Arendonk JAM, Woolliams JA (2004) Criteria to assess the degree of endangerment of livestock breeds in Europe. Livest Prod Sci 91:173–182

Ganguly I, Sharma A, Singh R, Deb SM, Singh DK, Mitra A (2008) Association of microsatellite (GT) polymorphism at 3′ UTR of NRAMP1 with the macrophage function following challenge with Brucella LPS in buffalo (*Bubalus bubalis*). Vet Microbiol 129:188–196

Gebremedhin KG, Ni H, Hillman PE (1997) Temperature profile and heat flux through irradiated fur layer. In: Proceedings 5th International Livestock Environment Symposium. American Society of Agricultural and Biological Engineers, St Joseph MI, vol 1, pp 226–233

Georgopoulos C, Welch WJ (1993) Role of the major heat shock proteins as molecular chaperones. Annu Rev Cell Biol 9:601–634

Gillespie JR (1997) Modern livestock and poultry production, 5th edn. Delmar Publisher, U S A

Goodland R, Daly H (1996) Environmental sustainability: universal and non-negotiable. Ecol Appl 6:1002–1017

Goodwin PJ, Gaughan J, Skele P, Josey M, Hall A, Young B (1997) Coat color and alleviation of heat load in Holstein-Friesian cows. In: Proceedings of 5th International Livestock Environment Symposium. American Society of Agricultural and Biological Engineers, St Joseph MI, vol 2, pp 923–927

Gorman GM, Park SDE, Hill WE, Meade KG, Coussens PM, Agaba M, Naessens J, Kemp SJ, MacHugh DE (2009) Transcriptional profiling of cattle infected with *Trypanosoma congolense* highlights gene expression signatures underlying trypanotolerance and trypano-susceptibility. BMC Genom 10:207. doi:10.1186/1471-2164-10-207

Gowane GR, Arora AL (2010) Performance evaluation of sheep in farmers' flock of eastern semi-arid region of Rajasthan. Indian J Small Rumin 16:87–91

Gowane GR, Chopra A, Prince LLL, Paswan C, Arora AL (2010) Estimates of (co)variance components and genetic parameters for body weights and first greasy fleece weight in Bharat merino sheep. Animal 4:425–431

Groeneveld LF, Lenstra JA, Eding H, Toro M A, Scherf B, Pilling D, Negrini R, Finlay EK, Jianlin H, Groeneveld E, Weigend S, The GLOBALDIV consortium (2010) Genetic diversity in farm animals—a review. Anim Genet 4:6–31

Haile-Mariam M, Malmfors B, Philipsson J (1998) Boran—indigenous African cattle with potential. Currents (Sweden) 17(18):38–43

Hall SJG, Ruane J (1993) Livestock breeds and their conservation—a global review. Conserv Biol 7(4):815–825

Hamadeh SK, Moussa Z, Abi Sa M, Barbour E (1997) Physiological indicators of adaptation in Awassi and Finn × Texel × Awassi sheep. Options méditerr 33:231–236

Hammond J (1947) Animal breeding in relationship to nutrition and environmental conditions. Biol Rev 22:195–213
Hammond PM (1994) Practical approaches to the estimation of the extent of biodiversity in speciose groups. Philos Trans R Soc London 345:119–136
Hanocq E, Bodin L, Thimonier J, Teyssier J, Malpaux B, Chemineau P (1999) Genetic parameters of spontaneous spring ovulatory activity in Mérinos d'Arles sheep. Genet Sel Evol 31:77–90
Hanotte O, Bradley DG, Ochieng JW, Verjee Y, Hill EW, Rege JEO (2002) African pastoralism: genetic imprints of origins and migrations. Science 296:336–339
Hansen PJ, Lander MF (1988) Coat color and heat stress in cattle. In: Intern Conference on Livestock in the Tropics. Gainesville, FL [mimeo]
Hansen PJ (1990) Effects of coat colour on physiological responses to solar radiation in Holsteins. Vet Rec 127:333–334
Hansen PJ (2004) Physiological and cellular adaptation of zebu cattle to thermal stress. Anim Reprod Sci 82–83:349–360
Harris DL (1970) Breeding for efficiency in livestock production: defining the economic objectives. J Anim Sci 30:860–865
Hatzimimaogiou J, Zervas NP, Boyazoglu JG (1990) Prolific dairy sheep breeds in Greece. Options Méditerr Sér 12:25–30
Hayes BJ, Bowman PJ, Chamberlain AJ, Savin K, van Tassell CP (2009) A validated genome wide association study to breed cattle adapted to an environment altered by climate change. PLoS ONE 4(8):e6676. doi:10.1371/journal.pone.0006676
Hedrick P, Kalinowski ST (2000) Inbreeding depression and conservation biology. Annu Rev Ecol Sys 32:139–162
Heitschmidt RK, Short RE, Grings EE (1996) Ecosystems, sustainability, and animal agriculture. J Anim Sci 74:1395–1405
Henderson ND, Turri MG, DeFries JC, Flint J (2004) QTL analysis of multiple behavioral measures of anxiety in mice. Behav Genet 34:267–293
Hiemstra SJ, deHass Y, Mäki-Tanila A, Gandini G (eds) (2010) Local cattle breeds in Europe development of policies and strategies for selfsustaining breeds. Wageningen Academic Publishers, Wageningen
Hillman PE, Lee CN, Carpenter JR, Back KS, Parkhurst A (2001) Impact of hair color on thermoregulation of dairy cows to direct sunlight. In: A.S.A.E. Paper No. 014031. American Society of Agricultural and Biological Engineers, St Joseph MI
Hoffmann I (2010) Climate change and the characterization, breeding and conservation of animal genetic resources. Anim Genet 41(1):32–46
Hoffmann I, Scherf B (2010) Implementing the global plan of action for animal genetic resources. Anim Genet Res 47:1–11
Hohenboken W, Jenkins T, Pollak J, Bullock D, Radakovich S (2005) Genetic improvement of beef cattle adaptation in America. In: Proceedings of the beef improvement federation's 37th Annual Research Symposium and Annual Meeting, pp 115–120
Homann S, Maritz HJ, Hülsebusch C, Meyn K, Valle Zárate A (2005) Boran and Tuli cattle breeds—origin, worldwide transfer, utilisation and the issues of access and benefit sharing. Verlag Grauer Beuren and Stuttgart, Germany
Horowitz M (2002) From molecular and cellular to integrative heat defence during exposure to chronic heat. Comp Biochem Physiol 131:475–483
International Commission for Thermal Physiology (ICTP) (2001) Glossary of terms for thermal physiology, 3rd edn. Jpn J Physio 51:245–280
Intergovernmental Panel on Climate Change (IPCC) (2002) Climate change and biodiversity. In: Gitay H, Suárez A, Watson RT, Dokken DJ (eds) IPCC Technical Paper V
International Union for the Conservation of Nature (IUCN) (1980) World conservation strategy International Union for the Conservation of Nature and Natural Resources. Gland, Switzerland

Janssen M, Weimann C, Gauly M, Erhardt G (2002) Associations between infections with *Haemonchus contortus* and genetic markers on ovine chromosome, vol 20. In: Proceedings of 7th WCGALP, pp 13–11

Jenkins TG (2004) Genetic adaptation of livestock: the power of change. In: Proceedings 1st Holt/Cat Symposium on Excellence in Ranch Management. 7–8 Oct 2004, Kingsville, pp 34–41

Jones HE, Warkup CC, Williams A, Audsley E (2008) The effect of genetic improvement on emission from livestock systems. In: EAAP Annual Conference, Vilnius

Joober R, Zarate JM, Rouleau GA, Skamene E, Boksa P (2002) Provisional mapping of quantitative trait loci modulating the acoustic startle response and prepulse inhibition of acoustic startle. Neuropsychopharmacology 27:765–781

Kadzere CT, Murphy MR, Silanikove N, Maltz E (2002) Heat stress in lactating dairy cows: a review. Livest Prod Sci 77:59–91

Khalifa HH, SHalaby T, Abdel-Khalek TMM (2005) An approach to develop a biometeorological thermal discomfort index for sheep and goats under Egyptian conditions. In: Proceeding of the 17th International Congress of Biometeorology (International Society of Biometeorology). Offenbach am Main, Garmisch-Partenkirchen, Germany, pp 118–122

Khazzaka A, Figwer P, Poirel MT, Serrar M, Franck M (2006) Hsp70 response in pigs is affected by their Halothane genotypes after heat stress. J Therm Biol 31:605–610

Klinedinst PL, Wichita DA, Hahn GL, Hubbard KG (1993) The potential effects of climate change on summer season dairy cattle milk production and reproduction. Clim Change 23:21–36

Klip A, Britt BA, Elliott ME, Walker D, Ramlal T, Pegg W (1986) Changes in cytoplasmic free calcium caused by halothane. Role of the plasma membrane and intracellular Ca^{2+} stores. Biochem Cell Biol 64(11):1181–1189

Kosgey IS, Baker RL, Udo HMJ, van Arendonk JAM (2006) Successes and failures of small ruminant breeding programmes in the tropics: a review. Small Rumin Res 61(1):13–28

Krätli S (2008) Time to outbreed animal science? A cattle-breeding system exploiting structural unpredictability: the WoDaaBe herders in Niger. In: STEPS Working Paper 7, Brighton STEPS Centre

Lai LP, Deng CL, Moss AJ, Kass RS, Liang CS (1994) Polymorphism of the gene encoding a human minimal potassium ion channel (minK). Gene 151(1–2):339–340

Lanari MR, Domingo E, Perez Centeno MJ, Gallo L (2005) Pastoral community selection and the genetic structure of a local goat breed in Pantagonia. Anim Genet Res Info 37:31–42

Lande (1981) The minimum number of genes contributing to quantitative variation between and within populations. Genetics 99(3–4):541–553

Legarra A, Ugarte E (2005) Genetic parameters of udder traits, somatic cell score, and milk yield in Latxa sheep. J Dairy Sci 88:2238–2245

Lilja IE, Wattrang E, Andersson L, Fossum C (2000) Mapping QTL for stress induced alteration in porcine leukocyte number and functions. Anim Genet 31:186–193

Lincoln GA, Lincoln CE, McNeilly AS (1990) Seasonal cycles in the blood plasma concentration of FSH, inhibin and testosterone, and testicular size in rams of wild, feral and domesticated breeds of sheep. J Reprod Fertil 88(2):623–633

Lindquist S (1986) The heat-shock response. Annu Rev Biochem 55:1151–1191

Liu YX, Zhou X, Li DQ, Cui QW, Wang GL (2010) Association of ATP1A1 gene polymorphism with heat tolerance traits in dairy cattle. Genet Mol Res 9(2):891–896

Liu X, Stancliffe D, Lee S, Mathur S, Gershenfeld HK (2007) Genetic dissection of the tail suspension test: a mouse model of stress vulnerability and antidepressant response. Biol Psychiatry 62(1):81–91

Loerch SC, Fluharty FL (1999) Physiological changes and digestive capabilities of newly received feedlot cattle. J Anim Sci 77:1113–1119

LPP, LIFE Network, IUCN–WISP, FAO (2010) Adding value to livestock diversity—marketing to promote local breeds and improve livelihoods. FAO Animal Production and Health Paper. No. 168. Rome

Marshall K, Maddox JF, Lee SH, Zhang Y, Kahn L, Graser HU, Gondoro C, Walkden-Brown SW, van der Werf JHJ (2009) Genetic mapping of quantitative trait loci for resistance to Haemonchus contortus in sheep. Anim Genet 40:262–272

Mathur PK, Horst P (1994) Genotype by environment interactions in laying hens based on relationship between breeding values of sires in temperate and tropical environments. Poultry Sci 73:1777–1784

Matijasevic Z, Snyder JE, Ludlum DB (1998) Hypothermia causes a reversible, p53-mediated cell cycle arrest in cultured fibroblasts. Oncol Res 10:605–610

Mavrogenis AP, Antoniades NY, Hooper RW (2006) The Damascus (Shami) goat of Cyprus. Anim Genet Resour Inf 38:57–65

McCrabb GJ, Hunter RA (1999) Prediction of methane emissions from beef cattle in tropical production systems. Austr Agric Res 50(8):1285–1339

McDowell RE (1972) Improvement of livestock production in warm climates. Freeman and Co, San Francisco

McManus C, Prescott E, Paludo GR, Bianchini E, Louvandini H, Mariante AS (2008) Heat tolerance in naturalized Brazilian cattle breeds. Livest Sci 120(3):256–264

Menendez Buxadera A, Mandonnet N (2006) The importance of the genotype × environment interaction for selection and breeding programmes in tropical conditions. In: CAB reviews: perspectives in agriculture, veterinary Science, nutrition and natural resources, vol 1, p 26

Meuwissen THE, Hayes BJ, Goddard ME (2001) Prediction of total genetic value using genome-wide dense marker maps. Genetics 157:1819–1829

Mitlohner FM, Morrow JL, Dailey JW, Wilson SC, Gaylean ML, Miller MF, McGlone JJ (2001) Shade and water misting effects on behavior, physiology, performance, and carcass traits of heat stressed feedlot cattle. J Anim Sci 79:2327–2335

Mitlohner FM, Gaylean ML, McGlone JJ (2002) Shade effects on performance, carcass traits, physiology, and behavior of heat-stressed feedlot heifers. J Anim Sci 80:2043–2050

Monin G, Larzul C, Le Roy P, Culioli J, Mourot J, Rousset-Akrim S, Talmant A, Touraille C, Sellier P (1999) Effects of the halothane genotype and slaughter weight on texture of pork. J Anim Sci 77:408–415

Morimoto RI (1998) Regulation of the heat shock transcriptional response: cross talk between a family of heat shock factors, molecular chaperones, and negative regulators. Genes Dev 12:3788–3796

Moyo S (1990) Relative productivity of indigenous, exotic and crossbred cattle at matopos research station. J Zimbabwe Soc Anim Prod 5:23–28

Mrode RA, Swanson GJT (1996) Genetic and statistical properties of somatic cell count and its suitability as an indirect means of reducing the incidence of mastitis in dairy cattle. Anim Breed Abs 64:847–857

Mulder HA, Bijma P (2006) Benefits of cooperation between breeding programs in the presence of genotype by environment interaction. J Dairy Sci 89:1727–1739

Mulder HA, Veerkamp RF, Ducro BJ, van Arendonk JAM, Bijma P (2006) Optimization of dairy cattle breeding programs for different environments with genotype by environment interaction. J Dairy Sci 89:1740–1752

Naylor RL, Goldburg RJ, Mooney H, Beveridge M, Clay J, Folke C, Kautsky N, Lubschenko J, Primavera J, Williams M (1998) Nature's subsidies to shrimp and salmon farming. Science (Wash DC) 282:883–884

Nienaber JA, Hahn GL, Eigenberg RA (1999) Quantifying livestock responses for heat stress management: a review. Int J Biomet 42:183–188

Notter DR (1999) The importance of genetic diversity in livestock populations of the future. J Anim Sci 77:61–69

Notter DR, Cockett NE (2005) Opportunities for detection and use of QTL influencing seasonal reproduction in sheep: a review. Genet Sel Evol 37(1):39–53

Ogilvie FB (1947) Cattle breeding in India: the development of milk production in two Indian breeds. J Hered 38:23–28

Ohnishi T, Wang X, Ohnishi K, Takahashi A (1998) Dependent induction of WAF1 by cold shock in human glioblastoma cells. Oncogene 16:1507–1511

Olesen I, Groen AF, Gjerde B (2000) Definition of animal breeding goals for sustainable production systems. J Anim Sci 78:570–582
Olson TA, Lucena C, Chase CC Jr, Hammond AC (2003) Evidence of a major gene influencing hair length and heat tolerance in *Bos taurus* cattle. J Anim Sci 81:80–90
Oseni S, Bebe O (2010) Climate change, genetics of adaptation and livestock production in low input systems. In: 2nd International Conference: climate, sustainability and development in semi-arid regions. Fortaleza-Ceará, Brazil
Otterbein LE, Choi AM (2000) Heme oxygenase: colors of defense against cellular stress. Am J Physiol Lung Cell Mol Physiol 279:029–037
Page TJ, Sikder S, Yang L, Pluta L, Wolfinger RD, Kodadek T, Thomas RS (2006) Genome-wide analysis of human HSF1 signaling reveals a transcriptional program linked to cellular adaptation and survival. Mol BioSyst 2:627–639
Parsell DA, Lindquist S (1993) The function of heat-shock proteins in stress tolerance: degradation and reactivation of damaged proteins. Annu Rev Genet 27:437–496
Payne WJA, Hodges J (1997) Tropical cattle: origins, breeds and breeding policies. Blackwell Science, Oxford, p 328
Pearson RE (1986) Economic evaluation of breeding objectives in dairy cattle: intensive specialized milk production in temperate zones. In: 3rd World Congress of Genetics Applied to Livestock Production, vol 9, pp 11–17
Pelletier J, Bodin L, Hanocq E, Malpaux B, Teyssier J, Thimonier J, Chemineau P (2000) Association between expression of reproductive seasonality and alleles of the gene for Mel1a receptor in the ewe. Biol Reprod 62(4):1096–1101
Petit E, Balloux F, Excoffier L (2002) Mammalian population genetics: Why not Y? Trends Ecol Evol 17:28–33
Philipsson J, Rege JEO, Okeyo AM (2006) Sustainable breeding programmes for tropical farming systems. In: Ojango JM, Malmfors B, Okeyo AM (eds) Animal genetics training resource, version 2, 2006. International Livestock Research Institute, Nairobi, Kenya, and Swedish University of Agricultural Sciences, Uppsala, Sweden
Pinedo PJ, Buergelt CD, Donovan GA, Melendez P, Morel L, Wu R, Langaee TY, Rae DO (2009) Candidate gene polymorphisms (BoIFNG, TLR4, SLC11A1) as risk factors for paratuberculosis infection in cattle. Pre Vet Med 91:189–196
Pinto J, Bonacic C, Hamilton-West C, Romero J, Lubroth J (2008) Climate change and animal diseases in South America. Rev Sci et Tec (Int Off Epizoot) 27(2):599–613
Pirkkala L, Nyka nen P, Sistonen L (2001) Roles of the heat shock transcription factors in regulation of the heat shock response and beyond. FASEB J 15:1118–1131
Prayaga KC, Barendse W, Burrow HM (2006) Genetics of tropical adaptation. In: 8th World Congress on Genetics Applied to Livestock Production, Belo Horizonte, Brasil. Theriogenology 60:1139–1151
Rathore JS (2004) Drought and household coping strategies: a case of Rajasthan. Indian J Agric Econ 59(4):689–708
Ravagnolo O, Misztal I (2002) Effect of heat stress on nonreturn rate in Holsteins: fixed-model analyses. J Dairy Sci 85:3101–3106
Rege JEO, Gibson JP (2003) Animal genetic resources and economic development: issues in relation to economic valuation. Ecol Econ 45:319–330
Regitano LC, Ibelli AM, Gasparin G, Miyata M, Azevedo AL, Coutinho LL, Teodoro RL, Machado MA, Silva MV, Nakata LC, Zaros L, Sonstegard TS, Silva AM, Alencar MM, Oliveira MC (2008) On the search for markers of tick resistance in bovines. Dev Biol 132: 225–230
Renaudeau D, Collin A, Yahav S, de Basilio V, Gourdine JL, Collier RJ (2010) Adaptation to tropical climate and research strategies to attenuate heat stress in livestock production. In: International Symposium of Sustainable Animal Production in the Tropics, Nov 2010, Guadeloupe. https://colloque.inra.fr/var/sapt2010/storage/fckeditor/file/1030%20RENAUDEAU.pdf
Rhoad AO (1940) A method of assaying genetic differences in the adaptability of cattle to tropical and subtropical climates. Emp J Exp Agric 8:190–198

Roberts AJ, Finn DA, Phillips TJ, Belknap JK, Keith LD (1995) Genetic analysis of the corticosterone response to ethanol in BXD recombinant inbred mice. Behav Neurosci 109(6): 1199–1208

Robertson A (1959) The sampling variance of the genetic correlation coefficient. Biometrics 15:469–485

Rosa HJD, Bryant MJ (2003) Seasonality of reproduction in sheep. Small Rumin Res 48:155–171

Rosenberg S (2001) Evolving responsively: adaptive mutation. Nat Rev Genet 2:504–515

Rosenzweig C, Karoly D, Vicarelli M, Neofotis P, Wu Q, Casassa G, Menzel A, Root TL, Estrella N, Seguin B, Tryjanowski P, Liu C, Rawlins S, Imeson A (2008) Attributing physical and biological impacts to anthropogenic climate change. Nature 453 (May). doi:10.1038/nature06937

Sansthan LP, Köhler-Rollefson I (2005) Indigenous breeds, local communities. Life Initiative, LPPS, p 66

Sarge KD, Murphy SP, Morimoto RI (1993) Activation of heat shock gene transcription by heat shock factor 1 involves oligomerization, acquisition of DNA-binding activity, and nuclear localization and can occur in the absence of stress. Mol Cell Biol 13:1392–1407

SASBA (2004) South African Livestock Breeder. South African Stud Book and Livestock Improvement Association, Somerset East

Standing Committee on Agriculture (SCA) (1990) Feeding standards for Australian livestock: ruminants. CSIRO Publications, East Melbourne

Schena M, Shalon D, Heller R, Chai A, Brown PO, Davis RW (1996) Parallel human genome analysis: microarray-based expression monitoring of 1,000 genes. Proc Natl Acad Sci U S A 93:10614–10619

Schwaiger FW, Gostomski D, Stear MJ, Duncan JL, McKellar QA, Epplen JT, Buitkamp J (1995) An ovine MHC DRB1 allele is associated with low fecal egg counts following natural, predominantly *O. circumcincta* infection. Int J Parasitol 25:815–822

Settar P, Yalcin S, Türkmut L, Özkan S, Cahanar A (1999) Season by genotype interaction related to broiler growth rate and heat tolerance. Poult Sci 78:1353–1358

Sharma AK, Rodriguez LA, Mekonnen G, Wilcox CJ, Bachman KC, Collier RJ (1983) Climatological and genetic effects on milk composition and yield. J Dairy Sci 66:119–126

Silanikove N (2000a) Effects of heat stress on the welfare of extensively managed domestic ruminants. Livest Prod Sci 67:1–18

Silanikove N (2000b) The physiological basis of adaptation in goats to harsh environments. Small Rumin Res 35:181–193

Simianer H (2005a) Decision making in livestock conservation. Ecol Econom 53:559–572

Simianer H (2005b) Using expected allele number as objective function to design between and within breed conservation offarm animal biodiversity. J Anim Breed Genet 53:559–572

Simm G, Bünger L, Villanueva B, Hill WG (2004) Limits to yield of farm species: genetic improvement of livestock. In: Sylvester-Bradley R, Wiseman J (eds) Yields of farmed species. Constraints and opportunities in the 21st century. Publ. Nottingham University Press, pp 123–141

Singh D, Swarnkar CP, Kumar Sushi, Prince LLL, Arora AL (2009) Performance of Avikalin sheep selected for resistance to *Haemonchus contortus*. J Vet Para 23(1):23–27

Small R, Hosking J (2010) Rural development programme funding for farm animal genetic resources: a questionnaire survey. In: UK National Standing Committee on Farm Animal Genetic Resources. www.defra.gov.uk/fangr

Smith C (1984) Estimated costs of genetic conservation in farm livestock. Animal genetic resources conservation by management, data banks and training. FAO Animal Production and Health Paper, vol 44. FAO, Rome, pp 21–30

Smith C, Banos C (1991) Selection within and across populations in livestock improvement. J Anim Sci 69:2387–2394

Solkner J, Nakimbugwe H, Valle Zárate A (1998) Analysis of determinants for success and failure of village breeding programmes. In: Proceedings of the 6th World Congress on

Genetics Applied to Livestock Production (12–16 Jan 1998), vol 25. Armidale, Australia, pp 273–280

Sonna LA, Fujita J, Gaffin SL, Lilly CM (2002) Invited review: effects of heat and cold stress on mammalian gene expression. J Appl Physiol 92:1725–1742

Stephens M (2006) Handbook of Australian livestock, 5th edn. Australian Meat and Livestock Corporation, Sydney

Strandberg E (2006) Analysis of the genotype by environment interaction using random regression models. In: 8th World Congress on Genetics Applied to Livestock Production. 13–18 Aug, Belo Horizonte, pp 25–05

Strydom PE (1994) Production and product characteristics of bulls of five Bonsmara breeding lines. In: Research report to Bonsmara Breeder Society of South Africa. Irene Animal Production Institute of the Agricultural Research Council

Tarricone BJ, Hingtgen JN, Belknap JK, Mitchell SR, Nurnberger JI Jr (1995) Quantitative trait loci associated with the behavioural response of BXD recombinant inbred mice to restraint stress: a preliminary communication. Behav Genet 25(5):489–495

Thifault S, Ondřej S, Sun Y, Fortin A, Skamene E, Lalonde R, Trembley J, Hamet P (2008) Genetic determinants of emotionality and stress response in AcB/BcA recombinant congenic mice and in silico evidence of convergence with cardiovascular genes. Human Mol Genet 17(3):331–344

Thornton PK, Jones PG, Owiyo TM, Kruska RL, Herrero M, Kristjanson P, Notenbaert A, Bekele N, Omolo A (2006) Mapping climate vulnerability and poverty in Africa. In: Report of the DFID. ILRI, p 2000

Thornton PK, van de Steeg J, Notenbaert A, Herrero M (2009) The impacts of climate change on livestock and livestock systems in developing countries: a review of what we know and what we need to know. Agric Sys 101:113–127

Torp Donner H, Juga J (1997) Sustainability—a challenge to animal breeding. Agric Food Sci Finl 6:229–239

Turner HG (1984) Variation in rectal temperature of cattle in a tropical environment and its relation to growth rate. Anim Prod 38:417–427

Turner JW (1980) Genetic and biological aspects of zebu adaptability. J Anim Sci 50(6): 1201–1205

Turri MG, Datta SR, DeFries J, Henderson ND, Flint J (2001) QTL analysis identifies multiple behavioral dimensions in ethological tests of anxiety in laboratory mice. Curr Biol 11: 725–734

Vallee A (2007) Analysis of the genotype by environment interaction for post weaning growth traits in tropical Creole cattle breed using the random regression model. Minor thesis, Institut National de la Recherche Agronomique, Guadeloupe (Unité de Recherches Zootechniques), Wageningen University and Research (Animal Breeding and Genomic Centre)

Vallejo RL, Bacon LD, Liu HC, Witter RL, Groenen MAM, Hillel J, Cheng HH (1998) Genetic mapping of quantitative trait loci affecting susceptibility to Marek's disease virus induced tumors in F2 intercross chickens. Genetics 148:349–360

Van den Bossche P, Coetzer JA (2008) Climate change and animal health in Africa. Rev Sci et Tech (Int Off Epizoot) 27(2):551–562

Van der Waaij EH (2004) A resource allocation model describing consequences of artificial selection under metabolic stress. J Anim Sci 82:973–981

Veerkamp RF, Simm G, Oldham JD (1995) Genotype by environment interaction experience from Langhill. In: Lawrence TLJ, Gordon FJ, Carson A (eds) Breeding and feeding the high genetic merit dairy cow. Edinburgh, United Kingdom, Br Soc Anim Sci, vol 19, pp 59–66

Vidal S, Tremblay ML, Govoni G, Gauthier S, Sebastiani G, Malo D, Skamene E, Olivier M, Jothy S, Gros P (1995) The *Ity/Lsh/Bcg* locus: natural resistance to infection with intracellular parasites is abrogated by disruption of the *Nramp1* gene. J Exp Med 182:655–666

Vidal SM, Malo D, Vogan K, Skamene E, Gros P (1993) Natural resistance to infection with intracellular parasites: isolation of a candidate gene for Bcg. Cell 73:469–485

Vitali A, Segnalini M, Bertocchi L, Bernabucci U, Nardone A, Lacetera N (2009) Seasonal pattern of mortality and relationships between mortality and temperature humidity index in dairy cows. J Dairy Sci 92:3781–3790
von Haartman L (1982) Two modes of clutch size determination in passerine birds. J Yamashina Inst Ornithol 14:214–219
Wall E, Bell MJ, Simm G (2008) Developing breeding schemes to assist mitigation. In: Rowlinson P, Steele M, Nefzaoui A (eds) Livestock and global climate change. BSAS, pp 44–47
Wall E, Simm G, Moran D (2010) Developing breeding schemes to assist mitigation of greenhouse gas emissions. Animal 4:366–376
Wallenbeck A, Rydhmer L, Lundeheim N (2009) GxE interactions for growth and carcass leanness: re-ranking of boars in organic and conventional pig production. Livest Sci 123:154–160
Warkup C (2007) Environmental footprint: is there scope to reduce emissions by improving genetics? Presentation for FABRE-TP. http://www.fabretp.org/images/071024fabretpenvironmentalfootprintgfwarkup.pdf
Warren JJ (1996) Breeding and genetics. In: Skinner JL (ed) American poultry history 1974–1993. Watt Publishing Co, IL, pp 12–24
Wickham BW, Banos G (1998) Impact of international evaluations on dairy cattle breeding programmes. In: Proceedings of 6th World Congress on Genetics Applied to Livestock Production, vol 23. Armidale, Australia, pp 315–322
Willer H (1967) Technische fortschritte und landwirtschaft. Verlag Paul Parey, Hamburg
Williams RW, Strom RC, Goldowitz D (1998) Natural variation in neuron number in mice is linked to a major quantitative trait locus on chr 11. J Neurosci 18(1):138–146
Wilson RT (2009) Fit for purpose—the right animal in the right place. Trop Anim Health Prod 41:1081–1090
Winkler DW, Dunn PO, McCulloch CE (2002) Predicting the effects of climate change on avian life-history traits. In: Proceedings of National Academy of Sciences, vol 99, issue 21, pp 13595–13599
Winter A, Alzinger A, Ruedi F (2007) Assessment of the gene content of the chromosomal regions flanking bovine DGAT1. Genomics 83:172–180
Wolfe DW, Ziska L, Petzoldt C, Seaman A, Chase L, Hayhoe K (2008) Projected change in climate thresholds in the Northeastern U.S.: implications for crops, pests, livestock, and farmers. Mitig Adapt Strat Glob Change 13:555–575
Yonash N, Cheng HH, Hillel J, Heller DE, Cahaner A (2001) DNA microsatellites linked to quantitative trait loci affecting antibody response and survival rate in meat-type chickens. Poult Sci 80:22–28
Zhu JJ, Lillehoj HS, Allen PC, Van Tassell CP, Sonstegard TS, Cheng HH, Pollock D, Sadjadi M, Min W, Emara MG (2003) Mapping quantitative trait loci associated with resistance to coccidiosis and growth. Poult Sci 82:9–16
Zimbelman RB, Rhoads RP, Rhoads ML, Duff GC, Baumgard LH, Collier RJ (2009) A re-evaluation of the impact of temperature humidity index (THI) and black globe humidity index (BGHI) on milk production in high producing dairy cows. In: Collier RJ (ed) Proceedings of the Southwest Nutrition Conference, pp 158–169
Zwald NR, Weigel KA, Fikse WF, Rekaya R (2003) Identification of factors that cause genotype by environment interaction between herds of Holstein cattle in seventeen countries. J Dairy Sci 86(3):1009–1018

Chapter 14
Genes Involved in the Thermal Tolerance of Livestock

Robert J. Collier, Kifle Gebremedhin, Antoni R. Macko and Kajal Sankar Roy

Abstract Although there is a good amount of knowledge about the physiological aspects, the effects of heat stress at the cellular and genetic level still remain unrevealed. Functional genomics research is providing new knowledge about the impact of heat stress on livestock production and reproduction. Using functional genomics to identify genes that are up- or down-regulated during a stressful event can lead to the identification of animals that are genetically superior for coping with stress and toward the creation of therapeutic drugs and treatments that target affected genes. Given the complexity of the traits related to adaptation to tropical environments, the discovery of genes controlling these traits is a very difficult task. With the development of molecular biotechnologies, new opportunities are available to characterize gene expression and identify key cellular responses to heat stress. These new tools will enable to improve the accuracy and the efficiency of selection for heat tolerance. Studies evaluating genes identified as participating in the cellular acclimation response from microarray analyses or genome-wide association studies have indicated that heat shock proteins are playing a major role

R. J. Collier (✉)
Department of Animal Sciences, University of Arizona, Tucson, AZ 85721, USA
e-mail: rcollier@ag.arizona.edu

K. Gebremedhin
Department of Biological and Environmental Engineering, Cornell University, Ithaca, NY 48824, USA

A. R. Macko
Department of Physiological Sciences, University of Arizona, Tucson, AZ 85721, USA

K. S. Roy
National Institute of Animal Nutrition and Physiology, Adugodi, Bangalore, Karnataka 560 030, India

V. Sejian et al. (eds.), *Environmental Stress and Amelioration in Livestock Production*, DOI: 10.1007/978-3-642-29205-7_14,

in adaptation to thermal stress. Additional genes of interest which two or more studies have identified are the genes for fibroblast growth factor, solute carrier proteins, interluekins, and tick resistance genes. Genes which have only been identified by microarray analysis but not by genome-wide association studies include genes associated with cellular metabolism (phosphofructo kinase, isocitrate dehydrogenase, NADH dehydrogenase, glycosyltransferase, transcription factor, and mitochondrial inositol protein). Other genes of importance were thyroid hormone receptor, insulin-like growth factor II, and annexin.

Keywords Acclimation · Thermal tolerance · Functional genomics · Heat stress

Contents

14.1 Interaction of Animals with Their Environment

Of importance in the livestock industry are the complex physiological adaptive responses of the stress and how they affect the ability of domestic animals to reproduce and produce on an economic basis (Scott 1981). These adaptive responses have been evaluated and measured using two different research approaches; field observations under different seasons, and laboratory experiments using only a given part of the complex climatic factors encountered in the natural environment. Both approaches are necessary for a better understanding of the optimal environmental conditions which yield the highest economic productivity of livestock. Not only are responses of livestock under continued investigations, but emphasis is given to the mechanisms controlling these responses. In addition, new attention is being directed toward modeling of the complex interactions between the physiological responses and the thermal environment. Many large gaps remain to be bridged, but important accomplishments have been made to ameliorate environmental impacts in improving livestock industry in regions where climatic stress prevails for a significant part of the year (Yousef 1984).

The physical factors that are important to productivity of livestock include air temperature, humidity, solar radiation, and wind speed. Therefore, the microclimate around an animal is thought of as a four-dimensional space (e.g., hot, humid, sunny, windy or hot, dry, sunny, and calm) in which the four independent environmental variables are acting simultaneously and are each time dependent. Mammals may survive, thrive, or die in an unfavorable thermal environment depending on their limit of tolerance and ability to utilize efficient physiological and behavioral mechanisms to maintain a heat balance between their bodies and the environment.

Thermal energy is exchanged between an animal and its environment by radiation, conduction, convection, and evaporation. A conceptual picture of the streams of energy flowing to and from a cow outdoors is shown in Fig. 14.1.

The outermost surface of skin and hair coat of the animal is the transducing surface across which the environment interacts with the internal physiology. As such, the energy flow is a function of the physical properties (hair length, hair-coat thickness and density, color) and optical properties (absorptivity, transmisivity) of the hair coat. Hair-coat and skin characteristics are under genetic and seasonal (acclimatization) control. Animals have developed coping mechanisms to minimize the impact of these environmental stressors on their biological systems. These responses are termed acclimation, acclimatization, and adaptation. Acclimation is defined as the coordinated phenotypic response developed by the animal to a specific stressor in the environment (Fregley 1996) while acclimatization refers to a coordinated response to several simultaneous stressors (e.g. temperature, humidity and photoperiod, Bligh 1976). Generally, there is hardly ever an example under normal environmental conditions where only one variable is changing. Therefore, typically an animal is undergoing acclimatization to a changing environment. Acclimation and acclimatization are induced by the environment and are considered phenotypic and not genotypic change and the responses decay if the stress is removed. Acclimation and acclimatization act to improve the fitness of the animal to the environment. In many cases the response is induced by sudden environmental change such as heat or cold stress. In other examples the acclimation response is driven by seasonal changes in photoperiod or other environmental cues such as the lunar cycle which permit the animal to "anticipate" the coming change in the environment leading to seasonal acclimation adjustments in insulation (coat thickness, fat deposition), feed intake or reproductive activity in advance of the actual environmental change. However, in every case, the process is driven by the endocrine system and is "homeorhetic" meaning metabolism is coordinated to support a specific physiologic state (Bauman and Currie 1980). In this case the specific physiologic state is the "acclimatized animal". If the environmental stressors are present for prolonged

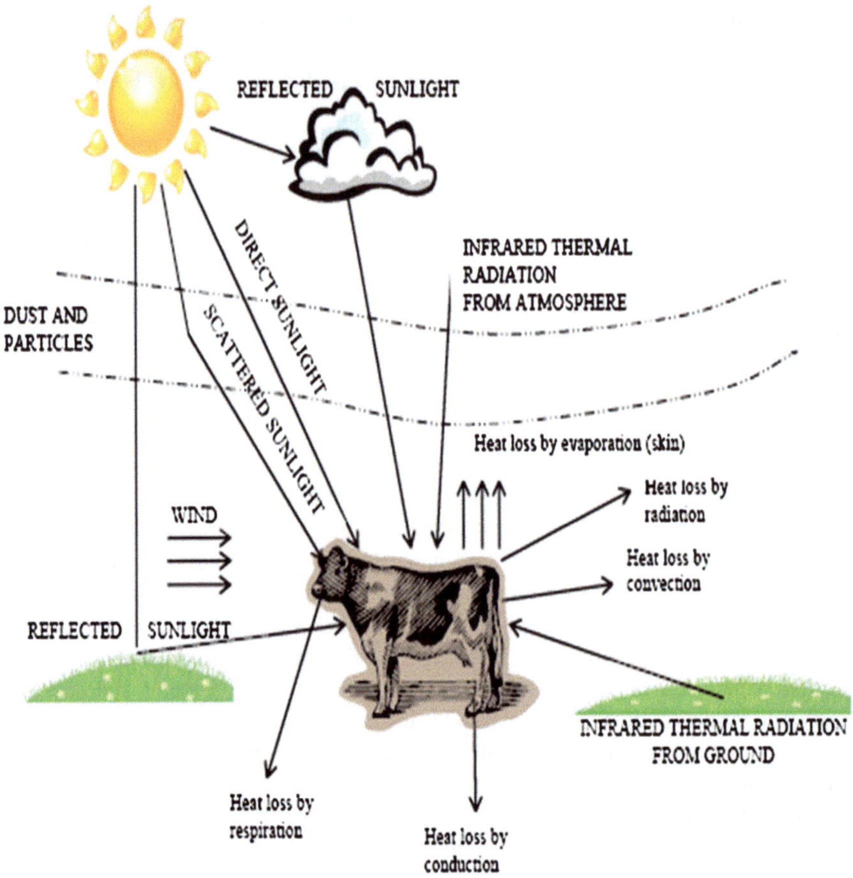

Fig. 14.1 Routes of heat exchange between a cow (outdoors) and its thermal environment include radiation, conduction, convection, and evaporation (from the skin and respiration) as the four basic modes of heat transfer

periods of time (e.g. years) these metabolic and physiologic adjustments can become "fixed genetically" and the animal is considered "adapted" to the environment. For example, *Bos indicus* breeds of cattle have evolved under conditions of high temperature and humidity and display a number of genetic differences which endow them with improved thermotolerance compared to *Bos taurus* breeds of cattle which evolved under temperate weather conditions. *B. indicus* cattle have greater thermoregulatory capability than *B. taurus*. As pointed out by Hansen 2004, *B. indicus* cattle produce less heat, have increased capacity to lose heat toward the environment, or a combination of both. This suggests that low metabolic rates resulting from reduced growth rates and milk yields of many indicus breeds is a major contributing factor to thermotolerance. The basal metabolic rate of *B. indicus* is infact lower than for *B. taurus*. (Finch 1985). The physiological and cellular basis for this difference have not been identified. One possibility for improved heat loss in *B. indicus* is that the density of arteriovenous anastomoses is higher in *B. indicus*. Since these structures have lower resistance to flow than vascular passages involving capillary networks they facilitate increased blood flow to the skin during heat stress (Hales et al. 1978).

The vascularity and degree of insulation of the skin and quality of the hair coat (hair and skin coat color, thickness and density of hair fibers) also contribute to the effectiveness of heat loss in cattle (Gebremedhin et al. 2008, 2010). All of these are affected by breed and contribute to well-known genotype x environment effects.

14.2 Heat Loss

14.2.1 Sensible Heat Loss

Heat loss from the animal is divided into sensible heat loss and evaporative heat loss. Sensible heat loss comprises heat loss by conduction, convection, and radiation.

Radiant energy exchange occurs between two surfaces as each surface emits energy at wavelengths that are dependent on the temperature of the emitting surface. Infrared radiation refers to radiation of wavelengths longer than those in the visible portion of the spectrum. Convection, however, is heat loss from a surface of an object by fluid (air) flowing over the surface. Convection could be natural (also called free convection) or forced. Free convection occurs when temperature difference in the boundary layer of air surrounding an object causes a movement of the air in response to changes in air density, but forced convection occurs when external pressure difference cause wind to blow past the object. Conduction heat transfer results from the exchange of energy from higher temperature to lower temperature by direct contact of the two surfaces. The heat loss by conduction would be very small when the animal is standing but the surface area in contact to the ground would be about 20% of the total surface area when lying down (Moen 1973). Cows are, however, more likely to stand in sun than lying down because their body temperature increases when lying down because of reduced effective surface area exposure for convective evaporative cooling (Gebremedhin et al. 2011).

Increasing ambient temperature lowers the temperature gradient between the animal and air and consequently decreases the sensible heat loss. Increased air velocity (between 0.5 and 3.0 m/s) over the cow's surface increases convective heat loss by reducing the insulation of the boundary layer, or by reducing the insulation of the hair coat if the velocity is greater than 3.0 m/s (Berman 2004). Higher velocities penetrate into the hair coat and thus increase the convective heat loss from the fur layer and skin surface.

14.2.2 Cutaneous Heat Loss

Various behavioral and autonomic thermoregulatory mechanisms are utilized by cattle to relieve heat stress. A distressed animal may seek shade, change its orientation to the sun, and increase water intake (Blackshaw and Blackshaw 1994). Sweating and panting are two of the primary autonomic responses exhibited by cattle under heat stress. Sweating leads to evaporative heat loss from the skin surface, whereas in panting, sensible heat is used to heat the water vapor and remove heat in the form of vaporized moisture from the lung.

At low ambient temperature, evaporative heat loss represents a small fraction of metabolic heat production. When air temperature is between 10 and 20°C, heat loss by cutaneous evaporation accounts for 20–30% of the total heat loss but when the temperature is greater than 30° C, cutaneous evaporation becomes the primary venue for heat loss, accounting for approximately 85% of the total heat loss, while the rest is lost by respiratory evaporation (Maia et al. 2005). When ambient temperature equals internal body temperature, sensible heat loss is zero and evaporative heat loss via sweating and panting becomes the only available venue for heat loss. Both sweating and panting have the undesirable side effect of depleting body-water reserves.

Cutaneous evaporation is affected by wind velocity, ambient temperature, relative humidity, and thermal and solar radiation. Other factors that affect the efficacy of evaporative cooling from the skin surface are hair-coat physical and optical properties such as hair density, hair length, hair-coat thickness, and hair color. Some of these properties such as hair and skin color may enhance solar absorption and thus increase solar heat load on the skin surface. Other properties such as hair-coat density, which is necessary to provide insulation during cold weather, obstruct free evaporation of water from the skin surface during hot weather resulting in heat stress (Gebremedhin et al. 2008).

Wind greatly increases evaporative heat loss from the skin surface. Wind penetrates the hair coat and reduces the effective thickness of the coat (McArthur 1987). Thermal resistance decreases linearly in hair coat with an increase in air velocity. Increasing air velocity over the hair-coat surface from about 0.2 m/s to about 0.9 m/s raised sweating rate from about 75 g/m^2-h to about 350 g/m^2-h and no further increase in sweating rate occurred as air velocity was increased to 2.2 m/s (Hillman et al. 2001).

Sweating moistens the skin surface and usually leaves the hair-coat layer dry. Wetting the hair coat with a mist of water adds water to the coat surface and penetrates deep down to the skin surface and wets it. The process of evaporative cooling is a complex interaction of humidity and temperature difference between air and skin, air velocity, and hair-coat characteristics such as density and depth. A model incorporating these variables has been developed, where evaporative heat loss resulting from wetting the hair coat as the more effective means of heat loss than sweating (Kimmel et al. 1991). Hillman et al. (2005) reported that heat-stressed cows actively stand under water spray to wet their hair coat. They further reported that this process lowered core-body temperature. A faster drop in core-body temperature was recorded when the hair coat was sprayed with water at short time intervals between sprays. The decrease in core-body temperature occurred at a faster rate when the spraying is combined with blowing air over the hair coat (Brouk et al. 2003; Hillman et al. 2001).

Gebremedhin et al. (2008) reported that cutaneous moisture production (sweating rate) ranged between 189 ± 84.6 and 522 ± 127.7 g/m^2-h for solar load exposure >500 W/m^2 (average 833 ± 132 W/m^2), average THI of 82.7 ± 1.64, air velocity between 0.8 and 1.2 m/s, and body (rectal) temperature >38.8°C (threshold for heat stress). The same report indicated that in a hot and dry environment, evaporative cooling was profoundly increased (from 68 g/m^2-h before wetting to 508 g/m^2-h after wetting the skin surface, exposed to 0.2 m/s air velocity, and without solar load), and the rate was further increased (from 296 g/m^2-h before wetting to 961 g/m^2-h after wetting) when air velocity over the wetted skin surface was increased to 0.9–1.0 m/s. In a hot and humid environment, however, the increase was relatively modest (from 258 g/m^2-h before wetting to 490 g/m^2-h after wetting) for an air velocity of 0.9–1.0 m/s and a solar load >600 W/m^2. Moisture production is higher in hot and dry conditions than in hot and humid because of higher moisture gradient between the skin surface and ambient air.

A prolonged (extended) exposure to a hot and dry condition or exposure to 3 h of 850 W/m^2 of solar load caused rectal temperature to rise above 40.0°C and respiration rate to rise above 125 breaths/min. Under these conditions, black or predominantly black cows were observed foaming in the mouth, sticking their tongues out, and drooling (Gebremedhin et al. 2010). During these events, immediate intervention with water spray of the body helps to alleviate heat stress. A physiological upper limit of moisture production, which is different for each cow, seems to exist. The maximum sweating rate was about 660 g/m^2-h for dairy cows and feedlot heifers, and the driving force for moisture production seems to be skin temperature (Gebremedhin et al. 2010).

14.2.3 Respiratory Heat Loss

Sweating and panting appear to be under independent control. Energy is lost by vaporizing water within the respiratory tract. The amount of water lost in grams per unit time when multiplied by the latent heat of vaporization of water at the appropriate temperature gives the rate of water loss in energy units.

Maia et al. (2008) developed a model for calculating the heat loss by respiratory evaporation (Q_R, W m^{-2}) and is expressed as:

$$Q_R = \frac{\lambda VF(w_E - w_A)}{60A} \tag{14.1}$$

where, $\lambda = 2500.7879 - 2.3737 * t_A$ is the latent heat of vaporization (Jg^{-1}); A = the body surface area; V (m^3 $breath^{-1}$) = tidal volume; F = respiratory rate (breaths min^{-1}); w_A, w_E (g m^{-3}) are absolute air humidity of the atmosphere and expired air, respectively, and they are given by:

$$w_A = \frac{2166.87 \times P_p\{t_A\}}{273.15 + t_A} \quad (14.2a)$$

$$w_E = \frac{2166.87 \times P_p\{t_E\}}{273.15 + t_E} \quad (14.2b)$$

where, $P_P\{t_A\}$ and $P_P\{t_E\}$ are the partial vapor pressures (kPa) of ambient air and expired air, respectively; t_A and t_E are temperatures (Celsius degree) of ambient and expired air, respectively.

Stowell (2000) suggested that a respiration rate of 80–90 breaths/min was a clear indication of a cow experiencing heat stress.

14.3 Modeling Heat Exchange Between an Animal and the Environment

It is impossible to describe mathematically all of the dynamic thermal relationships between a biological object and its thermal environment, but research on the thermal energy exchange of both plants and animals has provided insight into the mechanisms involved. The climatic energy demands (food and water) for an animal is inherently a coupled heat and mass transfer problem (Fig. 14.2). The energy balance is the diagonal in the figure, and the horizontal expressions represent the food ingested and water consumed (Gebremedhin et al. 1984).

It is obviously impractical to examine experimentally all the possible combinations of radiation, air temperature, wind velocity, animal size, metabolic rate, evaporative water loss, and hair-coat physical and optical properties. There are considerable experimental difficulties in measuring accurately animal thermal responses in their natural habitat. Therefore, mechanistic models have to be developed to simulate the energy exchange between the animal and its environment and be able to evaluate what if conditions.

The analysis given below is for a steady-state situation. However, much of the time, an animal is in a transient energy state as it moves about in the environment. While in transient states, the energy budget for an animal must average within the environmental limits permitted by the steady-state requirements for survival. An animal may search for food in an environment which is intolerable to it as a steady-state situation, provided it is in this environment for a short period of time compared to the body-time constant of the animal.

A basic concept of a one-dimensional mechanistic heat transfer model through artificial fur layer was described by Kowalski and Mitchell (1979) and through

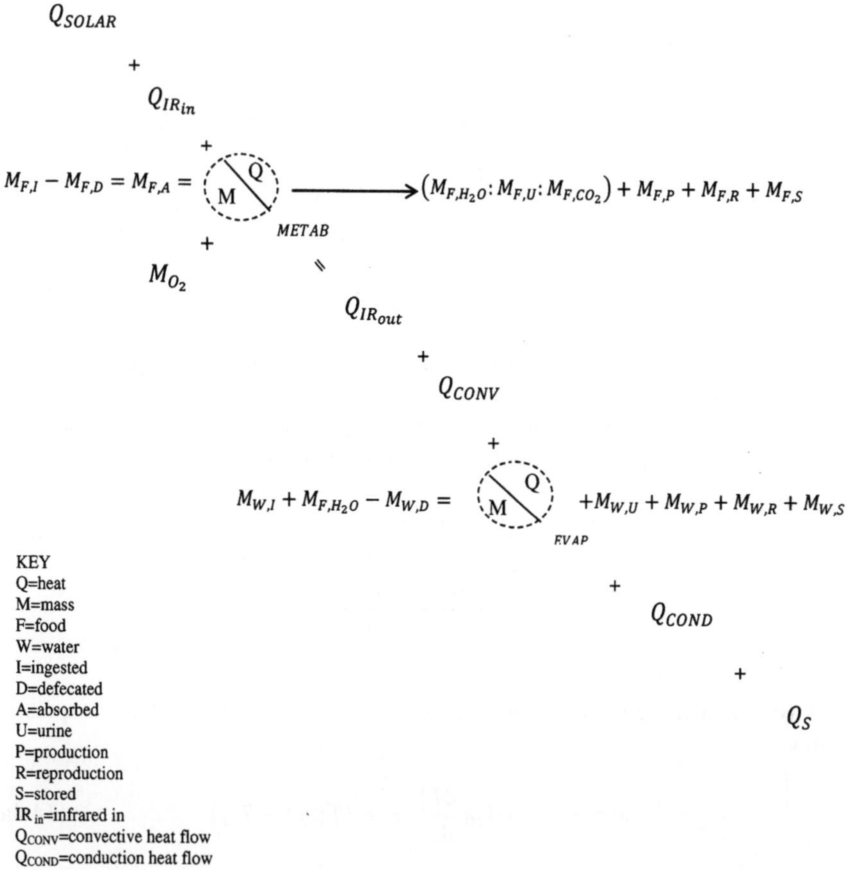

Fig. 14.2 Climatic energy demand (feed and water) for livestock (Gebremedhin et al. 1984)

animal hair coat by Gebremedhin et al. (1983). The model is a function of environmental parameters and hair-coat physical properties. The absorbed solar radiation, q''_{solar}, within the hair coat was incorporated into the model as an internal heat generation source. The model describes the energy transport through the hair coat and the solution provides the temperature profile through the hair coat. The model is developed from basic principles where conduction heat flux, q''_c, and radiation, q''_r heat flux enters into and leaves out of the element at both the top and bottom of a section of hair coat (Fig. 14.3). The effective conductivity, k_{eff}, accounts for the physical properties of the hair coat.

An energy balance on the incremental element yields the differential equation which describes the temperature profile through the hair coat. After solving for the temperature profile, the heat loss through the hair coat can be determined from an energy balance on the skin surface (Gebremedhin et al. 1983). Heat loss is directly

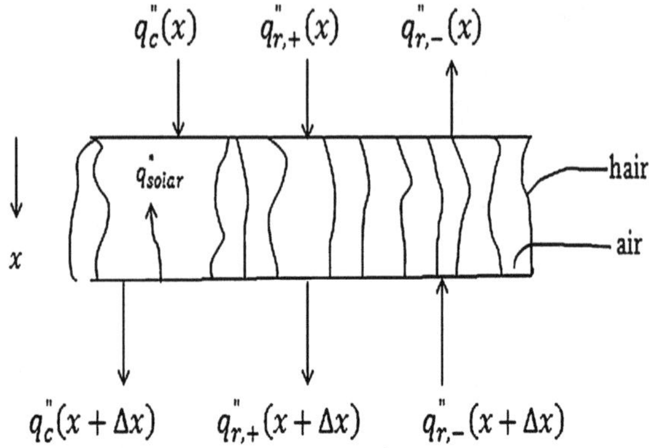

Fig. 14.3 Heat flow within an incremental element of hair coat

related to metabolism and, therefore, to feed requirements for survival, production (or growth), and reproduction potential. The energy balance is expressed as

$$\underbrace{k_{\text{eff}}\frac{\mathrm{d}^2T}{\mathrm{d}x^2}}_{\text{Conduction term}} - \underbrace{\underbrace{\frac{\mathrm{d}q''_{r,+}(x)}{\mathrm{d}x}}_{\text{Inward flux}} + \underbrace{\frac{\mathrm{d}q''_{r,-}(x)}{\mathrm{d}x}}_{\text{Outward flux}}}_{\text{Long-wave irradiation}} + \underbrace{q''_{\text{solar}}}_{\text{Solar}} = 0 \tag{14.3}$$

Two boundary conditions are needed to solve Eq. 14.3 and are expressed below:

$$\text{at } z = z_L \quad -k_{\text{eff}}\left.\frac{\mathrm{d}T}{\mathrm{d}z}\right|_{z_L} = h_c(T(z_L) - T_{\text{air}}) \tag{14.4a}$$

$$\text{at } z = 0 \quad T(0) = T_{\text{Skin}} \tag{14.4b}$$

In the first boundary condition, the energy conducted to the hair-coat-air interface equals the energy convected from that surface. This boundary condition couples the outer hair-coat surface to the prevailing wind conditions through the convective heat transfer coefficient, h_c, and ambient air temperature, T_{air}. The second boundary condition is the temperature at the skin surface, $z = 0$. It may be either specified or measured. More details of this model is given in Gebremedhin et al. (1983).

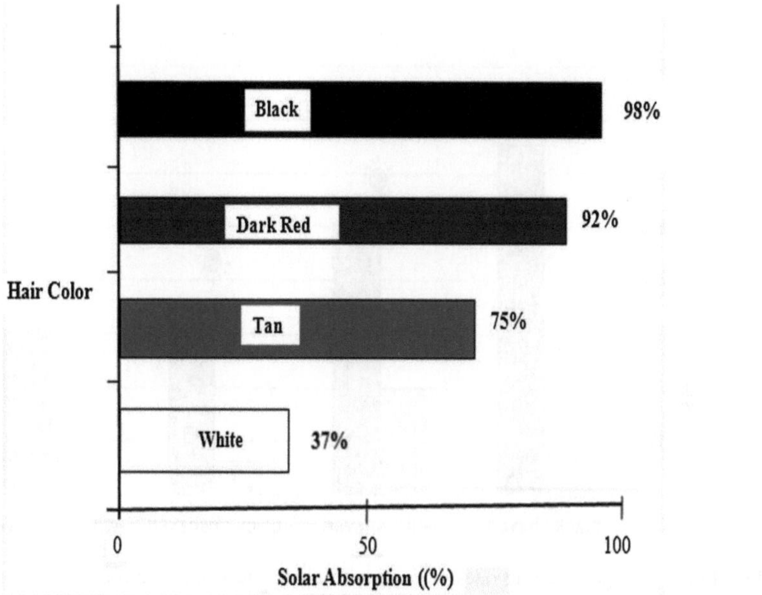

Fig. 14.4 Relationship between hair color and solar absorption for four breeds of heifers (*black* Angus, *dark-red* MARC III, *tan-colored* MARC I, and *white* Charolais) (Gebremedhin et al. 2011)

14.4 Effects of Hair Coat Physical and Optical Properties on Heat Exchange

Hair and feathers do not merely provide insulation from the cold, but from heat as well. They tend to ameliorate the thermal regime of the body proper by buffering the thermal variations.

The effect of direct solar radiation on the radiant temperature over the entire surface of an animal is not uniform because both the color of this surface and the angle of the rays striking the surface are important in determining just how much solar energy is absorbed. The relationship between hair color and solar absorption for four breeds of heifers is shown in Fig. 14.4.

In a study on the effect of hair color on thermoregulation, Hillman et al. (2001) reported that when black Holstein cows were exposed to direct sunlight, their surface temperature increased by 4.8°C, and by 0.7°C for white cows. The difference in temperature between black and white is because of higher solar absorption by black than white. They also reported that rectal temperature increased at a rate of 0.7°C/h for black cows and 0.3°C/h for white cows. In another study (Hillman et al. 2005), they reported an increase in sensible heat flux in the order of 26% for dark-red, 22% for black, 5% for tan, and 4% for white.

Da Silva et al. (2003) investigated radiative properties of the skin and hair coat of various breeds of cattle with respect to shortwave radiation. The study

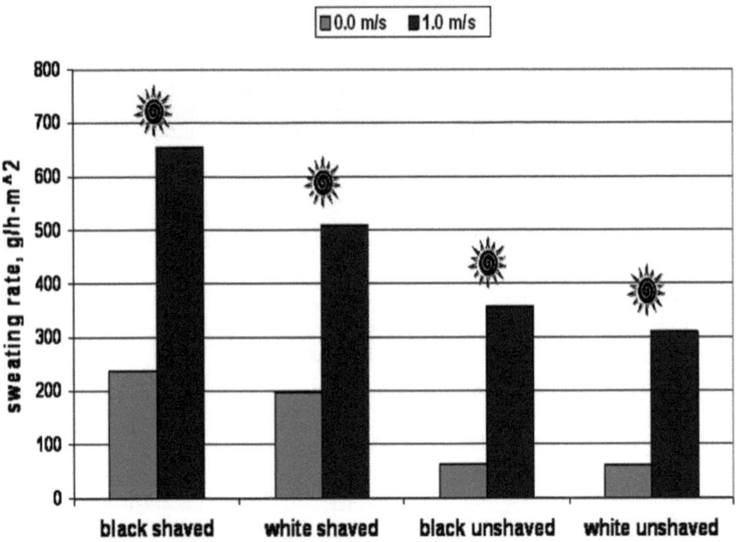

Fig. 14.5 Average sweating rates of cows with black and white hair coats shaved or unshaved under direct sunlight at zero air velocity, and 1.0 m/s (Gebremedhin et al. 2007)

concluded that light-hair coats exhibited much higher reflectivity than dark-hair coats for wavelengths ranging from 300 to 850 nm.

The presence of a hair coat conserves heat by entrapping air. The entrapped air serves as insulation in the case of cold environments but becomes an obstruction to evaporation of moisture from the skin surface in the case of hot environments. This was substantiated by measuring sweating rates from shaved and unshaved areas of a cow (Gebremedhin et al. 2007). Sweating rates from the shaved areas were higher than those from the unshaved areas, regardless of hair-coat color (Fig. 14.5). On the average, sweating rate from black shaved areas was 1.84 times higher (655 g/h-m^2) than that from unshaved black hair coat (356 g/h-m^2). Similarly, sweating rate from a white shaved area was 1.64 times higher (509 g/h-m^2) than that from unshaved white hair coat (310 g/h-m^2). On the shaved areas, the conversion of sensible heat to latent heat is unobstructed because sweat (water) on the skin surface is exposed directly to ambient air and solar radiation. On the unshaved areas, however, the presence of hair coat above the sweat glands might be acting as a moisture trap creating locally a more humid environment. Thus, the presence of moisture in the hair coat results in lowering the moisture gradient between the skin surface and the hair layer above it, causing less evaporation to take place, and consequently depressing heat loss from the skin surface. The presence of a hair coat, therefore, becomes a liability i.e., obstructing evaporative cooling of the skin surface in hot environments. That is perhaps why cows shed some of their hair during summer.

14.5 Genetics of Hair Coat

Several loci and some genes have been identified as having involvement in coat color and quality as well as heat dissipation, Olson et al. (2006). These changes are all adaptive in nature and are genetically fixed in their populations.

14.6 Acclimation and Adaptation to Thermal Stress

When animals are adapted, the physiologic differences between them and non-adapted animals do not disappear when the environment changes. This is not the case in acclimation where differences do disappear if the stress is removed. However, it is becoming apparent that the systems which are involved in acclimatization are the same systems which endow animals with thermotolerance or adaptation to heat. Therefore, obtaining a better definition of the gene networks up- and down-regulated in response to environmental stress will also lead us to those pathways which offer promise to improve thermotolerance.

Acclimatization is a homeorhetic process which requires several days to weeks to fully develop. There is a hormonal link between the central nervous system and the effector cell types involved and the end result of the hormonal change is to alter the responsiveness of the effector cells to environmental change (Blighs 1976). These key features are hallmarks of a homeorhetic process where metabolism of multiple tissues and organs are coordinated to support the new acclimatized state as contrasted to a homeostatic process, (Bauman and Currie 1980, Collier et al. 2005), where regulation is occurring around a set point. We then should consider the stages of acclimation, the hormones that are driving acclimation and what changes are occurring in effector tissues to accomplish development of the acclimatized state.

14.7 Stages of Acclimatization

Acclimatization is generally considered to occur in two stages; acute or short term and chronic or long term (Johnson and Vanjonack 1976; Horowitz 2002; Garrett et al. 2009). The acute phase includes the shock response at the cellular level, (Carper et al. 1987; Sonna 2002) and homeostatic endocrine, physiological, and metabolic responses at the systemic level while the chronic or long-term phase results in acclimation to the stressor and involves reprograming of gene expression and metabolism, (Horowitz 2002; Collier et al. 2006b).

In agricultural animals there is typically a loss in productivity as animals progress through the acute phase and some or even all of this productivity is restored as animals undergo acclimation to the stress.

14.7.1 Systemic Response

The systemic response to environmental stress is driven by two systems—(1) the central nervous system (CNS) and (2) peripheral nervous system and endocrine components (Charmandari et al. 2005). The central component is comprised of nuclei in the hypothalamus and the brainstem which releases corticotrophin-releasing hormone (CRH) and arginine vasopressin (AVP). The peripheral components of the stress system include the pituitary-adrenal axis, the efferent sympathetic-adrenomedullary system, and components of the parasympathetic system (Habib et al. 2001). However, relative to environmental stress and acclimation the initial phases of the response involve receptor systems at the periphery which drive autonomic and endocrine responses to the changing environment such as skin thermoreceptors and photoreceptors in the retina.

Sweating and panting are two of the primary autonomic responses exhibited by animals under heat stress. Sweating results in increased evaporative heat loss from the skin surface, whereas in panting, sensible heat from the body core is used to heat the water vapor and expel heat in the form of vaporized moisture from the lungs. However, these responses are likely driven more by surface than core-body temperatures, evaporative heat loss from skin and the respiratory tract is highly correlated with skin temperature. In fact, skin temperature is more highly correlated with these parameters than core temperature suggesting that thermal receptors in the skin initiate the autonomic systemic response to thermal stress. Another potential route of information flow from the surface to the whole system would be via secreted heat shock protein released from skin epithelium during heat stress which would act as an alarm system to assist in mobilizing the acute response to thermal shock. An examination of the relationship between skin temperature and expression of the gene for inducible heat shock protein 70 revealed that gene expression is increased several fold as skin temperature approaches 35°C which is below body temperature but represents the upper limit of the thermoneutral zone of cattle. Berman (2005) estimated that the stress response system in cattle would be activated at effective temperatures at and above 35°C. Previously, it has been demonstrated that evaporative heat loss and rectal temperature rise dramatically above an effective environmental temperature of 35°C, Collier et al. 2008a). It is now apparent that the heat shock response in bovine skin epithelial tissue is activated at effective environmental temperature of 35°C as well. Activation of the heat shock response in cells in many cases leads to secretions of HSP's into the extracellular space and plasma, Ireland et al. 2007)

Recently, secreted HSP was identified in plasma of cattle. Kristensen et al. (2004) and Gaughan and Bonner (2009) have demonstrated that secreted HSP concentrations rise in plasma when the effective environmental temperature exceeds 35°C.

Thus, activation of the heat shock response in cells also leads to secretion of Hsps into the extracellular space and plasma, (Ireland et al. 2007). It has been hypothesized that secreted heat shock protein acts as an alarm signal for the immune system and several measures of innate immunity are increased following increases in secreted heat shock protein in blood (Fleshner and Johnson 2005). Secreted heat shock protein has also been shown to improve survival of neural cells subjected to environmental and metabolic stressors, (Tytell 2005; Guzhova et al. 2001).

Thus, the acute response in cattle is initially driven by thermal receptors in skin which activate the central nervous system and subsequently, the endocrine system and the peripheral components of the autonomic system. This response is augmented by secreted heat shock proteins which rapidly rise in plasma and are believed to provide protective effects to a variety of cell types as well as activating the innate immune system. At a skin surface temperature of 35°C the respiration rate of cattle will reach or exceed half maximal which is about 60–70 breaths/min. At this point, the animal is entering the acute phase of the stress response.

During the acute phase there is rapid decline in productivity of domestic animals and this is especially true in high producing dairy cows. The decline in productivity begins on the day the stress is initiated but is not maximal until 48 h following the initiation of the stress (Collier et al. 1981). This suggests that there are intermediate events between the rise in body temperature and the reduction in milk yield. Rhoads et al. (2009) demonstrated that reduced feed intake only accounted for 40% of the decline in milk yield and that other factors were likely involved in the rapid decline in production due to severe heat stress. They postulated that reduced glucose availability could potentially reduce lactose synthesis rates and contribute to the reduced milk volume. Silanikove et al. (2009) reported that acute heat stress reduced milk secretion in lactating cows by up-regulating the activity of a milk-borne negative feedback regulatory system, specifically an n-terminal fragment of β-casein. They also reported that this fragment has an inhibitory activity on the mammary epithelial cell potassium channel. Identification of the exact mechanisms by which milk yield is reduced in response to heat stress offers potential in improving productivity of animals in warm climates.

After 5–7 days of continuous stress, animals enter the chronic "acclimation" phase of the stress response. During this phase there is a reprograming of metabolism resulting in altered responses to homeostatic signals. The overall impact of these changes is a reduction in the impact of the stress on the animal. The transition of animals from the acute to the chronic phase of the stress response has been extensively studied in laboratory models by Horowitz and coworkers (Horowitz 2002; Horowitz et al. 2004; Maloyan et al. 2005; Horowitz 2007). These changes are driven by the endocrine system and result in global changes in gene expression as well as post-translational alterations in protein function. Hormones which have been identified as being homeorhetic regulators are also linked to acclimation responses to thermal stress and changes in photoperiod related to season. These include somatotropin, prolactin, thyroid hormones, glucocorticoids, and mineralocorticoids. Several of these hormones are known to contribute to regulation of HSP gene expression.

The changes which occur at the cellular level which provide improved cytoprotection are described in the section on the cellular response to heat stress.

Table 14.1 Genes associated with heat shock response and/or acclimation to warm climates

Gene	References
Heat shock proteins	1, 2, 3
Interleukins	1, 3
Keratins	1, 3
Fibroblast growth factor	1, 2
CD antigen	3
Solute carrier family	1, 3
NADH dehydrogenase	1
Tick resistance genes	1, 3
Collagen, type VI	1
Kinesin family	1
Selenium binding protein	1
Annexin	1
Glycosyltransferase	1
Protein kinase C	1
Transcription factor	1
Bcl2 associated anthanogene 3	1
Zinc finger	1
Thyroid hormone receptor	1
Mitochondrial inositol Protein	1
Isocitrate dehydrogenase	1
Butyrophilin 3	1
Phosphofructo kinase	1
Insulin-like growth factor II	1

References: 1. Collier et al. 2006b; 2. Hayes et al. 2009; 3. Chan et al. 2010

At the systemic level, metabolism is coordinated to support a new physiological state. Some of the changes which occur include a lowering of the threshold for vasodilation and evaporative cooling (Roberts et al. 1977), reduced metabolic rate, increased resistance to thermal injury, and improved cardiac performance, (Horowitz 2002). Endocrine changes occurring with acclimation to heat stress in cattle include increased plasma prolactin, reduced glucocorticoid, somatotropin, and thyroxin concentrations, increased progesterone concentrations and decreased estrone sulfate concentrations in pregnant cattle (Collier et al. 2005). Somatotropin is a homeorhetic regulator and has been shown to be beneficial in improving evaporative heat loss and thermal balance in cattle during summer heat stress (Manulu et al. 1991).

Despite large reductions in feed intake and energy balance during heat stress there appears to be a tighter coupling of the somatotropin-IGF axis during summer resulting in higher IGF concentrations during summer months compared to winter months and only slight decreases in plasma IGF to severe heat stress even when somatotropin concentrations are reduced, (Collier et al. 2008b; Rhoads et al. 2009). This fact reinforces the importance of somatotropin in dealing with environmental stress, (Collier et al. 2005). Additionally, the somatotropin response to GRF is not affected by severe heat stress, (Rhoads et al. 2009). The seasonal variation in coupling of the growth hormone-IGF axis is also associated with effects of increased photoperiod on growth rate and milk yield in cattle, (Collier et al. 2006a).

14.7.2 Cellular Responses

Heat tolerance at the cellular level is directly related to the ability of the cell to maintain elevated levels of heat shock proteins (HSP's). As stated by Horowitz and Assadi (2010) "A hallmark of the acclimation process is the enhancement of cytoprotective networks-that of the heat shock proteins, anti-oxidative and apoptotic-and the stabilization of the Hypoxia Inducible Factor- Iα, the master regulator of oxygen homeostasis".

The heat shock response of bovine embryos and mammary epithelial cells to heat stress has been described (Edwards et al. 1997; Jousan and Hansen 2004; Collier et al. 2006b, 2008a). The dramatic effect of heat shock on mammary epithelial cell growth and structure has been well described (Collier et al. 2008a). Heat shock acutely downregulates DNA synthesis and adversely affects the ability of cells to maintain their cytoskeleton leading to a collapse of cell structure.

The transcriptome profile of heat-shocked bovine mammary epithelial cells indicated down-regulation of genes involved in cell structure, DNA synthesis, cell division, metabolism, biosynthesis, and intracellular transport while genes associated with cellular and protein repair and degradation were up-regulated. The heat shock response is induced by accumulation of mis-folded proteins in the cytoplasm and is mediated by heat shock transcription factors (HSF), (Voellmy and Boellmann 2007; Horowitz and Robinson 2007). There are four forms of HSF but HSF-1 is considered to be the primary transcription factor involved in the heat shock response, (Akerfelt et al. 2007). Regulation of HSF-1 activity has been reported to be largely controlled post-translationally and not at the level of synthesis/degradation of the transcription factor, (Voellmy and Boellmann 2007). Once activated the HSF-1 monomer trimerizes with other HSF-1 molecules which is essential for DNA binding (Sarge et al. 1993). The activated complex can then enter the nucleus and initiate transcription of heat shock proteins.

Genes currently identified as participating in the cellular acclimation response are shown in Table 14.1. These genes were either identified from microarray analysis (Collier et al. 2006b) or genome-wide association studies (Hayes et al. 2009; Chan et al. 2010). All three studies indicate that HSPs are playing a key role in acclimation and adaptation to thermal stress. Additonal genes of interest which two or more studies have identified are the genes for fibroblast growth factor, solute carrier proteins, interluekins, and tick resistance genes. Genes which have only been identified by microarray analysis but not by genome-wide association studies include genes associated with cellular metabolism (phosphofructo Kinase, isocitrate dehydrogenase, NADH dehydrogenase, glycosyltransferase, transcription factor, and mitochondrial inositol protein). Other genes of importance were thyroid hormone receptor, insulin-like growth factor II, and annexin.

Thermal acclimation and thermal adaptation is associated with increased basal levels of HSPs (Carper et al. 1987; Kregel 2002; Maloyan et al. 1999). Thermal acclimation and cyclopentenone prostaglandins have been shown to increase the DNA-binding activity of HSF-1 leading to increased HSP gene transcription (Amici et al. 1992; Strauss and Glass 2001; Ianaro et al. 2003; Buckley and Hofmann 2002). Several investigators have shown that heat acclimation also provides cross-tolerance against other types of stress such as hypoxia, ischemia, acidosis, and energy depletion, (Horowitz 2002; Kregel et al. 2002) and that HSF-1 is involved in this process.

Although there is little evidence for endocrine regulation of HSF-1 gene expression activity, there is substantial evidence that expression of heat shock proteins and other cytoprotective proteins are modulated by the endocrine system (Xu et al. 1996)

The separate evolution of *B. taurus*, *B. indicus*, and *Sanga* cattle has resulted in differing genotypes of *B. indicus* and *Sanga* cattle that confer improved thermotolerance compared to *B. taurus* cattle in both beef and dairy populations (Paula-Lopes et al. 2003; Hansen 2004). In addition, large genotype x environment interactions in dairy cattle for milk yield (Cerón-Muñoz et al. 2004; Ravagnolo et al. 2000; Bohmanova et al. 2006) of Holstein cattle indicates that there is considerable opportunity to improve thermal resistance and performance in dairy cattle. These differences include thermoregulatory capability, feed intake, and production responses and cellular differences in heat shock responses, (Hansen 2004; Collier et al. 2008a).

There are genetic differences in cellular resistance to elevated temperature in cattle. It is possible that the same gene or genes conferring cellular thermotolerance are present in Indicus, Senepol, and Romosinuano, especially because of the contribution of *B. indicus* genotypes to these two other breeds (Magee et al. 2002). An alternative explanation is that distinct thermotolerance genes are present in the different genotypes. Identification of the genes conferring cellular thermotolerance offers the possibility of transferring these genes to heat-sensitive breeds to improve reproduction and other physiological systems compromised by hyperthermia. Little is known regarding the molecular basis for the improved cellular resistance to elevated temperature in thermotolerant cattle. There were no detectable differences between Indicus, Senepol, and Angus in the amount of heat shock protein 70 (HSP70) in heat-shocked lymphocytes (Kamwanja et al. 1994) although the tendency for lower cellular concentration of HSP's in Brahman and Senepol may indicate that protein denaturation in response to elevated temperature (one of the signals for HSP70 synthesis; Shamovsky and Nudler 2008) is reduced in Brahman and Senepol. The capacity for transcription in response to elevated temperature seems to be important for expression of genetic differences because there were no differences between Indicus and Holstein embryos in resistance to elevated temperature at the two-cell stage, a time when the embryonic genome is largely inactive, (Hansen 2004). Also, in vitro effects of elevated temperature on spermatozoa were similar for Indicus, Indicus-influenced breeds, Angus, and Holstein (Block et al. 2002).

The cellular thermotolerance of crossbred embryos is dependent upon the genotype of the oocyte and not the spermatozoa. Embryos produced by insemination of Brahman oocytes with Angus spermatozoa were more thermotolerant than embryos produced by insemination of Holstein oocytes with Angus semen (Block et al. 2002). In contrast, there were no differences in thermotolerance between Indicus × Holstein embryos and Angus × Holstein embryos. These results indicate that either genes conferring thermotolerance are paternally imprinted (only the maternal allele being expressed) or that thermotolerance in embryos depends upon some genetically controlled factor produced in the oocyte. Regulation of body temperature is the most critical factor for genetic differences in reproductive function during heat stress since the depression in fertility per unit increase in body temperature is the same for *B. indicus* × *B. taurus* crossbred cows as for Hereford × Shorthorn cows (Hansen 2004).

New genomics tools are also beginning to provide information on specific gene networks associated with thermotolerance. Lillehammer et al. (2009) identified single nucleotide polymorphisms (SNPs) that were associated with gene by environment interactions for production traits in cattle. Hayes et al. (2009) also reported on a genome-wide association study aimed at identifying differences related to adaptation to environmental change. It is envisioned that, in the not too distant future, use of SNP markers will lead to greater progress in improving thermotolerance of high producing dairy breeds.

Acclimation is a homeorhetic process driven by the endocrine system which enables animals to respond to a stress. The resulting cellular, metabolic, and systemic changes associated with acclimation reduce the impact of the stress on the animal and allow it to function more effectively in the stressful environment. These changes are lost if the stress is removed, so the process is not based on changes in the genome. However, if the stressful environment is not removed over successive generations these changes will become "genetically fixed" and are referred to as adaptations. A better understanding of genetic differences between adapted animals will contribute useful information on the genes associated with acclimation. Likewise, study of gene expression changes during acclimation will assist in identifying genes associated with improved thermotolerance.

14.8 Reproduction

The primary objective of the reproductive process is to produce a viable and healthy offspring. Heat stress has been shown to negatively impact livestock reproduction. Over 50% of the world's livestock reside in tropical regions where temperature or more specifically, the temperature humidity index often exceeds thresholds reaching levels that have been shown to impair reproduction as well as other aspects of livestock performance. As compared to European breeds, zebu cattle experience less severe reduction in reproductive function (Hansen 2004).

However, differences exist between a multitude of genomic and environmental factors as well as their interactions determine animal performance by affecting several major components involved in male and female reproductive function. Heat stress, in general, has deleterious effects on the reproductive process across mammalian species. However, certain breeds have adapted genetically to cope with heat stress and experience less severe reductions in reproductive function due to elevations in temperature. Recent improvements in management and feeding strategies in livestock have stemmed from improved understanding and ability to regulate physical environment surrounding the animal and elucidation of physiological mechanisms that govern animal nutrition, metabolism, reproduction, and lactation. Emerging research in the field of genomics is uncovering additional, promising avenues by which to improve animal production, breeding

There is a general belief that the appendages of *B. indicus* cattle contribute to the superior thermoregulatory ability of these cattle since they increase the surface area per unit body weight as compared to *B. taurus*. The actual importance of these anatomical features is not likely to be crucial for thermoregulation; because surgical removal of the dewlap or hump of Red Sindhi bulls did not have a measurable impact on thermoregulatory ability (McDowell 1958). Additionally, differences in regulation of rectal temperature in response to heat stress were observed between Jersey and Red Sindhi × Jersey even though surface area per unit body weight or metabolic body weight was similar between the two genotypes.

14.8.1 Placenta

14.8.1.1 Ruminant Placenta

Heat stress has long been known to impair reproductive function in cattle, sheep, and swine species. These losses are largely attributed to failure of embryonic implantation and early embryonic loss as a result of heat stress during early pregnancy. However, it has become apparent that that heat stress during mid- and late-gestation also contributes significantly to reductions in fetal survival and birth weight which are associated with decreased placental weight. Although the exact mechanism by which heat stress reduces fetal survival and birth weight are not yet known, several studies have identified maternal, fetal, and placental molecular deficiencies and differences in gene expression that may serve as lead to the development of screening methods for genetic markers to improve breeding and other management practices by identifying and exploiting genes associated with thermal tolerance traits.

The placenta is a temporary organ, present only in eutherian animals during pregnancy, is comprised of cells from both maternal and fetal origin and facilitates fetal-maternal exchange of nutrients, gasses, hormones, and waste products. The placenta begins to form early during pregnancy and attaches to the uterus in a process referred to as implantation. The second trimester is characterized by rapid placental growth and development, which precedes and facilitates the subsequent rapid fetal growth phase that defines the third trimester. In mammals, 95% of birth weight is accumulated by the fetus during the second half of gestation.

Our current understanding of the ruminant placenta and its function has been gained from observations on the effects of environmental heat stress on livestock animal reproduction in production settings as well as in controlled experimental investigations utilizing various animal models of placental insufficiency and fetal growth restriction. The ruminant placenta is classified as cotyledonary and typically contains 60–100 placentomes in sheep and 70–120 placentomes in cattle. The placentome is composed of a maternal component, the caruncle, and a fetal component, the cotyledon, which together comprise a fetal-maternal exhchange unit, the placentome, which is characterized by an intricate capillary network that serves to facilitate fetal-maternal exchange of nutrients, electrolytes, gasses, hormones, and waste products, while serving as a barrier to prevent the passage of larger items such as blood cells, immune system components, and large molecules (Mossman 1987).

Fetal growth is directly related to placental mass (Reynolds et al. 2005; Myers et al. 1982) and importantly also to the proper development of the placental microvasculature (Kingdom and Kaufmann 1997; Grazul-Bilska et al. 2010, 2011). Heat stress exerts deleterious effects on placental mass, microanatomical development, and function of the placentome (Kingdom and Kaufmann 1997; Regnault et al. 2002). The degree to which heat stress impairs placental mass and function depends on the timing, duration, and severity of the heat stress (Hafez 1964). In the ewe, heat stress prior to breeding and during the first trimester is associated with increased embryonic loss and disruptions in the process of implantation (Dutt 1963) reflected by a reduction in the number of placentomes formed.

Heat stress initiated at the onset and limited in duration to the second trimester (approximately 40–95 days gestation in the sheep, term = 148 days), a period beginning after implantation has completed and ending prior to the onset of the rapid fetal growth, is sufficient to elicit a greater than 50% reduction in placental mass (Bell et al. 1989) and is associated with decreased placental mass, altered placentome morphology, and decreased placental transport function (Alexander and Williams 1971). Microscopic imaging of placentomes of heat-stressed pregnant ewes reveals increased tortuosity and sinusoidal structuring of the microvasculature (Hafez et al. 2010), which is thought to be a compensatory mechanism to increase surface area and decrease resistance to blood flow thereby increasing exchange efficiency of the smaller placenta.

Although reductions to placental mass due to heat stress occur during the second trimester, the effects on fetal growth restriction become apparent and progressively worsen over the course of the third trimester. It is thought that even a compromised placenta is capable of transporting an adequate supply of oxygen and nutrients to meet fetal demands for the first two trimesters, but becomes insufficient when challenged by rapid increases in fetal nutrient and oxygen demands that coincide with the fetal growth during the third trimester. Some studies suggest that fetal growth restriction is secondary to the effect of heat stress on placental mass while other studies in which heat stress was initiated during the second trimester and prolonged into the third trimester revealed additional fetal and placental growth restriction (Galan et al. 1999) demonstrating that the timing and duration of heat stress during gestation is a direct determinant of fetal growth restriction.

In addition to the impact on placental mass and morphology, heat stress alters the maternal endocrine profile and in cows this is associated with reduced calf birth weight and subsequent milk yield (Collier et al. 1982). Catecholamines are markedly elevated in the heat stress-induced PI-IUGR fetus and alter cotyledonary blood flow and fetal glucose metabolism (Yates et al. 2011). In addition to hemodynamic changes, heat stress induces alterations to the regulation of gene expression of angiogenic factors such as vascular endothelial growth factor (VEGF) and placental growth factor (PlGF) (Regnault et al. 2002) which play major roles in vascular development within the placentome (Ahmed et al. 2000).

During normal placental development, capacity for nutrient and gas exchange increases with gestation as fetal demands increase over the third trimester. This increase in placental capacity is achieved by increasing uterine blood flow (Christenson and Prior 1978), but also, importantly, by branching and expansion of the microvasculature within the placentome to increase capillary surface area for exchange (Leiser et al. 1997). Heat stress, in addition to decreasing uterine blood flow (Regnault et al. 2007), disrupts all major routes for placental nutrient and gas exchange, namely, diffusion, facilitated diffusion, and active transport. Diffusion capacity for oxygen across the placenta is markedly reduced in the ovine model of heat stress-induced placental insufficiency induced intrauterine growth restriction (PI-IUGR) (Regnault et al. 2007), which is reflected by fetal hypoxemia (Thureen et al. 1992) and this is associated with decreased expression of VEGFR-1 mRNA in the fetal cotyledon and increased expression of Growth Factor (PlGF) in the maternal caruncle (Regnault et al. 2003). Interestingly, when oxygen transport is normalized to fetal weight, no difference in oxygen transport between PI-IUGR and normal animals exists (Ansari et al. 2003) suggesting that oxygen availability is sensed by the fetus or placenta and may be an important determinant of fetal growth.

The mammalian fetus relies heavily on glucose as an energy source for growth and development (Battaglia and Meschia 1978). Fetal glucose supply is derived almost exclusively from maternal circulation via facilitated diffusion across the placenta via glucose transporter proteins (Olson and Pessin 1996). Fetal plasma glucose concentrations are reduced in PI-IUGR sheep fetuses compared to control fetuses, and the deficiency becomes increasingly pronounced as the third trimester progresses and fetal mass and demand for glucose increases (Bell and Ehrhardt 2002). The expression of glucose transporter protein isoforms GLUT-1 (Limesand unpublished data) and GLUT-8 (Limesand et al. 2004) is decreased in the placenta of heat-stressed ewes, representing specific mechanisms by which thermal stress alters placental function and fetal nutrition. Additionally, active transport processes have been shown to be impaired by heat stress as placental amino acid transport (De Vrijer et al. 2004), (Ross et al. 1996), and utilization rates (Regnault et al. 2005) are reduced in heat stress-induced PI-IUGR. In a rat model of nutrient restriction induced PI-IUGR, placental active transport of amino acids are associated with down-regulation of mRNA expression of a number of placental essential amino acid transporters (Jansson and Powell 1998; Norberg and Jansson 1998).

In summary, heat stress during gestation exerts deleterious effects on the ruminant placenta. Placental mass, transport function, and the consequential effects on fetal growth have been well documented. Recent studies have identified specific molecular mechanisms and altered gene expression induced by heat stress, representing potential targets to improve breeding and management practices.

14.9 Fetus

Offspring that are born small for gestational age tend to exhibit substandard performance in terms of growth and carcass qualities and are more likely develop metabolic complications postnatally (Barker et al. 1993), resulting in less accretion of lean soft tissue and more adipose tissue. Additionally, the effects of fetal growth restriction have been shown to be intergenerational as growth-restricted female offspring tends to produce smaller offspring of their own and both have an increased risk of cardiovascular and metabolic complications in adolescence and adulthood (Drake and Walker 2004).

Heat stress has negative impact on several aspects of livestock animal reproduction. However, within a species, breed differences in thermal tolerance and associated degrees of impairment to various aspects of male and female reproductive function highlight the importance of genetics. Animals that are adapted to warm climates generally exhibit less impairment of reproductive functions due to heat stress which is in part due to enhanced ability to regulate body temperature and dissipate heat during periods of increased temperature and humidity. Interestingly, cells isolated from these animals and studied in vitro continue to exhibit enhanced survival and function compared to cells from breeds not adapted to thermal stress despite isolation from total body thermoregulatory effects. This phenomenon indicates that at least one component of thermal tolerance is regulated at the cellular level independently of total body thermal regulatory functions. Current studies aim to elucidate the genetic basis and molecular mechanisms by which cells from certain breeds are exceptionally resistant to thermal stress while cells from less-adapted breeds experience high losses in cell function and survival. Additionally, heat stress during gestation has been shown to alter fetal gene expression, metabolism and subsequently growth. The effects of heat stress on fetal growth restriction have been shown to persist into adulthood, which, from a livestock production standpoint, translates to decreased animal performance and carcass qualities.

14.9.1 Fetal Gene Expression

The effects of heat stress on livestock reproduction in terms of conception, embryonic survival, and birth weight have long been known. Recent advances in the physiological understanding of processes that regulate animal reproduction highlight the importance of genetics as a target for strategies to enhance breeding programs and production management practices. Additionally, research utilizing facilities where environmental conditions such as temperature, humidity, and ultraviolet radiation can be experimentally manipulated have led to major advances in our understanding the effects of heat stress on various aspects of livestock reproduction and fetal development.

The increased metabolic expenditure associated with pregnancy generates copious amounts of heat in the ruminant. As a result, fetal temperature in a normal pregnancy is approximately 0.3–1.0°C higher than maternal body temperature (Power 1989). Under thermal neutral conditions, maternal thermoregulatory mechanisms adequately dissipate heat such that the additional heat generated by the fetus and placenta is effectively transferred to the environment and tolerable heat load is maintained. However, in hyperthermic environmental conditions, thermoregulatory mechanisms such as decreasing feed intake, and maternal evaporative and respiratory heat loss strategies, which rely on a temperature or humidity gradient, become compromised, maternal and fetal core-body temperatures become elevated (Hahn 1999). Modest increases in maternal core-body temperature of approximately 0.5–1.0°C for prolonged periods during mid-gestation are associated with placental insufficiency and intrauterine growth restriction (PI-IUGR) as reflected by 30–50% reductions in placental mass and up to 50% reduction to fetal weight near term (Leos et al. 2010; Limesand et al. 2006).

The mammalian fetus relies extensively on the availability of glucose and oxygen for growth (Owens 1991). PI-IUGR is characterized by an asymmetrical growth pattern (Galan et al. 1999) in which brain growth and nervous tissue function, which absolutely require glucose, are protected at the expense of somatic growth as reflected by increased fetal skull to abdominal diameter ratio, decreased fetal limb length, decreased skeletal muscle mass, and liver mass. Glucose transporter isoform GLUT-1 mRNA expression is increased in the brain of the PI-IUGR sheep fetus (Sadiq et al. 1999) while glucose utilization capacity is reduced in peripheral tissues. Fetal hypoglycemia and hypoxia both of which are present in the heat stress-induced PI-IUGR fetus are potent stimuli for fetal norepinephrine secretion from the chromaffin cells of the fetal adrenal medulla (Cohen et al. 1991). These cells become functional during mid-gestation (Comline and Silver 1966) and secrete predominantly norepinephrine due to relatively low fetal expression, of the enzyme phenylethanolamine N-methyltransferase (PNMT) which converts norepinephrine to epinephrine (Adams and McMillen 2000).

Norepinephrine is 4 to 5-fold elevated in the heat-stressed fetal sheep which is both hypoglycemic and hypoxic (Limesand et al. 1984). In the PI-IUGR sheep fetus, norepinephrine plays a major role in nutrient distribution (Green et al. 2011) by mediating widespread glucose sparing effects to promote fetal survival via three routes by; (1) increasing fetal hepatic gluconeogenic capacity, (2) mobilization of fuels such as fatty acids and amino acids that can be metabolized peripherally in lieu of glucose, and (3) suppressing insulin sercretion. Insulin is the major anabolic hormone that coordinates fetal nutrient availability with fetal growth (Gluckman and Liggins 1984). Fetal pancreatic ß-cells are nutrient sensing cells that secrete insulin in response to increases in plasma concentrations of glucose and amino acids (Fowden 1992). Catecholamines increase hepatic gluconeogenic capacity, which is associated with up-regulation of mRNA of key gluconeogenic enzymes phosphoenolpyruvate carboxy kinase (PEPCK), and glucose -6-phosphatase (G6P) as well as increased mRNA expression of peroxisome proliferator-activated receptor-γ coactivator-1α increased phosphorylation of cAMP response element binding protein (Thorn et al. 2009). Reductions in fetal glucose oxidation rates further promote gluconeogenesis by increasing lactate availability as a substrate for hepatic gluconeogenesis (Yates et al. 2011). Catecholamines bind β-adrenergic receptors on the plasma membrane of adipose cells and initiate intracellular signaling cascades that result in activation of hormone-sensitive lipase and mobilization of fatty acids. Chronic exposure to catecholamines upregulates mRNA expression of the $\beta 2$ adrenergic recepetor isoform and leads to adrenergic desensitization of perirenal adipose tissue in the fetal sheep and this defect persists postnatally leading to metabolic complications, excess fat accumulation, and unfavorable body composition. Norepinephrine suppresses fetal insulin secretion by binding to adrenergic receptor alpha-2 isoforms which are coupled to G-protein signaling pathways that suppress insulin secretion by decreasing intracellular cyclic AMP and/or calcium. Chronically elevated norepinephrine, as measured in the heat stress-induced PI-IUGR sheep fetus, is associated with increased mRNA expression of adrenergic receptor isoforms alpha1D, alpha2A, and alpha2B (Leos et al. 2010) explaining persistent impairment fetal β-cell function (Green et al. 2010). Furthermore, β-cell replication is diminished and therefore β-cell mass is reduced in PI-IUGR sheep fetuses (Limesand et al. 2005). Additionally, mRNA expression of insulin receptor (Thorn 2009) and glucose transporter isoforms GLUT-1 and GLUT-4 (Hay 2006) were found to be increased in the skeletal muscle of PI-IUGR sheep but intermediate insulin signaling elements phosphoinositide-3 kinase (p85) and protein kinase B (Akt2) were decreased (Thorn et al. 2009). Glucose utilization rates were found to be increased along with insulin sensitivity in fetal sheep with PI-IUGR (Limesand et al. 2007). This fetal programing adaptation which has been found to persist postnatally, may partially explain the observed phenomenon of "catch up" growth in PI-IUGR offspring.

While the metabolic adaptations described elsewhere by the frugal phenotype or thrifty phenotype may be advantageous for fetal survival in heat stress-induced

PI-IUGR, these adaptations have detrimental effects postnatally when nutrients and oxygen are no longer restricted. In this case animal exhibits "catch-up" growth and tends to accumulate disproportionally large amounts of fat compared to lean soft tissue which translates to lower carcass quality from a production standpoint. In summary, a better understanding of the genes affected by heat stress and the mechanisms by which their expression affects fetal and postnatal growth will elucidate opportunities improve practices to improve thermal tolerance and production in livestock.

References

Adams MB, McMillen IC (2000) Actions of hypoxia on catecholamine synthetic enzyme mRNA expression before and after development of adrenal innervation in the sheep fetus. J Physiol 529(3):519–531

Ahmed A, Dunk C, Ahmad S, Khaliq A (2000) Regulation of placental vascular endothelial growth factor (VEGF) and placenta growth factor (PlGF) and soluble Flt-1 by oxygen–a review. Placenta 21:S16–S24

Alexander G, Williams D (1971) Heat stress and development of conceptus in domestic sheep. J Agri Sci 76:53–72

Akerfelt MD, Trouillet V, Mezger , Sistonen L (2007) Heat shock factors at a crossroad between stress and development. Ann N Y Acad Sci 1113:15–27

Amici C, Sistonene L, Gabriella Santoro M, Morimoto RI (1992) Antiproliferative prostaglandins activate heat shock transcription factor. Proc Natl Acad Sci 89:6227–6231

Ansari T, Fenlon S, Pasha S (2003) Morphometric assessment of the oxygen diffusive conductance in placentae from pregnancies complicated by intra-uterine growth restriction. Placenta 24:618–626

Battaglia FC, Meschia G (1978) Principal substrates of fetal metabolism. Physiol Rev 58(2): 499–527

Barker DJ, Gluckman PD, Godfrey KM, Harding JE, Owens JA, Robinson JS (1993) Fetal nutrition and cardiovascular disease in adult life. Lancet 341:938–941

Bauman DE, Currie WB (1980) Partitioning of nutrients during pregnancy and lactation: a review of mechanisms involving homeostasis and homeorhesis. J Dairy Sci 63:1514–1529

Bell AW, Ehrhardt RA (2002) Regulation of placental nutrient transport and implications for fetal growth. Nutr Res Rev 15:211–230

Bell AW, McBride BW, Slepetis R, Early RJ, Currie WB (1989) Metabolites conceptus growth and maternal plasma hormones and chronic heat stress and prenatal development in sheep: I. J Anim Sci 67:3289–3299

Berman A (2004) Tissue and external insulation estimates and their effects on prediction of energy requirements and of heat stress. J Dairy Sci 87(5):1400–1412

Berman A (2005) Estimates of heat stress relief needs for Holstein dairy cows. J Anim Sci 83:1377–1384

Blackshaw JK, Blackshaw AW (1994) Heat stress in cattle and the effect of shade on production and behavior: a review. Aust J Exp Agric 34:285–295

Bligh J (1976) Introduction to acclimatory adaptation-including notes on terminology. In: Bligh J, Cloudsley-Thompson JL, Macdonald AG (eds) Environmental physiology of animals. Wiley, NY, pp 219–229

Block J, Chase CC, Hansen PJ (2002) Inheritence of resistance of bovine preimplantation embryos to heat shock: relative importance of the maternal versus paternal contribution. Mol Reprod Dev 63:32–37

Bohmanova J, Misztal I, Tsuruta S, Norman HD, Lawlor TJ (2006) Short communication: genotype by environment interaction due to heat stress. J Dairy Sci 91:840–846

Brouk MJ, Smith JF, Harner JP (2003) Effect of sprinkling frequency and airflow on respiration rate, body surface temperature and body temperature of heat stressed dairy cattle. In: Fifth international dairy housing proceedings. ASAE, Fort Worth, TX 29–31 Jan 2003, pp 263–268

Buckley BA, Hofmann GE (2002) Thermal acclimation changes DNA-binding activity of heat shock factor 1 (HSF1) in the goby Gillichthys mirabilis: implications for plasticity in the heat-shock response in natural populations. J Exper Biol 205:3231–3240

Carper SW, Duffy JJ, Gerner EW (1987) Perspectives in cancer: heat shock proteins in thermotolerance and other cellular processes. Cancer Res 47:5249–5255

Cerón-Muñoz M, Tonhati H, Costa CN, Rojas-Sarmiento D, Echeverri DM (2004) Factors that cause genotype by environment interaction and use of a multiple-trait herd-cluster model for milk yield of Holstein cattle from Brazil and Colombia. J Dairy Sci 87:2687–2692

Chan EKF, Nagaraj SH, Reverter A (2010) The evolution of tropical adaptation: comparing taurine and zebu cattle. Anim Genet 10:1365–2052

Charmandari E, Tsigos C, Chrousos G (2005) Endocrinology of the stress response. Annu Rev Physiol 67:259–284

Christenson RK, Prior RL (1978) Uterine blood flow and nutrient uptake during late gestation in ewes with different number of fetuses. J Anim Sci 46:189–200

Cohen WR, Piasecki GJ, Cohn HE, Susa JB, Jackson BT (1991) Sympathoadrenal responses during hypoglycemia, hyperinsulinemia, and hypoxia in the ovine fetus. AJP Endo 261(1): 95–102

Collier RJ, Eley RM, Sharma AK, Pereira RM, Buffington DE (1981) Shade management in subtropical environment for milk yield and composition in Holstein and Jersey cows. J Dairy Sci 64:844–849

Collier RJ, Doelger SG, Head HH, Thatcher WW, Wilcox CJ (1982) Effects of heat stress during pregnancy on maternal hormone concentrations, calf birth weight and postpartum milk yield of Holstein cows. J Anim Sci 54:309

Collier RJ, Baumgard LH, Lock AL, Bauman D (2005) Physiological limitations, nutrient partitioning. In: Sylvester-Bradley R, Wiseman J (eds) Yields of farmed species: constraints and opportunities in the 21st Century. pp 351–377

Collier RJ, Dahl GE, VanBaale MJ (2006a) Major advances associated with environmental effects on dairy cattle. J Dairy Sci Centenial 89:1244–1253

Collier RJ, Collier RP, Rhoads , Baumgard LH (2008a) Invited review: genes involved in the bovine heat stress response. J Dairy Sci 91(2):445–454

Collier RJ, Miller MA, McLaughlin CL, Johnson HD, Baile CA (2008b) Effects of Recombinant Bovine Somatotropin (rbST) and season on plasma and milk insulin-like growth factors I (IGF-I) and II (IGF-II) in lactating dairy cows. Domest Anim Endocrinol 35:16–23

Collier RJ, Stiening CM, Pollard BC, VanBaale MJ, Baumgard LH, Gentry PC, Coussens PM (2006b) Use of gene expression microarrays for evaluating environmental stress tolerance at the cellular level in cattle. J Anim Sci 84:1–13

Comline RS, Silver M (1966) Development of activity in the adrenal medulla of the foetus and new-born animal. British Med Bulletin 22(1):16–20

Da Silva RG, La Scala Jr N, Tonhati H (2003) Radiative properties of the skin and hair coat of cattle and other animals. Trans ASAE 46(3):913–918

De Vrijer B, Regnault TRH, Wilkening RB, Meshia G, Battaglia FC (2004) Placental uptake and transport of ACP, a neutral nonmetabolizable amino acid in an ovine model of fetal growth restriction. Am J Physiol Endocrinol Metab 287:1114–1124

Drake AJ, Walker BR (2004) The intergenerational effects of fetal pro- gramming: non-genomic mechanisms for the inheritance of low birth weight and cardiovascular risk. J Endocrinol 180:1–16

Dutt RH (1963) Critical period for early embryo mortality in ewes exposed to high ambient temperatures. J Anim Sci 22:713

Edwards JL, Ealy AD, Monterroso VH, Hansen PJ (1997) Ontogeny of temperature-regulated heat shock protein 70 synthesis in preimplantation bovine embryos. Mol Reprod Dev 48: 25–33

Finch VA (1985) Comparison of non-evaporative heat transfer in different cattle breeds. Aust J Agric Res 36:497–508

Fleshner M, Johnson JD (2005) Endogenous extra-cellular heat shock protein 72: releasing signal(s) and function. Int J Hyperthermia 21:457–471

Fowden AL (1992) The role of insulin in fetal growth. Early Hum Dev 29(1–3):177–181

Fregley MJ (1996) Adaptations: some general characteristics. In: Fregley MJ, Blatteis CM (eds) Handbook of physiology, section 4: environmental physiology. Oxford University Press, Oxford, pp 3–15

Galan HL, Hussey MJ, Barbera A, Ferrazzi E, Chung M, Hobbins JC, Battaglia FC (1999) Relationship of fetal growth to duration of heat stress in an ovine model of placental insufficiency. Amer J Ob Gyn 180:1278–1282

Gaughan JB, Bonner S (2009) Hsp70 expression in cattle exposed to hot climatic conditions. In: Proceedings the 4th international congress on stress responses in biology and medicine, Sapporo Japan, PP 231

Garrett AT, Goosens NG, Rehrer NG, Patterson MJ, Cotter JD (2009) Induction and decay of short-term heat acclimation. Eur J Appl Physiol 107:659

Gebremedhin KG, Porter WP, Cramer CO (1983) Quantitative analysis of the heat exchange through the fur layer of Holstein calves. Trans ASAE 26(1):188–193

Gebremedhin KG, Porter WP, Warner RG (1984) Heat flow through pelage of calves—a sensitivity analysis. Trans ASAE 27(4):1140–1143

Gebremedhin KG, Hillman PE, Lee CN, Collier RJ (2007) Sweating rate of dairy cows under shade and sunny environments. ASABE Paper No. 074083, presented at the 2007 ASABE 100th annual international meeting, Minneapolis Convection Center, Minneapolis, Minnesota, pp 17–20

Gebremedhin KG, Hillman PE, Lee CN, Collier RJ, Willard ST, Arthington J, Brown-Brandl TM (2008) Sweating rates of dairy cows and beef heifers in hot conditions. Trans ASABE 51(6):2167–2178

Gebremedhin KG, Lee CN, Hillman PE, Collier RJ (2010) Physiological responses of dairy cows during extended solar exposure. Trans ASABE 53(1):239–247

Gebremedhin KG, Lee CN, Hillman PE, Brown-Brandl TM (2011) Body temperature and behavioral activities of four breeds of heifers in shade and full sun. App Eng J Agri ASABE 27(6): 999–1006

Gluckman PD, Liggins GC (1984) Regulation of fetal growth. In: Beard RW, Nathanielsz PW (eds) Fetal Physiology and Medicine: the basis of perinatology, 2nd edn, vol VI. Butterworth, London, pp 511–557

Grazul-Bilska AT, Borowicz PP, Johnson ML, Minten MA, Bilski JJ, Wroblewski R, Redmer DA, Reynolds LP (2010) Placental development during early pregnancy in sheep: vascular growth and expression of angiogenic factors in maternal placenta. Reproduction 140(1): 165–174

Grazul-Bilska AT, Johnson ML, Borowicz PP, Minten M, Bilski JJ, Wroblewski R, Velimirovich M, Coupe LR, Redmer DA, Reynolds LP (2011) Placental development during early pregnancy in sheep: cell proliferation, global methylation, and angiogenesis in the fetal placenta. Reproduction 141:529–540

Green AS, Rozance PJ, Limesand SW (2010) Consequences of a compromised intrauterine environment on islet function. J Endocrinol 205(3):211–224

Green AS, Macko AR, Rozance PJ, Yates DT, Chen X, Hay WW Jr, Limesand SW (2011) Characterization of glucose-insulin responsiveness and impact of fetal number and sex difference on insulin response in the sheep fetus. Am J Physiol Endocrinol Metab 300(5): 817–823

Guzhova IK, Kislyakova O, Moskaliova I, Fridlanskaya M, Tytell M, Cheetham B (2001) In vitro studies show that Hsp70 can be released by glia and that exogenous Hsp70 can enhance neuronal stress tolerance. Brain Res 914:66–73

Habib KE, Gold PW, Chrousos GP (2001) Neuroendocrinology of stress. Endocrinol Metab Clin North Am 30:695–728

Hafez SA (1964) Effects of high temperature on reproduction. Int J Biometeor 7:223

Hafez SA, Borowicz P, Reynolds LP, Redmer DA (2010) Maternal and fetal microvasculature in sheep placenta at several stages of gestation. J Anatomy 216:292–300. doi:10.1111/j.1469-7580.2009.01184.x

Hales JR, Fawcett AA, Bennett JW, Needham AD (1978) Thermal control of blood flow through capillaries and arteriovenous anastomoses in skin of sheep. Pflugers Arch 378:55–63

Hahn GL (1999) Dynamic responses of cattle to thermal heat loads. J Anim Sci 77(2):10–20

Hansen PJ (2004) Physiological and cellular adaptations of zebu cattle to thermal stress. Anim Reprod Sci 82–83:349–360

Hay WW (2006) Placental-fetal glucose exchange and fetal glucose metabolism. Trans Am Clin Climatol Assoc 117:321–339

Hayes B, Bowman P, Chamberlain A, Goddard M (2009) A validated genome-wide association study to breed cattle adapted to an environment altered by climate change. PLoS ONE 4:6676

Hillman PE, Gebremedhin KG, Parkhurst AM, Fuquay J, Willard ST (2001) Evaporative and convective cooling of cows in a hot and humid environment. In livestock environment VI: Proceedings of the 6th International Symposium, ASABE, St. Joseph, MI, pp 343–350

Hillman PE, Lee CN, Willard ST (2005) Thermoregulatory responses associated with lying and standing in heat-stressed dairy cows. Trans ASAE 48(2):795–801

Horowitz M (2002) From molecular and cellular to integrative heat defense during exposure to chronic heat. Comp Biochem Physiol A Mol Integr Physiol 131(3):475–483

Horowitz M (2007) Heat acclimation and cross-tolerance against novel stressors: genomic-physiological linkage. Prog Brain Res 162:373–392

Horowitz M, Assadi H (2010) Heat acclimation-mediated cross-tolerance in cardioprotection: Do HSP70 and HIF-1a play a role? Ann N Y Acad Sci 1188:199–206

Horowitz M, Eli-Berchoer L, Wapinski I, Friedman N, Kodesh E (2004) Stress-related genomic responses during the course of heat acclimation and its association with ischemic-reperfusion cross-tolerance. J Appl Physiol 97:1496–1507

Horowitz M, Robinson SD (2007) Heat shock proteins and the heat shock response during hyperthermia and its modulation by altered physiological conditions. Prog Brain Res 162:433–436

Ianaro A, Ialenti A, Maffia P, Di Meglio P, Di Rosa M, Santoro MG (2003) Anti-Inflammatory Activity of 15-Deoxy-$D^{12,14}$-PGJ_2 and 2-Cyclopenten-1-one: Role of the Heat Shock Response. Mol Pharmacol 64:85–93

Ireland HE, Leoni F, Altaie O, Birch CS, Coleman RS, Hunter-Lavin C, Williams JJH (2007) Measuring the secretion of heat shock proteins from cells. Methods 43(176):183

Jansson T, Scholtbach V, Powell TL (1998) Placental transport of leucine and lysine is reduced in intrauterine growth restriction. Pediatr Res 44:532–537

Johnson HD, Vanjonack WJ (1976) Effects of environmental and other stressors on blood hormone patterns in lactating animals. J Dairy Sci 59:1603–1617

Jousan FD, Hansen PJ (2004) Insulin-like growth factor-I as a survival factor for the bovine preimplantation embryo exposed to heat shock. Biol Reprod 71(5):1665–1670

Kamwanja LA, Chase CC, Gutierrez JA, Guerriero V, Olson TA, Hammond AC (1994) Responses of bovine lymphocytes to heat shock as modified by breed and antioxidant status. J Anim Sci 72:438–444

Kimmel E, Arkin H, Broday D, Berman A (1991) A model of evaporative cooling in a wetted hide. J Agric Eng Res 49(3):227–241

Kingdom JCP, Kaufmann P (1997) Oxygen and placental villous development: origins of fetal hypoxia. Placenta 18:613–621

Kowalski GJ, Mitchell JW (1979) An analytical and experimental investigation of the heat transfer mechanism within fibrous media, ASME, New York, 79-WA/HT-40

Kregel KC (2002) Molecular biology of thermoregulation invited review: heat schock proteins: modifying factors in physiological stress responses and acquired thermotolerance. J Appl Physiol 92:2177–2186

Kristensen TN, Løvendahl P, Berg P, Loeschcke V (2004) Hsp 72 is present in plasma from Holstein-Friesian dairy cattle, and the concentration level is repeatable across days and age classes. Cell Stress Chaperones 9:143–149

Leiser R, Krebs C, Klisch K, Ebert B, Dantzer V, Schuler G, Hoffmann B (1997) Villosity and microvasculature of the bovine placentome in the second half of gestation. J Anat 191: 517–527

Leos RA, Anderson MJ, Chen X, Pugmire J, Anderson KA, Limesand SW (2010) Chronic exposure to elevated norepinephrine suppresses insulin secretion in fetal sheep with placental insufficiency and intrauterine growth restriction. American J Physiol 298(4):770–778

Lillehammer M, Hayes BJ, Meuwissen THE, Goddard ME (2009) Gene by environment interactions for production traits in Australian dairy cattle. J Dairy Sci 92:2008–4017

Limesand SW, Regnault TRH, Hay WW (2004) Characterization of glucose transporter 8 (GLUT8) in the ovine placenta of normal and growth restricted fetuses. Placenta 25(1):70–77

Limesand SW, Jensen J, Hutton JC, Hay WW (2005) Diminished beta-cell replication contributes to reduced beta-cell mass in fetal sheep with intrauterine growth restriction. Am J Physiol Regul Integr Comp Phyiol 288:1297–1305

Limesand SW, Rozance PJ, Zerbe GO, Hutton JC, Hay WW (2006) Attenuated insulin release and storage in fetal sheep pancreatic islets with intrauterine growth restriction. Endocrinology 147:1488–1497

Limesand SW, Rozance PJ, Smith D, Hay WW Jr (2007) Increased insulin sensitivity and maintenance of glucose utilization rates in fetal sheep with placental insufficiency and intrauterine growth restriction. Am J Physiol Endocrinol Metab 293:1716–1725

Maia ASC, daSilva RG, Loureiro CMB (2005) Sensible and latent heat loss from the body surface of Holstein cows in a tropical environment. Intl J Biometeor 50(1):17–22

Maia ASC, daSilva RG, Loureiro CMB (2008) Latent heat loss of Holstein cows in a tropical environment: a prediction model. R Bras Zootec 37(10):1806–9290

Manalu W, Johnson HD, Li R, Becker BA, Collier RJ (1991) Assessment of thermal status of somatotropin-injected lactating Holstein cows maintained under controlled-laboratory thermoneutral, hot and cold environments. J Nutr 121:2006–2019

McArthur AJ (1987) Thermal interaction between animal and microclimate–a comprehensive model. J Theor Biol 126(2):203–238

McDowell RE (1958) Physiological approaches to animal climatology. J Hered 49:52–61

Moen AN (1973) Wildlife ecology- an analytical approach. Freeman and Company WH, San Francisco, p 458

Magee DA, Meghen C, Harrison S, Troy CS, Cymbron T, Gaillard C, Morrow A, Maillard CG, Bradley DG (2002) A partial African ancestry for the creole cattle populations of the Caribbean. J Hered 93:429–432

Maloyan A, Palmon A, Horowitz M (1999) Heat acclimation increases basal HSP72 level and alters its produciton dynamics during heat stress. Am J Physiol Regul Integr Comp Physiol 276:1506–1515

Maloyan A, Eli-Berchoer L, Semenza GL, Gerstenblith G, Stern GD, Horowitz M (2005) HIF-1a-targeted pathways are activated by heat acclimation and contribute to acclimation-ischemic cross-tolerance in the heat. Physiol Genomics 23:79–88

Mossman HW (1987) Vertebrate fetal membranes. Rutgers University Press, New Brunswick

Myers SA, Sparks JW, Makowski EL, Meschia G, Battaglia FC (1982) Relationship between placental blood flow and placental and fetal size in guinea-pig. Am J Physiol 24:404–409

Norberg S, Powell TL, Jansson T (1998) Intrauterine growth restriction is associated with a reduced activity of placental taurine transporters. Pediatr Res 44:233–238

Olson AL, Pessin JE (1996) Structure, function, and regulation of the mammalian facilitative glucose transporter gene family. Annu Rev Nutr 16:235–256

Olson TA, Chase CC, Lucena C, Godoy E, Zuniga A, Collier RJ (2006) Effect of hair characteristics on the adaptation of cattle to warm climates. 8th World congress on genetics applied to livestock production, Belo Horizonte, MG, Brasil, 13–18 Aug 2006

Owens JA (1991) Endocrine and substrate control of fetal growth: placental and maternal influences and insulin-like growth factors. Reprod Fertil Dev 3(5):501–517

Paula-Lopes FF, Chase CC, Al-Katanani YM, Krininger CE, Rivera RM, Tekin S, Majewski AC, Ocon OM, Olson TA, Hansen PJ (2003) Genetic divergence in cellular resisitance to heat shock in cattle: differences between breeds developed in temperate versus hot climates in responses of preimplantation embryos, reproductive tract tissues and lymphocytes to increased culture temperatures. Reprod 125:285–294

Power GG (1989) Biology of temperature: the mammalian fetus. J Dev Physiol 12:295–304

Ravagnolo O, Misztal I, Hoogenboom G (2000) Genetic component of heat stress in dairy cattle, development of heat index function. J Dairy Sci 83:2120–2125

Regnault TRH, Friedman JE, Wilkening RB, Anthony RV, Hay WW (2005) Fetoplacental transport and utilization of amino acids in IUGR—a review. Placenta, pp 52–62

Regnault TRH, Orbus RJ, De Vrijer B, Davidsen ML, Galan HL, Wilkening RB, Anthony RV (2002) Placental expression of VEGF, PlGF and their receptors in a model of placental insufficiency–intrauterine growth restriction (PI-IUGR). Placenta 23(2–3):132–144

Regnault TRH, de Vrijer B, Galan HL, Davidsen ML, Trembler KA, Battaglia FC, Wilkening RB, Anthony RV (2003) The relationship between transplacental O2 diffusion and placental expression of PlGF, VEGF and their receptors in a placental insufficiency model of fetal growth restriction. J Physiol 550:641–656. doi:10.1113/jphysiol.2003.039511

Regnault TRH, de Vrijer B, Galan HL, Wilkening RB, Battaglia FC, Meschia G (2007) Development and mechanisms of fetal hypoxia in severe fetal growth restriction. Placenta 28:714–723

Reynolds LP, Borowicz PP, Vonnahme KA (2005) Placental angiogenesis in sheep models of compromised pregnancy. J Physiol 565:43–58

Rhoads ML, Rhoads RP, VanBaale MJ, Collier RJ, Sanders SRS, Weber WJ, Crooker BA, Baumgard LH (2009) Effects of heat stress and plane of nutrition on lactating Holstein cows: I. Production, metabolism, and aspects of circulating somatotropin. J Dairy Sci 92:1986–1997

Roberts MF, Wenger CB, Stolwijk JAJ, Nadel ER (1977) Skin blood flow and sweating changes following exercise training and heat acclimatization. J Appl Physiol 43:133–137

Ross JC, Fennessey PV, Wilkening RB, Battaglia FC, Meschia G (1996) Placental transport and fetal utilization of leucine in a model of fetal growth retardation. Am J Physiol Endocrinol Metab 270:491–503

Sadiq HF, Das UG, Tracy TF, Devaskar SU (1999) Intra-uterine growth restriction differentially regulates perinatal brain and skeletal muscle glucose transporters. Brain Res 823(1–2):96–103

Sarge KD, Murphy SP, Morimoto RI (1993) Activation of heat shock gene transcription by heat shock factor 1 involves oligomerization, acquisition of DNA-binding activity, and nuclear localization and can occur in the absence of stress. Mol Cell Biol 13:1392–1407

Scott GH (1981) What is animal stress and how is it measured? J Anim Sci 52:150

Shamovsky I, Nudler E (2008) New insights into the mechanism of heat shock response activation. Cell Mol Life Sci 65:855–861

Silanikove N, Shapiro F, Shinder D (2009) Acute heat stress brings down milk secretion in dairy cows by up-regulating the activity of the milk-borne negative feedback regulatory system. BMC Physiol 9(13):

Sonna LA, Fujita J, Gaffin SL, Lilly CM (2002) Molecular biology of thermoregulation invited review: effects of heat and cold stress on mammalian gene expression. J Appl Physiol 92:1725–1742

Stowell RR (2000) Heat stress relief and supplemental cooling. In: Dairy housing and equipment systems conference proceedings of the Natural Resource, Agriculture, and Engineering

Service (NRAES). Cornell University, Ithaca, New York, Department of Agricultural and Biological Engineering, pp 129

Straus DS, Glass CK (2001) Cyclopentenone prostaglandins: new insights on biological activities and cellular. Med Res Rev 21:185–210

Thorn SR, Regnault TRH, Brown LD, Rozance PJ, Keng J, Roper M, Wilkening RB, Hay WW, Friedman JE (2009) Intrauterine growth restriction increases fetal hepatic gluconeogenic capacity and reduces messenger ribonucleic acid translation initiation and nutrient sensing in fetal liver and skeletal muscle. Endocrinol 150(7):3021–3030

Thureen PJ, Trembler KA, Meschia G, Makowski EL, Wilkening RB (1992) Placental glucose transport in heat-induced fetal growth retardation. Am J Physiol Regul Integr Comp Physiol 263:578–585

Tytell M (2005) Release of heat shock proteins (Hsps) and the effects of extracellular Hsps on neural cells and tissues. Int J Hyperthermia 21:445–455

Voellmy R, Boellmann F (2007) Chaperone regulation of the heat shock protein response. Adv Exp Med Biol 594:89–99

Xu Q, Ganju L, Fawcett TW, Holbrook NJ (1996) Vasopressin-induced heat shock protein expression in renal tubular cells. Lab Invest 74:178–187

Yates DT, Green AS, Limesand SW (2011) Catecholamines mediate multiple fetal adaptations during placental insufficiency that contribute to intrauterine growth restriction: lessons from hyperthermic sheep. J Pregnancy 11:740408

Yousef MK (1984) Stress physiology in livestock, Vol I, basic principles. CRC Press Inc, Boca Raton, p 217

Part IV
Climate Change Impact and Adaptation

Chapter 15
Impact of Climate Change on Livestock Production

Lance H. Baumgard, Robert P. Rhoads, Michelle L. Rhoads, Nicholas K. Gabler, Jason W. Ross, Aileen F. Keating, Rebbeca L. Boddicker, Sangeeta Lenka and Veerasamy Sejian

Abstract Livestock production is the world's dominant land use, covering about 45% of the Earth's land surface, and much of it in harsh and variable environments that are unsuitable for other purposes. Climate change (CC) can impact the amount and quality of produce, reliability of production, and the natural resource base on which livestock production depends. Climate is an important factor of agricultural productivity and CC is expected to severely impact livestock production systems. Furthermore, global demand for animal protein will rise as populations become more affluent and eating habits change. Therefore, animal production plays (and will continue to do so) a key role in the food supply chain. While the increasing demand for livestock products offers market opportunities and income for small, marginal, and landless farmers, livestock production globally faces increasing pressure because of negative environmental implications, particularly because of greenhouse gas (GHG) emissions. Agriculture is one sector which is important to consider as it both impacts CC as well as is influenced by CC. Higher temperatures, potentially caused by GHG, would likely result in a decline in dairy production, reduced animal weight gain, reproduction, and lower feed-conversion efficiency in warm regions. Incidence of diseases among livestock and other

L. H. Baumgard (✉) · N. K. Gabler · J. W. Ross · A. F. Keating · R. L. Boddicker
Department of Animal Science, Iowa State University, Ames, IA 50011, USA
e-mail: baumgard@iastate.edu

R. P. Rhoads · M. L. Rhoads
Department of Animal Science, Virginia Tech, Blacksburg, VA 24061, USA

S. Lenka
Division of Soil Physics, Indian Institute of Soil Science, Nabibagh, Berasia Road, Bhopal, Madhya Pradesh 462038, India

V. Sejian
Division of Physiology and Biochemistry, Central Sheep and Wool Research Institute, Avikanagar, Jaipur, Rajasthan 304501, India

V. Sejian et al. (eds.), *Environmental Stress and Amelioration in Livestock Production*, DOI: 10.1007/978-3-642-29205-7_15,

animals are likely to be affected by CC, since most diseases are transmitted by vectors such as ticks and flies (development stages of ticks and flies are often dependent on ambient temperature). Cattle, goat, horses, and sheep are also vulnerable to an extensive range of nematode worm infections, most of which have their development stages influenced by climatic conditions. CC will have far-reaching consequences for dairy, meat, and wool production systems that rely primarily on grass and rangelands and this will likely detrimentally affect vulnerable pastoral communities which are engaged in extensive livestock production systems in drylands. Although the direct effects of CC on animals are likely to be small (as long as temperature increases do not exceed 3°C), CC will affect animals indirectly through physiological stress and thermoregulatory control, nutrition, and disease stress. Because livestock products are an incredibly important human food, and because animal farming is a significant source of income for millions of farmers, it is necessary to identify CC mitigation strategies and solutions.

Keywords Heat stress · Metabolism · Disease · Epigenetics

Contents

15.1 Introduction

Animal productivity is optimized within narrow environmental conditions. When the temperature is either below or above the threshold values for peak animal production, efficiency is compromised because nutrients are diverted to maintain euthermia as a means of maintaining body temperature which invariably takes priority over product synthesis such as milk, meat, fetus, etc. Heat stress negatively impacts a variety of productive parameters including milk yield, growth, reproduction, and carcass traits. In addition, heat load increases healthcare costs and animals can succumb to severe thermal stress (especially lactating cows and sows). Therefore, environmental heat stress is a significant financial burden to the industry given that about $900 million/year for dairy and over $300 million/year each in beef and swine are lost to heat stress in the U.S. alone (St. Pierre et al. 2003; Pollmann 2010). Advances in management (i.e., cooling systems, barn construction) have alleviated some negative impacts of thermal stress on animal agriculture (Armstrong 1994; Burgos et al. 2007; Stowell et al. 2009) although production is still compromised during the summer.

The detrimental effects of environmental heat stress on animal welfare and production will likely become more of an issue in the future if the Earth's climate continues to warm as some predict (IPCC 2007; Bernabucci et al. 2010). In addition, the human population continues to increase, especially in the tropical and subtropical areas of the globe (Roush 1994). Consequently animal agriculture in these warm areas will need to expand (Renaudeau et al. 2008) to keep pace with the global appetite for high quality protein. However, many countries in these geographical areas are still developing and may lack facilities and resources required for effective heat abatement strategies. Consequently, heat stress is likely one of the primary (if not the principal) factors limiting efficient animal protein production and may continue to compromise food security in developing countries.

Basal/metabolic heat production increases with enhanced production [i.e., enhanced milk yield (Spiers et al. 2004) and lean tissue accretion (Brown-Brandl et al. 2004)]. Therefore, genetic selection based upon traditional production traits may increase susceptibility of animals to thermal stress. Consequently, an animal or a breed's annual productivity needs consideration before introducing novel genetics into a particular geography. Understanding the biology and mechanisms of how heat stress jeopardizes animal performance, therefore, is critical in developing compatible approaches (i.e., genetic, managerial, nutritional, and pharmaceutical) to ameliorate current challenges facing animal production and future mitigating strategies to improve animal well-being, performance, and economics.

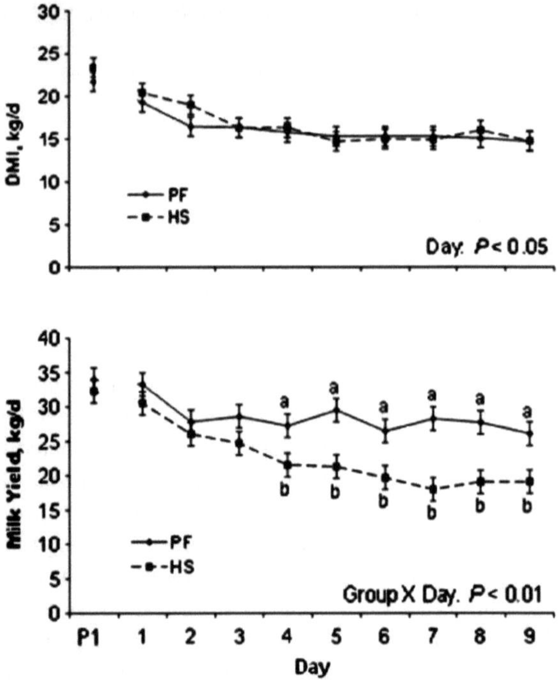

Fig. 15.1 Effects of heat stress (*HS*) or pair-feeding (*PF*) on **a** DMI and **b** milk yield in lactating Holstein cows. *Solid lines* with *diamonds* represent PF cows and *dashed lines* with *squares* represent HS cows. The mean value from day 1 to 9 of the TN ad libitum period (P1) was used as a covariate and is represented by P1 on the X axis. The day 1–9 results are from P2 when cows were exposed to HS or exposed to TN conditions and PF with the HS cows. Adapted from Rhoads et al. 2009

15.2 Animal Production

15.2.1 Lactation

Milk synthesis appears to be more sensitive (compared to growth models) to thermal stress as decreased yields of up to 35–40% are common (West 1999). Detailed description of how cows accumulate and dissipate heat and how, when, and where cows become heat stressed has already been published elsewhere (Berman 2003, 2004, 2005). It was traditionally thought that lactating cows become heat stressed when conditions exceed a temperature humidity index (THI) of 72 (Armstrong 1994). However, recent climate controlled experiments indicate that milk yield starts to decrease at a THI of 68 (Zimbleman et al. 2009) which is supported by field observations that evaluated the THI when cow standing time (a classic response to a thermal load) increases (Cook et al. 2007). The lower THI at which cows are thought to become heat stressed is consistent with the hypothesis that higher producing cows are more susceptible to a thermal load.

Heat stressed animals reduce feed intake and this is presumably a survival strategy as digesting (especially in ruminants) and metabolizing nutrients generates heat (i.e., thermic effect of feed). A popular opinion contends that inadequate feed intake caused by the thermal load was responsible for decreased milk production (Fuquay 1981; Beede and Collier 1986; Silanikove 2000; West 1999;

DeShazer et al. 2009). However, our recent experiment demonstrated disparate slopes in feed intake and milk yield responses to a cyclical heat load pattern (Shwartz et al. 2009). This led us to hypothesize that heat stress reduces milk synthesis by both direct and indirect (via reduced feed intake) mechanisms (Baumgard and Rhoads 2012). To examine this theory, we designed a series of pair-feeding experiments enabling us to evaluate thermal stress while eliminating the confounding effects of dissimilar nutrient intake. Employing this type of approach is required to differentiate between direct and indirect effects (e.g., reduced intake) of environmentally induced hyperthermia because both heat stressed and malnourished animals share common responses (i.e., reduced milk yield, growth, etc.). Our experiments demonstrate that reduced feed intake only explains approximately 35–50% of the decreased milk yield during environmental-induced hyperthermia (Fig. 15.1, Rhoads et al. 2009; Wheelock et al. 2010). These results agree with previous data (Bianca 1965), and suggest that when overall heat stress (extent and duration) exceeds a given threshold (as of yet unidentified) the cumulative thermal load disrupts the nutrient intake:milk production relationship and milk yield declines beyond expected levels.

15.2.2 Growth

15.2.2.1 Beef

In general, heat stress-induced production losses in beef cattle are not as severe as those for the dairy industry. It is not entirely clear why growing cattle tolerate higher THI conditions and exhibit a greater heat strain threshold than lactating dairy cows, but likely possibilities may involve: (1) increased surface area to mass ratio, (2) reduced rumen heat production (because of the mostly grain diet), and (3) reduced overall metabolic heat production (on a body weight basis). In addition, beef cattle will often experience compensatory gain after mild or short periods of heat stress (Mitlöhner et al. 2001). The combination of these factors translate into heat-related reduced gain that is typically less than 10 kg, which amounts to ~7 extra days on feed (St. Pierre et al. 2003). Furthermore, the impact of heat stress on reproductive indices is typically not as severe in beef cattle as in dairy cattle due to the seasonal nature of breeding programs (often occurring during the spring in the US).

15.2.2.2 Swine

Although accurately estimating the heat-induced economic loss is challenging, a recent assessment suggests poor sow performance alone (not including reduced growth, carcass quality, etc.) costs the US swine industry $330 million annually (Pollmann 2010). Even if optimal heat stress abatement strategies were implemented by all pig producers at all stages of production, losses to heat stress would still cost the American pig industry about $300 million annually (St. Pierre et al. 2003). The economic losses caused by a sustained thermal load include reduced growth and efficiency, increased healthcare costs, decreased carcass value (increased lipid and decreased protein), and increased mortality (especially sows and market hogs) and this typically coincides with the time of year that has the best market prices. Interestingly, the fact that pigs reared in heat stress conditions have reduced muscle mass and increased adipose tissue has been documented frequently over the past 40 years (Close et al. 1971; Verstegem et al. 1978; Stahly et al. 1979; Heath 1983, 1989; Bridges et al. 1998; Collin et al. 2001). This phenomenon is not unique to pigs, as heat stress also alters carcass composition similarly in rodents (Schmidt and Widdowson 1967; Katsumata et al. 1990) and growing poultry (Baziz et al. 1996; Geraert et al. 1996; Yunianto et al. 1997; Lu et al. 2007).

A dramatic reduction in feed intake (up to 50%) is an obvious sign of heat stress and is thought to be primarily responsible for the negative effects heat stress has on pig performance (Collin et al. 2001). It is counter-intuitive that heat stress stimulates a decrease in nutrient intake and depresses growth, yet increases carcass lipid accretion and decreases carcass nitrogen content. In thermal neutral (TN) conditions, pigs consuming a restricted diet will deposit protein at the expense of lipid accretion (i.e., the carcass lipid to protein ratio decreases, meaning the carcass becomes leaner) and the quantity of lipid deposited per unit of energy consumed decreases (Le Dividich et al. 1980; Van Milgen and Noblet 2003; Oresanya et al. 2008). The hallmark of heat stress in swine, therefore, is increased lipid deposition at the expense of protein. Hence, the reduced feed intake to body composition relationship is exactly opposite in pigs reared in heat stress conditions and is independent of plane of nutrition. Surprisingly, despite the enormous economic impact of heat stress on swine, little is known about how heat stress directly (not mediated by reduced nutrient intake) and indirectly alters metabolism and nutrient partitioning in pigs.

15.3 Reproduction

15.3.1 Swine

The swine industry suffers considerably due to impaired reproductive performance during periods of seasonal infertility, particularly during late summer and early autumn months (Pollmann 2010). The impact is quite visible in the US with day 28 pregnancy rates reaching their lowest levels in August to October and reduced farrowing rates in November and December. This phenomenon is not limited to specific regions and occurs internationally (Auvigne et al. 2010; Pollmann 2010). Several factors such as photoperiod and temperature can contribute to seasonal infertility and deciphering the precise contribution of each factor on swine reproductive performance is challenging. Despite that, heat stress has been repeatedly demonstrated to negatively impact reproductive efficiency in pigs by affecting germ cell development, pregnancy establishment, maintenance of gestation, and lactation performance.

15.3.1.1 Heat Stress During Oogenesis and Early Embryo Development in Pigs

The impact of heat stress during oocyte maturation and early embryonic development is evidenced in sows that are exposed to heat stress for 5 days following breeding have a reduced (38%) number of viable embryos after day 27 of gestation (Tompkins et al. 1967). In a related study, heat stress was administered following breeding, which generally occurs prior to ovulation and complete oocyte maturation, as pigs typically ovulate in the mid to latter half of estrus (Soede et al. 1992).

Difficult to study in vivo, characterization of heat stress during oocyte growth and maturation and early embryonic development in pigs has been demonstrated in in vitro oocyte maturation and embryo culture systems. Some evidence of in vitro heat stress models during the transition between germinal vesicle breakdown (GVBD) and the 4-cell stage of development demonstrates the susceptibility of this stage to heat stress. A 9 h culture of pig embryos at 42°C following porcine in vitro fertilization reduced blastocyst formation rate from 20·6 to 8·8% (Isom et al. 2007) and heat shock of 41·5°C for 4 h following in vitro maturation also reduced the developmental ability of parthenogenetically activated oocytes (Tseng et al. 2006). However, it has also been demonstrated that parthenogenetically activated pig oocytes that are exposed to 9 h of heat shock at 42°C following activation have accelerated MAPK dephosphorylation and improved developmental competency in vitro (Isom et al. 2009a, b).

The impact of in vitro heat stress during oocyte maturation and developmental competency has also been demonstrated (Wright and Ross, unpublished). Oocytes exposed to heat stress (41°C) during the first half (21 h) or complete duration (42–44 h) of maturation process in vitro demonstrated impaired ability to reach metaphase II arrest while oocytes exposed to heat stress during the second half (21 h) of in vitro maturation experienced normal maturation rate (Wright and Ross, unpublished data). In addition, following in vitro heat stress, metaphase II arrested oocytes had impaired developmental competency compared to oocytes matured at 39°C. We have subsequently used this model to demonstrate differences in gene expression in developing 4- to 8-cell embryos as a result of heat stress conditions during in vitro maturation (Wright and Ross, unpublished data).

15.3.1.2 Gestation

The effect of heat stress during pregnancy in pigs is variable, as different stages of gestation can be more severely affected than others. This is demonstrated by a study conducted by Omtvedt et al. (1971) as they exposed pregnant gilts to heat stress for 8 days during different stages of gestation. In this study, compared to sows in TN conditions, heat stress beginning either on day 0 or day 8 of gestation reduced the number of viable embryos at day 30 of gestation. Interestingly, the same conditions on days 53–61 did not affect farrowing performance while heat stress during late gestation (days 102–110) reduced the number of piglets born alive (Omtvedt et al. 1971). However, a more moderate, cyclic heat stress on bred gilts beginning on day 3 and extended to either day 24 or day 30 of gestation did not impact embryo survival (Liao and Veum 1994).

15.3.1.3 Lactation

Heat stress during lactation can also have a profound impact on production. Temperatures exceeding the evaporative critical temperature (ECT) during lactation reduces feed intake, and consequently, decreases milk production (Black et al. 1993). It was hypothesized that elevated core body temperature results in the redirection of blood flow from the mammary gland toward the skin in an effort to facilitate heat dissipation. In response, lactation and piglet growth during lactation are significantly reduced (McGlone et al. 1988; Black et al. 1993; Johnston et al. 1999). It is of interest to determine whether heat stress alters post-absorptive metabolism and lactation parameters similarly in pigs as it does cows (Wheelock et al. 2010). In addition to reduced lactation performance, heat stress during lactation can also reduce the number of sows returning to estrus within 15 days post weaning (Johnston et al. 1999).

15.3.1.4 Semen Quality

While the effects of heat stress on pig reproduction is notable, it is not clear whether heat stress affects boars differently from the way it does sows. Whereas it is clear that reproductive parameters in gilts and sows are negatively affected by heat stress. Exposure of boars to elevated ambient temperatures can also be detrimental to swine reproduction through impaired semen quality. Boars subjected to heat stress for 90 days had reduced motility and increased percentage of abnormal sperm within 2 weeks from the initiation of heat stress (Wettemann et al. 1976). Utilizing semen from heat stressed boars reduced the number of embryos on day 30 post insemination when compared to TN boars (Wettemann et al. 1976; Cameron and Blackshaw 1980).

15.3.2 Ruminant

The physiological effects of heat stress on productivity are financially devastating to animal production systems. During heat stress, dry matter intake (DMI) decreases and maintenance requirements increase as livestock attempt to dissipate excess heat load (West 1999). In addition, changes in blood flow and the production of various hormones ultimately result in decreased reproductive performance. During summer months, conception rates can decline 20–30% (Rensis and Scaramuzzi 2003). This reduced fertility is attributed to several factors, including a reduction in estrus detection rates, early embryonic death, inhibition of follicular dominance, and reduced ovarian steroidogenic output (Putney et al. 1988a, b; Wolfenson et al. 2000; Rensis and Scaramuzzi 2003). Thus, heat stress has a wide range of reproductive effects beginning with the development of follicle through early embryonic development. The biological mechanisms that mediate these effects, however, are not completely understood.

15.3.2.1 Reproductive Cycle

Inevitably, the decrease in DMI that occurs during periods of heat stress is accompanied by changes in circulating concentrations of several metabolic hormones. Consequently, these metabolic adaptations alter the production and secretion of the hormones controlling the reproductive cycle (Wolfenson et al. 2000). Such consequences are far-reaching and may involve detrimental effects on ovarian follicular development, oocyte competence, early embryonic development, and the maternal recognition of pregnancy.

During heat stress, the development of the dominant ovarian follicle is attenuated and circulating concentrations of estradiol are reduced. In addition, the luteal phase of heat stressed cattle is extended and follicular wave dynamics are altered (Wilson et al. 1998). These changes in ovarian function appear to be the result of decreased pulse amplitude of luteinizing hormone (LH) (Gilad et al. 1993). LH is directly involved in the processes of follicular growth, estradiol production, ovulation, and progesterone production. These changes in LH pulsatility during heat stress may simply be a consequence of inadequate nutrient consumption as decreased feed intake is associated with changes in circulating insulin, leptin, and ghrelin, all of which affect LH pulsatility in several species (Szymanski et al. 2007).

These heat stress-induced changes in ovarian dynamics ultimately result in the production of substandard oocytes and pose unique challenges in reproductive management. Many producers now rely on timed artificial insemination (AI) programs during periods of heat stress because estrus detection rates decrease significantly during the warm summer months. Indeed, mounting activity declines by nearly half and is likely the result of lower circulating estradiol concentrations (Hansen and Aréchiga 1999). Extended luteal phases during heat stress also make it more difficult to predict when individual animals will come into estrus. Use of timed protocols alleviates these challenges by allowing the producer to control the time of ovulation (de la Sota et al. 1998; de Rensis et al. 2002).

15.3.2.2 Oocyte and Embryo

The detrimental effects of heat stress on oocyte quality are not easily mitigated. Preovulatory oocytes can be damaged directly by heat stress, and indirectly by prolonged estrous cycles. These longer estrous cycles presumably result in the ovulation of an aged oocyte that has reduced potential for developmental competence. Oocytes contained within antral follicles appear to be the most susceptible to the damaging effects of heat stress (Roth 2008). As a result, conception rates remain depressed well into the fall as the oocytes that were damaged during the summer heat stress are cleared from the ovary via ovulation or degradation.

In particular, one management technique has shown promise for overcoming the oocyte-specific problems associated with heat stress. Conception rates are markedly increased by transferring fresh in vitro-produced embryos into heat stressed cattle (Al-Katanani et al. 2002; Stewart et al. 2011). This allows the producer to completely bypass the challenges associated with substandard oocyte quality. However, the advantage is only evident with the use of fresh embryos. Utilizing frozen embryos during heat stress yields conception rates similar to timed AI. This presents a logistical challenge since few producers have access and technical know-how for effective use of fresh in vitro-produced embryos. The source of the oocytes is also a concern if the offspring is needed as replacement animals. Two sources of oocytes which are readily available are genetically superior donor females (housed in a cool environment) and indiscriminate collection of oocytes from ovaries at the slaughterhouse. Depending on the geographical region, animals sent to the slaughterhouse vary widely in breed and genetics, and therefore would not be desirable as replacement animals. Future advances in in vitro embryo production and freezing will make this technique a viable alternative for use during periods of heat stress.

Clearly the effects of heat stress on reproduction are not limited to the estrous cycle and oocyte. Even with the use of timed embryo transfer, conception rates during periods of heat stress still lag behind herd averages during the cooler seasons. Most other effects of heat stress on fertility are attributed to a compromised uterine environment. Effects of temperature, blood flow, and hormone concentrations on uterine environment have been previously investigated (Lewis et al. 1984; Biggers et al. 1987; Putney et al. 1988a, 1988b). Each of these factors is capable of causing embryonic death and certainly lethal to the embryo when combined.

15.3.2.3 Semen Quality

Male fertility also suffers during periods of heat stress. Bull sperm concentration, motility, and morphology are all affected by increased testicular temperatures.

Morphological defects include various maladies affecting both the head and tail regions of the spermatozoa. The observed decrease in quality of the ejaculate occurs irrespective of testosterone concentrations, as both circulating and testicular concentrations of testosterone are unaffected by heat stress (Minton et al. 1981; Barth and Bowman 1994). The observed decrease in semen quality is, however, associated with an increase in the production of reactive oxygen species (ROS) during periods of heat stress (Nichi et al. 2006).

15.3.3 Plasma Urea Nitrogen and Fertility

Although poorly understood, heat stress increases circulating plasma urea nitrogen (PUN) concentrations (Shwartz et al. 2009; Rhoads et al. 2009). This phenomenon may be a consequence of increased skeletal muscle breakdown rather than sub-optimal rumen nitrogen efficiency. A better indicator of muscle catabolism is 3-methyl-histidine, which is reported to increase in heat stressed poultry and is not caused by heat-induced reduced DMI (Yunianto et al. 1997). Direct effects of heat stress on muscle breakdown (3-methyl-histidine or creatine) have been reported in exercising men (Febbraio 2001), rabbits (Marder et al. 1990), and lactating cows (Schneider et al. 1988; Kamiya et al. 2006). Other factors that may contribute to increased PUN are increased digestibility of the diet (due to decreased passage rate; West 1999) and decreased blood flow to the kidneys for urea excretion (Kenney and Musch 2004; Lee et al. 2005). The end result of these physiological changes is elevated PUN concentrations in heat stressed cows, irrespective of changes in diet formulation or declining DMI.

The conception rates of both dairy cows and heifers plummet dramatically in response to elevated PUN concentrations (Ferguson et al. 1988, 1993; Elrod and Butler 1993; Butler et al. 1996). Reproductive processes that are likely to be affected by excess PUN concentrations include follicular, luteal, and embryonic development. Circulating urea concentrations equilibrate within reproductive fluids and increase proportionally with increasing concentrations of PUN (Leroy et al. 2004; Hammon et al. 2005). Elevated urea concentrations within the uterus may affect embryonic development and survival by altering the uterine environment (especially secretory activity). For example, uterine luminal pH on day 7 after estrus is slightly lower in cows with high PUN concentrations (Elrod and Butler 1993; Elrod et al. 1993; Rhoads et al. 2004) During embryo culture, lowering the pH of the culture media to similar levels reduced cleavage rates and almost completely inhibited the development of embryos to the blastocyst stage (Ocon and Hansen 2003). When embryos were flushed from donor cows consuming high or low protein diets, no quantitative or qualitative differences were observed (Garcia-Bojalil et al. 1994; Rhoads et al. 2006). However, appearances can be deceiving: when those embryos were transferred, those collected from the high PUN donors resulted in fewer pregnancies than the embryos collected from moderate PUN donors (11 vs. 35% pregnancy rate). Interestingly, there were no differences in pregnancy rate based on the PUN concentrations of the recipient animals (Rhoads et al. 2006). Since pregnancy rates were only affected by the PUN concentrations of the donor animals, we can conclude that either the oocyte or embryo is damaged by high PUN concentrations on or before day 7 of pregnancy (the day that embryos were collected from donors), resulting in decreased long-term viability. Similar effects on embryonic development have been reported in ewes (Bishonga et al. 1996). Mechanistically, the embryos from ewes fed with high urea diets used more glucose and experienced up to a 2.8-fold increase in metabolic rate compared to embryos from ewes fed with low urea diets (McEvoy et al. 1997). The availability of nutrients to the embryo that would be needed to support such an accelerated metabolic rate may be a factor contributing to the observed decrease in embryo survival.

15.3.4 Heat Stress, Reproduction, and Heat Shock Proteins

The impact of increased environmental temperature on mammalian reproduction are well recognized and include detrimental effects on oogenesis (Putney et al. 1989), spermatogenesis (Kunavongkrit et al. 2005), puberty suppression (Kurowicka et al. 2006), ovulation (Rozenboim et al. 2007), conception (Christenson 1980; Putney et al. 1989), and embryogenesis (Putney et al. 1989; Hansen 2009). This section will highlight impacts of hyperthermia on signaling mechanisms and specifically heat shock proteins (HSP) in reproductive tissues. It is also important to mention detrimental effects of elevated body temperature on male reproductive function. The testis is located in a scrotal sacral pouch which is designed to maintain the temperature at a lower temperature than the body ~34–35°C. This reduced temperature is vital for spermatogenesis to occur (Martti 1967) because elevated body temperatures can have detrimental effects on male reproductive function. Damage to spermatocytes and spermatids occurs at temperatures of 43°C for 15 min, while a further two degree increase results in widespread damage and germ cell apoptosis (Jegou et al. 1984). Recovery of spermatogenesis takes approximately 50 days after thermal stress (Setchell 1998); however, if the primary germ cells are destroyed, the damage can be permanent. Hyperthermia can also impact male hormone production (Kuhn-Velten 1996).

It is proposed that ghrelin is thought to play a role in counteracting hyperthermia-induced testicular damage. For example, ghrelin is hypothesized to act as an endogenous antioxidant by increasing antioxidant enzyme activity and to reduce lipid peroxidation in the testis and ovary (Kheradmand et al. 2009a, b, c, 2010). In addition, ghrelin partially reverses the negative effects of heat stress on testicular and epididymal mass, testicular germ cells, and mitotic activity of seminiferous tubules (Kheradmand et al. 2011).

The heat shock response that mediates the detrimental effects of hyperthermia can also be activated in response to other stressors, such as chemical exposures. HSP are the central mediators in the heat shock response and have been classified into seven groups based on their functional role: (1) the molecular chaperones

(Ellis et al. 1989), (2) proteolytic enzymes, (3) DNA and RNA modifying enzymes alterations (Jantschitsch and Trautinger 2003), (4) metabolic enzymes (Voit and Radivoyevitch 2000; Malmendal et al. 2006), (5) regulatory proteins (Al Refaii and Alix 2009), (6) cellular structural proteins, e.g., cytoskeletal proteins, and (7) transport, detoxifying, membrane-modulating proteins (Welker et al. 2010). Across species, approximately 50–200 of these proteins are upregulated in response to thermal stress (Richter et al. 2010). The HSP genes that were upregulated are classified into the seven HSP groups described above. It is noteworthy that no genes involved in DNA/RNA repair were altered. It is also worth noting that these cells were a transformed lymphoma cell line, and therefore results may not accurately represent changes to HPS's that occur in normal, non-diseased cells. Additionally, this was a mild heat stress for a short period of time; however, it is notable that there were changes in 104 genes, and that many of these genes have a known function in mediating the heat shock response (Tabuchi et al. 2008). This family of proteins, which include several subgroups, is classified based on their molecular weights. A transcription factor, heat shock factor (Hsf1), is a critical regulator of the heat shock response (Wu 1984; Wu et al. 1986). Hsf1 binds to the heat shock element in target genes to initiate transcription.

15.3.4.1 Role of Heat Shock Protein in Reproduction

Hsp90: In Drosophila, HSP90 mutant alleles have a variety of ovarian defects, including blocked egg chambers, problems in transfer of nurse cell cytoplasm into the developing oocytes, decreased nuclear degeneration at end of dumping process, and smaller size of mature eggs (Pisa et al. 2009). HSP90 is thought to regulate the estrogen receptor (ER) (Ruden et al. 2005) and has been demonstrated to inhibit aromatase expression in bovine dominant follicles (Driancourt et al. 1999). Since aromatase is a critical enzyme in Estradiol synthesis, HSP90 may play a role in ovarian steroidogenesis. A role for auto-antibodies in HSP90 in premature ovarian failure (POF) patients has recently been demonstrated (Pires and Khole 2009). In cows, HSP90 has a widespread ovarian distribution (Velazquez et al. 2010), and has been shown to be unaffected by thermal stress in bovine follicular fluid (Guzeloglu et al. 2001). HSP90, therefore, is known to be important for ovarian function although the effect of thermal stress on its expression and activity remains unclear.

Hsp70: HSP70 plays an important function in refolding misfolded proteins and directing those misfolded proteins that are beyond repair toward apoptosis (Weber and Janz 2001). Conversely, HSP70 has an anti-apoptotic role (Mallouk et al. 1999; Beere 2004) and is increased in ovarian tertiary follicles relative to primary and secondary follicles (Velazquez et al. 2010). Bovine cystic follicles have increased expression of HSP70 relative to healthy follicles (Velazquez et al. 2010), and HSP70 was increased in heat shocked rat (Narayansingh et al. 2004) and quail (Sahin et al. 2009). In porcine ovaries, HSP70 protein is expressed in granulosa cells and is elevated in response to high temperatures (Sirotkin and Bauer 2011).

Further, thermal stress increases the stability of Hsp70 mRNA (Theodorakis and Morimoto 1987), and may represent a protective event in the ovary. In contrast to a protective ovarian role, elevated HSP70 is associated with complications during pregnancy (Molvarec et al. 2010). Increased circulating HSP70 is found in human patients with pre-eclampsia (Padmini and Lavanya 2011) and in those who experience preterm labor (Molvarec et al. 2010).

Mutation of the gene encoding HSP70 leads to defects in spermatogenesis and causes arrest at prophase I phase of meiosis, resulting in loss of spermatocytes by apoptosis (Mori et al. 1997). Also, loss of DNAJA1, which is a co-chaperone of HSP70, causes severe spermatogenesis defects (Terada et al. 2005).

Hsp60: HSP60 is localized to the human uterus and fallopian tube (Lachance et al. 2007). Additionally, the role of HSP60 in placental steroidogenesis has been determined (Olvera-Sanchez et al. 2011). In Drosophila, Hsp60C is expressed during oogenesis and lack of Hsp60C results in severe oogenesis defects and female sterility (Sarkar and Lakhotia 2008). In mice, lack of Hsp60 is embryonically lethal (Christensen et al. 2010), while heterozygosity for the mutant allele results in increased numbers of male offspring due to altered genotype of spermatogonia (Christensen et al. 2010).

In males, HSP60 is located in spermatogonia in the seminiferous tubule, interstitial cells, and in the mid-piece region of sperm cells (indicating a mitochondrial location) (Lachance et al. 2010). Interestingly, LPS/endotoxin exposure increases testicular HSP60 expression (Metukuri et al. 2010).

Small heat shock proteins (sHSPs): sHSPs including HSP27 have been studied in female reproductive tissues and is highly expressed in the corpus luteum during pregnancy in rats (Maizels et al. 1998). In the ovary, HSP27 is localized to the oocyte cytoplasm and nucleus. During oocyte maturation, HSP27 increases dramatically, and injection of HSP27 into mouse oocytes increased the rate of GVBD (Liu et al. 2010). Similar to HSP70, expression of HSP27 is increased in placentas from pre-eclamptic women (Alterovitz et al. 2010), and may have a pathogenic role during pregnancy.

HSPA4 is detected in many adult tissues with highest expression level in the spleen, testis, and ovary (Kaneko et al. 1997). Lack of Hspa4 results in impaired male fertility (Held et al. 2011) with reduced spermatozoa and testicular defects.

It is clear that a number of proteins in the HSP family play important roles in reproduction, regardless of hyperthermia. The role of individual HSPs during hyperthermia exposure remains to be clearly delineated.

15.4 Epigenetics

Literally translated to mean "above the genome", the study of epigenetics is the study of DNA modifications capable of impacting gene expression, and subsequently the cellular phenotype that is not a result of differences in DNA base pair sequence. Modifications which impact the 'epigenetic code' include DNA methylation, histone regulation, chromatin state, miRNA expression, etc. Prenatal and postnatal environmental factors can alter the epigenetic profile thereby altering future cell-specific responses to environmental stimuli and metabolic conditions. These changes may be transient or static and can be heritable for multiple generations. The type of epigenetic modification determines the duration and nature of the epigenetic response.

15.4.1 DNA Methylation

Methylation is a stable, long-lasting DNA modification that can be inherited by offspring or acquired throughout an animal's lifetime. DNA is methylated by the enzymatic transfer of a methyl group from the methyl donor, S-adenosyl methionine, through the action of DNA methyltransferases (DNMTs). Methylation often occurs by addition of a methyl group to the cytosine of CpG dinucleotides, the cytosine- and guanine-rich regions located upstream of gene promoters. Hypermethylation of CpG islands in promoter regions of genes typically suppresses gene expression through recruitment of DNA binding proteins that result in interference of transcription factors causing transcriptional repression. Alternatively, non-CpG methylation of DNA can confer more variable regulation patterns. Two classes of DNMTs exist with the distinct purposes of de novo or maintenance methylation (Klose and Bird 2006). Maintenance methylation is essential for copying existing methylation patterns during replication and development. De novo methylation, in contrast, introduces new methylation patterns to previously unmethylated DNA. De novo methylation may occur in response to environmental stress and is therefore a possible mechanism for potentiating a long-term or imprinted response to stress.

15.4.2 Histone Modifications

Histone modifications alter gene expression through regulation of histone interactions with DNA and control of histone structure. Histones modifications, unlike DNA methylation, can be reversible or static (Turner 2007). The primary post-translational modifications of histones are acetylation, methylation, and phosphorylation. Acetylation is catalyzed by histone acyltransferases (HATs) and generally acts to weaken the histone-DNA bond thereby promoting transcription. Acetylation primarily targets lysine residues and can be reversed, or deacetylated, by histone deacetylases (HDACs). HDACs restore the interaction between DNA and histones causing repression of transcription. Phosphorylation of histones targets serine and threonine residues and is controlled by the opposing action of kinases and phosphatases. It is currently unknown how phosphorylation alters histone function. Histone methylation of lysine and arginine is reversibly regulated by methyltransferases and demethylases. Methylated residues can be mono, di, or trimethylated. Less common histone modifications also modify histone function including ubiquitylation, sumoylation, and ADP ribosylation. Histone modifications can impact histone-DNA binding and regulate further histone modifications.

15.4.3 Chromatin State

The regulation of histone structure by the covalent modifications described above has a direct impact on chromatin state. Typically, chromatin exists in one of two states: the restricted state, heterochromatin or the relaxed state, euchromatin. While distinct associations have not been made between specific histone modifications and chromatin state, it is clear that chromatin state impacts gene expression.

15.4.4 miRNA

miRNA are 18–25 nucleotide non-coding RNAs that confer post-transcriptional gene regulation (PTGR) following binding with specific regions on the target mRNA which results primarily in mRNA degradation and/or translational repression. Although miRNA is not classically considered in an epigenetic event, miRNA is able to participate in epigenetic regulation through a feedback loop with the epigenetic regulators, DNMTs and HDACs (Sato et al. 2011). In this regulatory circuit, the expression of miRNAs is regulated epigenetically by DNMTs and HDACs. Subsequently, miRNAs regulate transcript expression of DNMTs and HDACs.

15.4.5 Genomic Imprinting

Following fertilization, both paternal and maternal genomes undergo global demethylation yielding a one cell totipotent zygote. As development progresses through a series of holoblastic cleavages, cells begin differentiation leading them toward specific cell lineages (Reik 2007). The mechanism controlling the molecular programming of these cells is not well understood but is largely attributed to differences in DNA methylation of CpG islands as well as alterations in histone methylation and acetylation patterns (Kelly and Trasler 2004). Continued differentiation and programming of cells during prenatal development is largely the result of continued modifications of DNA methylation patterns. During cellular differentiation, epigenetic modifications can be altered while others remain more permanent, such is the basis of numerous disease statuses as the result of abnormal epigenetic imprints. Functionally, epigenetic modifications can improve the plasticity of the genome to improve responses to specific environmental conditions. We hypothesize that epigenetic modifications established in utero in response to maternal heat stress can influence the postnatal development.

15.4.6 Epigenetic Response to Heat Stress

Heat stress during early postnatal development has been shown to cause histone modifications that extend into adulthood. This epigenetic conditioning then results in thermotolerance to heat stress later in life. This phenomenon has been best demonstrated in chickens, where heat stress at 3 days of age conferred protection against acute heat stress-related mortality during adulthood (Yahav and McMurtry 2001). This protective effect has been characterized by analysis of histone modifications and resulting epigenetic memory (Kisliouk et al. 2010). In chickens that were acutely heat stressed 3 days post hatching and again 1 week later, increases in H3K9 acetylation and H3K9 dimethylation were observed during heat stress exposure at developmental stage. Histone modifications in response to chronic heat stress have also been reported in rodents (Tetievsky and Horowitz 2010). In this study, histone H3 phosphorylation and subsequent H4 acetylation occurred in the promoters of Hsp-70 and Hsp-90 after heat acclimation. This resulted in a cytoprotective effect that was mediated through constitutively elevated expression of HSP-70 and increased HSP-90 protein in response to heat acclimation and reacclimation. This cytoprotective response to heat stress has been replicated in cell culture where a conditioning heat load increased survival of mouse fibroblasts in response to a lethal heat load (Luft et al. 2001).

15.4.7 Evidence of Epigenetically Programmed Response to Maternal Stress

15.4.7.1 Intrauterine Growth Restriction

Intrauterine growth restriction (IUGR) is caused by insufficient fetal nutrient and oxygen supply and placental development and has been associated with low birth weight and numerous adverse outcomes in affected offspring. The epigenetic programming that occurs in IUGR has been studied primarily in rodent models. Characterization of experimental-induced IUGR in rodents revealed global genome-wide hypomethylation and respective perturbations in methyl group metabolism in the liver (MacLennan et al. 2004). In pancreatic islets, experimentally induced IUGR caused alterations in the methylation profile and histone modification of genes responsible for insulin and glucose regulation (Raychaudhuri et al. 2008; Thompson et al. 2010). Many metabolic consequences have been reported in response to IUGR in rodents, humans, and agricultural species. IUGR and nutrient malnutrition prior to weaning caused impaired beta cell development in rats which is retained into adult life (Garofano et al. 1998). At 9 months of age, the effects of IUGR resulted in upregulation of hepatic lipogenic transcription factors and enzymes as well as upregulation of hepatic CRP expression in rats (Magee et al. 2008).

15.4.7.2 Metabolic and Hormonal Imprinting

Similar to IUGR, metabolic imprinting describes a lasting epigenetic imprint in response to the prenatal metabolic environment (Waterland and Garza 1999). This imprint is characterized by changes in organ and tissue structure, cell number, and differentiation. The phenomenon of metabolic imprinting is best illustrated by Csaba et al. (1984). In this study, the injection of parental rats with insulin altered the offspring's response to insulin. Metabolic and hormonal imprinting has since been demonstrated through the study of obesity-prone offspring born from diabetic mothers in rodents and humans (Poston 2011).

15.4.7.3 Maternal Heat Stress

Epigenetic programming and hormonal changes in IUGR animals is somewhat similar to that observed in animals exposed to prenatal heat stress. This change is likely a result of alterations in metabolism, uterine blood flow, and reproduction caused by heat stress. The effects of maternal heat stress on offspring may range from developmental deficits to an altered response to heat stress later in life. In mice and guinea pigs, prenatal exposure to heat stress resulted in reduced postnatal weight gain and smaller brain weights that lasted into maturation (Jonson et al. 1976; Shiota and Kayamura 1989). The cause of suboptimal postnatal performance in mice exposed to prenatal heat stress has been suggested to be due to interference with establishment of the hypothalamic pituitary axis. In mouse embryos, a gene-specific DNA methylation imprint of heat was observed following heat stress (Zhu et al. 2008). As we continue to expand our knowledge of the effects of maternal heat stress and its imprint on future generations, epigenetic regulation will likely be a major contributor.

15.4.8 Species-Specific Evidence of Maternal Stress-Induced Epigenetic Imprinting

In agricultural species, the postnatal effects of prenatal stresses have not been well characterized. The majority of research on epigenetic regulation in response to maternal stressors has been done in pigs and sheep. Although there is little direct evidence suggesting that heat stress in utero confers an epigenetically mediated response to heat in later life, there is ample evidence of stress-induced epigenetic changes in these species.

15.4.8.1 Swine

Heat stress in swine has significant impact on reproductive performance. In addition to decreased fertility, prenatal stress can also result in IUGR that can result in further losses as piglets born during periods of maternal stress have diminished performance in postnatal life.

The concept of IUGR impacting both fetal and postnatal development in pigs has been studied by Foxcroft et al. IUGR in pigs has lasting effects on growth potential and carcass quality (Foxcroft et al. 2006, 2009). In addition to IUGR, maternal diet has been demonstrated to impact postnatal performance of piglets. Supplementing gestating sow diets with omega three fatty acids improved glucose uptake in offspring (Gabler et al. 2009). Moreover, maternal stress may also result in an imprint on immune response. Exposure to maternal restraint stress caused a greater inflammatory response to endotoxin challenge in offspring (Collier et al. 2011).

15.4.8.2 Sheep

IUGR has also been studied in sheep, and it has been demonstrated that IUGR by nutrient restriction alters the offspring's response to similar undernutrition during gestation later in life (Kiani et al. 2008). Specifically, ewes that experienced IUGR in fetal life have an impaired ability to adjust energy expenditure in response to energy restriction during gestation. An impact of maternal heat stress has also been reported in sheep. Prenatal heat stress in lambs caused reduced uterine blood flow and resulted in decreased birth weights and tissue weights (Dreiling et al. 1991). In addition, seasonal differences in tissue development have been observed between lambs born in the spring vs. autumn (McCoard et al. 1997).

15.5 Gastrointestinal Tract

15.5.1 Barrier Function and Integrity

The main functions of the gastrointestinal tract (GIT) are to facilitate digestion and absorption of water and nutrients as well as maintain barrier integrity and immunity. The GIT achieves this by maintaining critical epithelial barriers which normally prevent passage of unwanted luminal contents and pathogens from entering the body, while allowing passage of ions, nutrients, and water. Thus, the intestines form a major physical barrier to prevent pathogens and toxic compounds

Table 15.1 Effects of heat stress on intestinal morphology measures in various species

Species	Villous height	Crypt depth	References
Pigs	Decreased	Decreased	Yu et al. (2010)
Pigs	Decreased	Decreased	Liu et al. (2009)
Broilers	No change	No change	Quinteiro-Filho et al. (2010)
Broilers	No change	Decrease	Burkholder et al. (2008)
Broilers	Decrease	–	Garriga et al. (2006)
Quail	Decrease	Decrease	Sandkc et al. (2004)
Fowl	Decrease	–	Mitchell and Carlisle (1992)

from entering the mucosa and circulation, while simultaneously activating the acquired and innate immune systems. This is largely dependent on junction complexes connecting enterocytes together via well-organized intercellular array of tight junctions, adhesion junctions, and desmosomes surrounding the apical region of epithelial cells. Cell to cell adhesion and tight junctions are regulated by the membrane spanning proteins claudin, occluding, zonula occudens-1 and 2 (ZO-1, ZO-2) and cingulin (Oswald 2006; Shen et al. 2011). Additionally, adhesion junction proteins such as E-cadherin, also contribute to gut integrity. Interestingly, these tight junctions undergo constant remodeling and are dynamic in nature (Shen et al. 2008).

Little is known about the direct or indirect effects of heat stress on tight junction complexes and barrier integrity in livestock. However, one of the most consistent effects of heat stress on livestock is the rapid damage and death of the intestinal epithelial surface and sloughing of the epithelial layer (Liu et al. 2009; Yu et al. 2010). As a result, heat stress is associated with leaky epithelial barrier and increased intestinal permeability, especially endotoxin or lipopolysaccharides (LPS) (Lambert 2004, 2009). Furthermore, as a result of epithelial damage, intestinal morphology is altered during heat stress with villi and crypts being shortened (Table 15.1). Mechanistically, in vitro studies in Caco-2 cells indicated that heat stress directly alters tight junction protein expression and localization, leading to increased barrier dysfunction (Dokladny et al. 2006, 2008). The increase in occludin expression was blunted when cells were heated and given a HSP inhibitor. Furthermore, the decrease in barrier integrity and increased permeability may be explained by the upregulation of myosin light chain kinase (MLCK) (Yang et al. 2007). MLCK directly phosphorylates the regulatory light chain of type 2 myosin (MLC) resulting in the contraction of the actin cytoskeleton in the tight junction complex (Turner 2006). As actin filaments regulate tight junction adhesion, contraction of these filaments ultimately leads to tight junction openings, aiding in paracellular flux (Yang et al. 2007). Therefore, it is hypothesized that heat stress activates MLCK and phosphorylates MLC resulting in the opening of tight junctions and intestinal permeability.

Further evidence suggest that during heat stress there is reduced blood flow to the GIT causing tissue hypoxia, ATP depletion, intracellular acidosis, and altered

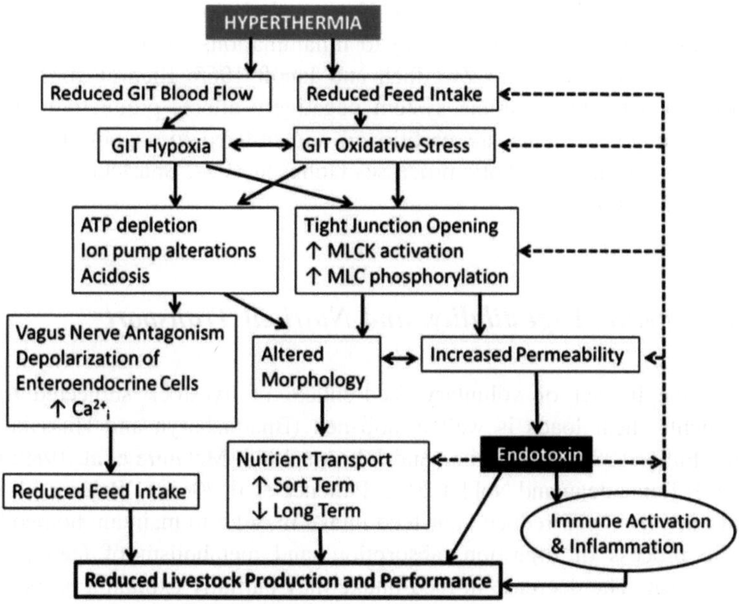

Fig. 15.2 Response of the gastrointestinal tract (*GIT*) to heat stress in livestock and events that lead to reduced production, performance, and health

ion pump activity (Hall et al. 1999). Depletion of ATP and intracellular acidosis jeopardizes tight junctions of the intestinal epithelium and this can result in bacterial and LPS translocation from the lumen into circulation. In addition, sepsis (Li et al. 2009; Zhang et al. 2010a, b), LPS (Albin et al. 2007; Chakravortty and Kumar 1999), hypoxia (Qi et al. 2011), and metabolic stress (Keller et al. 1992; Lewis and McKay 2009; Muckter et al. 1987) have all been shown to compromise intestinal epithelial barrier function via decreased transepithelial electrical resistance (TER), intestinal integrity, or junction protein reorganization. Interestingly, all of these factors are commonly found in animals undergoing heat stress or heat stroke and significantly contribute to altered intestinal function and integrity (Fig. 15.2).

Luminal bacteria present in the GIT serve as a major source of LPS/endotoxin (Ravin et al. 1960; Schweinburg and Fine 1960; Wiznitzer et al. 1960). As such, increased intestinal permeability due to heat stress augments circulating endotoxin concentrations in a variety of species (DuBose et al. 1983; Hall et al. 1999; Singleton and Wischmeyer 2006; Lambert 2008, 2009). However, little is known about whether heat stress increases circulating LPS in livestock species, but the consequences of LPS/endotoxin on animal performance and health is well documented. LPS is a potent immune stimulator (Frost and Lang 2002; Webel et al. 1997) that is recognized by the pattern recognition receptor, toll-like receptor 4 (TLR4), and the other proteins such as LPS binding protein (LBP),

CD14, and MD-2 (Hornef et al. 2003; Neal et al. 2006). Once recognized, LPS activates the immune system, leading to inflammation, septicemia, and eventually death (if severe enough; Zweifach and Janoff 1965; Rice et al. 2003). For livestock, an activated immune system negatively affects productivity because energy and other nutrients are partitioned and utilized to support the immune modulation rather than anabolic processes (Johnson 1997; Spurlock 1997; Gabler and Spurlock 2008).

15.5.2 Appetite, Digestibility, and Nutrient Transport

The negative impact of voluntary feed intake in livestock subjected to high environmental heat loads is well established (Bhattacharya and Hussain 1974; Dale and Fuller 1980; Mohammed and Johnson 1985; McGuire et al. 1989; Collin et al. 2001; Renaudeau and Noblet 2001; Patience et al. 2005; O'Brien et al. 2010). Heat stressed livestock reduce their feed intake in order to maintain homeothermy since the process of digestion, absorption, and metabolism of feed produces metabolic heat. The decrease in feed intake may partially explain a proportion of the weight loss or lack of body weight gain, reduced milk production, and decline in other performance parameters observed in thermally challenged livestock (Baumgard and Rhoads 2012). Feed intake is influenced by a number of factors and is very sensitive to environmental conditions. With the selection for more efficient and higher yielding genotypes, increasing nutrient intake to support high levels of production may render animals more sensitive to environmental conditions.

Compared to TN-housed pigs, apparent total track digestibility coefficients for energy, nitrogen, and dry matter are increased in pigs reared at 33°C (Collin et al. 2001). However, if corrected for equal feed intake (heat-stressed animals ate less) there was no effect of climate on digestibility. This is in agreement with other species in which no differences were observed in total track digestibility coefficients (once corrected for feed intake; laying hens, Koelkebeck et al. 1998; quail, Sahin and Kucuk 2003; dairy, Kim et al. 2010; beef, White et al. 1992) during heat stress. On the contrary, heat stressed sheep have reduced digesta passage rates, neutral detergent fiber (NDF), acid detergent fiber, and non-structural carbohydrates digestibility coefficients (Bernabucci et al. 2009). This may be due to changes in the cellulolytic and amylolytic rumen bacterial populations following exposure to climatic stress.

The GIT is the major site of nutrient uptake and intestinal nutrient absorption is highly susceptible to modification following exposure to environmental stress. Glucose and fructose are the major sugars present in the gut lumen. Glucose transport in the jejunum occurs via the apical sodium-dependent glucose co-transporter (SGLT1) and the basolateral glucose transporter (GLUT) 2, whereas fructose is absorbed from the lumen by the way of GLUT5 (Kellett 2001; Kellett and Brot-Laroche 2005). Similar to many amino acid transporters, SGLT1 is an active transporter that requires Na+ cation gradient to transport glucose (Ray et al. 2002). Na+/K+ ATPase pumps on the basolateral membrane restores and maintains these Na+ gradient and disruption of intestinal Na+/K+ ATPase can inhibit sugar absorption and glucose transporter mRNA in rodents (Ai et al. 2004). Interestingly, intestinal Na+/K+ ATPase activities are either unchanged (Garriga et al. 2006) or increased (Chen et al. 1994) due to heat stress in broilers. Heat stress may have two effects on an animal's GIT. Firstly, increased ion pump activity helps to maintain cellular osmolality and homeostasis. Secondly, increased ion pump activity helps to maintain membrane potential and active nutrient transport. Garriga et al. (2006) reported that heat stress increased intestinal glucose transport kinetics in broilers. The fact that they did not observe this in a pair-fed control group indicates this was mainly due to the thermal effect of feeding and not feed restriction. Additionally, the increase in SGLT-1 protein expression partially explains this phenomenon.

The nutrient transport function of the GIT decreases when the intestine is under inflammatory or metabolic stress. Studies show that glucose, lysine, arginine, and glutamine transport is reduced under LPS challenge in pigs and culture, respectively (Meng et al. 2005; Albin et al. 2007). While the animal is trying to fight the cause of the stress, the GIT reduces its ability to transport the necessary nutrients to mount an effective immune response. In turn there is increased skeletal muscle degradation and an upregulation of catabolic processes that are required to fuel the immune response (Webel et al. 1998). There are two proposed mechanisms through which the nutrient transport under an inflammatory stress might be controlled. First, the inflammatory cytokines secreted following immune challenge prevents food intake and nutrient transport by acting on the growth hormone axis (Johnson 1997, 1998). Second, LPS and other TLR4 ligands such as saturated fatty acids activate the enteroendocrine cells which act as a nutrient sensor in the intestine. This leads to the secretion of appetite regulating peptides such as CCK and GLP-1 from the enteroendocrine cells which acts on the satiety centers in the hypothalamus which ultimately results in reduced nutrient absorption from the intestine (Bogunovic et al. 2007; de Lartigue et al. 2011). While this increase in satiety peptide secretion has been shown in the cell lines, further research is needed to prove this theory in vivo. Apart from the direct damage to the intestinal epithelium, heat stress can lead to an increase in intracellular calcium ($[Ca^{2+}]i$) in myocytes (Chen et al. 2010). Importantly, and plausibly, an increase in $Ca^{2+}i$ leads to activation of different signaling cascades and cellular processes leading to augmented secretion of intestinal peptides such as glucagon-like peptide 1 (GLP-1), peptide YY (PYY), gastric inhibitory polypeptide (GIP), and cholecystokinin (CCK) (Chen et al. 2006; Sternini et al. 2008). Increased secretion of these peptides is associated with decreased gastric emptying, intestinal motility, and suppressed appetite. Therefore, the large rise in the rate of potassium flux due to hyperthermia could depolarize the cell and cause increased porosome secretion of appetite regulatory peptides. This could partially explain the reduction in feed intake in livestock species subjected to heat stress.

15.6 Impact of Climate Change on Disease occurrence in Livestock

The livestock sector accounts for 40% of the world's agricultural Gross Domestic Product (GDP) and 1.3 billion people depend upon livestock husbandry for their livelihood (FAO 2006). Moreover, global demand for foods of animal origin is steadily growing and it is apparent that the livestock sector will continue to expand accordingly (FAO 2009). Livestock productivity is heavily influenced by the animal's environment. Climate change (CC) impacts agricultural production because climate provides essential inputs (water, solar radiation, and temperature) needed for plant and animal growth. CC is a major threat to the survival of many species, ecosystems, and the sustainability of livestock production systems (Moss et al. 2000). The majority of animal species presumably prefer to be in their natural habitat so that they minimize the energy expense during the process of adaptation to harsh climatic conditions. The Intergovernmental Panel on Climate Change (IPCC) has declared that global warming is responsible for many environmental changes such as drought, flood, crop and animal farming, food production, health hazards, etc. CC affects livestock productivity by the following (1) price and availability of grain (2) production and quality of forage and pastures (3) changing patterns of livestock diseases, and (4) direct effect of weather on livestock health (Smit et al. 1996).

Global CC alters ecological construction which causes both the geographical and phonological shifts (Van den Bossche and Coetzer 2008; Slenning 2010). These shifts affect the efficiency and transmission pattern of the pathogen and increase their spectrum in the hosts (Brooks and Hoberg 2007). Increased spectrum of pathogen increase the disease susceptibility of the animal and thus CC supports the pathogenicity of the causative agent. Heavy rainfall causes flood and increases atmospheric humidity which results in favorable condition for the proliferation of pathogen, ticks, flies, and mosquitoes. These pathogens and insects may serve as vector or invader for the transmission of diseases in humans and livestock. Recent disease outbreaks are consistent with model projections that warmer and wetter conditions lead to greater transmission potential even at high altitudes and elevations. Mosquito-borne diseases are now reported at higher elevations than in the past at sites in Asia, Central Africa, and Latin America (Epstein et al. 1998). Environmental changes caused either by natural phenomenon or anthropogenic interference changes the ecological balance and context within which disease hosts or vectors and parasites breed, develop, and transmit disease (Patz et al. 2000). CC affects the occurrence and spread of disease by impacting the population size and range of hosts and pathogens, the length of the transmission season, and the timing and intensity of outbreaks (Epstein et al. 1998; Epstein 1999). Pathogens from terrestrial and marine texa are sensitive to hot temperature, heavy rainfall, and humidity. Climate warming can increase pathogen development and survival rates, disease transmission, and host susceptibility (Harvell et al. 2002). Understanding the spatial scale and temporal pattern of disease incidence is a fundamental prerequisite for the development

Strategies to counteract climate change effect on disease occurrences in livestock:

- Development of public awareness and health education programmes
- Appraisal of public health and socioeconomic impacts of zoonoses
- Strengthening of surveillance and disease investigation capacities
- Networking among epidemiological and laboratory units under public health and animal health sectors
- Developing prevention and control strategies for livestock diseases and incorporating them in the existing control programmes
- Development and harmonization of appropriate cross-border disease surveillance
- Use vaccinations (e.g., rabies vaccines) as a control measure where appropriate in regions where disease is endemic
- Limiting transportation of live animals
- Modification of veterinary and medical curricula with respect to epidemiology and public health aspects of zoonoses
- Involvement of medical and veterinary institutions NGOs, international professional associations and animal welfare organizations in zoonoses control activities.
- Identification of research needs on zoonoses and their promotion.
- Protocol should be designed to verify the absence or elimination of infection from a country
- Formulating appropriate environmental, energy and economic policies are necessary to control emerging infectious diseases
- Pilot projects should be developed for specific disease control and elimination

Fig. 15.3 Strategies to minimize CC effect on disease occurrences in livestock

of appropriate management and intervention strategies (Ward and Carpenter 2000; Koopmans et al. 2007; Rose et al. 2009). It is particularly important given the need to understand the elevated risks linked to CC (Rose and Wall 2011).

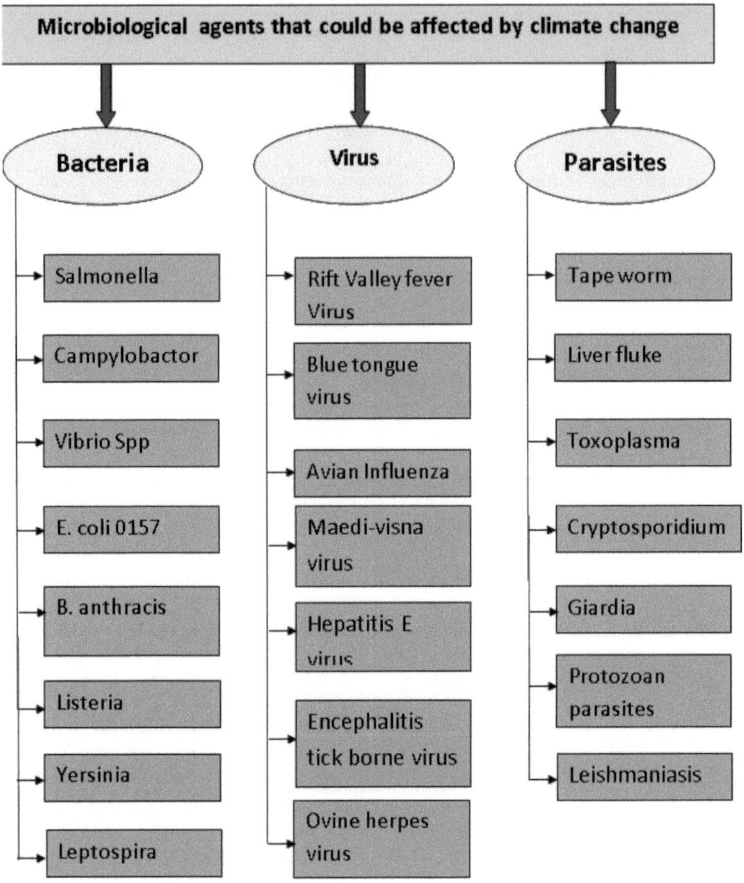

Fig. 15.4 Microbiological agents impacted by CC

15.6.1 Effects of Climate Change on Livestock Diseases

CC influences several factors that cause livestock disease outbreaks. These factors are mainly associated with temperature rise, host pathogen interaction, farming practice, number of vectors/host/reservoirs, environmental factors and microhabitats. Figure 15.3 describes various strategies that could be implemented to reduce the effect of changing climate on disease ocuurence.

15.6.1.1 Climate Change and Disease Occurrence in Livestock

Global warming, CC, and extreme weather events have an adverse effect on biodiversity, distribution of animals, and microflora, all of which may increase the likelihood of emerging zoonotic agents and infectious disease outbreaks (Sachan and Singh 2010). CC is expected to cause an increase in weather-related disasters and extreme weather events, such as droughts, heat waves, storms, desertification, and increases in insect infestations. Long-term changes in climate will jeopardize the future of all animals. CC may be responsible for the emergence and proliferation of many disease such as malaria and zoonotic parasitic diseases including Leishmaniasis, cryptosporidiosis, giardiasis, trypanosomiasis, schistosomiasis, filariasis, onchocerciasis, and loiasis (Patz et al. 2000). The most sensitive diseases are those that are indirectly transmitted especially those that require vehicle for transfer from host to host (water- and food-borne disease) or an intermediate host or vector as part of its life cycle.

Global CC predictions suggest that far-ranging effects might occur in the population dynamics and distributions of livestock parasites, provoking fears of widespread increases in disease incidence and production loss (Morgan and Wall 2009). Livestock are sensitive to temperature variations and it is the major of cause of economic loss due to CC (Reilly et al. 2003). Increased temperature is the major cause of reduction in livestock productivity. Economic losses per year in USA have been estimated to be $2.4 billion with minimal heat stress abatement. Climatic restrictions on vectors, environmental habitats, and disease causing agents are important for understanding the outbreak of several animal diseases (Stem et al. 1989). Changes in temperature and precipitation regimes may result in spread of disease and parasites in new regions or produce high incidence of diseases with concomitant decrease in animal productivity and increase in mortality (Baker and Viglizzo 1998). Baylis and Githeko (2006) evaluated the effect of CC on parasites, pathogens, disease hosts, and disease vectors on domestic livestock. The potential clearly exists for increased rate of development of pathogens and parasites due to early arrival of spring and warmer winters and such seasonal change allows greater proliferation and survivability of these organisms. Warming and changes in rainfall distribution may lead to changes in spatial or temporal distributions of diseases such as anthrax, blackleg, hemorrhagic septicemia, and vector-borne diseases (VBD) that thrive in the presence of moisture. Figure 15.4 gives the representation of different types of bacterial, viral, and parasitic organisms that gets affected due to CC.

15.6.1.2 Temperature and Rainfall

Variations in temperature and rainfall are the two most significant climatic variables affecting livestock disease outbreaks. Increase in temperature and fluctuations in rainfall patterns impact prevalence, profiles, and sustenance of disease pathogenic bacteria, viruses, parasites, and fungi (Tirado et al. 2010). Such changes also have an impact on microbial ecology and growth, plant and animal physiology, and host susceptibility which may result in the emergence, redistribution, and changes in the incidence and intensity of pest infestations, plant and animal diseases (FAO 2008). Warm temperatures tend to favor proliferation of exogenous bacteria and parasites whose survival and development depends on atmospheric temperature. Warmer temperatures tend to promote survivability, shorter development rates, and transmission better than colder temperatures. Insects such as mosquitoes and ticks that transmit disease agents may also benefit from CC (Bradley et al. 2005; Randolph et al. 2008; Sumilo et al. 2009). Recently, Wall and Ellse (2011) contended that "a generally warmer environment will result in higher parasite abundance and increased disease incidence". This argument is reasonable given that insects such as biting flies and ticks that transmit pathogenic microorganisms are more likely to survive a mild winter rather than a harsh one.

15.6.2 Relationship Between Climate Change and Disease Susceptibility

CC, in the present scenario, is an important driver for the emergence of infectious diseases in humans and animals (Wolfe et al. 2005; Chomel et al. 2007; Woolhouse and Gaunt 2007; Jones et al. 2008; Reperant 2010). Warmer global temperatures will allow expansion of geographic range within which mosquitoes and parasites could survive with sufficient abundance for sustained transmission. Model predictions indicate that a 3°C global temperature rise by 2100 could increase the number of annual malaria cases by 50–80 million (Martens et al. 1995). CC is widening viral disease among farm animals, expanding the spread of some microbes that are pathogenic to humans. According to OIE, several countries are indicating that CC has been responsible for at least one emerging or re-emerging disease occurring in their territory. Little data exist to evaluate the effects of CC on infectious diseases in animals. However, some data exist from other parts of the world (Khasnis and Nettleman 2005; Epstein 2001a; Kutz et al. 2005), and a recent report from the World Organization for Animal Health (OIE) addresses the impact of CC on the epidemiology and control of animal diseases (OIE 2008).

Sensitivity of vectors to CC could change the range, season, and incidence of many zoonotic diseases (CDC 2008). Some example are: (1) Increased night temperature could increase the flight activity of the vector for example malaria (Purse et al. 2005); (2) Night temperature could increase the replication and transmission cycle of viral pathogen (Baylis and Githeko 2006); (3) Cycle of massive humidity followed by drought provides breeding site of vector and pathogen facilitating them for disease outbreaks (Baylis and Githeko 2006); (4) Alteration in the precipitation range could help in the migration of arthropod vector to different landslides and convert the disease outbreaks from endemic to epidemic (Trape et al. 1996).

Table 15.2 List of VBD of animals

VBD	Occurrence	Vector	Hosts
African horse sickness	Africa, Middle East, Europe	*Culicoides imicola* and *C.bolitinos*	Equine
African swine fever	Sub-Saharan Africa, Europe	*Ornithodoros* tick species	Swine
Bluetongue	Worldwide	*Culicoides* midges	Ruminants
Crimean Congo-hemorrhagic fever	Africa, Europe, Balkans, South Africa, and Asia	*Argasid ixodid* ticks	Wild and domestic animals (zoonosis)
Equine encephalomyelitis (eastern and Western)	Canada, Caribbean, North, South, and Central America	Mosquito	Equine, birds (zoonosis)
Equine infecriousanaemia	Worldwide	Biting flies, mosquitoes	Equine
Japanese encephalitis	Asia, Australia, India	*Culex* mosquito	Pigs, birds (zoonosis)
Lumpky skin disease	Africa, South and North Africa, Isreal	Arthopods	Cattle, zebus, giraffe, impala
Nairobi sheep disease	East Africa	*Rhipicephalus* ticks	Sheep and Goats
Rift valley fever	Africa	Mosquito	Multiple species(zoonosis)
Venezuelan equine encephalomyelitis	North, Central, and South America	Mosquito	Equine (zoonosis)
Vesicular stomatitis	United States, Europe, South Africa	Arthropods	Mammals (zoonosis)
West Nile fever	Worldwide	Mosquito	Multiple species (zoonosis)
Bovine anaplamosis	Worldwide	Tick species	Cattle
Bovine babesiosis	Worldwide	Tick species	Cattle
Tularemia	North America, Europe, and Asia	Arthropods	Rodents, rabbits, hares
Equine piroplasmosis	Worldwide	Ixodid	Equine
Heartwater	Africa, West Indies	*Amblyomma* tick species	Ruminants
Leishmaniosis	Worldwide	*Phlebotomus* and *Lutzomyia* sandflies	Dogs, rodents, opossum (zoonosis)
Surra (*Trypanosoma evansi*)	Asia, Africa and South America	*Tabanus* spp. (biting flies)	Multiple species
Theileriois	Worldwide	*Ixodid* ticks	Bovidae and ruminants
Trypanosomosis	Africa, South, and Central America	*Glossina* spp. (*tse–tse*)	Cattle (zoonosis)

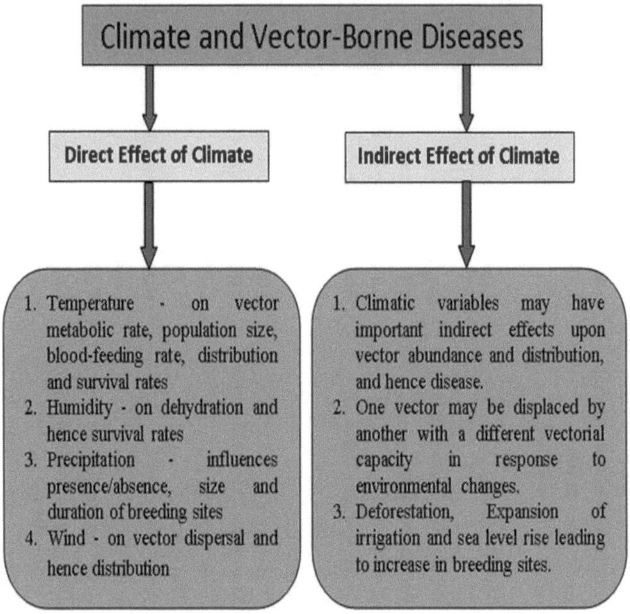

Fig. 15.5 Effect of climate on VBD. Changing climate has both direct and indirect effect on VBD

Table 15.3 Common livestock diseases due to CC

Sl. no	Type of disease
1	VBD
2	Bluetongue virus
3	Rift valley Fever
4	West Nile Virus
5	African horse sickness
6	Lumpy skin disease
7	Leishmaniasis
8	Epizootic haemorrhagic disease
9	Tick-borne diseases
10	Parasitic diseases (excluding tick-borne)
11	Pasteurellosis
12	Avian influenza
13	Anthrax
14	Blackleg
15	Rabies
16	Tuberculosis

15.6.3 Climate Change and Types of Disease Outbreaks

CC is considered a major threat to human health and well-being due to increasing association of emerging infectious diseases to CC (Dufour et al. 2006; Gale et al.

2009). Outbreaks of diseases such as foot and mouth disease (FMD) or Avian Influenza affect very large numbers of animals and contribute to further degradation of the environment and surrounding communities' health and livelihood.

15.6.3.1 Vector-Borne Diseases

Table 15.2 describes the different VBD of domestic livestock. CC impacts the distribution and abundance of arthropod vectors and interaction between the virus and its vector. CC can promote the spread of contagious diseases through increased contact between animals, or increased survival or availability of the agent or its intermediate host. The effect of CC on the distribution and prevalence of VBD is significant. In the context of increase in temperature and changes in other climatic variables are likely to have profound effects on the prevalence and transmission of VBD. Vectors are cold blooded and are influenced by climatic variables, especially temperature and humidity. The transmission of VBD depends upon three stages namely (1) Infectious agent (2) the Vector (3) the host. World Organization for Animal Health (OIE) categorized 66 diseases that affect ovine, equine, bovine, swine, and caprine of which 22 are considered to be VBDs. Effect of climate on VBD is depicted in Fig. 15.5.

15.6.3.2 Tick Borne Diseases

Tick born Encephalitis (TBE) is caused by an arbovirus of the family *Flaviviridae* and is transmitted by ticks, *Ixodes ricinus*, that act as both vector and reservoir. Similar to VBD this disease is also accelerated by the climatic changes mainly with increased day and night temperature. A common example of TBE is FMD. There are seven distinct types of FMD virus that are distributed in various geographical areas of the world. Common animal diseases which are caused or aggravated by CC are listed in Table 15.3.

15.6.3.3 Parasitic Diseases

Climatic variables are able to affect the prevalence, intensity, and geographical distribution of zoonotic helminthes. These climatic variables can influence free-living larval stages, other invertebrates, and vertebrate hosts. Fascioliasis, Schistosomiasis, and cercarial dermatitis caused by avian schistosomes have been most important (Singh 2010). Alveolar echinococcosis is currently the only cestode disease affected by CC. Nematodiases, including heterakiasis, different trichostrongyliases and protostrongyliases, anchylostomiases, and dirofilariases are the helminth disease most intensively analyzed with regard to CC.

15.7 Environmental Stress and Pasture/Forage Availability

Stress reduces crop growth (Solh 1993) and severely limits agricultural productivity (Boyer 1982). Perennial forages are grown in different environments and must endure stresses not normally encountered by annual crops. Most prominent of such stresses include repeated defoliation by machines and extremes in seasonal climatic conditions. Forage crops account for 60–90% of feedstuff input in animal production systems (Barnes and Baylor 1994).

Environmental stress which limit pasture and forage availability to livestock can be due to drought, high/low temperature, ozone, elevated carbon dioxide, soil water logging, and salinity. Perennial forages are grown commonly on degraded and barren soils with limited fertility, low water-holding capacity or infrequent irrigation, or high salt content. Furthermore, forages must endure subzero temperatures during winter or, in the case of tropical and subtropical forages, withstand chilling or infrequent frosts, and periodic defoliation (e.g., machine harvest). Collectively, stresses may reduce the harvested forage yield, alter its nutritive value, and change species composition of the sward. With the current societal emphasis on CC and its related impact on forage crop production, knowledge of how abiotic or environmental stresses limit forage production and the ability to mitigate negative impacts of these stresses are becoming increasingly important.

15.7.1 Climate Change and Related Environmental Stress

CC is expected to increase global temperature and alter precipitation regime and levels of atmospheric CO_2 concentrations. The overall effect of future CCs on pasture production is uncertain and likely to vary regionally, depending on the combination of changes in temperature, rainfall, and plant responses to elevated atmospheric CO_2 concentrations (Harle et al. 2007; Howden et al. 2008; McKeon et al. 2009). The direction and magnitude of the impacts will also depend on the specific cropping system as well as on regional conditions. CC impacts and adaptation in grasslands are of particular importance because grasslands are a major contributor to global food production, covering about 70% of the world's agricultural area (Soussana and Luscher 2007).

CC–induced increases in the coefficients of variation of grassland yields are in the range of 21 and 50%. This finding underpins that additional risk management strategies are needed to cope with CC impacts on grassland production (Finger and Calanca 2011). It is predicted that global CCs would significantly affect temperate grassland ecosystems, including their productivity, composition, and carbon exchange. This is because rainfall and seasonality would be altered and evapotranspiration would increase as a consequence of possible changes in temperature and water availability under doubled atmospheric CO_2 equilibrium conditions.

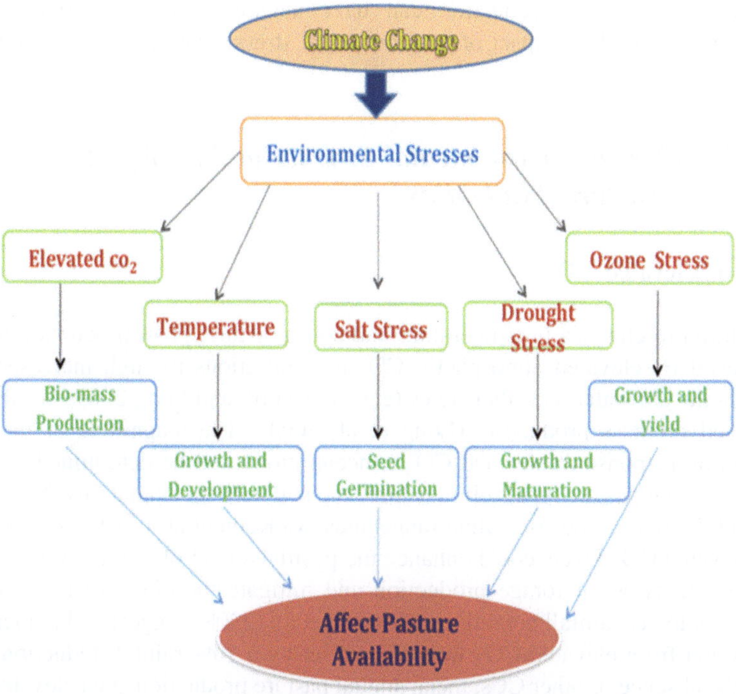

Fig. 15.6 Effect of CC on pasture availability. CC leads to different environmental stresses such as elevated CO_2, temperature stress, salt stress, drought stress, and ozone stress. All these stresses affect pasture availability by altering the different stages of growth of pastures

For example, Wang et al. (2007) reported that in *Leymus chinensis* meadow steppe (LCMS) net primary production (NPP) and soil organic carbon (SOC) are sensitive to climatic change and elevated CO_2. In the next 100 years and with an estimated doubled CO_2 concentration, if temperature increases from 2.7 to 3.9°C and precipitation increases by 10%, NPP and SOC will increase by 7–21% and 5–6%, respectively. However, if temperature increases by 7.5–7.8°C and precipitation increases by only 10%, NPP and SOC would decrease by 24 and 8%, respectively. Therefore, changes in the NPP and SOC of any production system are attributed mainly to the amount of temperature and precipitation change and the atmospheric CO_2 concentration in the future. Further the success of milk production in an out-of-doors grazing system depends largely on climatic conditions and other factors affecting pasture quality and growth. It is important to note that pasture characteristics influence the cost of production, the level and seasonal pattern of output. Seasonal variations in pasture growth lead to enforced use of dry herbage (composed of plants that have completed their life cycle or have become dormant) during certain period of the year. This herbage is usually fed to replacement stock and cows that are in non-lactating or dry period. Hence annual variations in pasture production can have negative impact on dairy farms. Figure 15.6 describes impact of environmental stresses on pasture availability.

15.7.2 Different Environmental Stresses and Its Effect on Pasture Availability

15.7.2.1 Elevated CO_2

In addition to changes in temperature and soil moisture, plant production will be influenced by elevated atmospheric CO_2 concentrations through increased photosynthetic and water use efficiencies (e.g. Ainsworth and Long 2005), resulting in increased biomass production (Long et al. 2004). The magnitude of the plant production response to elevated CO_2 concentrations will be determined by interactions among pasture type, soil moisture, and soil nutrient availability (Stokes and Ash 2007). Across the Australian rangelands, McKeon et al. (2009) demonstrated that elevated CO_2 levels could enhance the positive effect of higher rainfall future climate scenarios on forage production and mitigate the effect of reduced production in lower rainfall conditions. Howden et al. (2008) suggested that increased production from elevated CO_2 would be offset by a 10% rainfall reduction.

In the absence of other CCs, mean annual pasture production at an elevated CO_2 concentration of 550 ppm was predicted to be 24–29% higher than at 380 ppm CO_2 in temperate (C3) species-dominant pastures in southern Australia, with lower mean responses in a mixed C3/C4 pasture at Barraba in northern New South Wales (17%) and in a C4 pasture at Mutdapilly in south-eastern Queensland (9%). In the future climate scenarios at the Barraba and Mutdapilly sites in subtropical and subhumid climates, respectively, where climate projections indicated warming of up to 4.4°C, with little change in annual rainfall, the model predicted increased pasture production and a shift toward C4 species dominance.

The stimulatory effect of double ambient CO_2 on grassland above-ground ecosystem production averages about +17% in ecosystem-based experiments, although responses for particular systems and seasonal conditions can vary widely. There is a coupling between the stimulatory effect of CO_2 on growth and accompanying *long-term* changes in soil carbon, nitrogen and water availability resulting from elevated CO_2 (Lutze and Gifford 1998; Cannell and Thornley 1998; Campbell and Hunt 2000; Williams et al. 2000).

For water-limited systems, elevated CO_2 can result in greater water availability for longer in the growing season, especially if there is no increase in leaf transpiration surface per unit of ground area (Field et al. 1997; Owensby et al. 1997). These experiments suggest that the hydrological consequences of elevated CO_2 in water-limited systems can be as significant as the direct CO_2 fertilization effect on photosynthesis.

From the limited analyses that have been conducted to date in relation to temperature, it appears that warmer temperatures will increase the CO_2 response in warm-season grassland systems such as short grass steppe and tall grass prairie (Coughenour and Chen 1997). However, in temperate grasslands, a 3–4°C increase in temperature may either counterbalance the effect of CO_2 on productivity (Jones and Jongen 1996), or reduce productivity in summer but increase it in early spring and late autumn (Casella et al. 1996). In the latter case, a supplemental 3°C elevation in temperature with elevated CO_2 mitigated changes in the soil N cycle, without diminishing the sequestration of below-ground carbon (Casella and Soussana 1997). Apparently, this was due to a reduction in soil water content, which partly counterbalanced the positive effect of an increased air temperature on soil organic matter decomposition.

Therefore an increase in forage supply is expected to occur due to increase in atmospheric CO_2, with a greater increase in intensively managed farm systems. Reductions in forage supply are anticipated in drier or supra-optimal regions. Conversely, increases in forage supply are predicted if rainfall and temperature favor increased production. Other factors may constrain absolute changes (Campbell 2000). If atmospheric CO_2 increases, there will be a positive effect on pasture/forage yield and water use efficiency (WUE) although there could be a decrease in protein content. This boosting effect from higher CO_2 levels is stronger for temperate species than for sub-tropical species.

15.7.2.2 Temperature

Mean global temperature has increased by 0.76°C since 1850 owing to the emission of greenhouse gases (GHGs) and is predicted to rise by 1.8–4.0°C between 1990 and 2100 (IPCC 2007). Temperature is a crucial environmental factor that directly influences the growth and development of plants. Research has shown that effects of global warming can vary according to the temperature changes in different seasons and regions (Aber and Melillo 1991; Lamber et al. 1998; Shaver et al. 2000; Rustad et al. 2001; Rustad and Norby 2002; Post et al. 2008). Spring warming advances the phenophase of plants through accelerated leaf emergence and unfolding, while autumn warming lengthens the growing season by delaying leaf senescence; resulting in an increase in biomass production(Houghton et al. 2001; Wilmking et al. 2004). Summer plays an important role in biomass production by plants, because summer temperature is near the optimal range for normal plant growth and development (Lamber et al. 1998; Baldocchi et al. 2001). However, summer warming could potentially increase the risk of certain plants being exposed to their upper thermal limit. Some previous studies have shown that summer warming stimulates biomass production in high mountain or Arctic habitats (Post and Pedersen 2008); however, other studies conducted in temperate grasslands dominated by perennial grasses show a negative correlation between the aboveground net primary productivity (ANPP) and the mean summer temperature (Wan et al. 2009). Interestingly, belowground biomass accumulation responds to seasonal warming, decreasing during spring or autumn warming (Pedersen 2009), but changing unpredictably during summer warming (Chaves et al. 2002; De Boeck et al. 2007). In the northern steppe zone of China, and many other grassland types around the world, the plant community is dominated by rhizomatous, perennial grasses. These plants produce new shoots principally through vegetative propagation, by rhizomes or other perennial organs (Bond and Midgley 2001; Rogers and Hartnett 2001). The aboveground biomass of these species is determined by the contribution of the parent shoot and its daughter shoots which are regulated by the belowground bud bank dynamics (Harper 1977; Benson et al. 2004; Benson and Hartnett 2006). The belowground bud bank (meristem) has the potential to influence the pattern of NPP in grassland ecosystems (Knapp and Smith 2001). It has been predicted that the ANPP of the northern grassland ecosystem in China, which is dominated by *L. chinensis*, is especially sensitive to global warming (Gao et al. 1997; Piao et al. 2003). Both a field experiment manipulating ambient temperatures and wider observational data have shown that summer warming significantly or marginally decreases gross ecosystem productivity and the total aboveground grass biomass (Niu et al. 2008), and also decreases the ratio of grass to forb biomass (Xia et al. 2009).

Summer warming significantly decreased the biomass of both parent and daughter shoots in *L. Chinensis*, slightly increased the belowground biomass, and lead to a significant increase in root :shoot ratio. Warming significantly increased the total belowground bud number and decreased the daughter shoot number. Importantly, the proportions of each type of bud changed; vertical apical rhizome buds decreased, while horizontal rhizome buds increased in number (Wang et al. 2010). If temperatures increase, then there is likely to be more heat stress on cows (and pastures/crops) over summer, and perhaps more of the low quality subtropical grasses (such as paspalum and kikuyu) in pastures, but pasture growth in winter and early spring is likely to increase in response to warmer conditions.

Changes in temperature cause a pronounced effect on plant growth and productivity. Most temperate species are exposed to temperatures below the optimum for growth during winter months, and freezing or low temperatures limit persistence of tropical plants in northern latitudes or high altitudes. Plants that grow in suboptimum temperatures and acquire increased tolerance to freezing are "hardened." The hardening process takes several weeks to reach a maximum in many perennial forages. Different levels of freeze tolerance may develop depending on the genotype and hardening conditions. Dehardening may take less than one week and occurs without subsequent lowering of temperature or on reintroduction to warm temperatures. Low temperature constraints on pasture growth exist in small areas of the Eastern Highlands of China. Pastures here are dominated by temperate species, such as perennial ryegrass (*Lolium perenne*) and white clover (*Trifolium repens*) and dairying is limited to supplying local markets. And therefore dairying which means rearing of animals for milk production and products made out of milk is limited to supplying local markets.

15.7.2.3 Drought Stress

Drought is an environmental stress which is characterized by periods of limited or no water during the growing season and results in decrease in forage production for grazing and haymaking. Prolonged drought forces livestock and hay producers to better manage their fields to minimize recovery after the drought ends. To assimilate atmospheric CO_2 plants must encounter evapotranspirational water loss and some level of water stress. Some forages escape water stress by completing their life cycle prior to water becoming limiting {e.g, crested wheatgrass [*Agropryron cristutum* (L.) Beauv. ssp. *pectinarum* (Bieb.) Tzvel.]}. Most perennial forages resist water stress with a combination of avoidance and tolerance mechanisms. Plants employ avoidance mechanisms by making alterations in morphology that reduce evapotranspiration and conserve water. These may include deeper roots, leaf surface modifications (wax and pubescence), leaf orientation, and leaf senescence (Chaves 1991; Hsiao 1973; Jones and Corlett 1992). Dehydration avoidance in tall fescue and cocksfoot and summer dormancy in cocksfoot were the main strategies contributing to persistence under summer drought. Grasses such as *Chloris gayana* is one of the major tropical forage grasses that have vigorous root system that go up to 4.7 m down the Earth which confers reasonable drought resistance to the plant (Skerman and Riveros 1990).

Tolerance mechanisms enable plants to protect the water status of critical tissues, such as the apical meristem, and mainly include osmotic adjustment. Osmotic adjustment is the ability of the plant to accumulate solute molecules that reduce osmotic potential of the cell sap, and is an important physiological mechanism for dealing with water stress. Mutualistic relationships between plants and microbes modify the physiology of stress resistance in forages. Infection of tall fescue with an endophytic fungus (*Acremonium coenophialum*) reduces the weight gain and disrupts the metabolism of livestock feeding on the infected forage (Joost 1995). Ironically, infection with the fungus also confers resistance to several biotic and abiotic stresses to the plant and enables fescue to persist for many years.

When water is limiting, the efficiency with which water is used for agricultural production must be maximized, and in this context perennial forage species have a number of advantages when compared to annuals. Perennials can utilize water from autumn to spring, while annuals need to be planted from the seed bank annually. In addition, perennials can utilize water throughout the year and may improve soil macroporosity which could be beneficial to subsequent crops. Further restores soil fertility and enhance forage production, thereby contributing to reduce rangeland degradation and to increase sustainability of rainfed agricultural systems in southern European countries and North Africa (Lelievre and Volaire 2009). In addition, perennials can utilize water throughout the year and may improve soil macroporosity which could be beneficial to subsequent crops, could restore soil fertility, and enhance forage production, thereby contributing to reduce rangeland degradation and to increase sustainability of rainfed agricultural systems in southern European countries and North Africa (Lelievre and Volaire 2009).

In Mediterranean areas, WUE is mainly increased by maximizing crop growth during the rainy seasons. Under Mediterranean annual rainfall pattern, perennial plants must grow from autumn to spring and survive under summer aridity. WUE in autumn is highly correlated with sward recovery after drought. Seasonal and total WUE are also highly correlated with biomass production over the same period and with depth and density of the root system (Lelievre et al. 2011).

Water deficit frequently reduces forage yields; however, because the total value of a forage crop also depends on its quality, effects of water deficit stress on forage quality are of interest. Drought has been shown to reduce fiber concentrations and increase the digestibility of legumes and grasses. When plants encounter water deficit, there is a decline in photosynthesis. This may be due to reductions in C fixation per unit leaf area as stomata close or as photo-oxidation damages the photosynthetic mechanisms (Bruce et al. 2002). Forage produced during drought periods may be so stressed that it is unfit for livestock consumption. Several strategies have been employed to maintain healthy perennial pastures, hayfields, reduce loss of livestock to drought stressed forage consumption, and recovery practices following drought period. Repeated and severe drought, especially during summer, and limited precipitation limit forage production in the southern Great Plains. Gaps in forage availability exist in late autumn (prior to grazing of dual-purpose wheat) and early spring (grazing termination on wheat). Pitta et al. (2005) suggested that grazing willow fodder blocks can be used to maintain ewes, and is a useful supplementary feed under drought conditions.

In many regions of the world, despite generally favorable regional precipitation, periodic drought is the greatest obstacle to optimal uniform forage production throughout the growing season. With the adoption of intensive grazing systems, long-term pasture productivity is dependent on improved management systems. Improper management during drought can jeopardize the survival of desirable forage species and can allow 'weedy' invasions. Pasture characteristics and varieties differ in field crop agriculture because pasture plants are generally perennial. During severe drought, leaves senesce and new growth ceases until the arrival of precipitation. The period of active growth of pasture plants is longest in the wetter coastal areas where most dairy cattle are found, and least in the dry western parts of the State, where only xerophytic grasses survive and dairying is not undertaken.

15.7.2.4 Salt Stress

Scarcity of fresh water, competition for fresh water, and soil salinization have resulted in a need for grasses and legumes with increased salt tolerance, especially in irrigated regions of the world. Salts lower the solute potential of the external solution, presenting the plant with increased water potential gradients to overcome. Salinity reduces seed germination, stand establishment, and yield in forage grasses and legumes, but susceptibility of plants to salinity stress at different developmental stages varies among species. Increasing levels of salinity generally reduce seed germination and seedling emergence. Significant differences in germination over several saline solutions have been reported in perennial ryegrass (Horst and Dunning 1989), subterranean clover (Shannon and Noble 1995), and alfalfa (Al-Niemi et al. 1992). Plants cope with salinity stress by accumulating (primarily in tissues removed from meristems), extruding, or diluting ions, and by selective ion absorption (e.g., absorbing K in the presence of high Na).

Grasses and legumes vary within and among species for growth responses to salinity, often showing increased dry weight with moderate increases in Na levels (Shannon and Noble 1995). *Panicum coloratum* var. Bambatsi was more tolerant to salt than several tropical legumes; it sustained a 50% yield reduction at 16.4 dS m^{-1}, whereas the best tropical legume had a 50% yield reduction at 10.6 dS m^{-1} (Keating et al. 1986). Rhodesgrass, a halophytic forage grass, shows stimulated growth of single roots under NaCl concentrations that inhibit growth of the whole plant (Waisel 1985).

Grasses such as *C. gayana*, *Leptochloa fusca*, *Pennisetum clandestinum*, *P. coloratum*, *Cynodon dactylon*, *Sporobolus* genus are reported to be tolerant to saline conditions while the other grasses such as *Cenchrus ciliaris*, *Digitaria decumbens*, *Sorghum almum*, *Panicum maximum* are only moderately tolerant (Russell 1976; Maas and Hoffman 1977; Ahmad and Ismail 1992; Truog and Roberts 1992; Kumar 1998; Qureshi and Barrett-Lenard 1998). Saline ground water can also increase productivity of Panicum laevifolium and Panicum maximum if irrigation facilities are available in semi-arid regions of India (Tomar et al. 2003).

15.7.2.5 Ozone Stress

Tropospheric ozone concentrations have increased in the last decade and are predicted to remain at higher level. Tropospheric ozone is an air pollutant which is of most concern to agricultural crops, forests, and seminatural grasslands as well and is also a significant GHG. Ozone is an air pollutant which is of most concern to agricultural crops, forests, and seminatural grasslands as well as a significant GHG. Troposheric factors that favor climatic change may also favor ozone formation and ozone effects on vegetation and increase plant predisposition to climatic stressors. The relationship between ozone and CC are thus topical. Carbon assimilation, translocation, nutrient acquisition, and other physiological processes are inhibited by ozone exposure which may ultimately lead to suppressed plant growth and yield. The most detrimental effects of ozone probably occur after ozone enters the leaf through stomata and involve reactions in the gas and liquid phases. Most crops in developed and developing countries are grown in summer when O_3 concentrations are elevated and frequently are sufficiently high to reduce yields. Moderately elevated O_3 level reduces the productivity of intact grasslands by 23% during a 5-year exposure under real field conditions (Volk et al. 2006). Ozone (O_3) is known to reduce the yield of clover more than the yield of grass species, inhibit nitrogen fixation of clover, shiftsthe grass/clover yield ratio in favor of the grass, and, by implication, speed the loss of clover from grass-clover pastures. Ozone also reduces %N and % estimated digestibility of clover (Blum et al. 1983; Fernandez et al. 2008). Forage quality may be changed because of ozone effects on leaf chemistry. Long-term ozone exposure can lead to increased levels of phenolic acids, flavonoids, and related compounds in what forages (Booker and Miller 1998) that may negatively affect ruminant microorganisms and enzyme systems. One of the most common effects of ozone is to promote leaf senescence. Thus, in pastures or other types of grasslands exposed to ozone, the fraction of senescing tissue may be increased. Because of increased lignification, and a decreased leaf/stem ratio, forage digestibility would decline. Quantification of effects, however, is lacking, and the consequences for animal nutrition need to be studied in more detail. In grass-clover forage, white clover (*T. repens* L.) leaf in vitro dry matter disappearance (IVDMD) and N were decreased, and NDF increased by ambient ozone (50 nl l^{-1}) compared with charcoal-filtered air (Burns et al. 1997). Changes in forage quality may result from shifts in species composition. Differential ozone sensitivities between grasses and legumes have been found to cause a shift in the grass/legume ratio in favor of grasses, and, hence, in protein concentration and other quality traits relevant for animal nutrition (Rebbeck et al. 1988; Fuhrer 1997).

15.8 Conclusions

Global warming has generated intense speculation about how CC may adversely affect the livestock industry and how this will ultimately impact the production of high quality protein for human consumption. It is clear that CC and specifically heat stress will both directly and indirectly influence animal welfare and productivity. Recent discoveries indicate that heat stress directly alters post-absorptive metabolism, nutrient portioning, and eventually production. However, the indirect effects of heat stress on animal agriculture are numerous and probably far-outweigh the direct effects. Indirect effects of CC on animal productivity include reduced feed intake, altered immune systems, reduced feed, water availability, and quality, and increased environmental interaction with disease causing parasites and vectors.

The negative effects of CC on the livestock industry will increase if global warming continues. Furthermore, heat stress is already likely the largest impediment to efficient animal production in agriculturally rich regions and certainly one of the biggest obstacles to animal agriculture in the tropical and subtropical portions of the globe. Consequently, CC is a serious threat to food sustainability and this is especially apparent in many developing countries.

References

Aber JD, Melillo JM (1991) Terrestrial ecosystems. Saunders College Publishing, Philadelphia

Ahmad R, Ismail S (1992) Studies on selection of salt tolerant plants for food, fodder and fuel from world flora. In: Lieth H, Masoom Al (eds) Towards the rational use of high salinity tolerant plants. Agriculture and forestry under marginal soil water conditions, vol 2. Academic Publishers, Doedraft, The Netherlands, Kluwer, pp 295–304, 447

Ai J, Du J, Wang N et al (2004) Inhibition of small-intestinal sugar absorption mediated by sodium orthovanadate Na_3VO_4 in rats and its mechanisms. World J Gastroenterol 10:3612–3615

Ainsworth EA, Long SP (2005) What have we learned from 15 years of free air CO_2 enrichment (FACE)? Ameta-analytical review of the responses of photosynthesis, canopy properties and plant production to rising CO_2. New Phytol 165:351–372

Dufour B et al (2006) A method to rank the risks of infectious diseases development linked to global warming. In: Proceedings of the 11th international symposium on veterinary epidemiology and economics (www.sciquest.org. nz). Cairns, Australia, 6–11 Aug 2006

Al-Katanani YM, Drost M, Monson RL, Rutledge JJ, Krininger CE 3rd, Block J, Thatcher WW, Hansen PJ (2002) Pregnancy rates following timed embryo transfer with fresh or vitrified in vitro produced embryos in lactating dairy cows under heat stress conditions. Theriogenology 58:171–182

Al Refaii A, Alix JH (2009) Ribosome biogenesis is temperature-dependent and delayed in *Escherichia coli* lacking the chaperones DnaK or DnaJ. Mol Microbiol 71:748–762

Albin DM, Wubben JE, Rowlett JM et al (2007) Changes in small intestinal nutrient transport and barrier function after lipopolysaccharide exposure in two pig breeds. J Anim Sci 85: 2517–2523

Al-Niemi TS, Campbell WF, Rumbaugh MD (1992) Response of alfalfa cultivars to salinity during germination and post-germination growth. Crop Sci 32:976–980

Alterovitz G, Xiang M et al (2010) Ontology engineering. Nat Biotechnol 28:128–130

Armstrong DV (1994) Heat stress interaction with shade and cooling. J Dairy Sci 77:2044–2050

Auvigne V, Leneveu P, Jehannin C, Peltoniemi O, Salle E (2010) Seasonal infertility in sows: a five year field study to analyze the relative roles of heat stress and photoperiod. Theriogenology 74:60–66

Baker B, Viglizzo JF (1998) Chapter 9: Rangelands and livestock. In: Feenstra JF, Burton I, Smith JB, Tol RSJ (eds) Handbook of methods for climate change impact assessment and adaptation strategies. IVM/UNEP Version 2.0. http://130.37.129.100/ivm/pdf/handbook_range.pdf. Accessed 22 Sept 2002

Baldocchi DD, Falge E, Gu L, Olson R, Hollinger D, Running S, Anthoni P, Bernhofer C, Davis K, Evans R, Fuentes J, Goldstein A, Katul G, Law B, Lee X, Malhi Y, Meyers T, Munger W, Oechel W, Paw UKT, Pilegaard K, Schmid HP, Valentini R, Verma S, Vesala T, Wilson K, Wofsy S (2001) Fluxnet: a new tool to study the temporal and spatial variability of the ecosystem-scale carbon dioxide, water vapor, and energy flux densities. Bull Am Meteorol Soc 82:2415–2434

Barnes RF, Baylor J (1994) Forages in a changing world. In: Forages. An introduction to grassland agriculture, vol 1. Iowa State University Press, Ames, pp 3–13

Barth AD, Bowman PA (1994) The sequential appearance of sperm abnormalities after scrotal insulation or dexamethasone treatment in bulls. Can Vet J 35:93–102

Baumgard LH, Rhoads RP (2012) Chapter 6: Effects of environment on metabolism. In: Collier RJ, Collier JL (eds) Environmental physiology of livestock. Wiley, Inc., Ames, pp 81–100

Baylis N, Githeko AK (2006) The effects of climate change on infectious diseases of animals. T 7.3. Foresight. Infectious disease. Preparing for the future. Office and science and innovation. (www.foresight.gov.uk)

Baziz HA, Geraert PA et al (1996) Chronic heat exposure enhances fat deposition and modifies muscle and fat partition in broiler carcasses. Poult Sci 75:505–513

Beede DK, Collier RJ (1986) Potential nutritional strategies for intensively managed cattle during thermal stress. J Anim Sci 62:543–554
Beere HM (2004) the stress of dying: the role of heat shock proteins in the regulation of apoptosis. J Cell Sci 117:2641–2651
Benson EJ, Hartnett DC (2006) The role of seed and vegetative reproduction in plant recruitment and demography in tallgrass prairie. Plant Ecol 187:163–178
Benson EJ, Hartnett DC, Mann KH (2004) Below ground bud banks and meristem limitation in tallgrass prairie plant populations. Am J Bot 91:416–421
Berman A (2003) Effects of body surface area estimates on predicted energy requirements and heat stress. J Dairy Sci 86:3605–3610
Berman A (2004) Tissue and external insulation estimates and their effects on prediction of energy requirements and of heat stress. J Dairy Sci 87:1400–1412
Berman A (2005) Estimates of heat stress relief needs for holstein dairy cows. J Anim Sci 83:1377–1384
Bernabucci U et al (2009) Influence of different periods of exposure to hot environment on rumen function and diet digestibility in sheep. Int J Biometeorol 53:387–395
Bernabucci U, Lacetera N et al (2010) Metabolic and hormonal adaptations to heat stress in ruminants. Animal 4:1167–1183
Bhattacharya AN, Hussain F (1974) Intake and utilization of nutrients in sheep fed different levels of roughage under heat stress. J Anim Sci 38:877–886
Bianca W (1965) Reviews of the progress of dairy science. Physiology. Cattle in hot environment. J Dairy Sci 32:291–328
Biggers BG, Geisert RD, Wetteman RP, Buchanan DS (1987) Effect of heat stress on early embryonic development in the beef cow. J Anim Sci 64:1512–1518
Bishonga C, Robinson JJ et al (1996) Excess dietary urea intake in ewes and its effect on ovulation rate and embryo development. Jpn J Vet Res 44:139–151
Black JL, Mullan BP, Lorschy ML, Giles LR (1993) Lactation in the sow during heat stress. Livest Prod Sci 35:18
Blum U, Allen SH, Burns JC, Linthurst RA (1983) The effects of ozone on fescue-clover forage: regrowth, yield and quality. Environ Exp Bot 23:121–132
Bogunovic M et al (2007) Enteroendocrine cells express functional toll-like receptors. Am J Physiol Gastrointest Liver Physiol 292:G1770–G1783
Bond WJ, Midgley JJ (2001) Ecology of sprouting in woody plants: the persistence niche. Trends Ecol Evol 16:45–51
Booker FL, Miller JE (1998) Phenylpropanoid metabolism and phenolic composition of soybean [*Glycine max* (L.) Merr.] leaves following exposure to ozone. J Exp Bot 49:1191–1202
Boyer JS (1982) Plant productivity and environment. Science (Washington DC) 218:43–448
Bradley MJ, Kutz SJ, Jenkins E, O'Hara TM (2005) The potential impact of climate change on infectious diseases of arctic fauna. Int J Cir Health 64:468–477
Bridges TC, Turner LW, Gates RS (1998) Economic evaluation of misting-cooling systems for growing/finishing swine through modelling. Appl Eng Agric 14:425–430
Brooks DR, Hoberg EP (2007) How will global climate change affect parasite–host assemblages? Trends Prasitol 23:571–574
Brown-Brandl TM, Nienaber JA, Zin H, Gates S (2004) A literature review of swine heat production. Trans ASAE 47:259–270
Bruce WB, Emeades GO, Barker TC (2002) Molecular and physiological approaches to maize improvement for drought tolerance. J Exp Bot 53:13–25
Burgos R, Odens LJ et al (2007) Evaluation of different cooling systems in lactating heat stressed dairy cows in a semi-arid environment. Prof Anim Sci 23:546–555
Burkholder KM, Thompson KL et al (2008) Influence of stressors on normal intestinal microbiota, intestinal morphology, and susceptibility to Salmonella enteritidis colonization in broilers. Poult Sci 87:1734–1741
Burns JC, Heagle AS, Fisher DS (1997) Nutritive value of ozone sensitive and resistant ladino white clover clones after chronic ozone and carbon dioxide exposure. Advances in carbon

dioxide effects research. ASA Spec Publ, vol 61. American Society of Agronomy, Madison, pp 153–67

Butler WR, Calaman JJ et al (1996) Plasma and milk urea nitrogen in relation to pregnancy rate in lactating dairy cattle. J Anim Sci 74:858–865

Cameron RD, Blackshaw AW (1980) The effect of elevated ambient temperature on spermatogenesis in the boar. J Reprod Fertil 59:173–179

Campbell BD (2000) A synthesis of recent global change research on pasture and rangeland production: reduced uncertainties and their management implications. Agric Ecosyst Environ 82:39–55

Campbell BD, Hunt DY (2000) Global climate change effects on competition and succession in pastures. In: Tow P, Lazenby A (eds) Competition and succession in pastures. CAB International, Wallingford (in press)

Cannell MGR, Thornley JHM (1998) N-poor ecosystems may respond more to elevated [CO_2] than N-rich ones in the long term. A model analysis of grassland. Glob Change Biol 4:431–442

Casella E, Soussana JF (1997) Long-term effects of CO_2 enrichment and temperature increase on the carbon balance of a temperate grass sward. J Exp Bot 48:1309–1321

Casella E, Soussana JF, Loiseau P (1996) Long-term effects of CO2 enrichment and temperature increase on a temperate grass sward. I. Productivity and water use. Plant Soil 182:83–99

CDC (2008) Congressional testimony. Select committee on energy independence and global warming United States house of representatives climate change and public health. Statement of Howard Frumkin MD, Dr. Charron PH, Thomas MK, Waltner-Toews D, Armini JJ, Edge T, Kent RA, Maarouf AR, Wilson J (2004) Vulnerability of waterborne diseases to climate change in Canada. A review. J toxi Env Health 67: 1667–1677

Chakravortty D, Kumar KS (1999) Modulation of barrier function of small intestinal epithelial cells by lamina propria fibroblasts in response to lipopolysaccharide: possible role in TNFalpha in inducing barrier dysfunction. Microbiol Immunol 43:527–533

Chaves MM (1991) Effects of water deficits on carbon assimilation. J Exp Bot 42:1–16

Chaves MM, Pereira JS, Maroco J, Rodrigues ML, Ricardo CPP, Osorio ML, Carvalho I, Faria T, Pinheiro C (2002) How plants cope with water stress in the field? Photosynthesis and growth. Ann Bot 89:907–916

Chen CL, Sangiah S, Chen H et al (1994) Effects of heat stress on Na+, K(+)-ATPase, Mg(2+)-activated ATPase, and Na(+)-ATPase activities of broiler chickens vital organs. J Toxicol Environ health 41:345–356

Chen MC, Wu SV et al (2006) Bitter stimuli induce Ca2+ signaling and CCK release in enteroendocrine STC-1 cells: role of L-type voltage-sensitive Ca2+ channels. Am J Physiol Cell Physiol 291:C726–C739

Chen YC et al (2010) Heat stress responses modulate calcium regulations and electrophysiological characteristics in atrial myocytes. J Mol Cell Cardiol 48:781–788

Chomel BB, Belotto A, Meslin FX (2007) Wildlife, exotic pets and emerging zoonoses. Emerg Infect Dis 13:6–11

Christensen JH, Nielsen MN et al (2010) Inactivation of the hereditary spastic paraplegia-associated Hspd1 gene encoding the Hsp60 chaperone results in early embryonic lethality in mice. Cell Stress Chaperones 15:851–863

Christenson RK (1980) Environmental influences on the postpartum animal. J Anim Sci 51(Suppl 2):53–67

Close WH, Mount LE, Start IB (1971) The influence of environmental temperature and plane of nutrition on heat losses from groups of growing pigs. Anim Prod 13:285–294

Collier CT, Williams PN, Carroll JA, Welsh TH Jr, Laurenz JC (2011) Effect of maternal restraint stress during gestation on temporal lipopolysaccharide-induced neuroendocrine and immune responses of progeny. Domest Anim Endocrinol 40:40–50

Collin A, van Milgen J, Dubois S, Noblet J (2001) Effect of high temperature and feeding level on energy utilization in piglets. J Anim Sci 79:1849–1857

Cook NB, Mentink RL, Bennett TB, Burgi K (2007) The effect of heat stress and lameness on time budgets of lactating dairy cows. J Dairy Sci 90:1674–1682
Coughenour MB, Chen DX (1997) Assessment of grassland ecosystem responses to atmospheric change using linked plant–soil process models. Ecol Appl 7:802–827
Csaba G, Inczefi-Gonda A, O. Dobozy O (1984) Hereditary transmission to the F1-generation of hormonal imprinting (receptor memory) induced in rats by neonatal exposure to insulin Acta Physiol Hung 63:93–99
Dale NM, Fuller HL (1980) Effect of diet composition on feed intake and growth of chicks under heat stress. II. Constant vs. cycling temperatures. Poult Sci 59:1434–1441
De Boeck HJ, Lemmens CMHM, Gielen B, Malchair S, Carnol M, Merckx R, Van den Berge J, Ceulemans R, Nijs I (2007) Biomass production in experimental grasslands of different species richness during three years of climate warming. Biogeosciences Discuss 4:4605–4629
de la Sota RL, Burke JM, Risco CA, Moreira F, DeLorenzo MA, Thatcher WW (1998) Evaluation of timed insemination during summer heat stress in lactating dairy cattle. Theriogenology 49:761–770
de Lartigue G, de La Serre CB, Raybould HE (2011) Vagal afferent neurons in high fat diet-induced obesity; intestinal microflora, gut inflammation and cholecystokinin. Physiol Behav, Corrected Proof (in press)
de Rensis F, Marconi P, Capelli T, Gatti F, Facciolongo F, Franzini S, Scaramuzzi RJ (2002) Fertility in postpartum dairy cows in winter or summer following estrus synchronization and fixed time AI after the induction of an LH surge with GnRH or hCG. Theriogenology 58:1675–1687
DeShazer JA, Hahn GL, Xin H (2009) Chapter 1: Basic principals of the thermal environment and livestock energetics. In: DeShazer JA (ed) Livestock energetics and thermal environment management. ASABE, St. Joseph, MI
Dokladny K, Moseley PL, Ma TY (2006) Physiologically relevant increase in temperature causes an increase in intestinal epithelial tight junction permeability. Am J Physiol Gastrointest Liver Physiol 290:G204–G212
Dokladny K, Ye D et al (2008) Cellular and molecular mechanisms of heat stress-induced up-regulation of occludin protein expression: regulatory role of heat shock factor-1. Am J Pathol 172:659–670
Dreiling CE, Carman FS 3rd, Brown DE (1991) Maternal endocrine and fetal metabolic responses to heat stress. J Dairy Sci 74:312–327
Driancourt MA, Guet P et al (1999) Presence of an aromatase inhibitor, possibly heat shock protein 90, in dominant follicles of cattle. J Reprod Fertil 115:45–58
DuBose DA, Basamania K, Maglione L, Rowlands J (1983) Role of bacterial endotoxins of intestinal origin in rat heat stress mortality. J Appl Physiol 54:31–36
Ellis RJ, van der Vies SM et al (1989) The molecular chaperone concept. Biochem Soc Symp 55:145–153
Elrod CC, Butler WR (1993) Reduction of fertility and alteration of uterine pH in heifers fed excess ruminally degradable protein. J Anim Sci 71:694–701
Elrod CC, Van Amburgh M et al (1993) Alterations of pH in response to increased dietary protein in cattle are unique to the uterus. J Anim Sci 71:702–706
Epstein PR (1999) Climate and health. Science 285:347–348
Epstein PR (2001) Climate change and emerging infectious diseases. Microbes Infect 3:747–754
Epstein P, Diaz H, Elias S, Grabherr G, Graham N, Martens W, Thompson EM, Susskind J (1998) Biological and physical signs of climate change, focus on mosquito borne diseases. Bull Am Meteorol Soc 79:409–417
FAO (Food and Agricultural Organization of the United Nations) (2006) Livestock a major threat to the environment. Remedies urgently needed. Retrieved from http://www.fao.org/newsroom/en/news/2006/1000448/index.html
FAO (Food and Agriculture Organization of the United Nations) (2008) FAOSTAT

FAO (Food and Agriculture Organization of the United Nations) (2009) Submission to UNFCCC AWG LCA. Enabling agriculture to contribute to climate change. Available at http://unfccc.int/resource/docs/2008/smsn/igo/036.pdf

Febbraio MA (2001) Alterations in energy metabolism during exercise and heat stress. Sports Med 31:47–59

Ferguson JD, Blanchard T et al (1988) Infertility in dairy cattle fed a high percentage of protein degradable in the rumen. JAVMA 192:659–662

Ferguson JD, Galligan DT et al (1993) Serum urea nitrogen and conception rate: the usefulness of test information. J Dairy Sci 76:3742–3746

Fernandez IG, Bass D, Muntifering R, Mills G, Barnes J (2008) Impacts of ozone pollution on productivity and forage quality of grass clover swards. Atmos Environ 42:8755–8769

Field CB, Lund CP, Chiarello NR, Mortimer BE (1997) CO_2 effects on the water budget of grassland microcosm communities. Glob Clim Change 3:197–206

Finger R, Calanca P (2011) Risk management strategies to cope with climate change in grassland production: an illustrative case study for the Swiss plateau. Reg Environ Change 11:1–15

Foxcroft GR, Dixon WT, Novak S, Putman CT, Town SC et al (2006) The biological basis for prenatal programming of postnatal performance in pigs. J Anim Sci 84(Suppl):E105–E112

Foxcroft GR, Dixon WT, Dyck MK, Novak S, Harding JC et al (2009) Prenatal programming of postnatal development in the pig. Soc Reprod Fertil Suppl 66:213–231

Frost RA, Lang CH (2002) Lipopolysaccharide regulates proinflammatory cytokine expression in mouse myoblasts and skeletal muscle. Am J Physiol Regul Integr Comp Physiol 283:698–709

Fuhrer J (1997) Ozone sensitivity of managed pasture. In: Cheremisinoff PN (ed) Ecological advances and environmental impact assessment. Adv Environ Control Technol Ser. Gulf Publ, Houston (TX), pp 681–706

Fuquay JW (1981) Heat stress as it affects animal production. J Anim Sci 52:164–174

Gabler NK, Spurlock ME (2008) Integrating the immune system with the regulation of growth and efficiency. J Anim Sci 86:E64–E74

Gabler NK, Radcliffe JS, Spencer JD, Webel DM, Spurlock ME (2009) Feeding long-chain n-3 polyunsaturated fatty acids during gestation increases intestinal glucose absorption potentially via the acute activation of AMPK. J Nutr Biochem 20:17–25

Gale P et al (2009) The effect of climate change on the occurrence and prevalence of livestock diseases in Great Britain. A review. J Appl Microbiol 106:1409–1423

Gao Q, Yu M, Zhang XS, Guan F (1997) Dynamic modeling of northeast China transect responses to global change—a regional vegetation model driven by remote sensing information. Acta Bot Sinica 39:800–810

Garcia-Bojalil CM, Staples CR et al (1994) Protein intake and development of ovarian follicles and embryos of superovulated nonlactating dairy cows. J Dairy Sci 77:2537–2548

Garofano A, Czernichow P, Breant B (1998) Beta-cell mass and proliferation following late fetal and early postnatal malnutrition in the rat. Diabetologia 41:1114–1120

Garriga C et al (2006) Heat stress increases apical glucose transport in the chicken jejunum. Am J Physiol Regul Integr Comp Physiol 290:R195–R201

Geraert PA, Padilha JC, Guillaumin S (1996) Metabolic and endocrine changes induced by chronic heat exposure in broiler chickens: growth performance, body composition and energy retention. Br J Nutr 75:195–204

Gilad E, Meidan R, Berman A, Graber Y, Wolfenson D (1993) Effect of heat stress on tonic and GnRH-induced gonadotrophin secretion in relation to concentration of oestradiol in plasma of cyclic cows. J Reprod Fertil 99:315−321.

Guzeloglu A, Ambrose JD et al (2001) Long-term follicular dynamics and biochemical characteristics of dominant follicles in dairy cows subjected to acute heat stress. Anim Reprod Sci 66:15–34

Hall DM, Baumgardner KR, Oberley TD, Gisolfi CV (1999) Splanchnic tissues undergo hypoxic stress during whole body hyperthermia. Am J Physiol 276:G1195–G1203

Hammon DS, Holyoak GR et al (2005) Association between blood plasma urea nitrogen levels and reproductive fluid urea nitrogen and ammonia concentrations in early lactation dairy cows. Anim Reprod Sci 86:195–204

Hansen PJ (2009) Effects of heat stress on mammalian reproduction. Philos Trans R Soc Lond B Biol Sci 364:3341–3350

Hansen PJ, Aréchiga CF (1999) Strategies for managing reproduction in the heat-stressed dairy cow. J Anim Sci 77(Suppl 2):36–50

Harle KJ, Howden SM, Hunt LP, Dunlop M (2007) The potential impact of climate change on the Australian wool industry by 2030. Agric Syst 93:61–89

Harper JL (1977) Population biology of plants. Academic, London

Harvell CD, Mitchell CE, Ward JR et al (2002) Climate warming and disease risks or terrestrial and marine biota. Science 296:2158–2162

Heath ME (1983) The effects of rearing-temperature on body composition in young pigs. Comp Biochem Physiol 76:363–366

Heath ME (1989) Effects of rearing temperature and level of food intake on organ size and tissue composition in piglets. Can J Physiol Pharmacol 67:526–532

Held T, Barakat AZ et al (2011) Heat-shock protein HSPA4 is required for progression of spermatogenesis. Reproduction 142:133–144

Hornef MW, Normark BH, Vandewalle A, Normark S (2003) Intracellular recognition of lipopolysaccharide by toll-like receptor 4 in intestinal epithelial cells. J Exp Med 198: 1225–1235

Horst GL, Dunning NB (1989) Germination and seedling growth of perennial ryegrass in soluble salts. J Am Soc Hort Sci 114:338–342

Houghton JT, Ding Y, Griggs DJ, Noguer M, Vander Linden PJ, Dai X, Maskell K, Johnson CA (2001) Climate Change 2001: the scientific basis. Cambridge University Press, New York

Howden SM, Crimp SJ, Stokes CJ (2008) Climate change and Australian livestock systems: impacts, research and policy issues. Aust J Exp Agric 48:780–788

Hsiao TC (1973) Plant responses to water stress. Annu Rev Plant Physiol 24:519–570

IPCC (2007) IPCC WGI fourth assessment report. Climatic change: the physical science basis. Intergovernmental Panel on Climate Change, Geneva

Isom SC, Prather RS, Rucker EB 3rd (2007) Heat stress-induced apoptosis in porcine in vitro fertilized and parthenogenetic preimplantation-stage embryos. Mol Reprod Dev 74:574–581

Isom SC, Lai L, Prather RS, Rucker EB 3rd (2009a) Heat shock of porcine zygotes immediately after oocyte activation increases viability. Mol Reprod Dev 76:548–554

Isom SC, Prather RS, Rucker Iii EB (2009b) Enhanced developmental potential of heat-shocked porcine parthenogenetic embryos is related to accelerated mitogen-activated protein kinase dephosphorylation. Reprod Fertil Dev 21:892–900

Jantschitsch C, Trautinger F (2003) Heat shock and UV-B-induced DNA damage and mutagenesis in skin. Photochem Photobiol Sci 2:899–903

Jegou B, Laws AO et al (1984) Changes in testicular function induced by short-term exposure of the rat testis to heat: further evidence for interaction of germ cells, sertoli cells and leydig cells. Int J Androl 7:244–257

Johnson RW (1997) Inhibition of growth by pro-inflammatory cytokines: an integrated view. J Anim Sci 75:1244–1255

Johnson RW (1998) Immune and endocrine regulation of food intake in sick animals. Domest Anim Endocrinol 15:309–319

Johnston LJ, Ellis M, Libal GW, Mayrose VB, Weldon WC (1999) Effect of room temperature and dietary amino acid concentration on performance of lactating sows. NCR-89 committee on swine management. J Anim Sci 77:1638–1644

Jones HG, Corlett JE (1992) Current topics in drought physiology. J Agric Sci (Cambridge) 119:291–296

Jones M, Jongen M (1996) Sensitivity of temperate grasslands to elevated CO_2 and the interaction with temperature and water stress. Agric Food Sci Finland 5:271–283

Jones KE, Patel NG, Levy MA, Storeygard A, Balk D, Gittleman JL, Daszak P (2008) Global trends in emerging infectious diseases. Nature 451:990–993

Jonson KM, Lyle JG, Edwards MJ, Penny RH (1976) Effect of prenatal heat stress on brain growth and serial discrimination reversal learning in the guinea pig. Brain Res Bull 1:133–150

Joost RE (1995) Acremonium in fescue and ryegrass: boon or bane? A review. J Anim Sci 73:881–888

Kamiya M, Kamiya Y et al (2006) Effects of high ambient temperature and restricted feed intake on urinary and plasma 3-methylhistidine in lactating holstein cows. Anim Sci J 77:201–207

Kaneko Y, Kimura T et al (1997) Cloning of apg-2 encoding a novel member of heat shock protein 110 family. Gene 189:19–24

Katsumata M, Yano H et al (1990) Influence of a high ambient temperature and administration of clenbuterol on body composition in rats. J Nutr Sci Vitaminol 36:569–578

Keating BA, Strickland RW, Fisher MJ (1986) Salt tolerance of some tropical pasture legumes with potential adaptation to cracking clay soils. Aust J Exp Agric 26:181–186

Keller G, Ben-Shaul Y, Bacher A (1992) Influence of metabolic inhibitors on the degradation of tight junctions in HT29 cells. Exp Cell Res 200:16–25

Kellett GL (2001) The facilitated component of intestinal glucose absorption. J Physiol 531:585–595

Kellett GL, Brot-Laroche E (2005) Apical GLUT2: a major pathway of intestinal sugar absorption. Diabetes 54:3056–3062

Kelly TL, Trasler JM (2004) Reproductive epigenetics. Clin Genet 65:247–260

Kenney MJ, Musch TI (2004) Senescence alters blood flow responses to acute heat stress. Am J Physiol Heart Circ Physiol 286:H1480–H1485

Khasnis AA, Nettleman MD (2005) Global warming and infectious disease. Arch Med Res 36:689–696

Kheradmand A, Alirezaei M et al (2009a) Antioxidant enzyme activity and MDA level in the rat testis following chronic administration of ghrelin. Andrologia 41:335–340

Kheradmand A, Roshangar L et al (2009b) The role of ghrelin on the morphometry and intracellular changes in the rat testis. Tissue Cell 41:105–111

Kheradmand A, Taati M, Babaei H (2009c) The effects of chronic administration of ghrelin on rat sperm quality and membrane integrity. Anim Biol 59:159–168

Kheradmand A, Alirezaei M et al (2010) Ghrelin promotes antioxidant enzyme activity and reduces lipid peroxidation in the rat ovary. Regul Pept 162(1–3):84–89

Kheradmand A, Dezfoulian O et al (2011) Ghrelin attenuates heat-induced degenerative effects in the rat testis. Regul Pept 167:97–104

Kiani A, Chwalibog A, Tygesen MP, Nielsen MO (2008) Effect of undernutrition in foetal life on energy expenditure during gestation in ewes. Arch Anim Nutr 62:117–126

Kim KH et al (2010) Productivity and energy partition of late lactation dairy cows during heat exposure. Anim Sci J = Nihon chikusan Gakkaiho 81:58–62

Kisliouk T, Ziv M, Meiri N (2010) Epigenetic control of translation regulation: alterations in histone H3 lysine 9 post-translation modifications are correlated with the expression of the translation initiation factor 2B (Eif2b5) during thermal control establishment. Dev Neurobiol 70:100–113

Klose RJ, Bird AP (2006) Genomic DNA methylation: the mark and its mediators. Trends Biochem Sci 31:89–97

Knapp AK, Smith MD (2001) Variation among biomes in temporal dynamics of aboveground primary production. Science 291:481–484

Koelkebeck KW, Parsons CM, Wang X (1998) Effect of acute heat stress on amino acid digestibility in laying hens. Poult Sci 77:1393–1396

Koopmans MMM, Reusken C, van Maanen K (2007) West Nile virus in Europe: waiting for the start of the epidemic. In: Takken W, Knols BGJ (eds) Emerging pests and vector-borne diseases in Europe. Wageningen Academic Publishers, Wageningen, pp 123–151

Kuhn-Velten WN (1996) Rapid down-regulation of testicular androgen biosynthesis at increased environmental temperature is due to cytochrome P450c17 (CYP17) thermolability in leydig cells, but not in endoplasmic reticulum membranes. Exp Clin Endocrinol Diab 104:243–249
Kumar A (1998) Growing forages in salt-affected soils. In: Tyagi NK, Minhas PS (eds) Agricultural salinity management in India. CSSRI, Karnal, India, pp 489–498, 526
Kunavongkrit A, Suriyasomboon A et al (2005) Management and sperm production of boars under differing environmental conditions. Theriogenology 63:657–667
Kurowicka B, Gajewska A et al (2006) Effect of early thermal experience on pituitary-gonadal axis in female rats. Reprod Biol 6:63–77
Kutz SJ, Hoberg EP, Polley L, Jenkins EJ (2005) Global warming is changing the dynamics of Arctic host-parasite systems. Proc R Soc Lond B Biol Sci 272:2571–2576
Lachance C, Bailey JL et al (2007) Expression of Hsp60 and Grp78 in the human endometrium and oviduct, and their effect on sperm functions. Hum Reprod 22:2606–2614
Lachance C, Fortier M et al (2010) Localization of Hsp60 and Grp78 in the human testis, epididymis and mature spermatozoa. Int J Androl 33:33–44
Lamber H, Chapin FS, Pons TL (1998) Plant physiological ecology. Springer, New York
Lambert GP (2004) Role of gastrointestinal permeability in exertional heatstroke. Exerc Sport Sci Rev 32:185–190
Lambert GP (2008) Intestinal barrier dysfunction, endotoxemia, and gastrointestinal symptoms: the 'canary in the coal mine' during exercise-heat stress? Med Sport Sci 53:61–73
Lambert GP (2009) Stress-induced gastrointestinal barrier dysfunction and its inflammatory effects. J Anim Sci 87:E101–E108
Le Dividich J, Vermorel M et al (1980) Effects of environmental temperature on heat production, energy retention, protein and fat gain in early weaned piglets. Br J Nutr 44:313–323
Lee IY, Lee CC et al (2005) Sheng mai san, a Chinese herbal medicine, protects against renal ischaemic injury during heat stroke in the rat. Clin Exp Pharmacol Physiol 32:742–748
Lelievre F, Volaire F (2009) Current and potential development of perennial grasses in rainfed Mediterranean farming systems. Crop Sci 49:2371–2378
Lelievre F, Seddaiu G, Ledda L, Porqueddu C, Volaire F (2011) Water use efficiency and drought survival in Mediterranean perennial forage grasses. Field Crops Res 121:333–342
Leroy JL, Vanholder T et al (2004) Metabolic changes in follicular fluid of the dominant follicle in high-yielding dairy cows early postpartum. Theriogenology 62:1131–1143
Lewis K, McKay DM (2009) Metabolic stress evokes decreases in epithelial barrier function. Ann N Y Acad Sci 1165:327–337
Lewis GS, Thatcher WW, Bliss EL, Drost M, Collier RJ (1984) Effects of heat stress during pregnancy on postpartum reproductive changes in holstein cows. J Anim Sci 58:174–186
Li Q et al (2009) Disruption of tight junctions during polymicrobial sepsis in vivo. J Pathol 218:210–221
Liao CW, Veum TL (1994) Effects of dietary energy intake by gilts and heat stress from days 3 to 24 or 30 after mating on embryo survival and nitrogen and energy balance. J Anim Sci 72:2369–2377
Liu F et al (2009) Heat-stress-induced damage to porcine small intestinal epithelium associated with downregulation of epithelial growth factor signaling. J Anim Sci 87:1941–1949
Liu JJ, Ma X et al (2010) Downregulation of both gene expression and activity of Hsp27 improved maturation of mouse oocyte in vitro. Reprod Biol Endocrinol 8:47
Long SP, Ainsworth EA, Rogers A, Ort DR (2004) Rising atmospheric carbon dioxide: plants FACE the future. Annu Rev Plant Biol 55:591–628
Lu Q, Wen J, Zhang H (2007) Effect of chronic heat exposure on fat deposition and meat quality in two genetic types of chicken. Poult Sci 86:1059–1064
Luft JC, Benjamin IJ, Mestril R, Dix DJ (2001) Heat shock factor 1-mediated thermotolerance prevents cell death and results in G2/M cell cycle arrest. Cell Stress Chaperones 6:326–336
Lutze LJ, Gifford RM (1998) Carbon accumulation and water use of *D. richarsonii* swards in response to CO2 and nitrogen supply over four years growth. Glob Change Biol 4:851–861

Maas EV, Hoffman GJ (1977) Crop salt tolerance current assessment. J Irrig Drain Div 174:115–134

MacLennan NK, James SJ, Melnyk S, Piroozi A, Jernigan S et al (2004) Uteroplacental insufficiency alters DNA methylation, one-carbon metabolism, and histone acetylation in IUGR rats. Physiol Genomics 18:43–50

Magee TR, Han G, Cherian B, Khorram O, Ross MG et al (2008) Down-regulation of transcription factor peroxisome proliferator-activated receptor in programmed hepatic lipid dysregulation and inflammation in intrauterine growth-restricted offspring. Am J Obstet Gynecol 199(271):e271–e275

Maizels ET, Peters CA et al (1998) Heat-shock protein-25/27 phosphorylation by the delta isoform of protein kinase C. Biochem J 332:703–712

Mallouk Y, Vayssier-Taussat M et al (1999) Heat shock protein 70 and ATP as partners in cell homeostasis (review). Int J Mol Med 4:463–474

Malmendal A, Overgaard J et al (2006) Metabolomic profiling of heat stress: hardening and recovery of homeostasis in *Drosophila.* Am J Physiol Regul Integr Comp Physiol 291: R205–R212

Marder J, Eylath U et al (1990) The effect of heat exposure on blood chemistry of the hyperthermic rabbit. Comp Biochem Physiol A 97:245–247

Martens WJM, Jetten TH, Rotmans J, Niessen LW (1995) Climate change and vector-borne diseases: a global modelling perspective. Glob Environ Change 5:195–209

Martti K (1967) Development of rectum-testis temperature difference in the post-natal rat. J Reprod Fertil 14:427–437

McCoard SA, Peterson SW, McNabb WC, Harris PM, McCutcheon SN (1997) Maternal constraint influences muscle fibre development in fetal lambs. Reprod Fertil Dev 9:675–681

McEvoy TG, Robinson JJ et al (1997) Dietary excesses of urea influence the viability and metabolism of preimplantation sheep embryos and may affect fetal growth among survivors. Anim Reprod Sci 47:71–90

McGlone JJ, Stansbury WF, Tribble LF, Morrow JL (1988) Photoperiod and heat stress influence on lactating sow performance and photoperiod effects on nursery pig performance. J Anim Sci 66:1915–1919

McGuire MA et al (1989) Effects of thermal stress and level of feed intake on portal plasma flow and net fluxes of metabolites in lactating holstein cows. J Anim Sci 67:1050–1060

McKeon GM, Stone GS, Syktus JI, Carter JO, Flood NR, Ahrens DG, Bruget DN, Chilcott CR, Cobon DH, Cowley RA, Crimp SJ, Fraser GW, Howden SM, Johnston PW, Ryan JG, Stokes CJ, Day KA (2009) Climate change impacts on northern Australian rangeland livestock carrying capacity: a review of issues. Rangeland J 31:1–29

Meng Q et al (2005) Regulation of amino acid arginine transport by lipopolysaccharide and nitric oxide in intestinal epithelial IEC-6 cells. J Gastrointest Surg Off J Soc Surg Aliment Tract 9:1276–1285 (discussion 1285)

Metukuri MR, Reddy CM et al (2010) Bacterial LPS-mediated acute inflammation-induced spermatogenic failure in rats: role of stress response proteins and mitochondrial dysfunction. Inflammation 33:235–243

Minton JE, Wettemann RP, Meyerhoeffer DC, Hintz RL, Turman EJ (1981) Serum luteinizing hormone and testosterone in bulls during exposure to elevated ambient temperature. J Anim Sci 53:1551–1558

Mitchell MA, Carlisle AJ (1992) The effects of chronic exposure to elevated environmental temperature on intestinal morphology and nutrient absorption in the domestic fowl (*Gallus domesticus*). Comp Biochem Physiol A: Comp Physiol 101:137–142

Mitlöhner FM, Morrow JL et al (2001) Shade and water misting effects on behavior, physiology, performance, and carcass traits of heat-stressed feedlot cattle. J Anim Sci 79:2327–2335

Mohammed ME, Johnson HD (1985) Effect of growth hormone on milk yields and related physiological functions of holstein cows exposed to heat stress. J Dairy Sci 68:1123–1133

Molvarec A, Tamasi L et al (2010) Circulating heat shock protein 70 (HSPA1A) in normal and pathological pregnancies. Cell Stress Chaperones 15:237–247

Morgan ER, Wall R (2009) Climate change and parasitic disease, farmer mitigation? Trends Parasitol 25:308–313
Mori C, Nakamura N et al (1997) Morphological analysis of germ cell apoptosis during postnatal testis development in normal and Hsp 70–2 knockout mice. Dev Dyn 208:125–136
Moss AR, Jounany JP, Neevbold J (2000) Methane production by ruminants. Its contribution to global warming. Ann Zootech 49:231–253
Muckter H, Ben-Shaul Y, Bacher A (1987) ATP requirement for induced tight junction formation in HT 29 adenocarcinoma cells. Eur J Cell Biol 44:258–264
Narayansingh RM, Senchyna M et al (2004) Expression of prostaglandin G/H synthase (PGHS) and heat shock protein-70 (HSP-70) in the corpus luteum (CL) of prostaglandin F2 alpha-treated immature superovulated rats. Can J Physiol Pharmacol 82:363–371
Neal MD et al (2006) Enterocyte TLR4 mediates phagocytosis and translocation of bacteria across the intestinal barrier. J Immunol 176:3070–3079
Nichi M, Bols PE, Züge RM, Barnabe VH, Goovaerts IG, Barnabe RC, Cortada CN (2006) Seasonal variation in semen quality in *Bos indicus* and *Bos taurus* bulls raised under tropical conditions. Theriogenology 66:822–828
Niu SL, Wu MY, Han Y, Xia JY, Li LH, Wan SQ (2008) Water-mediated responses of ecosystem carbon fluxes to climatic change in a temperate steppe. New Phytol 177:209–219
O'Brien MD, Rhoads RP et al (2010) Metabolic adaptations to heat stress in growing cattle. Domest Anim Endocrinol 38:86–94
Ocon OM, Hansen PJ (2003) Disruption of bovine oocytes and preimplantation embryos by urea and acidic pH. J Dairy Sci 86:1194–1200
OIE (2008) Climate change, impact on epidemiology and control of animal diseases. www.oie.int
Olvera-Sanchez S, Espinosa-Garcia MT et al (2011) Mitochondrial heat shock protein participates in placental steroidogenesis. Placenta 32:222–229
Omtvedt IT, Nelson RE, Edwards RL, Stephens DF, Turman EJ (1971) Influence of heat stress during early, mid and late pregnancy of gilts. J Anim Sci 32:312–317
Oresanya TF, Beaulieu AD, Patience JF (2008) Investigations of energy metabolism in weanling barrows: the interaction of dietary energy concentration and daily feed (energy) intake. J Anim Sci 86:348–363
Oswald IP (2006) Role of intestinal epithelial cells in the innate immune defence of the pig intestine. Vet Res 37:359–368
Owensby CE, Ham JM, Knapp AK, Bremer D, Auen LM (1997) Water vapour fluxes and their impact under elevated CO_2 in a C4-tallgrass prairie. Glob Clim Change 3:189–195
Padmini E, Lavanya S (2011) Over expression of HSP70 and HSF1 in endothelial cells during pre-eclamptic placental stress. Aust N Z J Obstet Gynaecol 51:47–52
Patience JF, Umboh JF et al (2005) Nutritional and physiological responses of growing pigs exposed to a diurnal pattern of heat stress. Livest Prod Sci 96:205–214
Patz JA, Graczyk TK, Geller N, Vitto YA (2000) Effects of environmental change on emerging parasitic diseases. Int J Parasitol 30:1395–1405
Pedersen JK (2009) Above and belowground phenology in a heathland during future climate change. In: Climate change: global risks, challenges and decisions. IOP conference series, IOP publishing. Earth and environmental science, p 31
Piao SL, Fang JY, Chen AP (2003) Seasonal dynamics of terrestrial net primary production in response to climatic changes in China. Acta Bot Sinica 45:269–275
Pires ES, Khole VV (2009) A block in the road to fertility: autoantibodies to heat-shock protein 90-beta in human ovarian autoimmunity. Fertil Steril 92:1395–1409
Pisa V, Cozzolino M et al (2009) The molecular chaperone Hsp90 is a component of the cap-binding complex and interacts with the translational repressor cup during *Drosophila* oogenesis. Gene 432:67–74
Pitta DW, Barry TN, L′opez-Villalobos N, Kemp PD (2005) Effects on ewe reproduction of grazing willow fodder blocks during drought. Anim Feed Sci Technol 120:217–234
Pollmann DS (2010) Seasonal effects on sow herds: Industry experience and management strategies. Midwest American Society of Animal Science, Des Moines

Post E, Pedersen C (2008) Opposing plant community responses to warming with and without herbivores. Proc Nat Acad Sci USA 105:12353–12358

Post E, Pedersen C, Wilmers CC, Forchhammer MC (2008) Phenological sequences reveal aggregate life history response to climate. Ecology 89:363–370

Poston L (2011) Intergenerational transmission of insulin resistance and type 2 diabetes. Prog Biophys Mol Biol 106:315–322

Purse BV, Mellor PS, Rogers DJ, Samuel AR, Mertens PPC, Baylis M (2005) Climate change and the recent emergence of bluetongue in Europe. Nat Rev Microbiol 3:171–182

Putney DJ, Drost M et al (1988a) Embryonic development in superovulated dairy cattle exposed to elevated ambient temperatures between days 1 to 7 post insemination. Theriogenology 30:195–209

Putney DJ, Gross TS, Thatcher WW (1988b) Prostaglandin secretion by endometrium of pregnant and cyclic cattle at day 17 after oestrus in response to in vitro heat stress. J Reprod Fertil 84:475–483

Putney DJ, Drost M et al (1989) Influence of summer heat stress on pregnancy rates of lactating dairy cattle following embryo transfer or artificial insemination. Theriogenology 31:765–778

Qi H et al (2011) Involvement of HIF-1alpha in MLCK-dependent endothelial barrier dysfunction in hypoxia. Cell Physiol Biochem: Int J Exp Cell Physiol Biochem Pharmacol 27:251–262

Quinteiro-Filho WM et al (2010) Heat stress impairs performance parameters, induces intestinal injury, and decreases macrophage activity in broiler chickens. Poult Sci 89:1905–1914

Qureshi RH, Barrett-Lenard EG (1998) Saline agriculture for irrigated land in Pakistan: a handbook, monograph, vol 50. ACIAR, Canberra, p 142

Randolph SE, Asokliene L, Avsic-Zupanc T, Bormane A, Burri C, Gern L, Golovljova I, Hubalek Z, Knap N, Kondrusik M, Kupca A, Pejcoch M, Vasilenko V, Ygutiene M (2008) Variable spikes in tick-borne encephalitis incidence in 2006 independent of variable tick abundance but related to weather. Parasites Vectors 1: 10.1186/1756-3305-1-44

Ravin HA, Rowley D, Jenkins C, Fine J (1960) On the absorption of bacterial endotoxin from the gastro-intestinal tract of the normal and shocked animal. J Exp Med 112:783–792

Ray EC, Avissar NE, Sax HC (2002) Growth factor regulation of enterocyte nutrient transport during intestinal adaptation. Am J Surg 183:361–371

Raychaudhuri N, Raychaudhuri S, Thamotharan M, Devaskar SU (2008) Histone code modifications repress glucose transporter 4 expression in the intrauterine growth-restricted offspring. J Biol Chem 283:13611–13626

Rebbeck J, Blum U, Heagle AS (1988) Effects of ozone on the regrowth and energy reserves of a ladino clover tall fescue pasture. J Appl Ecol 25:659–681

Reik W (2007) Stability and flexibility of epigenetic gene regulation in mammalian development. Nature 447:425–432

Reilly J, Tubiello F, Mccarl B, Abler D, Darwin R, Fuglie K, Hollinger S, Izaurralde C, Jagtap S, Jones J, Mearns L, Ojima D, Paul E, Paustian K, Riha S, Rosenberg N, Rosenzweig C (2003) U.S. agriculture and climate change. New results. Clim Change 57:43–69

Renaudeau D, Noblet J (2001) Effects of exposure to high ambient temperature and dietary protein level on sow milk production and performance of piglets. J Anim Sci 79:1540–1548

Renaudeau D, Gourdine JL, Silva BAN, Noblet J (2008) Nutritional routes to attenuate heat stress in pigs. In: Rowlinson P, Steele M, Nefzaoui A (eds) Livestock and global change. Cambridge University Press, Cambridge

Rensis FD, Scaramuzzi RJ (2003) Heat stress and seasonal effects on reproduction in the dairy cow-a review. Theriogenology 60:1139–1151

Reperant LA (2010) Applying the theory of island biogeography to emerging pathogens, toward predicting the sources of future emerging zoonotic and vector-borne diseases. Vector Borne Zoonotic Dis 10:105–110

Rhoads ML, Gilbert RO et al (2004) Effects of urea infusion on the uterine luminal environment of dairy cows. J Dairy Sci 87:2896–2901

Rhoads ML, Rhoads RP et al (2006) Detrimental effects of high plasma urea nitrogen levels on viability of embryos from lactating dairy cows. Anim Reprod Sci 91:1–10

Rhoads ML, Rhoads RP et al (2009) Effects of heat stress and plane of nutrition on lactating holstein cows: I. production, metabolism and aspects of circulating somatotropin. J Dairy Sci 92:1986–1997

Rice JB et al (2003) Low-level endotoxin induces potent inflammatory activation of human blood vessels: inhibition by statins. Arterioscler Thromb Vasc Biol 23:1576–1582

Richter K, Haslbeck M et al (2010) The heat shock response: life on the verge of death. Mol Cell 40:253–266

Rogers WE, Hartnett DC (2001) Temporal vegetation dynamics and recolonization mechanisms on different-sized soil disturbances in tall grass prairie. Am J Bot 88:1634–1642

Rose H, Wall R (2011) Modelling the impact of climate change on spatial patterns of disease risk, sheep blowfly strike by *Lucilia sericata* in Great Britain. Int J Parasitol 41:739–746

Rose H, Learmount J, Taylor M, Wall R (2009) Mapping risk foci for endemic sheep scab. Vet Parasitol 165:112–118

Roth Z (2008) Heat stress, the follicle, and its enclosed oocyte: mechanisms and potential strategies to improve fertility in dairy cows. Reprod Domest Anim 43(Suppl 2):238–244

Roush W (1994) Population—the view from Cairo. Science 265:1164–1167

Rozenboim I, Tako E et al (2007) The effect of heat stress on ovarian function of laying hens. Poult Sci 86:1760–1765

Ruden DM, Xiao L et al (2005) Hsp90 and environmental impacts on epigenetic states: a model for the trans-generational effects of diethylstibesterol on uterine development and cancer. Hum Mol Genet 14 Spec No 1: R149–155

Russell JS (1976) Comparative salt tolerance of some tropical and temperate legumes and tropical grasses. Aust J Exp Agric Anim Husb 16:103–109

Rustad LE, Norby RJ (2002) Temperature increase: effects on terrestrial ecosystems. In: Mooney HA, Canadell JG (eds) The earth system: biological and ecological dimensions of global environmental change. Encyclopedia of global environmental change, vol 2. Wiley, Chichester, pp 575–581

Rustad LE, Campbell JL, Marion GM, Norby RJ, Mitchell MJ, Hartley AE, Cornelissen JHC, Gurevitch J (2001) A meta-analysis of the response of soil respiration, net nitrogen mineralization, and aboveground plant growth to experimental ecosystem warming. Oecologia 126:543–562

Sachan N, Singh VP (2010) Effect of climatic changes on the prevalence of zoonotic diseases. Vet World 13:519–522

Sahin K, Kucuk O (2003) Zinc supplementation alleviates heat stress in laying Japanese quail. J Nutr 133:2808–2811

Sahin N, Tuzcu M et al (2009) The effects of vitamin C and E supplementation on heat shock protein 70 response of ovary and brain in heat-stressed quail. Br Poult Sci 50:259–265

Sandkc M, Eren U, Onol AG, Kum S (2004) The effect of heat stress and the use of *Saccharomyces cerevisiae* or (and) bacitracin zinc against heat stress on the intestinal mucosa in quails. Rev Med Vet 155:552–556

Sarkar S, Lakhotia SC (2008) Hsp60C is required in follicle as well as germline cells during oogenesis in *Drosophila melanogaster*. Dev Dyn 237:1334–1347

Sato F, Tsuchiya S, Meltzer SJ, Shimizu K (2011) MicroRNAs and epigenetics. FEBS J 278:1598–1609

Schmidt P, Widdowson EM (1967) The effect of a low-protein diet and a cold environment on calorie intake and body composition in the rat. Br J Nutr 21:457–465

Schneider PL, Beede DK et al (1988) Nycterohemeral patterns of acid-base status, mineral concentrations and digestive function of lactating cows in natural or chamber heat stress environments. J Anim Sci 66:112–125

Schweinburg FB, Fine J (1960) Evidence for a lethal endotoxemia as the fundamental feature of irreversibility in three types of traumatic shock. J Exp Med 112:793–800

Setchell BP (1998) The parkes lecture. Heat and the testis. J Reprod Fertil 114:179–194

Shannon MC, Noble CL (1995) Variation in salt tolerance and ion accumulation among subterranean clover cultivars. Crop Sci 35:798–804

Shaver GR, Canadell J, Chapin FS III, Gurevitch J, Henry JHG, Ineson P, Jonasson S, Melillo J, Pitelka L, Rustad L (2000) Global warming and terrestrial ecosystems: a conceptual framework for analysis. Bioscience 50:871–882

Shen L, Weber CR, Turner JR (2008) The tight junction protein complex undergoes rapid and continuous molecular remodeling at steady state. J Cell Biol 181:683–695

Shen L, Weber CR et al (2011) Tight junction pore and leak pathways: a dynamic duo. Annu Rev Physiol 73:283–309

Shiota K, Kayamura T (1989) Effects of prenatal heat stress on postnatal growth, behavior and learning capacity in mice. Biol Neonate 56:6–14

Shwartz G, Rhoads ML et al (2009) Effects of a supplemental yeast culture on heat-stressed lactating holstein cows. J Dairy Sci 92:935–942

Silanikove N (2000) Effects of heat stress on the welfare of extensively managed domestic ruminants. Livest Prod Sci 67:1–18

Singh RK (2010) Impact of climate change on pattern of zoonotic diseases occurance and spread in livestock, In: Compendium on national symposium on climate change and livestock productivity in India, NDRI, Karnal, Haryana, India pp 38–46

Singleton KD, Wischmeyer PE (2006) Oral glutamine enhances heat shock protein expression and improves survival following hyperthermia. Shock 25:295–299

Sirotkin AV, Bauer M (2011) Heat shock proteins in porcine ovary: synthesis, accumulation and regulation by stress and hormones. Cell Stress Chaperones 16:379–387

Skerman PJ, Riveros F (1990) Tropical grasses. FAO plant production and protection series no. 23. FAO, Rome

Slenning BD (2010) Global climate change and implications for disease emergence. Vet Path 47:28–33

Smit B, McNabb D, Smithers J (1996) Agricultural adaptation to climatic change. Clim Change 33:7–29

Soede NM, Noordhuizen JP, Kemp B (1992) The duration of ovulation in pigs, studied by transrectal ultrasonography, is not related to early embryonic diversity. Theriogenology 38:653–666

Solh MB (1993) New approaches to breeding for stress environments-discussion. Int Crop Sci 1:579–581

Soussana JF, Luscher A (2007) Temperate grasslands and global atmospheric change: a review. Grass Forage Sci 62:127–134

Spiers DE, Spain JN, Sampson JD, Rhoads RP (2004) Use of physiological parameters to predict milk yield and feed intake in heat-stressed dairy cows. J Therm Biol 29:759–764

Spurlock ME (1997) Regulation of metabolism and growth during immune challenge: an overview of cytokine function. J Anim Sci 75:1773–1783

St. Pierre NR, Cobanov B, Schnitkey G (2003) Economic losses from heat stress by US livestock industries. J Dairy Sci 86:E52–E77

Stahly TS, Cromwell GI, Aviotti P (1979) The effect of environmental temperature and dietary lysine source and level on the performance and carcass characteristics of growing swine. J Anim Sci 49:1242–1250

Stem E, Mertz GA, Stryker JD, Huppi M (1989) Changing animal disease patterns induced by the greenhouse effect. In: Smith J, Tirpack DA (eds) The potential effects of global change on theUnited States. Appendix C—Agriculture, vol 2.0. US Environmental Protection Agency, Washington, D.C., pp 11–38

Sternini C, Anselmi L, Rozengurt E (2008) Enteroendocrine cells: a site of 'taste' in gastrointestinal chemosensing. Curr Opin Endocrinol Diabetes Obes 15:73–78

Stewart BM, Block J, Morelli P, Navarette AE, Amstalden M, Bonilla L, Hansen PJ, Bilby TR (2011) Efficacy of embryo transfer in lactating dairy cows during summer using fresh or vitrified embryos produced in vitro with sex-sorted semen. J Dairy Sci 94:3437–3445

Stokes C, Ash A (2007) Impacts of climate change on marginal tropical animal production systems. In: Newton PCD, Carran RA, Edwards GR, Niklaus PA (eds) Agroecosystems in a changing climate. CRC Press, Boca Raton, FL, pp 323–328. Subterranean clover. In: Proceedings of the XVIII international grasslands congress, pp 9.21–9.22

Stowell RR, Mader TR, Gaughan JB (2009) Chapter 8: Environmental management. In: JA DeShazer (ed) Livestock Energetics and Thermal Environment Management, ASABE, St. Joseph, MI.

Sumilo D, Bormane A, Vasilenko V, Golovljova I, Asokliene L, Zygutiene M, Randolph S (2009) Upsurge of tick-borne encephalitis in the Baltic States at the time of political transition, independent of changes in public health practices. Clin Microbiol Infect 15:75–80

Szymanski LA, Schneider JE et al (2007) Changes in insulin, glucose and ketone bodies, but not leptin or body fat content precede restoration of luteinising hormone secretion in ewes. J Neuroendocrinol 19:449–460

Tabuchi Y, Takasaki I et al (2008) Genes and genetic networks responsive to mild hyperthermia in human lymphoma U937 cells. Int J Hyperth 24:613–622

Terada K, Yomogida K et al (2005) A type I DnaJ homolog, DjA1, regulates androgen receptor signaling and spermatogenesis. EMBO J 24:611–622

Tetievsky A, Horowitz M (2010) Posttranslational modifications in histones underlie heat acclimation-mediated cytoprotective memory. J Appl Physiol 109:1552–1561

Theodorakis NG, Morimoto RI (1987) Posttranscriptional regulation of hsp70 expression in human cells: effects of heat shock, inhibition of protein synthesis, and adenovirus infection on translation and mRNA stability. Mol Cell Biol 7:4357–4368

Thompson RF, Fazzari MJ, Niu H, Barzilai N, Simmons RA et al (2010) Experimental intrauterine growth restriction induces alterations in DNA methylation and gene expression in pancreatic islets of rats. J Biol Chem 285:15111–15118

Tirado MC, Clarke R, Jaykus LA, McQuatters-Gollop A, Frank JM (2010) Climate change and food safety, a review. Food Res Int 43:1745–1765

Tomar OS, Minhas PS, Sharma VK, Gupta Raj K (2003) Response of nine forage grasses to saline irrigation and its schedules in a semi-arid climate of north-west India. J Arid Environ 55:533–544

Tompkins EC, Heidenreich CJ et al (1967) Effects of post-breeding thermal stress on embryonic mortality in swine. J Anim Sci 26:4

Trape JF et al (1996) The spread of tick-borne borreliosis in West Africa and its relationship to sub-Saharan drought. Am J Trop Med Hyg 54:289–293

Truog PN, Roberts MH (1992) Salt tolerance of some tropical and subtropical grass species grown in Queensland. In: Proceedings of National workshop on productive use of saline land. Proceeding no 42, ACIAR, Perth, Western Australia, 1–14 May, pp 36–44

Tseng JK, Tang PC, Ju JC (2006) In vitro thermal stress induces apoptosis and reduces development of porcine parthenotes. Theriogenology 66:1073–1082

Turner JR (2006) Molecular basis of epithelial barrier regulation: from basic mechanisms to clinical application. Am J Pathol 169:1901–1909

Turner BM (2007) Defining an epigenetic code. Nat Cell Biol 9:2–6

Van den Bossche P, Coetzer JA (2008) Climate change and animal health in Africa. Rev Sci Tech 27:551–562

Van Milgen J, Noblet J (2003) Partitioning of energy intake to heat, protein, and fat in growing pigs. J Anim Sci 81:E86–E93

Velazquez MM, Alfaro NS et al (2010) Heat shock protein patterns in the bovine ovary and relation with cystic ovarian disease. Anim Reprod Sci 118:201–209

Verstegem MWA, Brascamp EW, Van Der Hel W (1978) Growing and fattening of pigs in relation to temperature of housing and feeding level. Can J Anim Sci 58:1–13

Voit EO, Radivoyevitch T (2000) Biochemical systems analysis of genome-wide expression data. Bioinformatics 16:1023–1037

Volk M, Bungener P, Contat F, Montani M, Fuhrer J (2006) Grassland yield declined by aquater in 5 years of free air ozone fumigation. Glob Change Biol 12:74–83

Waisel Y (1985) The stimulating effects of NaCl on root growth of rhodes grass (*Chloris gayana*). Physiol Plant 64:519–522

Wall R, Ellse LS (2011) Climate change and livestock parasites: integrated management of sheep blowfly strike in a warmer environment. Glob Clim Change 17:1770–1777

Wan SQ, Xia JY, Liu WX, Niu SL (2009) Photosynthetic overcompensation under nocturnal warming enhances grassland carbon sequestration. Ecology 90:2700–2710

Wang Y, Zhou G, Wang Y (2007) Modeling responses of the meadow steppe dominated by *Leymus chinensis* to climate change. Clim Change 82:437–452

Wang J-F, Gao S, Lin J-X, Mu Y-G, Mu C-S (2010) Summer warming effects on biomass production and clonal growth of *Leymus chinensis*. Crop Pasture 61:670–676

Ward MP, Carpenter TE (2000) Analysis of time space clustering in veterinary epidemiology. Prev Vet Med 43:225–237

Waterland RA, Garza C (1999) Potential mechanisms of metabolic imprinting that lead to chronic disease. Am J Clin Nutr 69:179–197

Webel DM, Finck BN, Baker DH, Johnson RW (1997) Time course of increased plasma cytokines, cortisol, and urea nitrogen in pigs following intraperitoneal injection of lipopolysaccharide. J Anim Sci 75:1514–1520

Webel DM, Johnson RW, Baker DH (1998) Lipopolysaccharide-induced reductions in food intake do not decrease the efficiency of lysine and threonine utilization for protein accretion in chickens. J Nutr 128:1760–1766

Weber LP, Janz DM (2001) Effect of beta-naphthoflavone and dimethylbenz[a]anthracene on apoptosis and HSP70 expression in juvenile channel catfish (*Ictalurus punctatus*) ovary. Aquat Toxicol 54:39–50

Welker S, Rudolph B et al (2010) Hsp12 is an intrinsically unstructured stress protein that folds upon membrane association and modulates membrane function. Mol Cell 39:507–520

West JW (1999) Nutritional strategies for managing the heat-stressed dairy cow. J Anim Sci 77(Suppl 2):21–35

Wettemann RP, Wells ME, Omtvedt IT, Pope CE, Turman EJ (1976) Influence of elevated ambient temperature on reproductive performance of boars. J Anim Sci 42:664–669

Wheelock JB, Rhoads RP et al (2010) Effects of heat stress on energetic metabolism in lactating holstein cows. J Dairy Sci 93:644–655

White TW, Bunting LD et al (1992) Influence of fish meal and supplemental fat on performance of finishing steers exposed to moderate or high ambient temperatures. J Anim Sci 70:3286–3292

Williams MA, Rice CW, Owensby CE (2000) Carbon and nitrogen dynamics and microbial activity in tallgrass prairie exposed to elevated CO_2 for 8 years. Plant Soil (in press)

Wilmking M, Juday GP, Barber VA, Zald HSJ (2004) Recent climate warming forces contrasting growth responses of white spruce at tree line in Alaska through temperature thresholds. Glob Change Biol 10:1724–1736

Wilson SJ, Marion RS et al (1998) Effects of controlled heat stress on ovarian function of dairy cattle. J Dairy Sci 81:2124–2131

Wiznitzer T, Schweinburg FB, Atkins N, Fine J (1960) On the relation of the size of the intraintestinal pool of endotoxin to the development of irreversibility in hemorrhagic shock. J Exp Med 112:1167–1171

Wolfe ND, Daszak P, Kilpatrick AM, Burke DS (2005) Bushmeat hunting, deforestation, and prediction of zoonotic disease emergence. Emerg Infect Dis 11:1822–1827

Wolfenson D, Roth Z et al (2000) Impaired reproduction in heat-stressed cattle: basic and applied aspects. Anim Reprod Sci 60–61:535–547

Woolhouse M, Gaunt E (2007) Ecological origins of novel pathogens. Crit Rev Microbiol 33:231

Wu C (1984) Two protein-binding sites in chromatin implicated in the activation of heat-shock genes. Nature 309:229–234

Wu BJ, Kingston RE et al (1986) Human HSP70 promoter contains at least two distinct regulatory domains. Proc Nat Acad Sci USA 83:629–633

Xia JY, Niu SL, Wan SQ (2009) Response of ecosystem carbon exchange to warming and nitrogen addition during two hydrologically contrasting growing seasons in a temperate steppe. Glob Change Biol 15:1544–1556

Yahav S, McMurtry JP (2001) Thermotolerance acquisition in broiler chickens by temperature conditioning early in life–the effect of timing and ambient temperature. Poult Sci 80: 1662–1666

Yang PC, He SH, Zheng PY (2007) Investigation into the signal transduction pathway via which heat stress impairs intestinal epithelial barrier function. J Gastroenterol Hepatol 22:1823–1831

Yu J et al (2010) Effect of heat stress on the porcine small intestine: a morphological and gene expression study. Comp Biochem Physiol A: Mol Integr Physiol 156:119–128

Yunianto VD, Hayashi K et al (1997) Effect of environmental temperature on muscle protein turnover and heat production in tube-fed broiler chickens. Br J Nutr 77:897–909

Zhang Q, Li Q et al (2010a) Redistribution of tight junction proteins during EPEC infection in vivo. Inflammation 35:23–32

Zhang Q et al (2010b) Enteropathogenic Escherichia coli changes distribution of occludin and ZO-1 in tight junction membrane microdomains in vivo. Microb Pathog 48:28–34

Zhu JQ, Liu JH, Liang XW, Xu BZ, Hou Y et al (2008) Heat stress causes aberrant DNA methylation of H19 and Igf-2r in mouse blastocysts. Mol Cells 25:211–215

Zimbleman RB, Rhoads RP, Baumgard LH, Collier RJ (2009) Revised temperature humidity index (THI) for high producing dairy cows. J Dairy Sci 92(E-Suppl 1):347

Zweifach BW, Janoff A (1965) Bacterial endotoxemia. Annu Rev Med 16:201–220

Chapter 16
Global Climate Change: Enteric Methane Reduction Strategies in Livestock

Veerasamy Sejian, Indu Shekhawat, Victor Ujor, Thaddeus Ezeji, Jeffrey Lakritz and Rattan Lal

Abstract The greenhouse gas (GHG) emission from the agricultural sector is considered to be a key contributor to the climate change, accounting for about 25.5% of total global anthropogenic emission. While carbon dioxide (CO_2) receives the most attention as a factor, which causes global warming, methane (CH_4), nitrous oxide (N_2O), and chlorofluorocarbons (CFCs), also cause significant radiative forcing. With the relative global warming potential of 25 compared with CO_2, CH_4 is one of the most important GHGs. This chapter reviews the prediction models, estimation methodologies, and strategies for reducing enteric CH_4 emissions. Emission of CH_4 in ruminants differs between developed and developing countries, depending on factors like animal species, breed, pH of rumen fluid, ratio of acetate: propionate, methanogen population, composition of diet, and amount of concentrate fed. Among ruminants, cattle contribute the most toward greenhouse effect through methane emission, followed by sheep, goat and buffalo, respectively. The estimated CH_4 emission rate per cattle, buffalo, sheep, and goat in developed countries are 150.7, 137, 21.9, and 13.7 (g/animal/day) respectively. However, the estimated rates in

V. Sejian (✉) · I. Shekhwat
Adaptation Physiology Laboratory, Division of Physiology and Biochemistry, Central Sheep and Wool Research Insitute, Avikanagar, Rajasthan 304501, India
e-mail: drsejian@gmail.com

V. Ujor · T. Ezeji
Department of Animal Sciences, The Ohio State University-Wooster Campus, Wooster, OH 44691, USA

J. Lakritz
Department of Veterinary Clinical Sciences, The Ohio State University, Columbus, OH 43210, USA

R. Lal
Carbon Management and Sequestration Center, School of Environment and Natural Resources, The Ohio State University, Columbus, OH 43210, USA

V. Sejian et al. (eds.), *Environmental Stress and Amelioration in Livestock Production*, DOI: 10.1007/978-3-642-29205-7_16,

developing countries are significantly lower, at 95.9 and 13.7 (g/animal/day) per cattle and sheep, respectively. There is a strong interest in developing new and improving existing CH_4 prediction models that are effective in identifying mitigation strategies for reducing the overall CH_4 emissions. A careful examination of the literature suggests that the mechanistic models are superior to empirical models in accurately predicting the CH_4 emission from dairy farms. The latest development in prediction model is the integrated farm system model, which is a process-based whole farm simulation technique. Several techniques are used to quantify enteric CH_4 emissions starting from whole animal chambers to sulfur hexafluoride (SF_6) tracer techniques. The latest technology developed to estimate CH_4 more accurately is the micrometeorological mass difference technique. Understanding this basic information about enteric methane is very vital for formulating suitable mitigation strategies to curtail methane production. There are varieties of mitigation strategies available, which can be grouped under managemental, nutritional, and advanced molecular technologies. Strategies that are cost-effective, improve productivity, with potentially limited negative effects on livestock production are likely to be adopted by producers.

Keywords Enteric methane · Climate change · Global warming potential · Prediction models for enteric CH_4 · Respiratory chamber · SF_6 · Whole farm model · Defaunation · Immunization

Contents

16.1 Introduction

There is a growing interest in reducing the emission of greenhouse gases (GHGs) into the atmosphere, because of its potential effect on global warming (Moss et al. 2000). Agricultural activities contribute significantly to global GHG emissions. Examples include carbon dioxide (CO_2), methane (CH_4), nitrous oxide (N_2O), and ammonia (NH_3), which are major GHGs contributing to global warming (IPCC 2001). Increasing awareness of environmental problems associated with global warming has brought significant attention to the need to mitigate global warming and protect the environment (Meadows et al. 1992). Minimizing the contamination of air with CO_2, CH_4, NH_3, and N_2O and other GHGs that contribute to radiative forcing has been promoted as a possible route for mitigating global warming (Tamminga 1996). The consequences of increasing the atmospheric concentration of GHGs responsible for the radiative forcing are gradual elevation of average global temperatures, altered viability of plants, animals, insects, and microbes with numerous adverse consequences to human well-being. The degree to which these changes are projected to occur is dependent upon reliable GHG policy models with a range of scenarios for the levels of GHG emissions (Moss et al. 2000).

The scientific evidence of anthropogenic interference with the climate system through GHG emissions has led to worldwide research on assessing impacts that could result from potential climate change associated with GHG accumulation (Sejian 2012). As ecosystems are sensitive to climatic changes, it is necessary to examine the likely impacts of climate change on various components within ecosystems to develop a comprehensive understanding of these effects on climate change. While CO_2 receives the most attention, methane with the global warming potential of 25 and longer residence time in the atmosphere, is an important GHG (Wuebbles and Hayhoe 2002; Forster et al. 2007). The rising concentration of CH_4 is strongly correlated with increasing populations, and currently about 70% of its production arises from anthropogenic sources (Moss et al. 2000; IPCC 2007).

Increase in the global release of CH_4 from agricultural sources can be ascribed to growing rice paddies, fermentation of feed by ruminants (enteric CH_4), biomass burning, and animal wastes (Sejian et al. 2011). In recent times, CH_4 production has been aggravated by large-scale ruminant production (Beauchemin and McGinn 2005). Globally, ruminant livestock are responsible for about 85 Tg (1 Tg $= 10^{12}$ g $=$ 1 million metric ton) of the 550 Tg CH_4 released annually (McGinn et al. 2006). Ruminant animals, particularly cattle (*Bos taurus*), buffalo (*Bubalus bubalis*), sheep (*Ovis aris*), goat (*Capra hircus*), and camels (*Camalus camalis*) produce significant amounts of CH_4 via anaerobic digestion. This microbial fermentation process, referred to as 'enteric fermentation', produces CH_4 as a by-product, which is released through eructation, normal respiration, and small quantities as flatus (Lassey 2007; Chhabra et al. 2009).

Measurement of CH_4 production in animals requires complex and often expensive equipment. Some models have been developed to specifically predict CH_4 emissions from animals (Ellis et al. 2007). At present, mathematical models are used to estimate CH_4 emissions from enteric fermentation at a national and global level, based on the guidelines of the intergovernmental panel on climate change (IPCC 2006). However, the accuracy of these models has been widely challenged (Kebreab et al. 2006).

Understanding the relationship of diet to enteric CH_4 production is essential to reduce uncertainty in GHG emission inventories and to identify viable reduction strategies. As a result, quantifying and reducing emissions from livestock farms is important in reducing overall CH_4 emission. This article reviews the sensitivity of livestock production to climate change in general, while attempting to highlight some important progresses made in predicting accurately the enteric CH_4 emissions. As animal production systems are vulnerable to climate change, while also contributing to potential global warming through CH_4 and N_2O emissions, this review also collates and synthesises the literature on prediction models, estimation methodology, and strategies for reducing CH_4 emission from ruminants.

16.2 The Contribution of CH_4 to Global Warming/Climate Change

Increasing concentrations of GHGs in the atmosphere have contributed to an increase in the earth's atmospheric temperature, an occurrence known as global warming (FAO 2006). Indeed, average global temperatures have risen considerably, and the IPCC (2007) predicts increases of 1.8–3.9°C (3.2–7.1°F) by 2100. Based on the current trends, the earth's temperature may rise by 1.4–5.8°C by the end of this century. With rare unanimity, the scientific community warns of a more abrupt and greater climatic change in the future (Moss et al. 2000; Gleik et al. 2010).

The GHG emissions from the agriculture sector account for about 25.5% of the total global radiative forcing and over 60% of anthropogenic sources (FAO 2009). Animal husbandry accounts for 18% of GHG emissions that cause global warming. Emission of CH_4 is responsible for nearly as much radiative forcing as all other non-CO_2 GHG gases combined. While atmospheric concentrations of GHGs have risen by about 39% since the pre-industrial era, CH_4 concentration has more than doubled during this period (WHO 2009). Reducing the increase of GHG emissions from agriculture, especially livestock production, should therefore be a top priority, because it could curb warming considerably (McMichael et al. 2007).

The major global warming potential (GWP) of livestock production worldwide comes from the natural life processes of the animals. In fact, CH_4 is considered to be the largest potential contributor to the global warming phenomenon (Johnson et al. 2002; Steinfeld et al. 2006). It is an important component of GHG in the atmosphere (Leng 1993; Moss et al. 2000). The development of management strategies to mitigate CH_4 emissions from cattle is possible and desirable. Not only can the enhanced utilization of dietary carbon (C) improve feed efficiency and

Table 16.1 Global domesticated ruminant population (10^6) and their respective distribution across major continents (Adapted from FAO 2008)

Animal type	World	Africa	North America	South America	Asia	Europe	Oceania
Cattle	1347	270	111	315	431	127	38
Buffalo	181	5	–	1	174	0.33	0.0002
Sheep	1078	288	7	73	452	134	113
Goat	862	291	3	21	514	18	1

animal productivity, but a decrease in CH_4 emissions may also reduce the contribution of ruminant livestock to the global CH_4 inventory.

16.3 Sources of CH_4 Emission

CH_4 is emitted from a variety of anthropogenic and natural sources (Wuebbles and Hayhoe 2002; Rotz et al. 2010). The anthropogenic sources include fossil fuel production and exploitation, animal husbandry, including manure storage (Chianese et al. 2009), paddy rice cultivation, biomass burning, and waste management (Kumaraswamy et al. 2000; Mosier et al. 2004). Other sources of methane emission are predominantly natural sources, including wetlands, gas hydrates, permafrost, termites, oceans, freshwater bodies, non-wetland soils, volcanos, and wildfires (Breas et al. 2001).

16.4 Enteric Emission of CH_4 by Livestock

Domestic animal production increased sharply since the 1990s due to policies encouraging large-scale livestock production, improvements in feeding technology, and market requirements (Yang et al. 2003; Sejian and Saumya 2011). Improved animal feeding techniques have had a marked impact on livestock production over the past few decades. In ruminants, a significant amount of fermentation takes place in the rumen, resulting in relatively large CH_4 emissions per unit of feed energy consumed. Capturing and burning methane produced by methanogens in manure lagoons, and regulation of dairy diet to reduce enteric CH_4 emission are among effective strategies that could be employed to reduce enteric CH_4 production. Pseudo-ruminants (e.g. horses) and monogastric animals (e.g. pigs) do not support the same level of feed fermentation, and consequently emissions from these animals are relatively low. Synthesis in Table 16.1 depicts world ruminant livestock population.

Among farm animals, cattle population contributes the most toward enteric CH_4 production (Johnson and Johnson 1995). Enteric fermentation emissions for cattle are estimated by multiplying the emission factor for each species by the relevant cattle populations. The emission factors are an estimate of the amount of CH_4 produced (kg) per animal, and are based on animal and feed characteristics, average energy requirement of the animal, the average feed intake, and the quality of the feed consumed (Sejian et al. 2011a). The district on county level emission from enteric fermentation is computed as a product of the livestock population under each category and its emission coefficient (Chhabra et al. 2009; Sejian and Naqvi 2011a). However, emission coefficients are country-specific, and should conform to IPCC guidelines (IPCC 2007).

16.4.1 Enteric Fermentation: Process Description

Enteric fermentation is the digestive process in herbivore animals by which carbohydrates are broken down by microorganisms into simple molecules for absorption into the bloodstream (Sejian and Saumya 2011). CH_4 is produced as a waste product of this fermentation process. CH_4 production through enteric fermentation is of concern worldwide for its contribution to the accumulation of greenhouse gases in the atmosphere, as well as its waste of fed energy for the animal. CH_4 is produced in the rumen and hindgut of animals by a group of *Archaea* known collectively as methanogens, which belong to the phylum *Euryarcheota*. Among livestock, CH_4 production is greatest in ruminants, as methanogens are able to produce CH_4 freely through the normal process of feed digestion. Ruminant animals are the principal source of emissions, because they produce the maximum CH_4 per unit of feed consumed. What makes ruminant animals unique is their 'fore-stomach' or rumen, a large, muscular organ. The rumen is characterized as a large fermentation vat where approximately 200 species and strains of microorganisms are present. The microbes ferment the plant material consumed by the animal through a process known as enteric fermentation. Figure 16.1 describes the enteric CH_4 emission from different feed components by the various rumen methanogen species. The products of this fermentation provide the animal with the nutrients it needs for survival, enabling them to subsist on coarse plant material. CH_4 is produced as a byproduct of the fermentation and is expelled. Monogastric animals produce small amounts of CH_4 as the result of incidental fermentation that takes place during the digestion process. Non-ruminant herbivores produce CH_4 at a rate that is of between monogastric and ruminant animals. Although these animals do not have a rumen, significant fermentation takes place in the large intestine, allowing significant digestion and use of plant material.

CH_4 producing bacteria reside in the reticulo-rumen and large intestine of ruminant livestock. These bacteria, commonly referred to as methanogens, use a range of substrates produced during the primary stages of fermentation to produce CH_4, thus creating generated energy required for their growth. All methanogen species can utilize hydrogen ions (H_2) to reduce CO_2 in the production of CH_4 as this reaction is thermodynamically favorable to the organisms. Availability of H_2

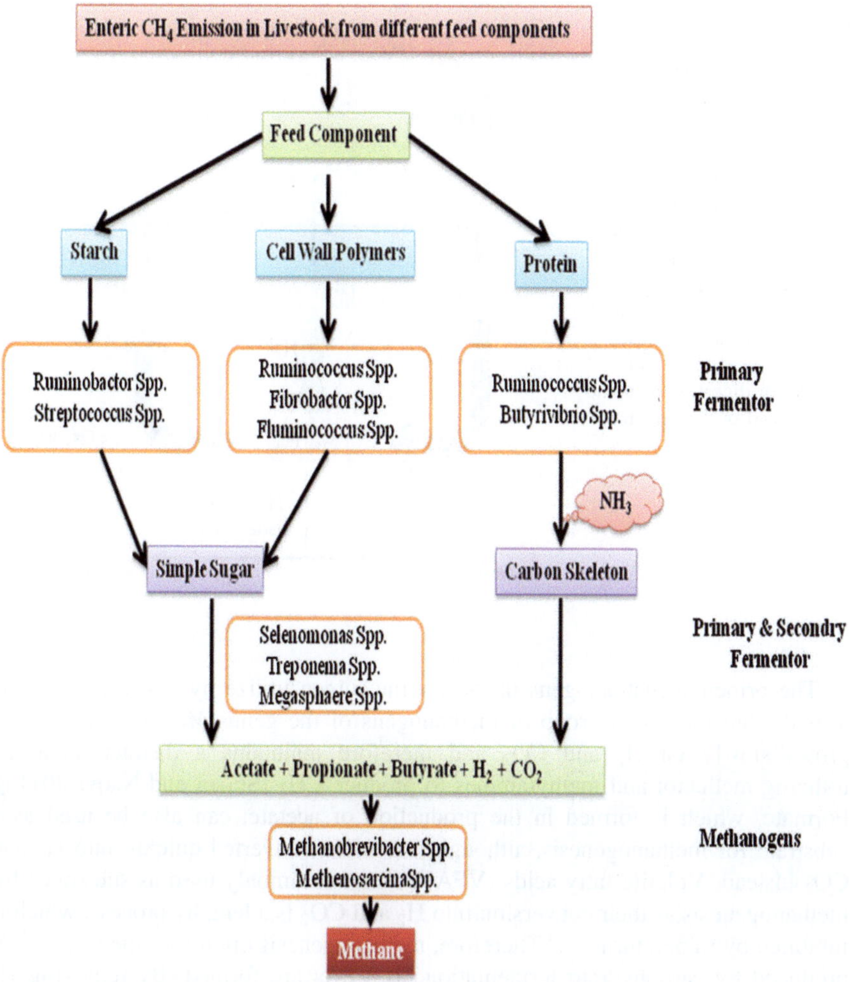

Fig. 16.1 Enteric methane emission from different feed components. Different feed components, such as starch, cell wall polymers, and proteins, are broken down to simpler sugars and carbon skeleton by primary fermentor microbes in the rumen. These simpler sugars are converted into volatile fatty acids by both primary and secondary fermentor microbes. This VFAs are acted upon by methanogen microbes in the rumen to produce methane

in the rumen is determined by the proportion of end products resulting from fermentation of the ingested feed. Processes that yield propionate and cell dry matter act as net proton-using reactions, whereas a reaction that yields acetate results in a net proton increase. Other substrates available to methanogens include formate, acetate, methanol, methylamines, dimethyl sulfide, and some alcohols; however, only formate has been documented as an alternative CH_4 precursor in the rumen. Figure 16.2 describes the mechanism of CH_4 production during digestion process in the rumen.

Fig. 16.2 Mechanism of enteric CH_4 production during digestion in the rumen. This figure describes the process of methane production in the rumen. From feed components, initially glucose is produced and from glucose VFAs are produced. Acetate and butyrate are hydrogen providers while propionate is hydrogen consumer. Based on the proportion of end product of digestion methane is produced in the rumen

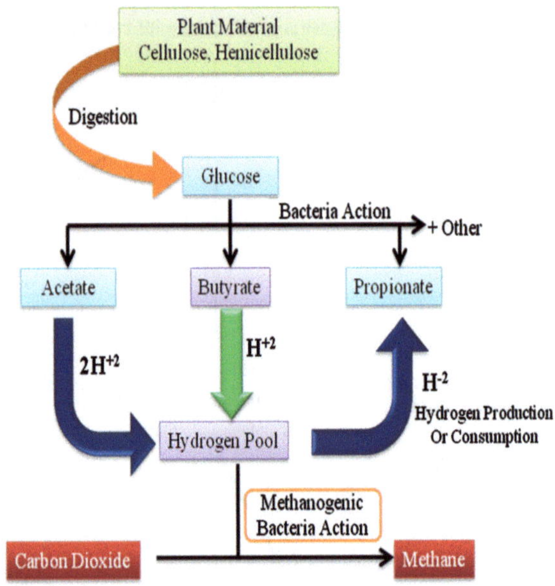

The principal methanogens in the bovine rumen utilize hydrogen and carbon dioxide, but there is a group of methanogens of the genus *Methanosarcina* that grows slowly on H_2 and CO_2 and therefore maintains a distinct niche by utilizing methanol and methylamines to produce CH_4 (Sejian and Naqvi 2011a). Formate, which is formed in the production of acetate, can also be used as a substrate for methanogenesis, although it is often converted quickly into H_2 and CO_2 instead. Volatile fatty acids (VFA) are not commonly used as substrates for methanogenesis as their conversion into H_2 and CO_2 is a lengthy process, which is inhibited by rumen turnover. Therefore, methanogenesis often uses the C and CO_2 produced by carbohydrate fermentation, as VFAs are formed. By removing H_2 from the ruminal environment as a terminal step of carbohydrate fermentation, methanogens allow the microorganisms involved in fermentation to function optimally and support the complete oxidation of substrates (Sejian and Naqvi 2011a). The fermentation of carbohydrates results in the production of H_2 and if this end product is not removed, it can inhibit metabolism of rumen microorganisms.

16.5 CH_4 Emission in Developed and Developing Countries

The global production and consumption of farm animals and their products are increasing significantly and are predicted to continue up until ~2050. This increase comes at a time when it is recognized that the world is facing a crisis about the impact of humans on the planet's climate (Sejian et al. 2011a). Animal products have a high

GWP per kg compared to most plant-derived foods (FAO 2008). In addition to enteric fermentation, which is the major source of CH_4 emission, meat and dairy production and consumption also contribute significantly to the emission of GHGs in developed countries (Eshel and Martin 2006; Ogino et al. 2007 Comparatively, due to intensive animal production, developed countries tend to record higher CH_4 emission than their developing counterparts do.

Thus, developed countries have the highest emissions factors per animal while Asia, North, and Sub-Saharan Africa have the lowest (USEPA 1994). Among dairy cows, the temperate mixed farming production system has the highest emissions factors, reflecting higher levels of milk production. In light of this, emissions per unit of product can be used to identify opportunities for reducing CH_4 emissions from enteric fermentation.

The emission of CH_4 is quantified by its conversion rate, which differs among developed and developing countries. The conversion rate in developed countries is 4% for feedlot cattle and 6% for other cattle (Lassey et al. 1997). Better-digested feeds with higher energy value result in lower conversion rates. The CH_4 conversion rate in developing countries is also 6% for dairy cows and young cattle, and 7% for non-dairy cattle with the exception of stall-fed animals (Lassey et al. 1997). Grazing cattle have a conversion rate of 6%, except for grazing cattle in tropical Africa, which have a rate of 7% because of the poor quality forage. These estimates are based on the general feed characteristics and production practices observed in many developed and developing countries.

The data on GHG emissions from animal production depend on the natural life processes of the animals. However, these processes are difficult to control without reducing productivity. Hence, overproduction of livestock is likely to have a negatively impact on overall food production given the accompanying damages to the environment, such as more frequent droughts, floods, storms, and harvest failures. This calls for better-planned and well-managed reduction in worldwide production and consumption of meat and milk products to mitigate global warming. Global trends indicate that overproduction and consumption of animal products (meat, milk, and eggs) in high-income, developed countries may be nearing its peak and further increase in production may be unsustainable due to population growth and environmental concerns (Gold and Porritt 2004). While low-income countries and some fast-developing countries are expected to continue their rapid growth in animal production and consumption, it is essential that emission of livestock-related GHGs is monitored and controlled in the short-term.

16.6 Factors Influencing CH_4 Production

Factors that influence enteric methane production outlined in Fig. 16.3 indicate that CH_4 production varies with microbial species and livestock breed. Consequently, breeding of livestock on the basis of CH_4 production in an effort to reduce enteric CH_4 emissions without compromising animal productivity is possible

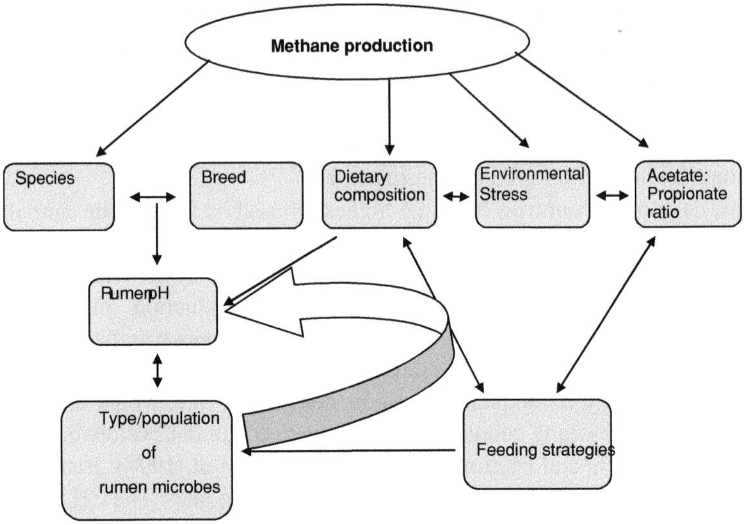

Fig. 16.3 Factors influencing methane output from enteric fermentation. These factors can be broadly categorized into animal, feed, rumen, and environmental factors

(Hegarty et al. 2007). Rumen pH plays a major role in determining the predominant microbial population, which has a direct bearing on CH_4 gas emission (Erfle et al. 1982). Reduction of ruminal pH can decrease ruminal CH_4, which can improve feed utilization in the ruminant animals (Lana et al. 1998). The ratio of ruminal acetate: propionate in vivo is highly influenced by the capacity of the bacteria to produce CH_4 in vitro. Cattle with low acetate: propionate ratios also have low ruminal pH values, and in vitro experiments corroborate the concept that pH has a major impact on CH_4 production and acetate: propionate ratio (Lana et al. 1998; Sejian et al. 2011a).

Methanogens that thrive on and within rumen ciliate protozoa may account for up to 37% of the rumen CH_4 emissions (Hegarty et al. 2007). In the absence of protozoa, rumen CH_4 emissions are reduced by up to 13% on the average depending on the diet. Decreased CH_4 emissions from protozoa-free rumen may be a consequence of reduced ruminal dry matter digestion, decreased methanogen population, altered pattern of volatile fatty acid production, decreased hydrogen availability, and increased partial pressure of oxygen in the rumen (Yoon and Stern 1995). Decline in methanogenesis associated with removal of protozoa is strung on high concentrate diets, because protozoa are relatively more important sources of hydrogen on starch diets, and many starch-fermenting bacteria do not produce hydrogen. Because protozoa also decrease the supply of protein available to the host animal, their elimination offers benefits in both decreasing GHG emissions and potentially increasing feed conversion ratio in livestock production (Hegarty 1999; Sejian et al. 2011a).

The composition of diet fed to the livestock is another important factor, which influences CH_4 emission (McCrabb et al. 1997), especially from enteric fermentation in lactating dairy cows. For instance, the proportion of forage in livestock diet and the source of grain influence enteric CH_4 production by ruminants (Beauchemin and McGinn 2005). The amounts of digestible nutrients especially the carbohydrate fraction is used to estimate CH_4 emission from livestock with high precision. Conversely, a diet rich in fat reduces CH_4 formation in the rumen (Jentsch et al. 2007). Puchala et al. (2005) identified the potential of condensed tannins in forages to reduce CH_4 emission by ruminants. The level of CH_4 emission is positively correlated with live weight, dry matter intake (DMI), milk yield (MY), and feeding level (Yan et al. 2006). To enhance productivity, manipulation of dietary composition of cattle feed is a nutritional management strategy with potential to reduce CH_4 production by dairy cows (Yan et al. 2006).

There are several feed supplements, which significantly alter the level of CH_4 emission (Leng 1991; Moss et al. 1995). However, additional feed supplement should be used with caution as they can reduce animal productivity when the wrong dosage is administered (Beauchemin and McGinn 2006; Foley et al. 2009). Incremental inclusion of malic acid in beef cattle diets results in linear reductions of both the total daily emission of CH_4 and its emissions expressed per unit of DMI. Nevertheless, the dietary inclusion of malate is also associated with a decline in DMI in both studies (Martin and Streeter 1995). This could potentially decrease animal performance, with consequent increases in lifetime CH_4 emission owing to extended slaughter age. Therefore, further in vivo research is required to clarify the long-term effects of malate supplementation on CH_4 emission, animal performance, and productivity (Foley et al. 2009). Beauchemin and McGinn (2006) demonstrated that canola oil (*Brassica campestris L*) can be used to reduce CH_4 emissions from cattle, but animal performance may be compromised due to lower feed intake and decreased fiber digestibility. Essential oils have no effect on CH_4 emissions, whereas fumaric acid causes potentially beneficial changes in ruminal fermentation although with no measurable reductions in its emissions. Certain fats and oils are potential natural CH_4 reducing feed compounds, and are effective even at common dietary proportions (Machmuller et al. 1998). McGinn et al. (2004) demonstrated that sunflower (*Helianthus annuus L*) oil, ionophores, and possibly some yeast products can be used to decrease the gross energy loss as CH_4 from cattle, however, fiber digestibility is impaired with oil supplementation.

Research involving in vitro and in vivo experiments indicates that high grain feeding reduces CH_4 emission in farm animals. (Lana et al. 1998; Hristov et al. 2001). There are two possible reasons for this viz., reduced methanogenesis and lower acetate: propionate ratio (Christophersen et al. 2008). Not only enhanced utilization of dietary C can improve feed utilization efficiency and animal productivity, but also a decrease in CH_4 emissions can reduce the contribution of ruminant livestock to the global CH_4 inventory (Johnson and Johnson 1995). Using best management practices (BMPs), the emission of CH_4 per unit of animal weight gain reduces significantly. Management-intensive grazing (MIG) is a BMP that offers the potential for more efficient utilization of grazed forage crops via controlled rotational grazing and more efficient conversion of forage into meat and milk (DeRamus et al. 2003).

As highlighted above, there are several factors, which play important roles in influencing enteric fermentation, hence CH_4 emission. There is an urgent need to understand these factors better, with a view to decreasing GHG emissions (Beauchemin and McGinn 2005).

16.7 Prediction of CH_4 Emission by Emission Models

CH_4 production is generally predicted on the basis of equations involving DMI, intake of carbohydrates, digestibility and intake of dietary energy, animal size, milk components, and digestibility of dietary components (Sejian et al. 2011b). The models developed for the prediction of CH_4 emission can be classified into two principal groups: (i) empirical (statistical) models, which directly relate nutrient intake to CH_4 output, and (ii) dynamic mechanistic models that attempt to simulate CH_4 emissions based on a mathematical description of ruminal fermentation biochemistry (Kebreab et al. 2006). Synthesis in Table 16.2 depicts the merits and demerits of the empirical and mechanistic models.

16.7.1 Empirical Models

Although many statistical models have been fairly successful in predicting CH_4 production, many variables in these models are not commonly measured, which may lead to difficulties in predicting CH_4 production outside the range of values used for model development (Johnson et al. 1996; Sejian and Rajni 2011). These problems may be addressed by using equations with common input variables and by developing models with minimum input variables from multiple sources. The limitation of using some of the extant models, such as the equation of Moe and Tyrrell (1979), is the difficulty in obtaining reliable model input variables, which might have compromised the predictive ability of the model in the study. Ellis et al. (2007) formulated the most widely accepted equations, which could be useful to the livestock industry for accurately predicting CH_4 production from a minimum set of inputs. Although the extant models evaluated performed well, the new equations developed in the study were more user-friendly and reliable than the extant models, as a result, it is a preferable model for generating national CH_4 emissions inventory. Synthesis in Table 16.3 depicts different types of regression/prediction equations for enteric methane emission.

Table 16.2 The merits and demerits of empirical and mechanistic prediction models for methane emission

Type of prediction models	Empirical models	Mechanistic models
Description	Statistical models that relate nutrient intake to CH_4 output directly	Dynamic models that attempt to simulate CH_4 emissions based on a mathematical description of ruminal fermentation biochemistry
Models	IPCC (2006) tier II model Moe and Tyrrell (1979) model	MOLLY (Baldwin (1995) and its current version MOLLY 2007), COWPOLL (rumen model of Dijkstra et al. 1992)
Merits	1. Simple uncomplicated regression equation based on feed characteristics	1. Predict CH_4 production from cattle without undertaking extensive and costly experiments 2. The advantage over empirical models is that mitigation options implemented at a farm or national level can be evaluated for effectiveness 3. Climate and management informations are included in these models 4. Take into account the methanogenic bacterial population and their methane production efficiency
Demerits	1. Predict CH_4 production from livestock by undertaking extensive and costly experiments 2. Do not take into account the influence of environmental effect and rumen microbial population on methane production 3. Cannot accurately predict CH_4 production under all perturbation conditions as it is based only on feed characteristics, and do not take other factors into account 4. Empirical models lack the biologic basis necessary to evaluate mitigation strategies and cannot be used to predict changes in methane emissions outside the range they were developed for	1. For effective application, these models should possess accurate values for input parameters, such as the chemical composition of the diet, degradation rates of feed components, and passage rates 2. One of the main limitations of mechanistic models is its lack of detailed and accurate data, in light of, the complexity of the system 3. Similarly, the rapid dynamic changes in metabolic flux during lactation, especially in late pregnancy and early lactation pose another major limitation

Source Sejian et al. (2011a)

Table 16.3 Examples of regression/prediction equations for methane emission

Prediction equations	Reference
Methane (Mcal/d) = (18 + 22.5 × DM1 (kg/d)) × .013184 (Mcal/g of methane)	Kriss (1930)
Methane (Mcal/d) = (17.68 + .04012 × digested carbohydrate (g/d)) × .013184 (Mcal/g of methane)	Bratzler and Forbes (1940)
Methane (Mcal/d) = −0.494 + 0.629 × DMI (kg/d) − 0.025 × DMI^2 (kg/d)	Axelsson (1949)
CH_4 (%GE) = 1.30 + .122 *D* + *L* (2.37 .05 *D*) GE—gross energy; *D*—digestibility of energy; *L*—intake relative to maintenance	Blaxter and Clapperton (1965)
Methane (MJ/d) = 3.41 + 0.51 NFC + 1.74 HC + 2.65 *C* For intake of carbohydrate fractions Methane (Mcal/d) = .814 + .122 × NFC (kg/d) + .415 × HC (kgId) + .633 × *C* (kg/d) NFC—non fiber carbohydrate; HC—hemicellulose; *C*—Cellulose	Moe and Tyrrell (1979)
F_{CH4} = 9.77 + .87 × CF F_{CH4} is methane emissions as a percentage of digestible energy and CF is crude fiber	Giger-Reverdin et al. (1992)
CH_4 (g/d) = SF_6 permeation rate (g/d) × [CH_4]/[SF_6]	Johnson et al. (1995)
CH_4 (g/d) = 55 + 4.5**M* + 1.2**W* (grass) CH_4 (g/d) = 26 + 5.1**M* + 1.8**W* (corn silage) *M* = milk yield; *W* = live weight	Kirchgessner et al. (1991)
CH_4 (g/d) = 63 + 79 CF + 10 NFE + 26 CP − 212 EE (all nutrients in kg/d) CF—crude fiber; NFE—nitrogen free extract; EE—ether extract	Kirchgessner et al. (1995)
CH_4 (MJ/d) = 1.36 + 1.21*Dm − 0.825*Dmc + 12.8*Nd (all) CH_4 (MJ/d) = −35.5 + 0.0216*N + 27.6*Sdm + 1.63*Gdm (silage) Dm = dry matter intake (kg); Dmc = dry matter concentrates (kg); Nd = ratio of NDF/Dm; *N* = *N* intake; Sdm = ratio of silage Dm/Dm; Gdm = ratio of gross energy/Dm	Yates et al. (2000)
CH_4 (MJ/d) = *E*max – *E*max*exp(−*c**MEI) *E*max = maximum value of CH_4 production; *c* = shape parameter; MEI = metabolizable energy intake	Mills et al. (2003)
Based on dry matter intake $Y = 164.18(\pm 39.39).\ X_1 + 8427.54(\pm 603.20)$ $Y = 82.3418(\pm 8.41).\ X_2 + 5227.35(\pm 596.12)$ where $y = CH_4$ energy (kJ); $x1$ = dry matter intake (g feed/kg BW) and $x2$ = dry matter intake (g feed/kg $W^{0.75}$)	Jentsch et al. (2007)

Source Sejian et al. (2011a)

16.7.2 Mechanistic Models

Synthesis in Table 16.4 describes the various emission models available to predict enteric methane emission for farm households. Yan et al. (2000) improved on the earlier representation of methanogenesis in the mechanistic model, and outlined some likely reasons for the differences between observed and predicted values for

Table 16.4 Examples of prediction models for methane emission

Prediction models	References
Statistical models/empirical model	Blaxter and Clapperton (1965)
Statistical models/empirical model	Moe and Tyrrell (1979)
Dynamic/mechanistic model	Dijkstra et al. (1992)
Dynamic/mechanistic model	Baldwin (1995)
MOLLY (mechanistic model)	Baldwin (1995)
Dynamic/mechanistic model	Benchaar et al. (1998)
Farm system simulation framework (whole farm model)	Neil et al. (1997), Sherlock et al. (1997), Sherlock and Bright (1999), Bright et al. (2000)
Mechanistic model refined with knowledge of the dietary components	Yan et al. (2000), Mills et al. (2003)
COWPOLL (mechanistic model)	Kebreab et al. (2004)
Tier II model (empirical model)	IPCC (2006)
Dynamic/mechanistic model	Ellis et al. (2007)
Integrated farm system model	Rotz et al. (2007)
Diet specific nonlinear mechanistic models	Kebreab et al. (2008)
Integrated farm system model refined with process-based whole farm simulation	Rotz et al. (2009)

Source Sejian et al. (2011a)

CH_4 production. One of these is the error attributable to dietary composition, not only in the analysis, but also due to variation in nutrient composition between samples of the same feedstock. This knowledge of the dietary components is a prerequisite for successful use of the models to compare the effect of different feeds on CH_4 production (Palliser and Woodward 2002). Nonlinear mechanistic model of CH_4 production provides a significant opportunity to enhance scientific capacity to estimate CH_4 emissions from cattle (Mills et al. 2003; Kebreab et al. 2008). In addition, researchers have identified that mechanistic models can be used to generate Ym values, which can be applied in national CH_4 emission inventory models. It was suggested that if incentives were introduced to mitigate CH_4 emissions at a farm level, mechanistic models would be excellent tools to make reliable estimates of enteric CH_4 emissions. The advantage of mechanistic models compared with empirical models is that mitigation options implemented at a farm or national level can be assessed for their effectiveness.

16.7.3 Whole Farm Model

Computer simulation can provide a cost-effective and efficient method of estimating CH_4 emissions from dairy farms and analyzing effects of management strategies on emissions. Invariably all whole farm models (WFM) are mechanistic models. A commonly used simulation is that proposed by Rotz et al. (2007). The model is an integrated farm system model (IFSM), which is a potential tool for simulating whole farm emissions of CH_4 and evaluating the overall impact of management strategies used to reduce CH_4 emissions. The IFSM was further refined into a process-based whole farm simulation including major components for soil processes, crop growth, tillage, planting, and harvest operations, feed storage, feeding, herd production, manure storage, and economics (Rotz et al. 2009). Incorporation of the CH_4 module with IFSM in addition to modules simulating N_2O emissions, provides an important tool for evaluating the overall impact of management strategies used to reduce GHG emissions in dairy farms.

Farm system simulation framework (FSSF) is another type of WFM, which uses pasture growth and cow metabolism for predicting CH_4 emissions in dairy farms. Also included in the WFM is climate and management information. Other examples of WFMs include those developed by Neil et al. (1997), Sherlock et al. (1997), Sherlock and Bright (1999), and Bright et al. (2000); however, these models are adequate only for predicting CH_4 production by non-lactating Holstein cows. Prediction rates for lactating cows are less accurate and WFMs currently described in the literature seem inappropriate (IPCC 1997). Hence, development of WFMs is required for the prediction of nutrient and GHG emissions and to get better estimate of enteric CH_4 production (Sejian and Singh 2011). Currently available WFMs may incorrectly estimate CH_4 emission levels, because they cannot broadly predict enteric CH_4 emissions as affected by DMI and diet. The low prediction accuracy of CH_4 equations in current WFMs may introduce substantial errors in the calculation of GHG emissions, thereby leading to incorrect mitigation recommendations. If regression equations examined here and elsewhere continue to explain only a small fraction of the variations in observed values, moving toward regression equations including more nutritional informations and details on a subanimal level, or toward a dynamic mechanistic description of enteric CH_4 emission, will improve predictions (Ellis et al. 2010; Sejian et al. 2011b).

16.8 Estimation Methodologies for Enteric CH_4 Emission

There is a growing interest in changing the management strategies to reduce enteric CH_4 production without negatively influencing animal productivity. This theme has been the focus of much research to improve environmental sustainability from an agricultural standpoint. However, accurate CH_4 measurements are required for identifying mitigation strategies that can discriminate among treatments relevant to on-farm conditions (Lassey 2008). Several techniques are used to quantify enteric CH_4 emissions including whole animal chambers (Grainger et al. 2007), and SF_6 tracer technique (Pinares-Patiño et al. 2008; McGinn et al. 2009).

Measurement of CH_4 emissions from individual animals has traditionally been made with open-circuit respiration chambers, which are highly accurate and reliable for animals offered indoor diets (Sejian and Saumya 2011). However, these chambers are not as suitable for evaluating emissions for grazing animals. A new technique that makes use of an inert gas (SF_6) has recently been developed for determining CH_4 emissions from cattle and sheep under grazing conditions. The SF_6 tracer technique enables the determination of enteric CH_4 emissions from both individual as well as on a large number of animals (Vlaming et al. 2007). This technique is based on the use of a controlled release bolus containing SF_6 gas, which is inserted into the animal's rumen (DeRamus et al. 2003). This technique slowly samples the mixed eructated and respired air from the animal, generally over a 24-h period. This air sample is then analyzed for the ratio of CH_4:SF_6 concentration, and the ratio is multiplied by the known release rate of SF_6 emitted from a permeation tube placed in the rumen (Pinares-Patiño et al. 2008). This technique is extremely useful for examining grazing management influences on enteric CH_4 emissions (Pinares-Patiño et al. 2007). In addition, the authors concluded that the SF_6 tracer technique can be used with a reasonable degree of accuracy for inventory purposes and for evaluating the effects of mitigation strategies on CH_4 emissions.

A more flexible technique for quantifying emissions is to model the dispersion of a target gas from the source (Flesch et al. 2004), so that a downwind concentration of gas can establish the emission rate. This "inverse-dispersion" technique has the advantage of simplicity, as it requires only a single gas concentration measurement and basic wind information (McGinn et al. 2006). However, several important factors must be considered to get accurate results from this technique. These factors include: (1) ambient conditions including landscape with clearly defined wind regime as well as not having other nearby emission sources; (2) duration of measurement; (3) period specific measurements (while ignoring periods known to be problematic for inverse-dispersion calculations), and (4) lactation specific concentration measurement.

A micrometeorological mass difference technique is used to measure CH_4 production by cattle in pasture and feedlot conditions (Harper et al. 1999). Measurements are made continuously under field conditions, semiautomatically for several days. The method permits a relatively large number of cattle to be sampled. These techniques do not infringe on the measurement being made, and are generally nonintrusive (Laubach and Kelliher 2004). Limitations include light winds, rapid wind direction changes, and high precision CH_4 gas concentration measurement. The mass difference method provides a useful tool for 'undisturbed' measurements on the influence of feedstuffs and nutritional management practices on CH_4 production from animals and for developing improved management practices for enhanced environmental quality (Harper et al. 1999).

McGinn et al. (2006) used a backward Lagrangian Stochastic (bLS) dispersion technique where plume gas concentrations are measured several hundred meters downwind of a dairy farm. The bLS dispersion technique is a useful approach for measuring whole farm emissions (e.g., dairies and feedlots) and emissions from well-defined point sources within the farm. In addition, this technique is useful in evaluating mitigation strategies, is nonintrusive, less labor-intensive, and is much easier to implement than most techniques. Furthermore, using the bLS dispersion technique in conjunction with global positioning system (GPS) may have application for monitoring enteric CH_4 emissions from grazing cattle herds in large pad docks (McGinn et al. 2009).

16.9 Mitigation Strategies for Reducing CH_4 Production

Any reduction strategies must be confined to the following general framework viz., development priority, product demand, infrastructure, livestock resource, and local resources (Sejian et al. 2011a). The most attractive emissions mitigation projects must balance the needs in all of these areas, so that, no one factor creates a constraint on continued improvement in production efficiency, and the resulting CH_4 emissions reductions. Within this framework, CH_4 emissions mitigation options for enteric fermentation can encompass a wide range of activities across these areas (Sejian and Indu 2011). However, underlying these activities must be specific options for improving the production efficiency of the livestock. Without these options, CH_4 emissions cannot be reduced. A range of reduction strategies to reduce enteric methane production outlined in Fig. 16.4 are briefly discussed below.

16.9.1 Reduction of Livestock Numbers

Reducing livestock numbers may be the best possible way for countries that hold large livestock population and are committed to reducing GHG emission from animal sources. Nevertheless, countries, which are heavily dependent on their livestock industries for generation of national income, imposition of regulations aimed at reducing livestock numbers would be economically unacceptable. Reducing livestock numbers through normal market processes can be effective. For example, in New Zealand, sheep farming has become less profitable since ~1990, and farmers have reduced sheep numbers and used the land for alternative enterprises, such as forestry (Ulyatt and Lassey 2001). Sheep populations have been reduced from 57.9 million in 1990 to 45.2 million in 2000, while dairy cattle and beef cattle population have increased slightly. The net outcome was a decline in ruminant CH_4 emission from 1.45 to 1.31 Tg/year from 1990 to 2000. This change in stock numbers, predominantly a reflection of the profitability of sheep farming, implies that New Zealand is in a position to meet its commitments to the UNFCCC. However, livestock population may respond positively to improved economic conditions. However, if sheep farming become more profitable an increase in stock numbers and thus CH_4 emission become most probable.

16.9.2 Management Strategies

Adoption of the basic livestock management principles offers the best opportunity for improving production efficiency while also reducing emissions (Sejian and Naqvi 2011b). Within the context of these livestock management principles, specific techniques for improving production efficiency and reducing CH_4

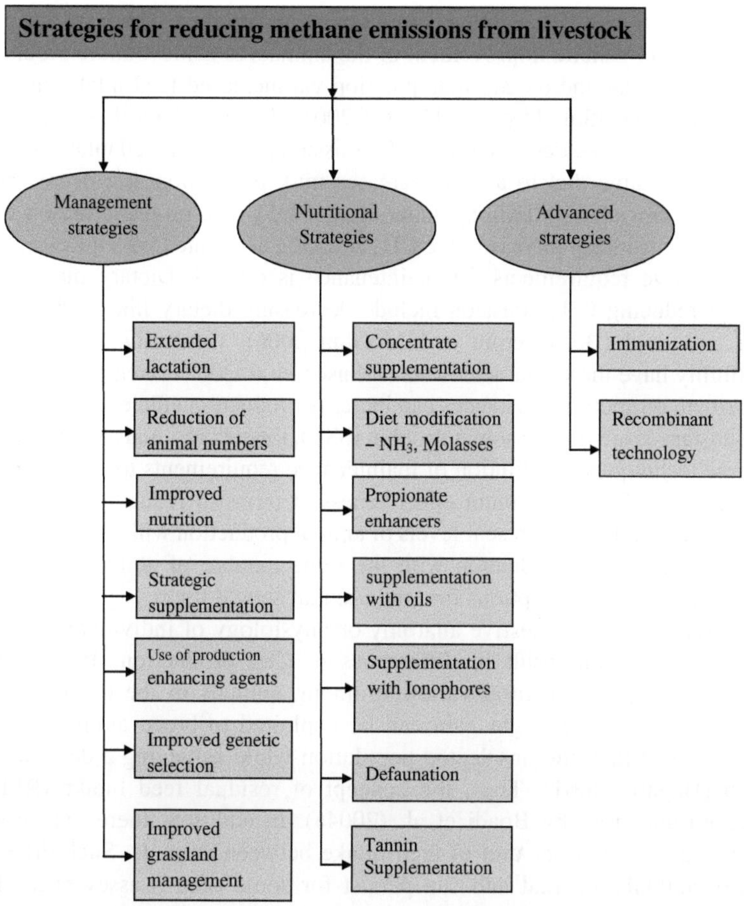

Fig. 16.4 Various strategies to reduce enteric methane production. These mitigation strategies can be grouped into three broader category-managemental, nutritional and advanced strategies (*Source* Sejian et al. 2011b)

emissions include the following: (1) enhanced nutrition through mechanical and chemical feed processing, (2) balanced nutrition through strategic supplementation, (3) intake of production enhancing agents, (4) improved production through enhanced genetic characteristics, (5) increased production efficiency through improved reproduction, and (6) improved grassland and rangeland management (USEPA 1993).

There is a wide range of management practices that improve animal productivity, with concomitant reduction in CH_4 emission per unit animal product (Naqvi and Sejian 2011). Genetic selection of animals that consume less feed or produce less CH_4 per unit of feed is a management strategy that may be used to reduce enteric CH_4 emissions. Improving the efficiency of ruminant animal performance will generally lead to a reduction of CH_4 emitted per unit of animal product. There are two aspects of this practice: genetic improvement of the animals to achieve more products per unit of feed intake and dietary manipulation via increased feed intake and appropriate feed composition (Ulyatt and Lassey 2001). These authors also suggested that increasing feed intake decreases the CH_4 emission per unit of feed intake. Increased intake of the same diet to a cow increases milk production, but decreases CH_4 emitted per unit of milk (Kirchgessner et al. 1995). The probable reason for this could be that as intake increases, the CH_4 emission associated with the essential but non-productive requirements for maintenance is diluted. Dietary manipulations aimed at reducing CH_4 emission include decreasing dietary fiber and increasing starch and lipid (Beauchemin and McGinn 2006). Generally, diets of higher digestibility have these characteristics (Johnson et al. 2002). Treatment of animals with growth promoting substances can increase production efficiency. Examples of such substances are bovine somatotrophin (bST), ionophores, and anabolic steroids. All these techniques use dilution of maintenance requirements to achieve reduced CH_4 emission. Their maximum effectiveness in terms of reducing CH_4 emission would be in maintaining present levels of animal production with fewer animals, or in increasing animal production with the same number of animals. This would provide the farmer with options for land use that should improve productivity.

Differences in the digestive anatomy or physiology of individual animals, or between breeds can result in differences in CH_4 production (Robertson and Waghorn 2002). The natural variation among animals in the quantity of feed eaten per unit of liveweight gain can be exploited to breed animals that consume less feed than the unselected population while achieving a desired rate of growth (Hegarty 2001). Thus, the concept of residual feed intake (RFI) was developed and used by Boadi et al. (2004a). In addition, there are large differences in emission per unit of feed intake between animals. Such differences between animals are real and can persist for some time (Lassey et al. 1997). Implementing proper grazing management practises to improve the quality of pastures increases animal productivity and has a significant effect on reducing CH_4 emission from fermentation in the rumen. Management-intensive grazing (MIG) is an effective form of grazing BMPs. As a result, the CH_4 emissions per unit of product as well as total emissions into the atmosphere are reduced (DeRamus et al. 2003; Naqvi and Sejian 2011).

There are many potential opportunities for mitigating CH_4 through microbial intervention in the rumen, such as: targeting methanogens with antibiotics, bacteriocins, or phage; removing protozoa from the rumen; development of alternative sinks for H_2 such as reductive acetogenesis (Ulyatt and Lassey 2001). There is considerable evidence that the rumen can function satisfactorily in the absence of methanogens (Joblin 1999). Given that CH_4 results from microbial activity, the animal can only have an impact on this variation through interactions with the microbes directly, through the diet selected, or through control of the fermenter (rumen) conditions (Ulyatt and Lassey 2001). Animal effects on fermentation could include saliva, feed processing, or flow rate through the rumen. It is possible that the animal impact on fermentation is genetically determined and if this is the case, it may be possible to create market opportunities that can be used to select low CH_4 emitters. Several CH_4 oxidizing bacteria have been isolated from the rumen and when added to the rumen fluid in vitro, decrease CH_4 accumulation in the rumen (Boadi et al. 2004a). In the long term, however, CH_4 oxidizers from gut sources could be screened for their activity in the rumen to reduce the proportion of CH_4.

16.9.3 Nutritional Strategies

Diet modification is one way in which the cattle industry can reduce its contribution to GHG emissions (Beauchemin and McGinn 2006; Sejian et al. 2011c). It has been accepted generally that digestibility is maximized when optimum levels of NH_3 is maintained in the rumen (Bowman et al. 1992). Supplying NH_3 can, therefore, greatly increase digestive efficiency and utilization of available energy. NH_3 can be supplied via urea, chicken manure, or soluble protein that degrades in the rumen. In addition to NH_3, there are some nutrients that must be present in the diet to support the microbe population in the rumen. The most common nutrients required are sulfur (S) and phosphorus (P), although this may vary greatly with region. In addition, molasses provides the energy needed to realize the improved microbial growth that can result from enhanced ammonia levels. Molasses nutrient blocks (MNBs) have been used as a supplement in many countries with encouraging results of increased milk yield by 20–30%; increased growth rate of 80–200% and increased reproductive efficiency. Based on these results, CH_4 emissions per unit product are expected to be reduced by up to 40% (Leng 1991). Bowman et al. (1992) reported that strategic supplementation of dairy animals with MNBs may reduce CH_4 emissions by 25%.

Improving the nutritive value of the feed given to grazing animals by balancing the diet with concentrates, or by breeding improved pasture plants, may reduce CH_4 emission. The proportion of concentrate within the diet is negatively correlated with CH_4 emissions (Yan et al. 2000; Lovett et al. 2005). Increased concentrate supplementation decreases enteric CH_4 production, increases proportion of propionate in rumen VFA and therefore reduces hydrogen for CH_4 synthesis. Indeed increasing the digestibility of pasture for grazing ruminants is the most practical means of reducing their CH_4 emissions (Hegarty 1999). If animal numbers do not decrease in response to the improved productivity, however, then emissions from the sector may increase rather than decrease (Hegarty 2002). Waghorn et al. (2002) observed that the impact of condensed tannins on rumen methanogenesis is small, although significant. Furthermore, legumes reduce CH_4 emissions when fed at comparable intake levels (Beever et al. 1985).

16.9.4 Chemical Inhibitors of Methanogenesis

Several mechanisms influence the availability of H_2 in the rumen and subsequent production of enteric CH_4 emissions by cattle. Processes that yield propionate act as net proton-using reactions while those that yield acetate result in a net increase in protons. That is, the proportion of volatile fatty acids, specifically acetate, propionate; produced as a consequence of microbial fermentation in the rumen has a significant influence on CH_4 production. If precursors of propionate are added to the diet, they should reduce CH_4 production by removing some of the H_2 produced during ruminal fermentation (O'Mara 2004). Another mechanism by which CH_4 production may be reduced during the rumen fermentation process is through the provision of alternative hydrogen acceptors or sinks.

A wide range of chemicals are available that may reduce rumen methanogenesis (Johnson and Johnson 1995; Mathison 1997). Organic acids, such as malate, fumarate, citrate, succinate, etc., are propionate precursors. Experiments conducted in vitro (Martin and Streeter 1995; Ulyatt and Lassey 2001) and in vivo (Newbold et al. 2002) show that addition of organic acids to the diet reduces CH_4 production, in a dose dependent manner (Martin and Streeter 1995). Thus, a feed additive or ingredient that reduces CH_4 emissions from cattle fed high-forage diets could have an important impact on reducing the emissions from the livestock sector (Beauchemin and McGinn 2006). Adding 4.6% of canola oil, a source of unsaturated fat, to a high-forage diet is an effective suppressant of CH_4, with daily emissions decreased by 32% and emissions as a percentage of gross energy intake decreased by 21%. McGinn et al. (2004) also reported reduced CH_4 production after supplementing the feed with sunflower oil and monensin. Halogenated CH_4 analogs, such as, chloroform, CCl_4, chloral hydrate, bromochloromethane, and bromoethanesulphonic acid are potent CH_4 inhibitors (McCrabb et al. 1997).

There is an increasing interest in exploiting natural products as feed additives to manipulate enteric fermentation and possibly reduce CH_4 emissions from livestock (Wenk 2003). Essential oils can interact with microbial cell membranes and inhibit the growth of some gram-positive and gram-negative bacteria. As a result of such inhibition, the addition of some plant extracts to the rumen results in an inhibition of deamination and methanogenesis, resulting in lower ammonia N, CH_4, and acetate, and in higher propionate and butyrate concentrations (Boadi et al. 2004b; Calsamiglia et al. 2007). Data from in vitro and in vivo experiments involving essential oil indicate that these are antimethanogenic agents leading to reduction in CH_4 production (Busquet et al. 2006; Castillejos et al. 2006). Supply of lipids from linseed (*Linum usitatissimum*) significantly decreased the amount of CH_4 emitted by dairy cows. Linseed fatty acid offers a promising dietary means to depress ruminal methanogenesis. The form of linseed fatty acid greatly influences CH_4 output from dairy cows (Martin et al. 2008). Polyunsaturated fatty acid decreases CH_4 through a toxic effect on microorganisms involved in fiber digestion and hydrogen production, such as protozoa (Doreau and Ferlay 1995) and cellulolytic bacteria (Nagaraja et al. 1997).

Ionophores, such as monensin and lasalocid also reduce CH_4 emission (Johnson and Johnson 1995). In the rumen, ionophores increase the proportion of gram-positive bacteria, resulting in a shift in fermentation acids from acetate and butyrate to propionate, consequently CH_4 production is reduced (NRC 2001). However, the rumen microbes adapt to the additives within two weeks (Saa et al. 1993; O'Mara 2004). Monensin supplementation can reduce CH_4 production by 25% (Tedeschi et al. 2003).

Defaunating agents, like manoxol, teric, alkanate $3SL_3$ and sulphosuccinate can reduce CH_4 emission (Mathison et al. 1998). They act by disrupting the close symbiotic relationship between methanogenic bacteria and protozoa. One method by which defaunation can be brought about is the addition of certain oils/fats (Machmuller et al. 1998; Hegarty 1999). The magnitude of reduction in CH_4 output following dietary supplementation of fats/oils is source-dependent, with coconut (*Cocos nucifera*) oil identified as being very effective (Lovett et al. 2005; Jordan et al. 2004). However, the use of defaunation to decrease CH_4 production from ruminants would have to be balanced against the effects on fiber and protein metabolism in the rumen (Gworgwor et al. 2006).

The main problems with chemical additives are the following: (i) some are toxic to the animal and rumen microflora thereby reducing digestion and food intake, (ii) some have short-lived effects, because the rumen microbes adapt, they are mainly volatile hence, difficult to administer, and (iii) some are expensive, and may not meet consumer product acceptance.

16.9.5 Immunization

The most noteworthy achievement with regards to reducing CH_4 production in livestock is the development of a vaccine containing an antigen derived from methanogenic bacteria (Gworgwor et al. 2006) and an immunogenic preparation which reduces the activity of rumen protozoa (Baker et al. 1997). Such vaccines are potent in providing a cost-effective treatment for reducing CH_4 emission and enhancing animal production. In light of these, Baker (1995) and Shu et al. (1999) proposed that it may be possible to immunize ruminants against their own methanogens with associated decrease in CH_4 output and that such an approach could reduce the members of streptococci and lactobacilli in the rumen.

16.10 Conclusions

This chapter addresses enteric CH_4 production from livestock and its mitigation strategies. It describes factors influencing CH_4 production, prediction models, and estimation methodologies for identifying different strategies to reduce such emissions. There are urgent needs to understand the various factors affecting variability in enteric CH_4 production to decrease GHG emission inventories and to identify viable GHG reduction strategies. Considerable progress has been made in the design of simulation models for predicting CH_4 emissions and the latest integrated farm system models offer greater capacity to predict more accurately GHG emissions with the incorporation of climatic and management information. Although the reduction in GHG emissions from livestock industries is of high priority, strategies for reducing emissions should not reduce the economic viability of the livestock industry.

The reduction of enteric CH_4 emissions from livestock by the selection of more feed-efficient animals based on their estimated breeding value offers a novel way of reducing the CH_4 emissions without compromising animal growth rate. Decreasing feed losses to CH_4 emission in the cattle industry can represent an improvement in feed efficiency. Therefore, mitigating CH_4 losses from cattle have both environmental and economic benefits. Improved knowledge of quantitative nutrition provides powerful tools to develop concepts to undertake a wide range of problem-oriented research with the goal of curbing emissions by livestock farms. Many authors have recommended on-farm practises, such as genetic selection for production traits, feed testing, and ration balancing, using CH_4 inhibitors to reduce enteric CH_4 emissions with concomitant reduction in feed costs. Several other CH_4 reduction strategies are at various stages of investigation, such as the use of feed additives, ionophores, defaunation, and vaccination. Pending an improved understanding of the global impact of these mitigation strategies on livestock, there are grounds for optimism that in the medium term, more effective strategies may become available to supplement those presently available. Finally, more consideration must be given to total farm GHG emissions from enteric fermentation. As low-income countries and some fast-developing countries are expected to continue their rapid growth in animal production, it is essential that the growth of global livestock-related GHG emissions is restricted in the short term to reduce the effects of CH_4 on global temperatures.

16.11 The Scope of Future Research

There is new need for further concerted research on methane emission by livestock and its mitigation. For instance, there are several new and more advanced CH_4 mitigation options, including the addition of probiotics, acetogens, bacteriocins, archaeal viruses, organic acids, and plant extracts to the diet, as well as immunization and genetic selection of animals. Although these new strategies are promising, more research is needed for validation purposes and to assess in vivo their effectiveness in reducing CH_4 production in dairy animals. In addition, there is a need to improve the efficacy of current strategies both economically, for livestock production and increasing their capacity to limit emissions. Vaccine development against methanogens is promising and calls for concerted research efforts if this strategy is to become a reality. The development of biomarkers to identify low CH_4 emission animals or low CH_4-producing bacteria also merits further investigation. Future developments in the area of modeling must accompany any improved understating of the underlying rumen biology. Furthermore, the need to develop simpler and more accurate models compatible with current trends in computer technology cannot be overemphasized.

References

Axelsson J (1949) The amount of produced methane energy in the European metabolic experiments with adult cattle. Ann R Agric Coll Sweden 16:404–419

Baker SK (1995) Method of improving utilization of nutrients by ruminants or ruminant like animals. International patent, WO9511041, USA

Baker SK, Gnanasampanthan G, Purser DB, Hoskinson RM (1997) Immunogenic preparation and method for improving the productivity of ruminant animals. Patent Application WO 97/00086

Baldwin RL (1995) Modeling ruminant digestion and metabolism. Chapman and Hall, London

Beauchemin KA, McGinn SM (2005) Methane emissions from feedlot cattle fed barley or corn diets. J Anim Sci 83:653–661

Beauchemin KA, McGinn SM (2006) Methane emissions from beef cattle: effects of fumaric acid, essential oil, and canola oil. J Anim Sci 84:1489–1496

Beever DE, Thomson DJ, Ulyatt MJ, Camell SB, Spooner MC (1985) The digestion of fresh perennial ryegrass (Lolium perenne L. cv.Melle) and white clover (Trifolium repens L. cv. Blanca) by growing cattle fed indoors. Br J Nutr 54:763–775

Benchaar C, Rivest J, Pomar C, Chiquette J (1998) Prediction of methane production from dairy cows using existing mechanistic models and regression equations. J Anim Sci 76:617–627

Blaxter KL, Clapperton JL (1965) Prediction of the amount of methane produced by ruminants. Br J Nutr 19:511–522

Boadi D, Benchaar C, Chiquette J, Massé D (2004a) Mitigation strategies to reduce enteric methane emissions from dairy cows: update review. Can J Anim Sci 84:319–335

Boadi D, Wittenburg KM, Scott SL, Burton D, Buckley K, Small JA, Ominski KH (2004b) Effect of low and high forage diets on enteric and manure pack GHG emissions from a feedlot. Can J Anim Sci 84:445

Bowman RL, Croucher JC, Picard MT (1992) Assessment of the prefeasibility of strategic supplementation as an opportunity for reducing methane emissions in Gujarat, India, A.T, International, prepared for the Global Change Division, U.S. Environmental protection agency, Washington DC

Bratzler JW, Forbes EB (1940) The estimation of methane production by cattle. J Nutr 19(6):611–613

Breas O, Guillou C, Reniero F, Wada E (2001) The global methane cycle: isotopes and mixing ratios, sources and sinks. Isot Environ Health Stud 37(4):257–379

Bright KP, Sherlock RA, Lile J, Wastney ME (2000) Development and use of a whole farm model for dairying, applied complexity: from neural nets to managed landscapes. NZ Institute for Crop and Food Research, Christchurch, pp 382–389

Busquet M, Calsamiglia S, Ferret A, Kamel C (2006) Plant extracts affect in vitro rumen microbial fermentation. J Dairy Sci 89:761–771

Calsamiglia S, Busquet M, Cardozo PW, Castillejos L, Ferret A (2007) Invited review: essential oils as modifiers of rumen microbial fermentation. J Dairy Sci 90:2580–2595

Castillejos L, Calsamiglia S, Ferret A (2006) Effect of essential oils active compounds on rumen microbial fermentation and nutrient flow in vitro systems. J Dairy Sci 89:2649–2658

Chhabra A, Manjunath KR, Panigrahy S, Parihar JS (2009) Spatial pattern of methane emissions from Indian livestock. Curr Sci 96(5):683–689

Chianese DS, Rotz CA, Richard TL (2009) Whole farm GHG emissions: a review with application to a Pennsylvania dairy farm. Appl Eng Agric 25(3):431–442

Christophersen CT, Wright ADG, Vercoe PE (2008) In vitro methane emission and acetate: propionate ratio are decreased when artificial stimulation of the rumen wall is combined with increasing grain diets in sheep. J Anim Sci 86:384–389

DeRamus HA, Clement TC, Giampola DD, Dickison PC (2003) Methane emissions of beef cattle on forages: efficiency of grazing management systems. J Environ Qual 32:269–277

Dijkstra J, Neal HD, Beever DE, France J (1992) Simulation of nutrient digestion, absorption and outflow in the rumen: model description. J Nutr 122:2239–2256

Doreau M, Ferlay A (1995) Effect of dietary lipids on nitrogen metabolism in the rumen: a review. Livest Prod Sci 43:97–110

Ellis JL, Kebreab E, Odongo NE, McBride BW, Okine EK, France J (2007) Prediction of methane production from dairy and beef cattle. J Dairy Sci 90:3456–3467

Ellis JL, Bannink A, France J, Kebreabz E, Dijkstra J (2010) Evaluation of enteric methane prediction equations for dairy cows used in whole farm models. Glob Change Biol. doi:10.1111/j.1365-2486.2010.02188.x

Erfle JD, Boila RJ, Teather RM, Mahadevan S, Sauer FD (1982) Effect of pH on fermentation and protein degradation by rumen microorganism in vitro. J Dairy Sci 65:1457–1464

Eshel G, Martin PA (2006) Diet, energy and global warming. Earth Interact 10:1–17

Food and Agriculture Organization of the United Nations (2006) Livestock a major threat to the environment: remedies urgently needed. Available: http://www.fao.org/newsroom/en/news/2006/1000448/index.html

Food and Agriculture Organization of the United Nations (2008) Accessed 24 Mar 2008. FAOSTAT

Food and Agriculture Organization of the United Nations (2009) Submission to UNFCCC AWG LCA, enabling agriculture to contribute to climate change. Available: http://unfccc.int/resource/docs/2008/smsn/igo/036.pdf).

Flesch TK, Wilson JD, Harper LA, Crenna BP, Sharpe RR (2004) Deducing ground-to-air emissions from observed trace gas concentrations: a field trial. J Appl Meteorol 43:487–502

Foley PA, Kenny DA, Callan JJ, Boland TM, O'Mara FP (2009) Effect of DL-malic acid supplementation on feed intake, methane emission, and rumen fermentation in beef cattle. J Anim Sci 87:1048–1057

Forster P, Ramaswamy V, Artaxo P, Berntsen T, Betts R, Fahey DW, Haywood J, Lean J, Lowe DC, Myhre G, Nganga J, Prinn R, Raga GMS, Van Dorland R (2007) Changes in atmospheric constituents and in radiative forcing. In: Solomon S et al (eds) Climate change 2007: the physical science basis. Contribution of working group I to the fourth assessment report of the intergovernmental panel on climate change. Cambridge University Press, Cambridge

Giger-Reverdin S, Vermorel M, Sauvant D (1992) Factors of variation of methane energy-loss in ruminants given compound feeds. Ann Zootech 41:37–38

Gleik PH, Adams RM, Amasino RM (2010) Climate change and the integrity of science. Science 328:689–691

Gold M, Porritt J (2004) The global benefits of eating less meat. A report for compassion in world farming trust, CIWF Trust, Petersfield, Hampshire, GU32 3EH, UK pp 1–76

Grainger C, Clarke T, McGinn SM, Auldist MJ, Beauchemin KA, Hannah MC, Waghorn GC, Clark H, Eckard RJ (2007) Methane emissions from dairy cows measured using the sulfur hexafluoride (SF_6) tracer and chamber techniques. J Dairy Sci 90:2755–2766

Gworgwor ZA, Mbahi TF, Yakubu B (2006) Environmental implications of methane production by ruminants: a review. J Sustain Dev Agr Env 2(1):1–14

Harper LA, Denmead OT, Freney JR, Byers FM (1999) Direct measurements of methane emissions from grazing and feedlot cattle. J Anim Sci 77:1392–1401

Hegarty RS (1999) Reducing rumen methane emissions through elimination of rumen protozoa. Aust J Agric Res 50:1321–1327

Hegarty RS (2001) Strategies for mitigating methane emissions from livestock—Australian options and opportunities. In: Proceeding of the 1st international conference on GHGes and animal agriculture. Obihiro, Hokkaido, pp 31–34

Hegarty RS (2002) Strategies for mitigating methane emissions from livestock—Australian options and opportunities. In: Takahashi J, Young BA (eds) GHGes and animal agriculture. Elsevier Science

Hegarty RS, Goopy JP, Herd RM, McCorkell B (2007) Cattle selected for lower residual feed intake have reduced daily methane production. J Anim Sci 85:1479–1486

Hristov AN, Ivan M, Rode LM, McAllister TA (2001) Fermentation characteristics and ruminal ciliate protozoal populations in cattle fed medium- or high-concentrate barley-based diets. J Anim Sci 79:515–524

Intergovernmental Panel on Climate Change (1997) Guidelines for national GHG inventories: workbook and reference manual. Agriculture (Chapter 4) and Waste (Chapter 6). OECD, Paris

Intergovernmental Panel on Climate Change (2001) Climate change 2001: the scientific basis. Cambridge University Press, Cambridge

Intergovernmental Panel on Climate Change (2006) IPCC guidelines for national GHG inventories. IGES, Hayama. http://www.ipccnggip.iges.or.jp

Intergovernmental Panel on Climate Change (2007) Climate change: synthesis report; summary for policymakers. Available: http://www.ipcc.ch/pdf/assessment-report/ar4/syr/ar4_syr_spm.pdf

Jentsch W, Schweigel M, Weissbach F, Scholze H, Pitroff W, Derno M (2007) Methane production in cattle calculated by the nutrient composition of the diet. Arch Anim Nutr 61(1):10–19

Joblin KN (1999) Ruminal acetogens and their potential to lower ruminant methane emissions. Aust J Agric Res 50:1307–1313

Johnson KA, Johnson DE (1995) Methane emissions from cattle. J Anim Sci 73:2483–2492

Johnson DE, Ward GW, Ramsey JJ (1996) Livestock methane: current emissions and mitigation potential. Lewis Publishers, New York

Johnson DE, Phetteplace HW, Seidl AF (2002) Methane, nitrous oxide and carbon dioxide emissions from ruminant livestock production systems. In: Takahashi J, Young BA (eds) GHGes and animal agriculture. Proceeding of the 1st international conference on GHGes and animal agriculture, Obihiro, November 2001, pp 77–85

Jordan E, Lovett DK, Hawkins M, O'Mara FP (2004) The effect of varying levels of coconut oil on methane output from continental cross beef heifers. In: Proceeding of GHG emissions from agriculture—mitigation options and strategies, Leipzig

Kebreab E, Mills JAN, Crompton LA, Bannink A, Dijkstra J, Gerrits WJJ, France J (2004) An integrated mathematical model to evaluate nutrient partition in dairy cattle between animal and environment. Anim Feed Sci Technol 112:131–154

Kebreab E, Clarke K, Wagner-Riddle C, France J (2006) Methane and nitrous oxide emissions from Canadian animal agriculture—a review. Can J Anim Sci 86:135–158

Kebreab E, Johnson KA, Archibeque SL, Pape D, Wirth T (2008) Model for estimating enteric methane emissions from United States dairy and feedlot cattle. J Anim Sci 86:2738–2748

Kirchgessner M, Windisch W, Muller HL (1995) Nutritional factors for the quantification of methane production. In: Von Engelhardt W, Leonhard-Marek S, Breves G, Giesecke D (eds) Ruminant physiology: digestion, metabolism, growth and reproduction proceedings of the eighth international symposium on ruminant physiology. Stuttgart Ferdinand Enke Verlag, pp 333–348

Kirchgessner M, Windisch HJ, Muller HL, Kreuzer M (1991) Release of methane and of carbon-dioxide by dairy-cattle. Agribiol Res Agrarbiol Agrikulturchemie Okol 44:91–102

Kriss M (1930) Quantitative relations of the dry matter of the food consumed, the heat production, the gaseous outgo and the insensible loss in body weight of cattle. J Agric Res 40:283–295

Kumaraswamy S, Rath AK, Ramakrishnan B, Sethunathan N (2000) Wetland rice soils as sources and sinks of methane: a review and prospects for research. Biol Fertil Soils 31:449–461

Lana RP, Russell JB, Van Amburgh ME (1998) The role of pH in regulating ruminal methane and ammonia production. J Anim Sci 76:2190–2196

Lassey KR, Ulyatt MJ, Martin JR, Walker CF, Shelton DI (1997) Methane emissions measured directly from grazing livestock in New Zealand. Atmos Environ 31:2905–2914

Lassey KR (2007) Livestock methane emission: from the individual grazing animal through national inventories to the global methane cycle. Agric For Meteorol 142:120–132

Lassey KR (2008) Livestock methane emission and its perspective in the global methane cycle. Aust J Exp Agric 48:114–118

Laubach J, Kelliher FM (2004) Measuring methane emission rates of a dairy cow herd by two micrometeorological techniques. Agric For Meteorol 125:279–303

Leng RA (1991) Improving ruminant production and reducing methane emissions from ruminants by strategic supplementation, United States Environmental Protection Agency, Office of Air and Radiation, Washington DC, p 105

Leng RA (1993) Quantitative ruminant nutrition—a green house science. Aust J Agri Res 44:363–380

Lovett DK, Stack LJ, Lovell S, Callan J, Flynn B, Hawkins M, O'Mara FP (2005) Manipulating enteric methane emissions and animal performance of late-lactation dairy cows through concentrate supplementation at pasture. J Dairy Sci 88:2836–2842

Machmuller A, Ossowski DA, Wanner M, Kreuzer M (1998) Potential of various fatty feeds to reduce methane release from rumen fermentation in vitro (Rusitec). Anim Feed Sci Technol 77:117–130

Martin C, Rouel J, Jouany JP, Doreau M, Chilliard Y (2008) Methane output and diet digestibility in response to feeding dairy cows crude linseed, extruded linseed, or linseed oil. J Anim Sci 86:2642–2650

Martin SA, Streeter MN (1995) Effect of malate on in vitro mixed ruminal microorganism fermentation. J Anim Sci 73:2141–2145

Mathison G (1997) Effect of canola oil on methane production in steers. Can J Anim Sci 77:545–546

Mathison GW, Okine EK, McAllister TA, Dong Y, Galbraith J, Dmytruk OIN (1998) Reducing methane emission from ruminant animals. J Appl Anim Res 14:1–28

Meadows DH, Meadows DL, Randers J (1992) Beyond the limits. Earthscan Publications Ltd, London

McCrabb GJ, Berger KT, Magner T, May C, Hunte RA (1997) Inhibiting methane production in Brahman cattle by dietary supplementation with a novel compound and the effects on growth. Austr J Agric Res 48:323–329

McGinn SM, Beauchemin KA, Coates T, Colombatto D (2004) Methane emissions from beef cattle: effects of monensin, sunflower oil, enzymes, yeast, and fumaric acid. J Anim Sci 82:3346–3356

McGinn SM, Flesch TK, Harper LA, Beauchemin KA (2006) An approach for measuring methane emissions from whole farms. J Environ Qual 35:14–20

McGinn SM, Beauchemin KA, Flesch TK, Coates T (2009) Performance of a dispersion model to estimate methane loss from cattle in pens. J Environ Qual 38:1796–1802

McMichael AJ, Powles JW, Butler CD, Uauy R (2007) Food, livestock production, energy, climate change, and health. The Lancet 370(9594):1253–1263

Mills JAN, Kebreab E, Yates CM, Crompton LA, Cammell SB, Dhanoa MS, Agnew RE, France J (2003) Alternative approaches to predicting methane emissions from dairy cows. J Anim Sci 81:3141–3150

Moe PW, Tyrrell HF (1979) Methane production in dairy cows. J Dairy Sci 62:1583–1586

Mosier A, Wassmann R, Verchot L, King J, Palm C (2004) Methane and nitrogen oxide fluxes in tropical agricultural soils: sources, sinks and mechanisms. Environ Dev Sustain 6:11–49

Moss AR, Givens DI, Garnsworthy PC (1995) The effect of supplementing grass silage with barley on digestibility, in sacco degradability, rumen fermentation and methane production in sheep at two levels of intake. Anim Feed Sci Tech 55:9–33

Moss AR, Jounany JP, Neevbold J (2000) Methane production by ruminants: its contribution to global warming. Ann Zootech 49:231–253

Nagaraja TG, Newbold CJ, Van Nevel CJ, Demeyer DI (1997) Manipulation of rumen fermentation. In: Hobson PN, Stewart CS (eds) The rumen microbial ecosystem. Blackie Academic and Professional Press, London, pp 523–632

Naqvi SMK, Sejian V (2011) Managemental strategies to reduce enteric methane emission. In: Sejian V, Naqvi SMK, Bhatt RS, Karim SA (eds) NAIP sponsored national training manual on carbon sequestration, carbon trading and climate change., Division of physiology and biochemistryCentral Sheep and Wool Research Institute, Avikanagar, pp 248–256

National Research Council (2001) Nutrient requirements of dairy cattle, 7th revised edn. National Academy Press, Washington

Neil PG, Bright KP, Sherlock RA (1997) Integrating legacy subsystem components into an object-oriented model, MODSIM 97—Proceedings of the international conference on modelling and simulation, modelling and simulation society of Australia, Hobart, pp 1133–1138

Newbold CJ, Ouda JO, Lopez S, Nelson N, Omed H, Wallace RJ, Moss AR (2002) Propionate precursors as possible alternative electron acceptors to menthane in ruminal fermentation. In: Takahashi J, Young BA (eds) GHGes and animal agriculture. Proceeding of the 1st international conference on GHGes and animal agriculture, Obihiro, November 2001, pp 151–154

Ogino A, Orito H, Shimada K, Hirooka H (2007) Evaluating environmental impacts of the Japanese beef cow–calf system by the life cycle assessment method. Anim Sci J 78(4):424–432

O'Mara F (2004) GHG production from dairying: reducing methane production. Adv Dairy Technol 16:295–309

Palliser CC, Woodward SL (2002) Using models to predict methane reduction in pasture fed dairy cows. Proceedings integrating management and decision support. Coordinated by Susan M. Cuddy (CSIRO, Australia) 482(1):162–167

Pinares-Patiño CS, Hour PD, Jouany JP, Martin C (2007) Effects of stocking rate on methane and carbon dioxide emissions from grazing cattle. Agric Ecosyst Environ 121:30–46

Pinares-Patiño CS, Machmüller A, Molano G, Smith A, Vlaming JB, Clark H (2008) The SF_6 tracer technique for measurements of methane emission from cattle—effects of tracer permeation rate. Can J Anim Sci 88:309–320

Puchala R, Min BR, Goetsch AL, Sahlu T (2005) The effect of a condensed tannin-containing forage on methane emission by goats. J Anim Sci 83:182–186

Robertson LJ, Waghorn GC (2002) Dairy industry perspectives on methane emissions and production from cattle fed pasture or total mixed rations in New Zealand. Proc NZ Soc Anim Prod 62:213–218

Rotz CA, Corson MS, Coiner CU (2007) Integrated farm system model: reference manual. Pasture systems and watershed management research unit, USDA Agricultural Research Service. University Park, PA. Available: http://www.ars.usda.gov/Main/docs.html?docid=8519

Rotz CA, Corson MS, Chianese DS, Coiner CU (2009) The integrated farm system model: reference manual. USDA-ARS Pasture Systems and Watershed Management Research Unit. University Park, PA. Available: www.ars.usda.gov/SP2UserFiles/Place/19020000/ifsmreference

Rotz CA, Montes F, Chianese DS (2010) The carbon footprint of dairy production systems through partial life cycle assessment. J Dairy Sci 93(3):1266–1282

Saa CF, Hill TM, Johnson DE (1993) Persistance of methane suppression by ionophore and glycopeptide in steers fed a brome hay diet. CSU Beef Programe Report

Sejian V (2012) Climate change, green house gas emission and sheep production. In: Shinde AK, Swarnkar CP, Prince LLL (eds) Status papers on future research in sheep production and production development, 50 years research contributions (1962–2012). Director, Central Sheep and Wool Research Institute, Avikanagar, pp 155–178

Sejian V, Saumya B (2011) Enteric methane emissions in livestock: contributors, prediction, estimations and repercussion. In: Sejian V, Naqvi SMK, Bhatt RS, Karim SA (eds) NAIP sponsored national training manual on "Carbon sequestration, carbon trading and climate change"., Division of physiology and biochemistryCentral Sheep and Wool Research Institute, Avikanagar, pp 68–80

Sejian V, Rajni C (2011) Modeling of green house gases in livestock farms. In: Sejian V, Naqvi SMK, Bhatt RS, Karim SA (eds) NAIP sponsored national training manual on "Carbon sequestration, carbon trading and climate change"., Division of physiology and biochemistryCentral Sheep and Wool Research Institute, Avikanagar, pp 178–190

Sejian V, Singh AK (2011) In vitro gas production model (bioreactor model) and dairy GHG model to predict GHG emission from livestock farm. In: Sejian V, Naqvi SMK, Bhatt RS, Karim SA (eds) NAIP sponsored national training manual on "Carbon sequestration, carbon trading and climate change"., Division of physiology and biochemistryCentral Sheep and Wool Research Institute, Avikanagar, pp 295–301

Sejian V, Naqvi SMK (2011a) Enteric methane emissions. In: Training manual on NAIP national training on "Climate change carbon sequestration and carbon credits" at Indian Institute of Soil Science (ICAR), Nabi Bagh, Bhopal, pp 73–89

Sejian V, Naqvi SMK (2011b) Mitigation strategies to reduce methane production from livestock. In: Training manual on NAIP national training on "Climate change carbon sequestration and carbon credits" at Indian Institute of Soil Science (ICAR), Nabi Bagh, Bhopal, pp 90–105

Sejian V, Indu S (2011) Salient mitigation strategies to reduce enteric methane emission from livestock. In: Sejian V, Naqvi SMK, Bhatt RS, Karim SA (eds) NAIP sponsored national training manual on "Carbon sequestration, carbon trading and climate change"., Division of physiology and biochemistryCentral Sheep and Wool Research Institute, Avikanagar, pp 118–131

Sejian V, Naqvi SMK, Bhatt RS, Karim SA (2011a) Carbon sequestration, carbon trading and climate change. Director, Central Sheep and Wool Research Institute, Avikanagar, pp 1–324

Sejian V, Lal R, Lakritz J, Ezeji T (2011b) Measurement and prediction of enteric methane emission. Inter J Biometeorol 55:1–16

Sejian V, Rotz A, Lakritz J, Ezeji T, Lal R (2011c) Modeling of green house gas emissions in dairy farms. J Anim Sci Adv 1(1):12–20

Sejian V, Lakritz J, Ezeji T, Lal R (2011d) Forage and flax seed impact on enteric methane emission in dairy cows. Res J Vet Sci 4(1):1–8

Sherlock RA, Bright KP, Neil PG (1997) An object-oriented simulation model of a complete pastoral dairy farm, MODSIM 97—Proceedings of the international conference on modelling and simulation, Modelling and Simulation Society of Australia, Hobart, pp 1154–1159

Sherlock RA, Bright KP (1999) An object-oriented framework for farm system simulation, MODSIM 99—Proceedings of the international conference on modelling and simulation, Modelling and Simulation Society of Australia and New Zealand, Hamilton, pp 783–788

Shu Q, Gill HS, Hennessy DW, Leng RA, Bird SH, Rowe JB (1999) Immunisation against lactic acidosis in cattle. Res Vet Sci 67:65–71

Steinfeld H, Gerber P, Wassenaar T, Castel V, Rosales M, de Haan C (2006) Livestock's long shadow: environmental issues and options. Food and Agriculture Organization of the United Nations, Rome

Tamminga S (1996) A review on environmental impacts of nutritional strategies in ruminants. J Anim Sci 74:3112–3124

Tedeschi LO, Fox DG, Tylutki TP (2003) Potential environmental benefits of ionophores in ruminant diets. J Environ Qual 32:1591–1602

Ulyatt MJ, Lassey KR (2001) Methane emissions from pastoral systems: the situation in New Zealand. Arch Latinoam Prod Anim 9(1):118–126

United States Environmental Protection Agency (1993) Anthropogenic methane emissions in the United States. Estimates for 1990: report to congress. Atmospheric Pollustatestion Prevention Division, Office of Air and Radiation, Washington DC

United States Environmental Protection Agency (1994) International anthropogenic methane emissions: estimates for 1990, office of policy planning and evaluation, Washington DC, EPA 230-R-93-010.94

Vlaming JB, Brookes IM, Hoskin SO, Pinares-Patiño CS, Clark H (2007) The possible influence of intra-ruminal sulfur hexafluoride release rates on calculated methane emissions from cattle. Can J Anim Sci 87:269–275

Waghorn GC, Tavendale MH, Woodfield DR (2002) Methanogenesis from forages fed to sheep. Proc NZ Grassl Assoc 64:167–171

Wenk C (2003) Herbs and botanicals as feed additives in monogastric animals. Asian Austral J Anim Sci 16:282–289

World Health Organization (2009) Department of health statistics and health information systems, world health statistics

Wuebbles DJ, Hayhoe K (2002) Atmospheric methane and global change. Earth Sci Rev 57:117–210

Yan T, Agnew RE, Gordon FJ, Porter MG (2000) Prediction of methane energy output in dairy and beef cattle offered grass silage based diets. Livest Prod Sci 64:253–263

Yan T, Mayne S, Porter MG (2006) Effects of dietary and animal factors on methane production in dairy cows offered grass silage-based diets. Int Cong Ser 1293:123–126

Yang SS, Liu CM, Liu YL (2003) Estimation of methane and nitrous oxide emission from animal production sector in Taiwan during 1990–2000. Chemosphere 52:1381–1388

Yates CM, Cammell SB, France J, Beever DE (2000) Prediction of methane emissions from dairy cows using multiple regression analysis. Proc Br Soc Anim Sci 2000:94

Yoon IK, Stern MD (1995) Influence of direct-fed microbials on ruminal fermentation and performance of ruminants: a review. Asian Austral J Anim Sci 8:533–555

United States Environmental Protection Agency (1994) International anthropogenic methane emissions: estimates for 1990. Office of policy planning and evaluation, Washington DC, EPA 230-R-93-010

[illegible]

Part V
Research and Development Priorities

Chapter 17
Conclusions and Researchable Priorities

Veerasamy Sejian

Abstract This chapter provides the salient findings established by various researchers in their field of specialization and also elaborates on the future research priorities that are available before the researchers in the field of environmental stresses and livestock production. Among the numerous variables affecting livestock production, heat stress has been identified to be the cause of severe economic consequences. In view of changing climate, it is envisaged that apart from heat and nutritional stress the livestock are also subjected to locomotory stress as a result of long distance walking in search of pastures. The changing climatic conditions have contributed to the multiple stress phenomenons, particularly heat stress, nutritional stress, and walking stress which severely hampers livestock production. Different ameliorative strategies have been identified to counter such environmental extremes and the most noteworthy being the managemental and nutritional strategies. Further, the role of pineal gland in combating environmental stress has been established. Various aspects including genetic, neuroendocrine, cellular, and molecular mechanisms involved in livestock adaptation have also been dealt in detail. However, research priorities need to be set to further our understanding of adaptive strategies at cellular and molecular level in order to develop suitable adaptive and mitigation strategies to counter environmental stresses. The integration of new technologies into the research and technology transfer systems potentially offers many opportunities to take forward the development of climate change adaptation strategies.

Keywords Breeding · HSP · Management · Multiple stresses · Nutrition · Simulation model

V. Sejian (✉)
Adaptation Physiology Laboratory, Division of Physiology and Biochemistry,
Central Sheep and Wool Research Institute, Avikanagar, Jaipur,
Rajasthan 304501, India
e-mail: drsejian@gmail.com

V. Sejian et al. (eds.), *Environmental Stress and Amelioration in Livestock Production*, DOI: 10.1007/978-3-642-29205-7_17,

Contents

17.1 Conclusions

The introductory chapter provides the overall vision of the book. It presents an insight of the environmental stresses affecting livestock production and its amelioration. There are five chapters (from Chaps. 2 to 6) in the first part of this book. This part particularly addresses in detail the various environmental stresses influence on livestock production and reproduction. Chapter 2 addresses the range of factors which play a major role in determining the livestock production. Authors of this chapter have identified and grouped the factors influencing livestock production under three major categories: climate change (CC), health, and nutrition. Apart from these major factors, also are described the influences of livestock selection policy, and other anthropogenic factors such as religion, cultural beliefs, economic constraints, infrastructure, and marketing choices to play a significant role in affecting livestock production. The single most important factor that influences livestock production across the globe is the heat stress. With the changing climatic conditions worldwide, heat stress seems to be playing the most detrimental role in diminishing the economy of livestock owners. Chapter 3 presents in detail, the impact of heat stress on livestock production. High temperature, combined with high humidity and radiant energy, are the major factors that contributes to severity of heat stress.

Heat stress is the single largest factor affecting livestock production. It reduces the feed intake which is reflected in the lower body weight gain per unit of feed intake. Further, heat stress reduces milk production, egg production, and reproductive efficiency of livestock species leading to severe economic losses. Walking stress is another most important factor affecting livestock species reared particularly in the tropical and subtropical regions of the world. Chapter 4 addresses in detail the significant influence of walking stress on livestock production. Livestock reared in hot tropical environment during summer months need to walk long distances for grazing and water. The authors of Chap. 4 presented their experimental findings to prove the hypothesis that walking stress influences livestock performance by altering their adaptive capability so as to deviate most of their energy reserves in trying to adapt to the locomotory stress during walking. As a result, there is a severe energy crisis in these animals and hence the productive performance of these animals is at stake. These studies indicate that animals under walking stress need to be supplemented with more energy feed to compensate for the locomotory activity as well as to maintain production.

Chapter 5 targets environmental stress influence on livestock reproduction. Specifically, this chapter addresses the heat stress as the major environmental factor influencing livestock reproduction. Reproductive performance is one of the important productive parameters that are threatened by the heat stress in domestic livestock. Clearly, heat stress exhibits profound effects on most of the aspects of reproductive functions including male and female gamete formation and function, embryonic development, and fetal growth and development. The various reproductive processes in female such as follicular development, estrus activity, and embryonic development are adversely affected by heat stress. Similarly in male, heat stress has a deleterious effect on fertility through reduced sperm production and increasing the abnormal spermatozoa percentage in the ejaculates. The authors of Chap. 5 highlight that scarcity of feed resources during summer months is the principal contributor to reduced reproductive efficiency of livestock. This chapter identifies endocrine system involving hypothalamus, anterior pituitary, and gonads to play the leading role and contribute to the reduced reproductive efficiency. The effect of thermal environment on reproduction may be a direct action upon the reproductive tissues or be an indirect one due to lower nutrients intake, impairment of hypothalamic, pituitary, gonadal, and endometrial secretions. Further, this chapter emphasizes that glucocorticoids, the principle stress relieving hormones in livestock species, are paramount in mediating the inhibitory role to reproduction. The final chapter of part I (i.e., Chap. 6) addresses the concept of multiple stresses affecting livestock production. Authors of Chap. 6 conclude that, under the changing climatic condition, it is not only the heat stress that hampers the livestock production but also several other stress which occurs simultaneously in arid and semi-arid tropical environment. The changing climatic conditions have contributed to the multiple stress phenomenons, particularly heat stress, nutritional stress, and walking stress. Discussions in Chap. 6 vividly indicate that livestock subjected to heat, nutritional, and walking stress separately had less detrimental effects on their reproductive performance. In comparison with the heat stress, however, nutritional stress had less significant effect on the productive and reproductive performance of sheep. However, when both stresses are coupled, together these severely jeopardize important productive and reproductive parameters. The discussions in Chap. 6 support the conclusion that when nutrition is not a limiting factor, livestock species are able to better cope with heat stress. It is imperative, therefore, that researchers aiming to target improving livestock production under the changing climatic conditions should address the effects of multiple stresses simultaneously rather than heat stress alone.

Part II of this book deals with the managerial strategies to alleviate and mitigate environmental stresses. These strategies are presented in Chaps. 7–9, and cover this part with detailed emphasis on various strategies and options available to ameliorate environmental stresses which affect the livestock. Good management ideally seeks well-being, comfort, and high productive and reproductive efficiency maintenance of the animals. The management practices in hot climate should involve modification of the environment, reduction of the animal's heat production, and increasing its heat loss by dissipating the heat load. Alleviation of heat stress in animals can be achieved by physical, physiological, and nutritional techniques. Chapter 7 addresses a range of options available for physical modification of livestock's micro environment to specifically counter the heat stress. These strategies include proper livestock housing, different options for providing shade, using different cooling methods available, general feeding managements, and ideal water requirements to combat environmental stresses. This chapter also elaborates on the concept of cold diet during the hot climatic conditions. In addition, this chapter also provides information pertaining to advanced biotechnological strategies available to ameliorate the heat stress in livestock. Chapter 8 highlights the importance of optimum nutrition to counter environmental extremes. Under the changing climatic conditions, livestock need to be insulated against environmental stresses by providing optimum nutrition, proper managemental practices, and health care. Adverse environments can increase the nutritional requirements of animals directly, or they may reduce the supply of quality feed. Under these circumstances, a concerted effort must be made to harmonize the welfare of animals by reducing environmental stress of food animals by nutritional manipulation and managemental practices. A new theme of research, signifying the importance of pineal gland to combat heat stress in livestock is discussed in Chap. 9. Using a modeling technique, this chapter highlights the establishment of pineal-adrenal-immune system relationship to combat heat stress in goat. The authors of this chapter elaborate on their series of experimental findings and demonstrate the functional relationship among pineal, adrenal, and immune system, and how these relationships modulate stress and non-specific immune response for the well-being of animals (goats) under thermal stress. Such an empirical relationship was established by testing the hypothesis involving the intercommunication of these endocrine glands by their hormonal secretions. Given the significance of thermal stress hampering animal productivity, these findings could provide a useful mitigation measure in improving economy of farm households as well as poor farmers. Livestock being major source of food security need special attention during the era of changing climate. The capacity building is essential to sustain livestock productivity. Besides the other physical approaches to mitigate heat stress of livestock, physiological interventions may have a good promise which is to be explored and well-devised approaches need to be developed and practiced, though with full attention and cautions so as not to disturb the basic physiology of animal.

Part III of this book deals with the adaptation mechanisms of livestock to adverse environmental conditions. Disccusions in Chaps. 10–14 cover this section involving basic principles, neuro endocrine regulations, molecular mechanisms, genetic adaptability, and the various genes which are responsible for tolerance to thermal stresses. The mechanisms underlying the adaptive capability of livestock to different environmental stresses are complex in nature. The stress response initiated by a particular stressor integrates a variety of homeostatic mechanisms ranging from physiological, immune to behavioral adaptations. Neuroendocrine adaptations are important to conserve energy, promote thermogenesis, and help use energy sources effectively promoting long-term survival of the animal. As a result of deviation of this energy for survival, their production capacity is at stake. Several systems of the body respond differently in combating environmental stresses in livestock. There are genetic variations and breed differences in determining the adaptive capability of an individual. The immediate response by which the livestock counteract heat stress is the physiological response while endocrine response is the one which plays a major role in combating stress. At cellular level, eukaryotic cells respond to stresses by the production of a specific set of proteins called heat shock or stress proteins (HSPs). Upregulation of HSPs under elevated temperature corresponds with a concomitant downregulation of the transcription and translation of the genes involved in the growth, metabolism, and differentiation. These HSPs are stress proteins which are thought to help the cell to survive stress. Thus, animals acquire thermo-tolerance by exposure to a conditioning body heat dose and hyperthermia induces the synthesis of these stress proteins. These wide ranges of HSPs are the confirmatory physiological biomarkers for stress determination. Scientific advances have indicated specific molecular mechanisms and altered gene expression induced by heat stress, representing targets which can be exploited to improve breeding and management practices.

Promotion of sustainable agriculture and livestock rearing is vital to ensuring an effective adaptation to CC by rural communities. Important among these techniques are rearing of animals which are more sturdy, heat tolerant, disease resistant, and relatively adaptable to the adverse conditions. In such a situation, some of the indigenous breeds may be able to cope better than the crossbreds as crosses containing higher exotic inheritance have a lower survival compared to the indigenous breeds. This part of the book clearly emphasizes that using adapted animals can yield better results in terms of improving the economy of farm households than infusing 'heat tolerance' genes into a non-adapted breed.

Part IV of the book covers both impact of CC on livestock production and contribution of livestock to adapt to CC and its mitigation. Livestock are one sector which is important to consider as it both impacts and is influenced by the CC. Hence, it is crucial to address these issues to provide an ideal platform for improving livestock production under the changing climate scenario. The CC is projected to exacerbate weather-related disasters and extreme events, such as droughts, heat waves, storms, floods, desertification, and increases in incidence of pests and pathogens. The CC effects livestock productivity directly as well as obliquely through changes in the availability of fodder and pastures, and influences animal welfare and productivity. The CC affects livestock productivity by reducing the quality and quantity of feed, reducing the forage and pasture availability, increasing extent and severity of drought stress, and changing patterns of livestock diseases. Higher temperatures, potentially caused by CC, would reduce dairy production, decrease animal weight gain, adversely affect reproduction, and lower feed-conversion efficiency in warm regions.

Chapter 16 describes the role of livestock to CC, and addresses in detail enteric methane (CH_4) emission and mitigation strategies in ruminant livestock. CH_4 is one of the most important greenhouse gases (GHGs), and livestock enteric emission is one of the principal sources of CH_4. Mitigation of rumen CH_4 emission is essential and can be effectively achieved by strategies which improve the efficiency of animal production. Because the conditions under which animals are managed vary greatly among countries, CH_4 emission reduction strategies must be tailored to country-specific circumstances. The dietary/nutritional strategies, which improve productivity with no potential negative effects on livestock health and production and are cost-effective, have a better chance of being adopted. It is also important to evaluate CH_4 mitigation strategies in terms of the total GHG budget and to consider the economics of various strategies. Although reduction in GHG emissions from livestock industries are of a high priority, strategies for reducing emissions should not reduce the economic viability of enterprises. As low income countries and also some fast-developing economies are expected to achieve a rapid growth in animal production, it is essential that the growth of global livestock-related GHG emissions is restricted in the short term to reduce the radiative forcing of enteric and other livestock-related emissions.

17.2 Researchable Priorities

Responding to the challenges of global warming necessitates a paradigm shift in the practice of agriculture and in the role of livestock within the farming system. Science and technology are lacking in thematic issues, including those related to climatic adaptation, dissemination of new understandings in rangeland ecology, and a holistic understanding of pastoral resource management. Livestock farmers should have key roles in determining what adaptation and mitigation strategies they support if these have to sustain livestock production in changing climate. The integration of new technologies into the research and technology transfer systems potentially offers many opportunities for further development of CC adaptation strategies. Figure 17.1 describes the various future research areas to target for ameliorating environmental stresses in livestock.

17.2.1 Early Warning System

Future research must develop new techniques of cooling systems such as thermo-isolation, concentrating more than in the past on techniques requiring low energy

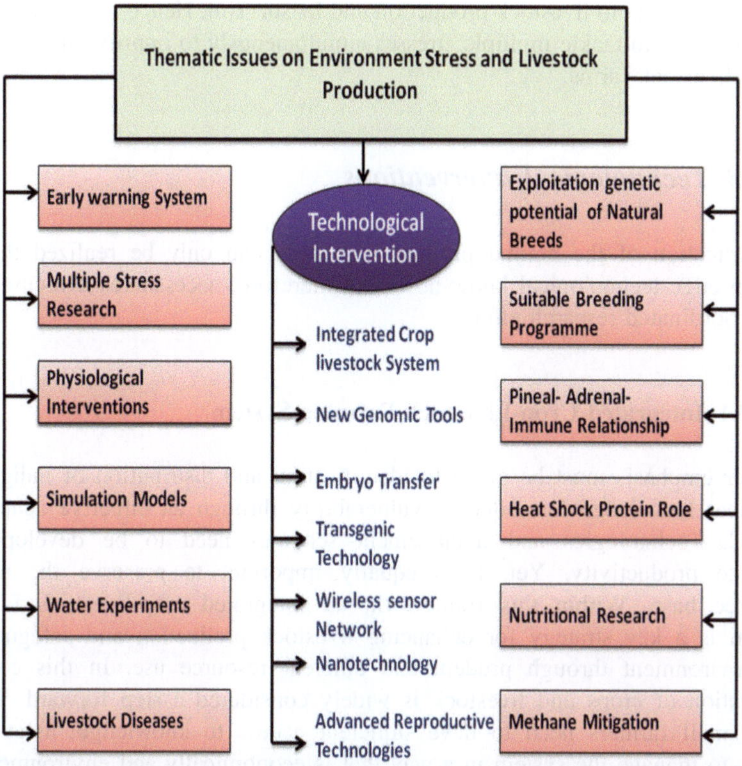

Fig. 17.1 Researchable priorities to target for ameliorating environmental stresses in livestock

expenditure. New indices that are more complete than temperature-humidity index (THI) to evaluate the climatic effects on each animal species must be developed and weather forecast reports must also be developed with these indices, to inform the farmers in advance. Above all to beat the CC or in any case not to let the climate beat livestock systems, researchers must be well aware of technologies that are available before them. Meteorological services need to be strengthened to provide timely weather and climate forecast/information to provide an early warning system which will help farmers to plan for the change well in advance.

17.2.2 Research on Multiple Stresses

Livestock researchers aiming at improving livestock production under changing climatic scenario must aim at counteracting multiple stresses as these stresses are of a common occurrence in most ecological zones where livestock are reared. Under the changing climatic scenario, the concept of multiple stresses emerges as a potential threat to livestock production and its survival. Hence research needs to be prioritized to tackle multiple stresses simultaneously to improve the economy of the livestock farms.

17.2.3 Technological Interventions

Intensification of the animal production system can only be realized through cutting edge technological know-how, and therefore, necessitates a focused and well-coordinated research effort.

17.2.3.1 Integrated Crop-Livestock Farming System

Greater emphasis must be given to identification and distribution of indigenous technical knowledge for reducing vulnerability through an effective adaptation to CC. Technologies and management schemes need to be developed to enhance productivity. Yet, it is equally important to preserve the natural resource base. Within this framework, an integrated crop-livestock farming system is a key strategy for enhancing livestock production and safeguarding the environment through prudent and efficient resource use. In this context, integration of crops and livestock is widely considered a step forward. Therefore, small farmers need to have sufficient access to knowledge, assets, and inputs to manage the system in a way that is economically and environmentally sustainable over the long term.

17.2.3.2 New Genomics Tools

Genomic tools are also useful in providing information on specific gene networks associated with thermotolerance. It is envisioned that, in the not too distant future, use of single nucleotide polymorphism (SNP) markers will lead to greater progress in improving thermotolerance of high producing dairy breeds. An important attempt that needs an immediate attention is the identification of genes involved in key regulatory/metabolic pathways for thermal resistance/sensitivity using genomic advancements such as whole genome arrays, RNA sequencing etc., in different native breeds of livestock. The elucidation of molecular basis of thermotolerance in native breeds and comparing with other livestock species would be key in developing strategies to minimize the effect of heat stress on productivity. Genome-wide association studies to detect panel of DNA markers associated with the sensitivity of production (milk/meat) and reproduction (semen quality/fertility) traits to heat stress and other environmental stresses is another priority. Research advances can facilitate selecting animals of high productive/reproductive capacity under anticipated CC scenarios. The identified SNPs can potentially be utilized as marker resource, to screen the variation in animal response (resistance/sensitivity) to heat stress and once validated can be utilized in the breeding programme. Such efforts can create opportunities to introgress desirable genes/alleles and drift herds toward superior tolerant ability against environmental stresses. In addition, systematic assessment and recording of phenotypes related to adaptive traits in response to different environmental stresses in breeds of livestock to undertake accurate phenotype-genotype association studies is vital for identifying the most appropriate site-specific breed.

17.2.3.3 Embryo Transfer

The embryo transfer can be an important future strategy for managing heat stress especially, if sources of affordable and high-quality embryos become available.

17.2.3.4 Transgenic Technology

The transgenic technology allows for the stable introduction of exogenous genetic information into livestock genomes. With its ability to enhance existing or introduce entirely novel characteristics at unprecedented magnitude and speed, the emerging technology can have a profound impact on the genetic improvement of livestock in the future. The continual advances in animal genomics toward the identification of genes that influence livestock production traits and impact human health can increase its ability and versatility for the purposeful modification of livestock to enhance their welfare, produce superior quality food and biomedical products, and reduce the environmental impact of animal husbandry.

17.2.3.5 Wireless Sensor Networks

Wireless sensors are enabling applications which can depict the full picture of the state of the animals and their environment on a continuous basis regarding parameters of animal health, animal behavior, and animal performance. This can be used for finding hidden patterns in animal behavior when exposed to environmental stresses, and this is valuable for developing decision support system.

17.2.3.6 Nanotechnology

The next few decades may well see nanotechnology applied in sustainable animal management. Nanosized, multipurpose sensors are already being developed that can report on the physiological status of animals. Nanoparticles may impact nutrient uptake and induce more efficient utilization of nutrients for milk production.

17.2.3.7 Advanced Reproductive Technologies

Applications of molecular technology will provide new ways to evaluate reproductive potential and the basic physiological mechanisms that limit reproductive performance. These technologies may also provide new tools for managing and monitoring livestock fertility. Animal reproductive biology must involve a multidisciplinary approach to address costly reproductive problems and improve reproductive efficiency. Understanding factors that control reproduction provides methodologies for improving reproduction. Research focus is needed to evolve new information on puberty, ovarian function and cycles, gamete formation and maturation, fertilization, establishment and maintenance of pregnancy, and placental function, including maternal-fetal interactions, fetal development and growth, and parturition. Some of the advanced reproductive technologies include controlling the sex of offspring through sperm separation, methods for early diagnosis of pregnancy, fundamental studies to determine molecular, cellular, and metabolic mechanisms regulating reproduction, new programs for efficient reproductive management and for identifying needed research through computer modeling, and embryonic stem cell technology to revolutionize the livestock breeding.

17.2.4 Physiological Interventions

The physiological mechanisms underlying the action of heat stress on the decline of production performance of domestic animals have not been fully investigated. They require further research for the elucidation of the mechanisms which may facilitate the adoption of comprehensive preventive and control measures of combating heat stress in domestic animals. The cellular and molecular mechanisms of the heat acclimation, as well as those which link the short- and long-term adaptability to heat stress, are clearly the area of future research, which may pave way for designing mitigatory strategies. Research in this area includes studies of the fundamental physiological processes within the animal at the organism, organ system, cellular, and molecular level. Neural, hormonal, or other chemical messengers that serve as regulators of physiologic processes and perform integrative functions in the animal are also needed to be elucidated for better understanding of these physiological processes for formulating suitable mitigation strategies.

17.2.5 Simulation Models

Animal-based research to compare total production systems is scanty because of high cost and resource requirements. There is a general lack of simulation models for assessing adaptation of livestock to CC. The development of computer technology, analytic methods, and simulation models for animal production systems provide a methodology for critically evaluating alternative production systems and management decisions. The models of animal production systems that allow comparisons among alternative management components and decisions are need of the hour to help launching nationwide mitigation strategies to counter the CC effects. The geographic information systems (GIS) also need to be strengthened to monitor the level of stress and the spread of CC-induced pest and pathogens. Predictive modeling systems are needed to predict the probability of an outcome. These projections by the simulation models need to be validated through laboratory and field research so that scientists and producers can predict and plan for threats by pests and pathogens.

17.2.6 Exploiting the Genetic Potential of Native Breeds

It is important to concentrate not only on breeds that may be of immediate value in current production systems, but also on the potential long-term value of preserving unique allelic diversity across the local breeds that may not have immediate commercial value. Strong investments are needed now to identify and preserve the unique traits in the indigenous germplasm. Several molecular tools are now available to identify valuable traits in indigenous breeds. This pool of information is crucial in increasing and sustaining future productivity.

Adaptation research, particularly for developing and upscaling strategies to cope with CC is an important researchable priority. Economic evaluation of adaptation technologies, appraisal of farmer's responses under different resource conditions, and change in regional, national, and international policies relating to farm level adaptation and assessment of the investment requirements for coping with the challenges of CC are among critical themes in which research efforts in social sciences and the human dimensions need to be intensified.

17.2.7 Developing Suitable Breeding Program

One of the most important researchable issues is the maintenance of the global animal genetic diversity. The choice of the most appropriate breed or breeds to use in a given environment or production system should be the first step when initiating a breeding program, and the due attention must be given to the adaptive performance. The appropriate strategy for any breeding program would, therefore, be to set suitable selection goals, which match the production system rather than ambitious performance objectives that cannot be reached under the prevailing environment. The present scenario calls for using the available genetic resources for identifying the genes/genomic regions associated with thermoregulation and adaptive biology, and develop further functional gene resource for native livestock with inherent thermotolerant ability as a model for better understanding of thermoregulation pathways/molecular mechanism of heat stress. A strong foundation for exploitation of molecular markers may be in place for selection of animals with superior adaptive ability. New developments in molecular biology and the emergence of mapping the human genome have led to the development of research to map and understand the genome of agriculturally important animal species. A more complete understanding of animal genome is needed to generate fundamental information important to genetics, physiology, nutrition, and related sciences relevant to animal production.

17.2.8 Research on Pineal-Adrenal-Immune System Relationship

There exists an elaborate interplay between the pineal-adrenal-immune system axis, the details of which need to be experimentally proven. Thus, detailed studies are required to fully understand the mechanisms of interactions between these glands (pineal-adrenal, pineal-adrenal-immune system etc.) in combating heat stress. It is still not apparent whether melatonin controls thermal stress by acting directly on adrenal cortex or by acting on hypothalamus or anterior pituitary. In addition, there are some unanswered questions on whether there is a direct link between glucocorticoids and melatonin production and secretion. If one attempts to establish this multi-endocrine gland relationship, research and experiments/trials should be conceptualized to unveil the precise mechanism of the interactions among pineal, adrenal, and immune system.

17.2.9 Heat Shock Proteins

The cellular and molecular mechanisms of the heat acclimation as well as the mechanistic link between the short- and long-term adaptability to heat stress are clearly the researchable priorities, which may pave way for designing mitigatory strategies. Genetic markers and/or whole genome sequencing may be useful to improving livestock fecundity. Additional research will be needed to establish the physiological mechanisms by which these heat shock proteins (HSP) polymorphisms alter livestock fertility and to evaluate genotype environment interactions. Research is needed to establish the physiological mechanisms regulating the relationships among HSP genotypes, serum and tissue concentrations of HSP, and livestock fertility. Understanding the genetic mechanisms associated with HSP may lead to genetic tools required to improve livestock tolerance of stressors.

17.2.10 Water Experiment

Research and development are needed on water as a limiting resource to increase animal production, and management practices to increase the use efficiency. Specific research is needed to understand the livestock–water interactions, a holistic approach to site-specific interventions, and to ensure that livestock production contributes to sustainable and productive use of scarce water resources. The climatic and environmental conditions being highly variable, temporal and spatial distributions of rainfall with consequent large fluctuations in the quantity and quality of feed and variations in ambient temperatures and water availability result in variations in body composition with a profound influence on the productivity of animals in a specific environment. It is, therefore, necessary to undertake studies on water metabolism and requirements for livestock under representative environmental conditions and physiological states. Water quality for drinking water for animals and for management practices also determine the water resources and is an important researchable issue. Moreover, research should provide information on improving the environment and management techniques to increase water use efficiency, and raise animal productivity as water productivity.

17.2.11 Nutritional Research

The efficiency with which animals convert feedstuffs to human food and other products varies among species, animal products produced, and types of diets. Enhancing the efficiency of nutrient utilization for animal productivity requires fundamental knowledge on a wide range of scientific processes such as molecular and cellular biology, digestion, metabolic processes, and feed processing technology. Applications of new informations through functional and nutrigenomics studies can revolutionize the way agriculturalists and animal scientists think about livestock production. These scientific informations are relevant to studies involving complex metabolic interactions that determine fertility and influence productive and reproductive efficiency. Nutrigenomic studies are becoming important as we develop an understanding of the relationship between nutrition, genetics, and fertility. Such scientific advances can provide new tools that can be used to more clearly understand how nutritional management can be applied to address diseases, performance issues, and fertility-related constraints in livestock. Development of nutritional strategies to manage heat stress in dairy animals is a researchable priority. It is widely accepted that livestock better copes heat stress when nutrition is not a limiting factor. Thus, additional research is needed to enhance the understanding of these associations at a mechanistic level to fully exploit the potential of nutritionally manipulated reproduction in livestock under tropical environment. This approach can be valuable in gaining a thorough understanding of how to adjust the nutrient requirement to deal with specific environmental constraints and will therefore provide decision-making managerial tools to optimize livestock productivity.

17.2.12 Livestock Diseases

Vaccination is the principal means to protect livestock against the outbreaks of climatically driven diseases. Thus, research is needed to strengthen the already available immunological protocols. Provision of trained manpower and resources to implement disease-control strategies in the event of outbreaks are an essential requirement which can help reduce outbreaks quickly. Regular surveillance can provide up-to-date information about changes in pathogen populations. Data based on laboratory and field research can illuminate how CC influences pathogen characteristics and models so that researchers and producers can predict and plan for pathogen threats. Establishment of nationwide network of research facilities and ground breaking pathogen research demands cross-field and transnational cooperation. Surveillance of infectious diseases should be increased and special emphasis must be given to vector borne and viral diseases where threat of epidemiology is the highest. Globally, improved disease surveillance and monitoring systems should be implemented with use of GIS and modern veterinary and animal health production systems. Priorities need to be set for research pertaining to understanding the interrelationships among environment, genetics, and infectious agents in the etiology of diseases and emphasis should be given to identify appropriate methods of diagnosis, prevention, treatment, control, and eradication of diseases, including development of equipment.

17.2.13 Advances in CH_4 Mitigation

Mitigating GHG emissions from livestock is increasingly recognized as a necessary part of meeting the worldwide CC obligations. There is a need to develop strategies to mitigate CH_4 emissions by identifying plant and feed additives that can interrupt the methanogenesis process, or manipulating CH_4-producing microbes in the guts of ruminant animals. There is renewed need for further concerted research on CH_4 emission by livestock and its mitigation. For instance, there are several new and more advanced CH_4 mitigation options, including the addition of probiotics, acetogens, bacteriocins, archaeal viruses, organic acids, and plant extracts to the diet, as well as immunization and genetic selection of animals. Although these new strategies are promising, more research is needed for validation purposes and to assess in vivo their effectiveness in reducing CH_4 production in dairy animals. Vaccine development against methanogens is promising and calls for concerted research efforts if this strategy is to become a reality. The development of biomarkers to identify low CH_4 emission animals or low CH_4-

producing bacteria also merits further investigation. Future developments in the area of modeling must accompany any improved understating of the underlying rumen biology. Furthermore, the need to develop simpler and more accurate models compatible with the current trends in computer technology cannot be overemphasized.

Just as concentration of active compounds in plants vary widely depending upon the cultivator, growing conditions, or processing methods so do those of elements in animal system. For consistency and accuracy, researchers should report the concentration of primary active components and recommended doses must be established. Further, there is a strong need to explore other potential effects on rumen fermentation, such as effects on biohydrogenation of fatty acids and activity against pathogenic microbes. The adaptation of microorganisms to plant secondary metabolites (PSM) in the rumen is a constraint, which diminishes the efficacy of compounds over time. These PSM should be included in diets on rotation to overcome this limitation. Although many of the PSM are generally recognized as safe for inclusion in diets, their toxicity in animals and the presence of residues in the animal products (meat or milk) must be tested.

It is also important to assess the scope for the application of breeding and genetic tools to reduce CH_4 emission from livestock system, while maintaining or improving animal well-being, food safety, quality, and biodiversity. It is therefore inevitable to understand the genetic components of unique characters of indigenous livestock species and create mechanisms to better exploit them in response to changing environments. Integrated research investigating animal, plant, microbe, and nutrient level strategies might offer a long-term solution of CH_4 production. At the animal level, genetic selection is the area of research with a best chance of identifying a solution. At the microbe level, vaccination and probiotics are the promising approaches to future research. Scientific advances through rumen metagenomic projects and the utilization of new technologies are needed to broaden the understanding of the mechanisms involved in methanogenesis and other metabolic hydrogen consuming and releasing processes to identify new targets for CH_4 mitigation.

17.2.14 Final Remarks

Immediate and far-reaching changes in current animal agriculture practices and consumption patterns are needed to mitigate adverse effects of CC. Indigenous knowledge is an integral part of the culture and history of a community. Thus, there is a need to learn from local communities, build upon the traditional knowledge, and enrich the development process. Considering the understanding of homeorhetic regulation of thermoregulation and advancement of technological innovations, it looks promising to mitigate environmental stresses to livestock even in the CC scenario. Such an approach can ultimately lead to sustainable security on livestock production, and food and health security to human. However, as physiological phenomena are very precisely and finely adjusted to maintain interior milieu of the body, any irrational modulation may disturb the physiological mechanism of the body. Therefore, it requires utmost care and vigilance to improve livestock production under CC scenario while simultaneously keeping pace with scientific developments.

Glossary

3-beta-hydroxysteroid dehydrogenase (3*β* HSD) 3*β* HSD is an enzyme that catalyses the synthesis of progesterone from pregnenolone. It is the only enzyme in the adrenal pathway of corticosteroid synthesis that is not a member of the Cytochrome P450 family.

Abiotic stress Abiotic stress is defined as the negative impact of non-living factors on the living organisms in a specific environment.

Acclimatization A long-term adaptive physiological adjustment which results in an increased tolerance to continuous or repeated exposure to complex climatic stressors (normally produced under field conditions).

Acidosis Acidosis is an increased acidity in the blood and other body tissue (i.e., an increased hydrogen ion concentration).

ACTH Adrenocorticotropic hormone (ACTH), also known as 'corticotropin', is a polypeptide tropic hormone produced and secreted by the anterior pituitary gland. It is an important component of the hypothalamic-pituitary-adrenal axis and is often produced in response to biological stress under the influence of corticotropin-releasing hormone from the hypothalamus. Its primary action is to stimulate the adrenal glands to produce corticosteroids such as cortisol, which acts to reduce stress and inflammation in the body.

Ad libitum Food available at all times with the quantity and frequency of consumption being the free choice of the animal. Ad libitum is also used in psychology and biology to refer to the "free-feeding".

Adaptation (biological) The morphological, anatomical, physiological, biochemical, and behavioral characteristics of the animal which promote welfare and favor survival in a specific environment.

V. Sejian et al. (eds.), *Environmental Stress and Amelioration in Livestock Production*, DOI: 10.1007/978-3-642-29205-7,

Adaptation (genetic) The heritable animal characteristics alterations which favor survival of a population in a particular environment. This may involve evolutionary changes over many generations (selected by nature) or acquiring specific genetic properties (selection by man).

Adaptation (physiological) The capacity and process of adjustment of the animal to itself, to other living material and to its external physical environment.

Adaptation Adaptation is the evolutionary process whereby an organism becomes better able to live in its habitat or habitats.

Adaptive capability Adaptive capability is the capacity of a system to adapt to the environment where the system exists is changing. It is applied to e.g., ecological and human social systems.

Adipose tissue Adipose tissue is specialized connective tissue that functions as the major storage site for fat in the form of triglycerides. Adipose tissue is found in mammals in two different forms: white adipose tissue and brown adipose tissue. The presence, amount, and distribution of each varies depending upon the species.

Adrenal cortex Cortical part of the adrenal gland. Situated along the perimeter of the adrenal gland, the adrenal cortex mediates the stress response through the production of glucocorticoids and mineralocorticoids.

Adrenal corticosteroids Adernal Corticosteroids are a class of chemicals that includes steroid hormones naturally produced in the adrenal cortex of vertebrates.

Adrenal medulla It is the innermost part of the adrenal gland. It is located at the center of the gland, being surrounded by the adrenal cortex. It is the innermost part of the adrenal gland, consisting of cells that secrete epinephrine (adrenaline) and norepinephrine (noradrenaline).

Agro-ecological zones Agro-ecological Zone refers to the division of an area of land into smaller units, which have similar characteristics related to land suitability, potential production, and environmental impact.

Alanine amino transferase An enzyme that catalyzes the reversible transfer of an amino group from alanine to α-ketoglutarate to form pyruvate and glutamate.

Aldosterone Aldosterone is a steroid hormone (mineralocorticoid family) produced by the outer-section (zona glomerulosa) of the adrenal cortex in the adrenal gland, and acts mainly on the distal tubules and collecting ducts of the nephron, the functioning unit of the kidney, to cause the conservation of sodium, secretion of potassium, increased water retention, and increased blood pressure.

Allostasis Allostasis is the process of achieving stability, or homeostasis, through physiological or behavioral change. This can be carried out by means of alteration in HPA axis hormones, the autonomic nervous system, cytokines, or a number of other systems, and is generally adaptive in the short term.

Ambient temperature Ambient temperature simply means "the temperature of the surroundings" or temperature of the immediate environment.

Anabolic steroids Anabolic steroids are drugs that mimic the effects of testosterone and dihydrotestosterone in the body. They increase protein synthesis within cells, which results in the buildup of cellular tissue (anabolism), especially in muscles. Anabolic steroids also have androgenic and virilizing properties, including the development and maintenance of masculine characteristics such as the growth of the vocal cords, testicles, and body hair (secondary sexual characteristics).

Anthropogenic emission The term is used in the context of green house gas emissions that are produced as a result of human activities but applies broadly to all major human impacts on the environment.

Antigen Any substance capable of inducing a specific immune response and of reacting with the products of that response, i.e., with specific antibody or specifically sensitized T lymphocytes, or both.

Anti-inflammatory Anti-inflammatory refers to the property of a substance or treatment that reduces inflammation. Anti-inflammatory drugs make up about half of analgesics, remedying pain by reducing inflammation as opposed to opioids, which affect the central nervous system.

Antioxidants Antioxidants are substances that may protect cells from the damage caused by unstable molecules known as free radicals. Free radicals are highly reactive chemicals that attack molecules by capturing electrons and thus modifying chemical structures. Antioxidants interact with and stabilize free radicals and may prevent some of the damage free radicals might otherwise cause. Examples of antioxidants include beta-carotene, lycopene, vitamins C, E and A, and other substances.

Apoptosis Apoptosis is the process of programmed cell death (PCD) that may occur in multicellular organisms. Biochemical events lead to characteristic cell changes (morphology) and death.

Appendages The main function of appendages is to increase the surface area of the skin therefore specifically contributing to the already mentioned lowering of the metabolic rate and the increase in the presence of sweat glands, hence reducing the amount of heat produced by the body and increasing heat dissipation.

Arid climate Generally, any extremely dry climate.

Aromatase Aromatase is an enzyme responsible for a key step in the biosynthesis of estrogens. It is a member of the cytochrome P450 superfamily, which are monooxygenases that catalyze many reactions involved in steroidogenesis. In particular, aromatase is responsible for the aromatization of androgens into estrogens.

Artificial insemination Artificial insemination, or AI, is the process by which sperm is placed into the reproductive tract of a female for the purpose of impregnating the female by using means other than sexual intercourse or natural insemination.

Aspartate amino transferase AST catalyzes the reversible transfer of an α-amino group between aspartate and glutamate and, as such, is an important enzyme in amino acid metabolism. AST is found in the liver, heart, skeletal muscle, kidneys, brain, and red blood cells, and it is commonly measured clinically as a marker for liver health.

Average daily gain Measurement of daily body weight change in animal on a feed test.

Bacteriocins Bacteriocins are proteinaceous toxins produced by bacteria to inhibit the growth of similar or closely related bacterial strain(s). They are typically considered to be narrow spectrum antibiotics. They are analogous to yeast and paramecium killing factors, and are structurally, functionally, and ecologically diverse.

Basal metabolism The minimum amount of energy required to maintain vital functions in an organism at complete rest, measured by the basal metabolic rate in a fasting individual who is awake and resting in a comfortably warm environment.

Behavioral adaptations A characteristic or modification in an animal's body that helps it survive in its habitat.

Best management practices (BMPs) BMPs are practices that you can use on your land to address concerns such as erosion, drainage, mud, and manure. By successfully managing these issues you can drastically improve the health of your property and local natural resources. In addition to supporting conservation, installing BMPs at your livestock facility can improve chore efficiency, benefit animal health and safety facility, enhance land aesthetics, improve neighborhood relationships, and even increase property value.

Bioclimatology The branch of climatology which deals with the relations of climate with animals and plants.

Biodiversity Biodiversity is the degree of variation of life forms within a given ecosystem, biome, or an entire planet. Biodiversity is a measure of the health of ecosystems. Biodiversity is in part a function of climate.

Biomarker A biomarker, or biological marker, is, in general, a substance used as an indicator of a biological state. It is a characteristic that is objectively measured and evaluated as an indicator of normal biological processes, pathogenic processes, or pharmacologic responses to atherapeutic intervention.

Biomass production Biomass production refers to the rate of generation of biomass in an ecosystem. It is usually expressed in units of mass per unit surface (or volume) per unit time, for instance grams per square metre per day. The mass unit may relate to dry matter or to the mass of carbon generated.

Biomass Biomass is biological material from living, or recently living organisms. As an energy source, biomass can either be used directly, or converted into other energy products such as biofuel.

Biotic Stress Biotic stresses include living disturbances such as pathogenic microorganism, whereas abiotic stress factors, or stressors, are naturally occurring, often intangible, factors such as intense sunlight or wind that may cause harm to the plants and animals in the affected area.

Blastocyst Blastocyst is a structure formed in the early embryogenesis of mammals, after the formation of the morula. It is a specifically mammalian example of a blastula. It possesses an inner cell mass (ICM), or embryoblast, which subsequently forms the embryo, and an outer layer of cells, or trophoblast, which later forms the placenta.

Blizzared A severe weather condition characterized by low temperatures and by strong winds bearing a great amount of snow.

Body condition scores An assessment of relative proportions of muscle and fat in farm animals. The assessment is made by palpation of amount of tissue cover between the points of the hip, over the transverse processes of the lumbar vertebrae, the cover over the ribs and the pin bones below the tail to determine the thickness of fat cover in these areas.

Body Growth It is an irreversible positive change in the measured dimensions of the body.

Bovine somatotrophin (bST) Bovine somatotrophin (bST) is a peptide hormone produced by the cow's pituitary gland. Like other hormones, it is produced in small quantities and is used in regulating metabolic processes.

Brucellosis Brucellosis is a bacterial disease caused by members of the Brucella genus that primarily infects livestock. Symptoms of the disease include intermittent fever, sweating, chills, aches, and mental depression. The disease can become chronic and recur, particularly if untreated.

Capillary permeability Capillary permeability characterizes the capacity of a blood vessel wall to allow the flow of small molecules (ions, water, nutrients) or even whole cells (lymphocytes on their way to the site of inflammation) in and out of the vessel.

Catabolism A metabolic process in which complex molecules are broken down into simple ones with the release of energy.

Catecholamines Catecholamines are molecules that have a catechol nucleus consisting of benzene with two hydroxyl side groups, and a side-chain amine. They include adrenaline, noradrenaline, and dopamine released by the adrenal medulla of the adrenal glands in response to stress.

Cell-mediated immune response Cell-mediated immune response is an immune response that does not involve antibodies but rather involves the activation of macrophages, natural killer cells (NK), antigen-specific cytotoxic T-lymphocytes, and the release of various cytokines in response to an antigen.

Chromatin A complex of nucleic acids and proteins, primarily histones, in the cell nucleus that stains readily with basic dyes and condenses to form chromosomes during cell division.

Chronic stress Chronic stress is the response to emotional pressure suffered for a prolonged period over which an individual has no control. It involves an endocrine system response which occurs with the release of corticosteroids.

Circadian rhythm/cyclic photic input A circadian rhythm, popularly referred to as **body clock**, is an endogenously driven (non-reliant on environmental cues), roughly 24-hour cycle in biochemical, physiological, or behavioral processes. Circadian rhythms have been widely observed in plants and animals.

Climate The long-term (some 30 years) average condition of the meteorological variable in a given region. The pattern or cycle of weather conditions such as temperature, wind, rain, snowfall, humidity, clouds, including extreme, or occasional ones, over a large area, averaged over many years.

Climatic changes Climatic changes are significant and lasting change in the statistical distribution of weather patterns over periods ranging from decades to millions of years. Climate change may be limited to a specific region or may occur across the whole Earth.

Climatography A thorough, quantitative description of climate, particularly with reference to the tables and charts which show the characteristic values of climatic elements at a station or over an area.

Climatology The scientific study of climate presentation of climatic data (climatography), analysis of the causes of differences of climate (physical climatology) and the application of climatic data to the solution of specific design or operational problems (applied climatology).

Cloud A visible collection of minute water and ice particles at some distance above the earth's surface. Clouds may be classified according to their shape, formation, height, and composition.

Comfort zone The ranges of indoor temperature, humidity, and air movement, under which the animal shows mental and physical well-being.

Conception rate The definition of conception rate is the percentage of matings that result in conception. Conception is important to allow reproduction to occur and high rates of conception to allow effective breeding.

Conditioning The transfer of an existing response to a new stimulus.

Conduction Conduction is the transfer of heat between materials that contact each other. Heat passes from the warmer material to the cooler material. For example, an animal's skin can transfer heat to a contacting surface if that surface is cooler, and vice versa.

Convection Convection is the transfer of heat in a moving fluid. Air flowing past the body can cool the body if the air temperature is cool. On the other hand, air that exceeds 35°C (95°F) can increase the heat load on the body.

Corticotropin releasing hormone (CRH) Also known as Corticotropin-releasing factor (CRF) is a family of related neuropeptides in vertebrates. CRH is secreted by the paraventricular nucleus (PVN) of the hypothalamus in response to stress. Its main function is to stimulate ACTH synthesis in anterior pituitary in response to stress.

Cortisol (hydrocortisone) Cortisol (hydrocortisone) is a steroid hormone, more specifically a glucocorticoid, produced by the zona fasciculata of the adrenal cortex. It is released in response to stress. Its primary functions are to increase blood sugar through gluconeogenesis suppressing the immune system; and aid in fat, protein, and carbohydrate metabolism.

Crossbreds An organism that is the offspring of genetically dissimilar parents or stock especially offspring produced by breeding plants or animals of different varieties or breeds or species.

Cud Cud is a portion of food that returns from a ruminant's stomach in the mouth to be chewed for the second time. More accurately, it is a bolus of semi-degraded food regurgitated from the reticulorumen of a ruminant. Cud is produced during the physical digestive process of rumination.

Cytoskeleton The cytoskeleton is a network of fibers throughout the cell's cytoplasm that helps the cell maintain its shape and gives support to the cell.

Data loggers A data logger (also datalogger or data recorder) is an electronic device that records data over time or in relation to location either with a built-in instrument or sensor or via external instruments and sensors. Increasingly, but not entirely, they are based on a digital processor (or computer).

Dehydration Excessive loss of water from the body or from an organ or body part, as from illness or fluid deprivation. It is a common symptom of livestock exposed to heat stress.

Dietary energy The dietary energy supply is the food available for animal consumption, usually expressed in kilocalories per animal per day. It gives an overestimate of the total amount of food consumed as it reflects both food consumed and food wasted.

Dietary fat Dietary fat is one of the three macronutrients, along with protein and carbohydrates that provide energy for your body. Fat is essential to your health because it supports a number of your body's functions. Some vitamins, for instance, must have fat to dissolve and nourish body.

DNA methylation DNA methylation is a biochemical process that is important for normal development in higher organisms. It involves the addition of a methyl group to the 5 position of the cytosine pyrimidine ring or the number 6 nitrogen of the adenine purine ring.

DNA sequence The **sequence** or **primary structure of a nucleic acid** is the composition of atoms that make up the nucleic acid and the chemical bonds that bond those atoms. Because nucleic acids, such as DNA and RNA, are unbranched polymers, this specification is equivalent to specifying the sequence of nucleotides that comprise the molecule.

Domestic animal A domesticated animal is any animal that depends on a human for food, water, and shelter. This includes farm animals such as cattle, horses, sheep, chickens, goats, dogs, and cats.

Dopamine (DA) Dopamine is a simple organic chemical in the catecholamine family, which plays a number of important physiological roles in the bodies of animals. Its name derives from its chemical structure, which consists of an amine group (NH_2) linked to a catechol structure called dihydroxyphenylalanine (acronym DOPA). In the brain, dopamine functions as a neurotransmitter—a chemical released by nerve cells to send signals to other nerve cells.

Drought A period of abnormally dry weather sufficiently prolonged for the lack of water to cause a serious hydrologic imbalance (i.e. croup damage, water-supply shortage, etc) in the affected area.

Dry bulb (DB) temperature Dry bulb (DB) temperature is measured by a thermal sensor, such as ordinary mercury-in-glass thermometer, that is shielded from direct radiant energy sources

Dry matter Dry matter can refer to the dry portion of animal feed. A substance in the feed, such as a nutrient or toxin, can be referred to on a dry matter basis to show its level in the feed (e.g., ppm). Considering nutrient levels in different feeds on a dry matter basis (rather than an as-is basis) makes a comparison easier because feeds contain different percentages of water.

Ecological region Sometimes called a **bioregion** is an ecologically and geographically defined area that is smaller than an ecozone and larger than an ecosystem. Ecoregions cover relatively large areas of land or water, and contain characteristic, geographically distinct assemblages of natural communities and species.

Ecological Zone A zone or an area that serves as a conserved natural habitat where plants and animals can thrive.

Ecosystem Ecosystem is a biological environment consisting of all the living organisms or biotic component, in a particular area, and the nonliving, or abiotic component, with which the organisms interact, such as air, soil, water, and sunlight.

Electroejaculation Electroejaculation is a procedure used to obtain semen samples from sexually mature male mammals. The procedure is applied for breeding programs and research purposes in various species, as well as in the treatment of an ejaculatory dysfunction in human males.

Embryo implantation Embryo implantation is the stage at which the embryo adheres to the wall of the uterus, believed by some to be the beginning of pregnancy. At this stage of prenatal development, the embryo is a blastocyst. It is by this adhesion that the fetus receives oxygen and nutrients from the mother to be able to grow.

Embryogenesis Embryogenesis is a developmental process that usually begins once the egg has been fertilized. It involves multiplication of cells (by mitosis) and their subsequent growth, movement, and differentiation into all the tissues and organs.

Endemic An infection that is maintained in a population without the need for external inputs.

Endocrine system The system of ductless glands in the body that secrete different hormones directly into the bloodstream to regulate the body. The endocrine system is in contrast to the exocrine system, which secretes its chemicals using ducts. It derives from the greek words "endo" meaning inside, within, and "crinis" for secrete.

Endogenous opioids An opioid is a psychoactive chemical that works by binding to opioid receptors, which are found principally in the central and peripheral nervous system and the gastrointestinal tract.

Endotoxin Endotoxins, which occur in the outer membrane of certain gram-negative bacteria, are not secreted but are released only when the cells are disrupted or destroyed. Endotoxins are complex polysaccharide molecules that elicit an antigenic response, resulting in fever and altered resistance to bacterial infections. Exposure may cause toxic hemorrhagic shock and severe diarrhea.

Enteric fermentation Enteric fermentation is a digestive process by which carbohydrates are broken down by microorganisms into simple molecules for absorption into the bloodstream of an animal. It is one of the important sources for methane emissions.

Enteric Methane It is the process by which methane is produced during enteric fermentation.

Environmental factors Environmental factors represent the external factors affecting an animal. These factors may be other living organisms (biotic factors) or nonliving variables (abiotic factors), such as water, soil, climate, light, and oxygen. All interacting biotic and abiotic factors together make up an ecosystem.

Enzymes Enzymes are protein catalysts that facilitate the catabolism (breakdown) and anabolism (synthesis) of organic compounds.

Epigenetic code The epigenetic code is hypothesised to be a defining code in every eukaryotic cell consisting of the specific epigenetic modification in each cell. It consists of histone modifications defined by the histone code and additional epigenetic modifications such as DNA methylation. The base for the epigenetic code is a system above the genetic code of a single cell. While in one individual the genetic code in each cell is the same, the epigenetic code is tissue and cell specific.

Epinephrine Also known as adrenaline is a hormone and a neurotransmitter. It increases heart rate, constricts blood vessels, dilates air passages and participates in the fight-or-flight response of the sympathetic nervous system.

Epithelial cells Membranous tissue composed of one or more layers of cells separated by very little intercellular substance and forming the covering of most internal and external surfaces of the body and its organs.

Eradication Eradication is the reduction of an infectious disease's prevalence in the global host population to zero.

Estivation A state of torpidity induced in some animals by the heat and dryness of the summer.

Estrogen Receptor Estrogen Receptor refers to a group of receptors that are activated by the hormone 17β-estradiol (estrogen). Two types of estrogen receptor exist: ER, which is a member of the nuclear hormone family of intracellular receptors, and the estrogen G protein-coupled receptor GPR30 (GPER), which is a G protein-coupled receptor.

Estrus periods The periodic state of sexual excitement in the female of most mammals, excluding humans, that immediately precedes ovulation and during which the female is most receptive to mating.

Etiology The cause or origin of a disease, condition, or constellation of symptoms or signs, as determined by medical diagnosis or research.

Evaporation Evaporation can be defined as the process by which liquid water is converted into a gaseous state. Evaporation can only occur when water is available. It also requires that the humidity of the atmosphere be less than the evaporating surface (at 100% relative humidity there is no more evaporation). The evaporation process requires large amounts of energy.

Evaporative cooling Process in which the heat is removed from an object by evaporation of a liquid coolant; also, the process in which outside air is pre-cooled before passing through a space.

Exons An exon is a nucleic acid sequence that is represented in the mature form of an RNA molecule either after portions of a precursor RNA (introns) have been removed by cis-splicing or when two or more precursor RNA molecules have been ligated by trans-splicing.

Fat rump A class of sheep, mostly Russian, similar to fat-tailed sheep but the large depot of fat is on the rump instead of the tail and although most of them are carpet wool sheep some are of merino type with fine wool.

Fat tail Fat-tailed sheep are characterized by a deposition of fat at the level of the hind quarters. This group of breeds is particularly adapted to extensive production systems in semi-arid environments e.g. Fat-tailed Damara sheep.

Feed Conversion ratio Also known as feed conversion efficiency (FCE) is a measure of an animal's efficiency in converting feed mass into increased body mass.

Feedback Mechanism It is the principal means by which hormone output is controlled in an animal body. The rate of hormone biosynthesis and secretion is often regulated by a homeostatic negative feedback control mechanism. Such a mechanism depends on factors that influence the metabolism and excretion of hormones. Thus, higher hormone concentration alone cannot trigger the negative feedback mechanism. Negative feedback must be triggered by overproduction of an "effect" of the hormone.

Fermentation Typically it is the conversion of carbohydrates to alcohols and carbon dioxide or organic acids using yeasts, bacteria, or a combination thereof, under anaerobic conditions. Fermentation in simple terms is the chemical conversion of sugars into ethanol.

Folliculogenesis Folliculogenesis is the maturation of the ovarian follicle, a densely packed shell of somatic cells that contains an immature oocyte. Folliculogenesis describes the progression of a number of small primordial follicles into large preovulatory follicles that enter the estrus cycle.

Forestomach The first division of the stomach of a ruminant animal, in which most food collects immediately after being swallowed and from which it is later returned to the mouth as cud for thorough chewing.

Gastrointestinal tract The tube that extends from the mouth to the anus in which the movement of muscles and release of hormones and enzymes digest food. The gastrointestinal tract starts with the mouth and proceeds to the esophagus, stomach, duodenum, small intestine, large intestine (colon), rectum and, finally, the anus. Also called the alimentary canal, digestive tract and, perhaps most often in conversation, the GI tract.

Genetic assimilation Genetic assimilation is a process by which the effect of an environmental condition, such as exposure to a teratogen, is used in conjunction with artificial selection or natural selection to create a strain of organisms with similar changes in phenotype that are encoded genetically. Despite superficial appearances, this does not require the inheritance of acquired characters, although epigenetic inheritance could potentially influence the result. Genetic assimilation is merely a method of overcoming the barrier to selection imposed by genetic canalization of developmental pathways.

Genetic biomarkers Measurable and quantifiable biological parameters which serve as indices for health—and physiology-related assessments. Biomarker analytes are DNA, gene expression profiles, proteins, protein expression, metabolites, cells, or combinations of these can constitute biomarkers.

Genetic diversity Genetic variation between and within species, which is measured by determining the proportion of polymorphic loci across the genome, or by the number of heterozygous individuals in a population.

Genetic drift Genetic drift or allelic drift is the change in the frequency of a gene variant (allele) in a population due to random sampling.

Genome It represents the organism's hereditary information. It is encoded either in DNA or, for many types of virus, in RNA. The genome includes both the genes and the non-coding sequences of the DNA/RNA.

Genotypes The genotype refers to the entire set of genes in a cell, an organism, or an individual. A gene for a particular character or trait may exist in two allelic forms. One is dominant (e.g. A) and the other is recessive (e.g. a). Based on this, there could be three possible genotypes for a particular character: AA (homozygous dominant), Aa (heterozygous), and aa (homozygous recessive).

Geographic information system Geographic information system is a system designed to capture, store, manipulate, analyze, manage, and present all types of geographically referenced data. The acronym GIS is sometimes used to mean geographical information science or geospatial information studies, these latter terms refer to the academic discipline or career of working with geographic information systems. In the simplest terms, GIS is the merging of cartography, statistical analysis, and database technology.

Germ cell A germ cell is any biological cell that gives rise to the gametes of an organism that reproduces sexually. In many animals, the germ cells originate near the gut of an embryo and migrate to the developing gonads.

Germinal cell The eggs and sperm are the germ cells: the reproductive cells.

Germplasm A germplasm is a collection of genetic resources for an organism.

Gestation Gestation is the carrying of an embryo or fetus inside a female viviparous animal.

Gibb's free energy In thermodynamics, the Gibbs free energy is a thermodynamic potential that measures the "useful" or process-initiating work obtainable from a thermodynamic system at a constant temperature and pressure (isothermal, isobaric).

Glial cells Glial cells, sometimes called neuroglia or simply glia are non-neuronal cells that maintain homeostasis, form myelin, and provide support and protection for neurons in the brain, and for neurons in other parts of the nervous system such as in the autonomic nervous system.

Global positioning system (GPS) GPS is a space-based satellite navigation system that provides location and time information in all weather, anywhere on or near the Earth, where there is an unobstructed line of sight to four or more GPS satellites.

Global warming potential (GWP) GWP is a relative measure of how much heat a greenhouse gas traps in the atmosphere. It compares the amount of heat trapped by a certain mass of the gas in question to the amount of heat trapped by a similar mass of carbon dioxide.

Global warming Global warming refers to the current rise in the average temperature of Earth's atmosphere and oceans. The warming is widely predicted to occur due to an increase in the greenhouse effect resulting especially from pollution that would lead to the production of more carbon dioxide, the main gas responsible for global warming.

Globe temperature Globe temperature is the temperature inside a blackened, hollow, thin copper globe.

Glucocorticoids (GC) GC are classes of steroid hormones secreted from adrenal cortex that bind to the glucocorticoid receptor (GR), which is present in almost every vertebrate animal cell. The name glucocorticoid (glucose + cortex + steroid) derives from their role in the regulation of the metabolism of glucose, their synthesis in the adrenal cortex, and their steroidal structure.

Gluconeogenesis Gluconeogenesis is a metabolic pathway that results in the generation of glucose from non-carbohydrate carbon substrates such as lactate, glycerol, and glucogenic amino acids.

Glutathione peroxidase (GPx) GPx is the general name of an enzyme family with peroxidase activity whose main biological role is to protect the organism from oxidative damage. The biochemical function of glutathione peroxidase is to reduce lipid hydroperoxides to their corresponding alcohols and to reduce free hydrogen peroxide to water.

Glutathione reductase (GR) GR is an enzyme that reduces glutathione disulfide to the sulfhydryl form GSH, which is an important cellular antioxidant.

Glycogenolysis Glycogenolysis is the conversion of glycogen polymers to glucose monomers. Glycogen is catabolized by removal of a glucose monomer through cleavage with inorganic phosphate to produce glucose-1-phosphate. This derivative of glucose is then converted to glucose-6-phosphate, an intermediate in glycolysis.

Gonadotropin-releasing hormone (GnRH) GnRH is a trophic peptide hormone responsible for the release of follicle-stimulating hormone (FSH) and luteinizing hormone (LH) from the anterior pituitary. GnRH is synthesized and released from neurons within the hypothalamus.

Gross Domestic Product (GDP) GDP refers to the market value of all final goods and services produced within a country in a given period. GDP per capita is often considered an indicator of a country's standard of living.

Habituation A gradual quantitative change of response which may lead to a loss of response, as a result of repeated stimulation.

Hatchability Percentage of fertilized eggs that hatch under incubation.

Heart rate Heart rate is the number of heartbeats per unit of time, typically expressed as beats per minute (bpm). Heart rate can vary as the body's need to absorb oxygen and excrete carbon dioxide changes.

Heat shock proteins Heat shock proteins are a class of functionally related proteins involved in the folding and unfolding of other proteins. Their expression is increased when cells are exposed to elevated temperatures or other stress. This increase in expression is transcriptionally regulated. The dramatic upregulation of the heat shock proteins is a key part of the heat shock response and is induced primarily by heat shock factor (HSF).

Heat stress Heat stress can be defined as a group of conditions due to over exposure to or over exertion in excess environmental temperature. The condition includes heat cramp, heat exhaustion, and heat stroke.

Heme oxygenase enzyme Heme oxygenase enzyme is an enzyme that catalyzes the degradation of heme. This produces biliverdin, iron, and carbon monoxide.

Hemodilution It refers to increase in fluid content of blood, resulting in lowered concentration of formed elements. In other way this represents an increase in the volume of plasma, resulting in a reduced concentration of red blood cells in blood.

Histamine Histamine is an organic nitrogen compound involved in local immune responses as well as regulating physiological function in the gut and acting as a neurotransmitter. Histamine triggers the inflammatory response. As part of an immune response to foreign pathogens, histamine is produced by basophils and by mast cells found in nearby connective tissues. Histamine increases the permeability of the capillaries to white blood cells and some proteins, to allow them to engage pathogens in the infected tissues.

Histones Histones are highly alkaline proteins found in eukaryotic cell nuclei that package and order the DNA into structural units called nucleosomes. They are the chief protein components of chromatin, acting as spools around which DNA winds, and plays a role in gene regulation.

Homeostasis Homeostasis refers to the tendency to maintain a balanced or constant internal state that is optimal for functioning. Animal have a specific "balanced" or "normal" body temperature. When there is a problem with the internal functioning animal body, this temperature may increase, signaling and imbalance. As a result, body attempts to solve the problem and restore homeostasis for normal body temperature.

Homeostatic equilibrium The ability of the body or a cell to seek and maintain a condition of equilibrium or stability within its internal environment when dealing with external changes.

Homeothermic animals Animals with a constant, persistent body temperature that is almost independent of the temperature of the environment. Birds and mammals are homeothermic. Characteristic features of homeothermic animals are mechanisms for chemical thermoregulation (regulation of heat production in the organism) and physical thermoregulation (regulation of heat loss to the environment).

Homeothermy Homeothermy is thermoregulation that maintains a stable internal body temperature regardless of external influence. This temperature is often, though not necessarily, higher than the immediate environment.

Homeoviscous adaptation An increase in unsaturated fatty acids at lower growth temperatures and an increase in saturated fatty acids at higher temperatures. This compositional adaptation of membrane lipids, called homeoviscous adaptation, serves to maintain the correct membrane fluidity at the new conditions.

Homogeneity Homogeneity is the state of being homogeneous. Pertaining to the sciences, it is a substance where all the constituents are of the same nature; consisting of similar parts, or of elements of the like nature.

Human chorionic gonadotrophin Human chorionic gonadotrophin is a glycoprotein hormone produced during pregnancy that is made by the developing embryo after conception and later by the syncytiotrophoblast (part of the placenta).

Humidity A measure of the vapor content of the atmosphere.

Humoral immune responses (HIR) The immune response involving the transformation of B cells into plasma cells that produce and secrete antibodies to a specific antigen

Hydrocortisone Hydrocortisone is a steroid hormone, more specifically a glucocorticoid, produced by the zona fasciculata of the adrenal gland. It is released in response to stress. Its primary functions are to increase blood sugar through gluconeogenesis, suppress the immune system and aid in fat, protein and carbohydrate metabolism.

Hyperthermia Hyperthermia is an elevated body temperature due to failed thermoregulation. Hyperthermia occurs when the body produces or absorbs more heat than it can dissipate. When the elevated body temperatures are sufficiently high, hyperthermia is a medical emergency and requires immediate treatment to prevent disability or death.

Hypogonadism Hypogonadism is a medical term for decreased functional activity of the gonads, thus resulting in lower amounts of sex steroids.

Hypo-osmolality A decrease in the osmolality of the body fluids, body fluid volume increases and solute volumes usually decrease.

Hypophysial portal Hypophysial portal is the system of blood vessels that link the hypothalamus and the anterior pituitary in the brain. It allows endocrine communication between the two structures. It is part of the hypothalamic-pituitary-adrenal axis. The anterior pituitary receives releasing and inhibitory hormones in the blood. Using these, the anterior pituitary is able to fulfill its function of regulating the other endocrine glands.

Hypothalamic-pituitary-adrenal (HPA) axis Also known as the **limbic-hypothalamic-pituitary-adrenal axis** (LHPA axis) and, occasionally, as the **hypothalamic-pituitary-adrenal-gonadotropic axis**, is a complex set of direct influences and feedback interactions among the hypothalamus, the pituitary gland, and the adrenal glands.

Hypothalamus The hypothalamus is located below the thalamus, just above the brain stem. In the terminology of neuroanatomy, it forms the ventral part of the diencephalon. All vertebrate brains contain a hypothalamus. The hypothalamus is responsible for certain metabolic processes and other activities of the autonomic nervous system. It synthesizes and secretes certain neurohormones, often called hypothalamic-releasing hormones, and these in turn stimulate or inhibit the secretion of pituitary hormones. The hypothalamus controls body temperature, hunger, thirst, fatigue, sleep, and circadian cycles.

IgG IgG is the most common type of antibody, comprising about 80% of the body's total. It is equally divided between the blood and interstitial fluid. IgG antibodies represent a large vocabulary of antigen recognition molecules. There are four subgroups, currently labeled with number suffixes (IgG1 to 4). In some mucosal tissues (e.g. mammary glands of ruminants), the IgG1 class of immunoglobulin-producing cells predominate while IgG4 is the major circulating antibody which enters tissues freely and participates in diverse immune events. IgG (as well as IgM) activates the complement system.

Immune response A bodily defense reaction that recognizes an invading substance (an antigen: such as a virus or fungus or bacteria or transplanted organ) and produces antibodies specific against that antigen.

Immune system Immune system is a system of biological structures and processes within an organism that protects against disease. In order to function properly, an immune system must detect a wide variety of agents, from viruses to parasitic worms, and distinguish them from the organism's own healthy tissue.

Immunity Immunity is a biological term that describes a state of having sufficient biological defenses to avoid infection, disease, or other unwanted biological invasion. In other words, it is nothing but the capability of the body to resist harmful microbes from entering the body.

Immunosuppression Immunosuppression involves an act that reduces the activation or efficacy of the immune system. Some portions of the immune system itself have immuno-suppressive effects on other parts of the immune system, and immunosuppression may occur as an adverse reaction to treatment of other conditions.

In vitro fertilization In vitro fertilization is a process by which egg cells are fertilized by sperm outside the body. IVF is a major treatment in infertility when other methods of assisted reproductive technology have failed.

In vitro In vitro is used in science to refer to state of being in an artificial environment outside the living organism.

In vivo In vivo is experimentation using a whole, living organism as opposed to a partial or dead organism, or an in vitro controlled environment. Animal testing and clinical trials are two forms of in vivo research. In vivo testing is often employed over in vitro because it is better suited for observing the overall effects of an experiment on a living subject.

Infertility Primarily refers to the biological inability of an animal to contribute to conception. Infertility may also refer to the state of a female animal which is unable to carry a pregnancy to full term.

Innate immune response The **innate immune system**, also known as **non-specific immune system** and first line of defense, comprises the cells and mechanisms that defend the host from infection by other organisms in a non-specific manner. This means that the cells of the innate system recognize and respond to pathogens in a generic way, but unlike the adaptive immune system, it does not confer long-lasting or protective immunity to the host. Innate immune systems provide immediate defense against infection, and are found in all classes of plant and animal life.

Innate immunity Immunity that occurs naturally as a result of an animal's genetic constitution or physiology and does not arise from a previous infection or vaccination.

Insensible heat Heat transfer to a material that does not result in a change in a temperature of this material, which is lost through evaporation.

Intake energy The energy intake is defined as energy content per unit mass of the food multiplied with the amount of food per unit time. Depending on the energy category obtained from the energy partition model, EI can be expressed as gross (GEI), digestible (DEI), metabolisable (MEI), and net energy (NEI), respectively. Sometimes it is useful to attribute the EI to BM (kg) or metabolic body mass (0.75 kg).

Integrated Stress Response (ISR) Is a cellular stress response common to all eukaryotes. ISR can be caused by cellular stresses and is often triggered by the activation of an eIF-2 kinase.

Intergovernmental Panel on Climate Change(IPCC) IPCC is a scientific intergovernmental body first established in 1988 by two United Nations organizations, the World Meteorological Organization (WMO) and the United Nations Environment Programme (UNEP), and later endorsed by the United Nations General Assembly through Resolution 43/53. Its mission is to provide comprehensive scientific assessments of current scientific, technical, and socio-economic information worldwide about the risk of climate change caused by human activity, its potential environmental and socio-economic consequences, and possible options for adapting to these consequences or mitigating the effects.

Interleukins Interleukins are a group of cytokines (secreted proteins/signaling molecules) that were first seen to be expressed by white blood cells (leukocytes). The term interleukin derives from (inter-) "as a means of communication", and (-leukin) "deriving from the fact that many of these proteins are produced by leukocytes and act on leukocytes".

Introns An intron is any nucleotide sequence within a gene that is removed by RNA splicing to generate the final mature RNA product of a gene.

Ionic channels A protein that acts as a pore in a cell membrane and permits the selective passage of ions (such as potassium ions, sodium ions, and calcium ions), by means of which electrical current passes in and out of the cell. Ion channels also serve many other critically important functions including chemical signaling, transcellular transport, regulation of pH, and regulation of cell volume.

Ionophores An ionophore is a lipid-soluble molecule usually synthesized by microorganisms to transport ions across the lipid bilayer of the cell membrane.

Isobar A line on a weather map of points of equal or constant pressure.

Isotherm A line on a weather map of equal or constant temperature.

Isozymes Isozymes (also known as isoenzymes) are enzymes that differ in amino acid sequence but catalyze the same chemical reaction. These enzymes usually display different kinetic parameters (e.g. different K_M values), or different regulatory properties.

Koppen classification The Koppen climate classification system is one of the most widely used climate classification systems. It uses six letters to divide the world into six major climate regions, based on average annual precipitation, average monthly precipitation, and average monthly temperature. The letters are a,b,c,d,e,h. A is for tropical humid, b is for dry, c is for mild mid-latitude, d is for severe mid-latitude, e is for polar, and h is for highland. Each category is further divided into sub-categories based on precipitation and temperature.

Learning The acquisition of a new response, or a qualitative change of an existing response, or an inhibition or facilitation of an existing response by a new stimulus.

Leucocytes Leucocytes are white blood cells of the immune system involved in defending the body against both infectious disease and foreign materials.

Granulocytes, monocytes, and lymphocytes are type of leukocytes.

Livestock Livestock refers to one or more domesticated animals raised in an agricultural setting to produce commodities such as food, fiber, and labor.

Luteinizing hormone Luteinizing hormone (LH, also known as lutropin) is a hormone produced by the anterior pituitary gland. In females, an acute rise of LH called the LH surge triggers ovulation and development of the corpus luteum. In males, where LH had also been called interstitial cell-stimulating hormone (ICSH), it stimulates Leydig cell production of testosterone. It acts synergistically with FSH.

Luteolysis Luteolysis is the structural and functional degradation of the corpus luteum (CL), which occurs at the end of the luteal phase of both the estrus and menstrual cycles in the absence of pregnancy.

Lymphocyte A mononuclear, non-granular leukocyte having a deeply staining nucleus containing dense chromatin and a sparse, pale-blue-staining cytoplasm. It participates in immunity. There are two main types of lymphocytes: B cells and T cells.

Macroclimate The climate of a large area as distinguished from that of a small area. The general large-scale climate of the open atmosphere in a large area or country, as distinguished from the mesoclimate and microclimate.

Management-intensive grazing Management-intensive grazing is a system of grazing in which ruminant and non-ruminant herds are regularly and systematically moved to fresh pasture with the intent to maximize the quality and quantity of forage growth.

Medial preoptic area The medial preoptic nucleus is bounded laterally by the lateral preoptic nucleus, and medially by the preoptic periventricular nucleus. It releases gonadotropin-releasing hormone, controls copulation in males, and is larger in males than in females.

Mediterranean basin Mediterranean basin refers to the lands around the Mediterranean Sea that have a Mediterranean climate, with mild, rainy winters and hot, dry summers, which supports characteristic Mediterranean forests, woodlands, and scrub vegetation.

Mediterranean climates The areas of the world that have a Mediterranean style climate are found at about 35 degrees north and south of the equator, on the western sides of continents. The distinctive feature of this climate is the low level of rain in the summer. Overall, the climate has warm winters and hot summers and a moderate amount of rainfall.

Meiotic process Meiotic process is a special type of cell division necessary for sexual reproduction. The cells produced by meiosis are gametes or spores. The animals' gametes are called sperm and egg cells.

Melanocortin The melanocortins are a group of peptide hormones which include adrenocorticotropic hormone (ACTH) and the different forms of melanocyte-stimulating hormone (MSH).

Mesoclimate The climate of small areas of the earth's surface which may not be respective of the general climate of the district.

Metabolic heat Rate of transformation of chemical energy into heat in an organism, usually expressed in terms of unit area of the total body surface

Metabolic heat Metabolic heat is a by-product of the body's activity.

Metabolic rate Two types of Basal Metabolic Rate (BMR), and the closely related resting metabolic rate (RMR), is the amount of daily energy expended by animals at rest. Rest is defined as existing in a neutrally temperate environment while in the post-absorptive state.

Metabolism Metabolism is the sum of many interconnected reaction sequences that interconvert cellular metabolites. Each sequence is regulated so as to provide what the cell needs at a given time and to expend energy only when necessary.

Metaphase Metaphase is a stage of mitosis in the eukaryotic cell cycle in which condensed and highly coiled chromosomes, carrying genetic information, align in the middle of the cell before being separated into each of the two daughter cells. Metaphase accounts for approximately 4% of the cell cycle's duration.

Methanogen Methanogen are microorganisms that produce methane as a metabolic by-product in anoxic conditions in the rumen of ruminant livestock. They are classified as archaea, a group quite distinct from bacteria.

Metyrapone Metyrapone is a drug used in the diagnosis of adrenal insufficiency and occasionally in the treatment of Cushing's syndrome (hypercortisolism).

Microclimate The climate condition directly surrounding the animal.

Mineralocorticoids A group of steroid hormones produced by the adrenal cortex that are important in maintaining electrolyte balance.

miRNA miRNA are post-transcriptional regulators that bind to complementary sequences on target messenger RNA transcripts (mRNAs), usually resulting in translational repression or target degradation and gene silencing

Mitosis Mitosis is the process by which an eukaryotic cell separates the chromosomes in its cell nucleus into two identical sets, in two separate nuclei. It is generally followed immediately by cytokinesis, which divides the nuclei, cytoplasm, organelles, and cell membrane into two cells containing roughly equal shares of these cellular components.

Molecular chaperones In molecular biology, molecular chaperones are proteins that assist the non-covalent folding or unfolding and the assembly or disassembly of other macromolecular structures, but do not occur in these structures when the structures are performing their normal biological functions having completed the processes of folding and/or assembly.

Monogastric animals A monogastric organism has a simple single-chambered stomach, compared to a ruminant organism which has a four-chambered complex stomach.

Monsoon climate A dry winter season and a rainy summer season, usually occurring near the tropics.

Morbidity A diseased condition or state.

Mortality The relative frequency of deaths in a specific population.

Natural selection Natural selection is the gradual, nonrandom process by which biological traits become either more or less common in a population as a function of differential reproduction of their bearers. It is a key mechanism of evolution.

Natural wet bulb (NWB) temperature Natural wet bulb (NWB) temperature is measured by exposing a wet sensor, such as a wet cotton wick fitted over the bulb of a thermometer, to the effects of evaporation and convection. The term natural refers to the movement of air around the sensor.

Neuroendocrine system The major center of neuroendocrine integration in the body is found in the hypothalamus and the pituitary gland. Here hypothalamic neurosecretory cells release factors to the blood. Some of these factors, released at the median eminence, control the secretion of pituitary hormones, while others (the hormones oxytocin and vasopressin) are released directly to the peripheral circulation.

Neutral detergent fiber Neutral detergent fiber is the most common measure of fiber used for animal feed analysis, but it does not represent a unique class of chemical compounds. NDF measures most of the structural components in plant cells (i.e. lignin, hemicellulose, and cellulose), but not pectin.

Non esterified fatty acids (NEFA) NEFA are the major component of triglycerides (the fat stores in the body), which consist of three fatty acids linked to a glycerol backbone. NEFAs can be used as an energy source by many tissues including skeletal muscle and hepatocytes.

Norepinephrine A catecholamine with multiple roles including as a hormone and a neurotransmitter. Also known as nor-adrenaline.

Normothermia Normal temperature, also known as normothermia or euthermia, is a concept that depends upon the place in the body at which the measurement is made, and the time of day and level of activity of the animal.

Osmoregulation Osmoregulation is the active regulation of the osmotic pressure of an organism's fluids to maintain the homeostasis of the organism's water content that is it keeps the organism's fluids from becoming too diluted or too concentrated.

Osmosis Diffusion of fluid through a semi-permeable membrane from a solution with a low solute concentration to a solution with a higher solute concentration until there is an equal concentration of fluid on both sides of the membrane.

Osmotic stress Osmotic shock or osmotic stress is a sudden change in the solute concentration around a cell, causing a rapid change in the movement of water across its cell membrane.

Oxidative stress Oxidative stress represents an imbalance between the production and manifestation of reactive oxygen species and a biological system's ability to readily detoxify the reactive intermediates or to repair the resulting damage.

Oxytocin Oxytocin is a mammalian hormone that acts primarily as a neuromodulator in the brain.

Packed cell volumes The hematocrit or packed cell volume or erythrocyte volume fraction is the percentage of the concentration of red blood cells in blood.

Panting Panting, a method of cooling, used by many mammals, most birds, and some reptiles, accomplished by means of the evaporation of water from internal body surfaces. As the animal's body temperature rises, its respiration rate increases sharply cooling results from the evaporation of water in the nasal passages, mouth, lungs, and (in birds) air sacs. Like other forms of evaporative cooling (e.g., perspiration), panting expends large amounts of water, which must be replaced if the animal is to maintain effective heat regulation.

Paraventricular Nucleus (PVN) The paraventricular nucleus is a neuronal nucleus in the hypothalamus. It contains multiple subpopulations of neurons that are activated by a variety of stressful and/or physiological changes.

Pathogenicity Pathogenicity is the ability of a pathogen to produce an infectious disease in an organism.

Pathogens A pathogen or infectious agent - in colloquial terms, a germ—is a microbe or microorganism such as a virus, bacterium, prion, or fungus that causes disease in its animal or plant host. There are several substrates including pathways whereby pathogens can invade a host the principal pathways have different episodic time frames, but soil contamination has the longest or most persistent potential for harboring a pathogen.

Peripheral vasodilation Peripheral vasodilation is the dilation of the veins and arteries of the periphery. This will lower blood pressure and provide less resistance for the heart to beat against.

pH The concentrations of hydrogen ions and indirectly hydroxide ions are given by a pH number. pH is defined as the negative logarithm of the hydrogen ion concentration. The equation is: $pH = -\log [H^+]$

Phenotypes The physical and biochemical characteristics of an organism as determined by the interaction of its genetic constitution and the environment

Phenotypic variation Phenotypic variation (due to underlying heritable genetic variation) is a fundamental prerequisite for evolution by natural selection. It is the living organism as a whole that contributes (or not) to the next generation, so natural selection affects the genetic structure of a population indirectly via the contribution of phenotypes.

Photoperiodism Photoperiodism is the physiological reaction of organisms to the length of day or night. It occurs in plants and animals.

Physiological responses A psychological and physical response of the body that occurs whenever animal must adapt to changing conditions, whether those conditions be real or perceived.

Physiological stress Physiological stress represents a wide range of physical responses that occur as a direct effect of a stressor causing an upset in the homeostasis of the body. Upon immediate disruption of either psychological or physical equilibrium the body responds by stimulating the nervous, endocrine, and immune systems. The reaction of these systems causes a number of physical changes that have both short and long-term effects on the body.

Pinealocytes Pinealocytes are the main cells of the pineal gland. They produce and secrete melatonin. Pinealocytes have an organelle called the synaptic ribbon this is considered to be a specific marker for pinealocytes. Some of the enzymes of the pinealocytes include 5-HydrooxyTryptaine, N-acetyl transferase, and 5-hydroxyindole-O-methyltransferase which are used to convert serotonin to melatonin.

Placental development Development of the placenta is a highly regulated process that is essential for normal fetal growth and development, and for maintenance of a healthy pregnancy. The placenta fulfills several critical roles as the interface between mother and fetus: it prevents rejection of the fetal allograft, enables respiratory gas exchange, transports nutrients, eliminates fetal waste products, and secretes peptide and steroid hormones.

Polymorphism Polymorphism is common in nature it is related to biodiversity, genetic variation and adaptation it usually functions to retain variety of form in a population living in a varied environment. The most common example is sexual dimorphism, which occurs in many organisms.

Precipitation Any or all of the forms of water particles, whether liquid or solid, that falls from the atmosphere and reach the ground. It is distinguished from cloud and virga in that it must reached the ground. Precipitation includes drizzle rain, snow, snow pellets, snow grains, ice crystal, ice pellets, and hail.

Pre-pubertal periods Before puberty, the period during which secondary sex characteristics start to develop and the capability for sexual reproduction is attained.

Probiotics Are live microorganisms thought to be beneficial to the host organism. Probiotics are commonly consumed as part of fermented foods with specially added active live cultures such as in yogurt, soy yogurt, or as dietary supplements.

Prolactin A hormone secreted by the pituitary gland. Prolactin stimulates lactation (milk production). It also has many other functions, including essential roles in the maintenance of the immune system.

Promoter In genetics, a promoter is a region of DNA that facilitates the transcription of a particular gene. Promoters are located near the genes they regulate, on the same strand and typically upstream.

Propranolol Propranolol is a sympatholytic non-selective beta blocker. Sympatholytics are used to treat hypertension, anxiety, and panic. It was the first successful beta blocker developed.

Proteomic Proteomic is the large-scale study of proteins, particularly their structures and functions. Proteins are vital parts of living organisms, as they are the main components of the physiological metabolic pathways of cells.

Pseudo-ruminants A pseudo-ruminant is a classification of an animal based on its digestive tract. Pseudoruminants differ from ruminants in that they do not have the compartment known as a reticulum. The main feedstuffs of pseudo-ruminants are forages.

Psychrometric chart A monograph for graphical representation of obtaining relative humidity, absolute humidity, and dew point from wet- and dry-bulb thermometer readings.

Psychrometry The art of determining the amount of moisture in the air by the use of psychrometer.

Radiation Radiation is the transfer of heat energy through space. An animal whose body temperature is greater than the temperature of the surrounding surfaces radiates heat to these surfaces. Hot surfaces and infrared light sources radiate heat that can increase the body's heat load.

Rain Precipitation of drops of water that have diameters larger than 0.5 mm.

Rangeland Rangelands are vast natural landscapes in the form of grasslands, shrublands, woodlands, wetlands, and deserts. Types of rangelands include tallgrass and shortgrass prairies, desert grasslands and shrublands, woodlands, savannas,chaparrals, steppes, and tundras.

Reactive oxygen species (ROS) ROS are chemically reactive molecules containing oxygen. Examples include oxygen ions and peroxides. Reactive oxygen species are highly reactive due to the presence of unpaired valence shell electrons. ROS form as a natural by-product of the normal metabolism of oxygen and have important roles in cell signaling and homeostasis. However, during times of environmental stress (e.g., UV or heat exposure), ROS levels can increase dramatically.

Relative humidity (humidity) A ratio of vapor pressure of the air to the saturation vapor pressure. The ratio of the actual vapor pressure of the air to the saturation vapor pressure. The corresponding ration of specific humidity or of mixing ratio give approximations of sufficient accuracy for many purposes in metrology. The relative humidity is usually expressed in percent, and can be computed from psychrometric data.

Resilient adaptation Resilient adaptation refers to the idea of an individual's tendency to cope with stress and adversity. This coping may result in the individual "bouncing back" to a previous state of normal functioning, or using the experience of exposure to adversity to produce a "steeling effect" and function better than expected. Resilience is most commonly understood as a process, and not a trait of an individual.

Respirometry chamber Respirometry chamber is a general term that encompasses a number of techniques for obtaining estimates of the rates of metabolism of vertebrates, invertebrates, plants, tissues, cells, or microorganisms via an indirect measure of heat production (calorimetry).

Ruminal fermentation Ruminal fermentation is a digestive process by which carbohydrates are broken down by microorganisms into simple molecules for absorption into the bloodstream of an animal. It is one of the important sources for increased methane emissions.

Ruminants A **ruminant** is a mammal of the order Artiodactyla that digests plant-based food by initially softening it within the animal's first compartment of the stomach, principally through bacterial actions, then regurgitating the semi-digested mass, now known as cud, and chewing it again. The process of rechewing the cud to further break down plant matter and stimulate digestion is called "ruminating".

Rumination activity Time spent chewing is directly related to saliva secretion, which helps buffer the rumen environment and optimizes fiber digestion. Rumination activity is critical for a ruminating animal's health.

Saliva The thin, watery, slightly viscid fluid secreted by the salivary glands: it serves as an aid to swallowing and digestion by moistening and softening food, and contains enzymes which convert starch to dextrin and maltose.

Season A division of the year according to some regularly recurrent event, usually climate or astronomic.

Sebaceous glands The sebaceous glands are microscopic glands in the skin that secrete an oily/waxy matter, called sebum, to lubricate and waterproof the skin and hair of mammals.

Serotonin Serotonin is a monoamine neurotransmitter. Biochemically derived from tryptophan, serotonin is primarily found in the gastrointestinal (GI) tract, platelets, and in the central nervous system (CNS) of animals including humans.

Solar radiation The total electromagnetic radiation emitted by the sun. About one-half of the total energy in the solar beam is contained within the visible spectrum from 0.4 to 0.7 μ, and most of the other half lies in the near infrared, a small additional portion lying in the ultraviolet.

Somatostatin Somatostatin is a peptide hormone that regulates the endocrine system and affects neurotransmission and cell proliferation via interaction with G-protein-coupled somatostatin receptors and inhibition of the release of numerous secondary hormones.

Spermatogenesis Spermatogenesis is the process of sperm cell development. Rounded immature sperm cells undergo successive mitotic and meiotic divisions (spermatocytogenesis) and a metamorphic change (spermiogenesis) to produce spermatozoa.

Stress Stress is defined as a state of endangered homeostasis or disharmony which is counteracted by a complex range of physiologic and behavioral responses that re-establish homeostasis.

Superoxide dismutase Superoxide dismutase are a class of enzymes that catalyze the dismutation of superoxide into oxygen and hydrogen peroxide. As such, they are an important antioxidant defense in nearly all cells exposed to oxygen.

Surrounding environment The totality of circumstances surrounding an organism or group of organisms, especially the combination of external physical conditions that affect and influence the growth, development, and survival of organisms.

Sustainability Sustainability describes how biological systems remain diverse and productive over time.

Sweat glands Sweat glands, or sudoriferous glands, are small tubular structures of the skin that produce sweat. There are two kinds of sweat glands (1) Eccrine sweat glands (2) Apocrine sweat glands.

Sweat Sweat is a fluid consisting primarily of water as well as various dissolved solids (chiefly chlorides) that is excreted by the sweat glands in the skin of mammals.

Sympathetic nervous system (SNS) SNS is one of the three parts of the autonomic nervous system, along with the enteric and parasympathetic systems. Its general action is to mobilize the body's nervous system fight-or-flight response. It is, however, constantly active at a basal level to maintain homeostasis.

T cells T cells are T lymphocytes belonging to a group of white blood cells known as lymphocytes, and play a central role in cell-mediated immunity. They can be distinguished from other lymphocytes, such as B cells and natural killer cells (NK cells), by the presence of a T cell receptor (TCR) on the cell surface. They are called T cells because they mature in the thymus, and thus, can also be called thymocytes.

Telemetry Telemetry is a technology that allows measurements to be made at a distance, via radio wave or IP network transmission and reception of the information.

Temperate climate Very generally, the climate of the "middle" latitudes the variable, climate between the extreme of tropical climate and polar climate.

Temperate zones In geography, temperate latitudes of the globe lie between the tropics and the polar circles. The changes in these regions between summer and winter are generally relatively moderate, rather than extreme hot or cold.

Temperature humidity index (THI) THI is a measure that has been used since the early 1990s. It accounts for the combined effects of environmental temperature and relative humidity, and is a useful and easy way to assess the risk of heat stress.

Thermal comfort zone Thermal comfort zone is a range in ambient temperature in which body temperature is possible and an animal needs not change the metabolic rate.

Thermo neutral zone (TNZ) The range of environmental temperatures over which the heat produced by a 'warm-blooded' animal remains fairly constant. Hence, it is the range in which the animal is 'comfortable', having neither to generate extra heat to keep warm nor expend metabolic energy on cooling mechanisms, such as panting. Animals adapted to cold environments tend to have broader thermoneutral zones than ones living in hot environments. The TNZ ranges from lower critical temperature (LCT) to upper critical temperature (UCT) and depends on age, species, feed intake, diet composition, previous state of temperature acclimation or acclimatization, production, specific housing and pen conditions, tissue insulation (fat, skin), external insulation (coat), and behavior of an animal.

Thermoneutrality The heat-neutral temperature for a given animal: the temperature at which it does not need to regulate its body temperature activity at thermoneutrality.

Thermoregulation Thermoregulation is the ability of an organism to keep its body temperature within certain boundaries, even when the surrounding temperature is very different.

Thyroid Gland Thyroid Gland is one of the largest endocrine glands. The thyroid gland is found in the neck, below (inferior to) the thyroid cartilage (which forms the laryngeal prominence, or "Adam's apple"). It secretes thyroxine and tri-iodo-thyronine which controls metabilc activity in an animal.

Thyrotropin releasing hormone (TRH) Also called thyrotropin-releasing factor (TRF), thyroliberin or protirelin, is a tropic, tripeptidal hormone that stimulates the release of thyroid-stimulating hormone and prolactin by the anterior pituitary.

Transcription Transcription is the process of creating a complementary RNA copy of a sequence of DNA. Both RNA and DNA are nucleic acids, which use base pairs of nucleotides as a complementary language that can be converted back and forth from DNA to RNA by the action of the correct enzymes.

Traumatic stress Traumatic stress is a commonly used term describing reactive anxiety (and depression). This includes subtypes of anxiety, depression, and disturbance of conduct and combinations of these symptoms. It results from events that are less threatening and distressing than the events that lead to posttraumatic stress disorder.

Tropical climates Tropical climate refers to zones in a range of latitudes between 5/10° and 35°. The temperatures remain high all over the year and show an annual wide change in precipitations with wet and dry seasons.

Ubiquitin-like protein Ubiquitin-like protein is a small regulatory protein that has been found in almost all tissues (ubiquitously) of eukaryotic organisms. Among other functions, it directs protein recycling.

Vaccination Vaccination is the administration of antigenic material (a vaccine) to stimulate the immune system of an individual to develop adaptive immunity to a disease. Vaccines can prevent or ameliorate the effects of infection by many pathogens.

Vaccines A preparation of killed microorganisms, living attenuated organisms, or living fully virulent organisms that are administered to produce or artificially increase immunity to a particular disease.

Vasodilation Vasodilation refers to the widening of blood vessels resulting from relaxation of smooth muscle cells within the vessel walls, particularly in the large veins, large arteries, and smaller arterioles. The process is essentially the opposite of vasoconstriction, which is the narrowing of blood vessels.

Vasopressin Vasopressin is a neurohypophysial hormone found in most mammals, including humans responsible for increasing water absorption in the collecting ducts of the kidney nephron.

Vasotocin Vasotocin is an oligopeptide hybrid of oxytocin and vasopressin found in all non-mammalian vertebrates including birds, fish, amphibians, and in fetal mammals. In mammals it appears to have similar biological properties to both oxytocin (stimulating reproductive tract contraction as in egg laying or birth) and vasopressin (diuretic and antidiuretic effects.

Vector A vector is an agent used to carry genes into another organism. Specific examples of natural vectors include plasmids or viruses.

Volatile fatty acids(VFAs) VFAs are fatty acids with a carbon chain of six carbons or fewer. They are now usually referred to as short-chain fatty acids. They can be created through fermentation in the intestine.

Vulnerability Vulnerability refers to the inability to withstand the effects of a hostile environment. A window of vulnerability is a time frame within which defensive measures are reduced, compromised or lacking.

Weather The short-term day to day fluctuations of the meteorological variables, as distinguished from climate, which is the long-term manifestation of the weather.

Whole-animal metabolic rates The metabolism of an animal is estimated by determining rates of carbon dioxide production (VCO_2) and oxygen consumption (VO_2) of individual animals, either in a closed or an open-circuit respirometry system.

Zoonosis A disease that can be transmitted from animals to people or, more specifically, a disease that normally exists in animals but that can infect humans. There are multitudes of zoonotic diseases.

Index

V. Sejian et al. (eds.), *Environmental Stress and Amelioration in Livestock Production*, DOI: 10.1007/978-3-642-29205-7,

H

I

N

O

P

R

S

MIX
Papier aus verantwortungsvollen Quellen
Paper from responsible sources
FSC® C105338

If you have any concerns about our products,
you can contact us on
ProductSafety@springernature.com

In case Publisher is established outside the EU,
the EU authorized representative is:
Springer Nature Customer Service Center GmbH
Europaplatz 3, 69115 Heidelberg, Germany

Printed by Libri Plureos GmbH
in Hamburg, Germany